The Vital Force:
A STUDY OF BIOENERGETICS

The Vital Force:
A STUDY OF BIOENERGETICS

FRANKLIN M. HAROLD

Department of Molecular and Cellular Biology
National Jewish Center for Immunology
and Respiratory Medicine
Denver, Colorado

Department of Biochemistry, Biophysics and Genetics
University of Colorado Health Sciences Center
Denver, Colorado

W. H. FREEMAN AND COMPANY
NEW YORK

Library of Congress Cataloging-in-Publication Data

Harold, Franklin M.
   The vital force: a study of bioenergetics.

   Bibliography: p.
   Includes index.
   1. Bioenergetics.   I. Title.
QH510.H37   1986   574.19′121   85-13640
ISBN 0-7167-1734-4
ISBN 0-7167-1736-0 (pbk. : international student ed.)

Printed in the United States of America

1 2 3 4 5 6 7 8 9 0   HA   4 3 2 1 0 8 9 8 7 6

*To my father, who held that the object of
life is to seek wisdom and lived accordingly.*

Contents

Preface

The time is short and the work is great and the labourers are sluggish and the reward is much and the master of the house is urgent. It is not thy task to complete the work, but neither art thou free to desist from it.

Mishnah

Should you ask a scientist, "What is the purpose of science?" he or she would probably reply, "To understand Nature" or some variation thereon. Newton held that the purpose of science was to read the mind of God, and the meaning of today's formula seems much the same. Few will doubt that the blend of reason, empiricism, and intuition that we call science has in fact given us a better understanding of how the universe works, but at times the sheer mass of the scientific enterprise swamps its meaning. Our very success has been creating a body of knowledge so vast and rich that no one can hope to master more than a small fraction. Outside of one's own subject, it becomes ever harder to discern order in the welter of detail, and even the secondary literature is often accessible only to specialists. This book was written in the belief that bioenergetics has become too important, and too interesting, to be relegated to bioenergeticists.

Today much of biology is dominated by the viewpoint of molecular biology. Historically, molecular biology stems from the fusion of two quite separate roots: one, the study of the structure of macromolecules, proteins and nucleic acids in particular; the other, the search for the mechanism of biological heredity. Its conceptual framework was complete by the 1960s, and today schoolchildren learn to recognize such landmarks as chromosomes, genes, DNA, and the double helix. No account of life that is not informed by the discoveries of molecular biology can be taken seriously. Yet anyone who peers through a microscope at the riot of creatures in a drop of pond water (not to mention the bird watcher or mushroom hunter) is likely to feel uneasy and dissatisfied with the molecular view of life. Too much is going on that is not adequately accounted for. Molecular biology, as it is usually presented, seems to deal more successfully with molecules than with cells and organisms, with the structure of things rather than with their activities.

What breathes life into the jumble of complex molecules is expressed in deceptively familiar words: energy, work, information, order. These terms describe the characteristic ability of living things to act—to respond to their environment, to manipulate molecules large and small, to move, grow, reproduce. In a word, bioenergetics is the scientific study of vitality, not a discipline so much as a point of view, complementary to that of molecular biology. Bioenergetics is by no means a novel concern. Its roots reach back to the discovery of oxygen by Lavoisier and Priestley, and much of modern biochemistry grew out of studies on the generation and utilization of biological energy. Nevertheless, bioenergetics has but recently attained a degree of integration comparable to that of molecular genetics, and its insights have yet to be fully felt in our conception of the nature of life.

The central issue in bioenergetics, at least in its cellular and molecular dimensions, is embodied in the phrase "energy coupling." It has been clear for a century that the familiar activities of living things almost invariably require the performance of work, be it chemical, mechanical, osmotic, or even electrical. Work must be done just to maintain the organization of the cell in the face of the forces of decay. The requisite energy is supplied by metabolism, particularly by the great highways of respiration, fermentation, and photosynthesis. The crux of the matter is the linkage between energy-yielding and energy-consuming processes: How is the energy released by metabolism captured and harnessed to the performance of the many kinds of work that living things find useful?

The first conceptual unification of bioenergetics, achieved in the 1940s, is inextricably bound up with the name of Fritz Lipmann. Its central feature was the recognition that adenosine triphosphate, ATP, serves as a universal energy currency much as money serves as economic currency. In a nutshell, the purpose of metabolism is to support the synthesis of ATP, and the hydrolysis of ATP in turn pays for the performance of work. This clear and simple framework made bioenergetics comprehensible and served to focus attention on the molecular mechanisms by which ATP is generated and utilized. It guided much of biochemical research during the past half century, and for some purposes it remains adequate today.

The great resurgence of bioenergetics during the past two decades began with the discovery that living things employ a second, and very different, energy currency: in many instances, the linkage of metabolism to work is effected by a current of ions passing from one side of a membrane to the other. The principle of energy coupling by ion currents has multiple historical sources, but it first found a clear and general expression in the chemiosmotic theory developed by Peter Mitchell between 1961 and 1966. Mitchell proposed novel answers to the intertwined riddles of oxidative phosphorylation, photosynthesis, and active transport, sparking a revolution in cell biology that continues today. A second seminal idea, the sliding-filament model of muscle contraction, was developed by A. F. Huxley and H. E. Huxley between 1956 and 1969. It is the foundation on which a clear conception of cellular movement and mechanical work is now being built. With the convergence of these disparate streams of research it is now possible, at least in principle, to give a reasonably coherent account of how cells generate useful energy and perform work. That is the object of this book.

For whom is this book intended? At one level, for my colleagues in bioenergetics research, who may find it useful as an overview of our collective achievement over the past quarter of a century. But it is chiefly addressed to students and researchers in biochemistry, physiology, microbiology, and cell biology who seek the wider perspective on their particular subject that may come from an appreciation of biological energetics. For insofar as bioenergetics is the study of the vital activities of living things, it is an essential element in any integrated view of life—whether we seek to understand it at the molecular, cellular, organismic, or even ecological level. I have therefore attempted to present neither an encyclopedic account of the entire field nor a technical scrutiny of the latest controversies but a synoptic view of the shape of bioenergetics. Since many biologists, including myself, find the language of mathematics excessively abstract, I have tried to express everything important in words and diagrams; formulas state relations quantitatively, but I have not relied on them to convey ideas. This approach inevitably entails a certain loss of rigor and precision; I can but hope that there has been a compensatory gain in clarity. The subject is inherently difficult, and if it seems at times that I am struggling to explain things to myself, so I am; that, after all, is what research is about.

I have come to think of this book as being a little like the mountain walks that, next to science, are my chief avocation. The hills are seamed with tracks and routes—a few well trodden, most barely marked, some gentle in the valleys, others steep along the ridges—each with its own charms and vistas. This book, the harvest of many rambles, is one man's map of the terrain called bioenergetics. Like all maps it is a projection of reality, not the thing itself, and it is circumscribed by the mapmaker's knowledge and by his moment in time. As the historian Lynn White once put it, speaking of what scientists mean by truth, it is "not a citadel of certainty to be defended against error; it is a shady spot where one eats lunch before tramping on."

Acknowledgments

This book grew out of a combination of research, teaching, and wide reading over half a lifetime. So many minds have influenced mine that colleagues who think they find here a cherished thought or phrase of their own, perchance without acknowledgment, may well be correct. I ask their indulgence, for to list all those from whom I have learned or borrowed, at first or at second hand, would have demanded thousands of references and made the text unreadable. To allow the narrative to stand by itself, I have sought to handle the mountainous literature by compromising between glibness and pedantry. Review articles, especially recent ones, have been cited whenever possible. Research papers were chosen for their historical significance and to give readers access to areas of current controversy or to the latest information available at the time of writing; responses to the literature extend into 1984. No attempt has been made to document every statement, to assign priority of discovery, or to cover all points of view, but readers who delve into the references can dig these out for themselves. To those who feel that their contributions have been insufficiently noted, I offer apologies and the reminder that it is the way of science for the best of our labors to endure in substance but to vanish individually, like raindrops falling into a pond.

But some debts I owe cannot go without explicit acknowledgment. To Albert Lehninger, whose little masterpiece *Bioenergetics* introduced me to the field. To Peter Mitchell, who made chemical sense of bioenergetics and patiently helped me over the hurdles that he had leaped. To the National Jewish Center for Immunology and Respiratory Medicine, which has for two decades provided me with salary, laboratory, and time to reflect with a minimum of distractions. To Allan Hamilton, my host at the University of Aberdeen during a sabbatical leave in 1981–1982, when this book first took shape. And to my friends and colleagues Martin Pato and Howard Rickenberg for questioning everything that seemed to me self-evident. The Institute of Allergy and Infectious Diseases of the National

Institutes of Health and the National Science Foundation have been liberal patrons of my laboratory; this book would certainly not have been written without the stimulus of prolonged immersion in laboratory research.

Colleagues from several nations reviewed chapters of the book during its gestation, some at my request and some at the publisher's. Philip Bragg, Clark Bublitz, Philip Laris, Robert Simoni, and Howard Rickenberg read the entire opus; sections were reviewed by Adolf Abrams (Chapters 9 and 10), Bradley Amos (11, 12), Susan Brawley (14), Graham Gooday (14), Allan Hamilton (1 to 6), Ruth Harold (1 to 6, 11 to 14), Peter Hinkle (7), Eva Kashket (1 to 6), Janos Lanyi (6), Ben Leichtling (13), Richard McCarty (8), Edward McGuire (11, 12), Richard McIntosh (11, 12), Harold Morowitz (1), Margaret Neville (9, 10), Martin Pato (13), Roger Prince (4, 8), James Prosser (14), Kenneth Robinson (11 to 14), Andrew Staehelin (8), Merna Villarejo (1 to 10) and Yu-Li Wang (11, 12). I thank them all for their counsel, even where I did not follow it, and absolve them of all blame for whatever errors of fact or interpretation remain in the text. These would be far more numerous but for the help that I have received. Special thanks are due to John Caldwell, who guided the work of my laboratory during the periods of partial or total neglect that attended this writing.

Production of a book is ultimately the work of many hands. This one would not exist but for the help of Sylvia Mackenzie, Susan Walker, Betty Woodson and especially Carol Breibart, who cheerfully typed and typed again an interminable stream of successive versions. John H. Staples of W. H. Freeman and Company provided editorial guidance and encouragement in the early stages of this project, Jim Dodd during the final ones. Mike Suh designed the book, and artists at Vantage Art, Inc., drew the illustrations. I am especially indebted to Karen McDermott, who, with skill and restraint, turned the manuscript into a book of which both author and publisher can be proud.

Finally, my thanks go to my wife, Ruth, collaborator in both the art of living and the art of science; for the innumerable times when I depended upon, and received, kindness and forbearance, criticism and support, companionship and love.

Symbols

E_0	Standard redox potential at pH 0 (V or mV).
E_0'	Standard redox potential at pH 7 (V or mV). In the bioenergetics literature the term E_m, 7 is preferred.
E_h	Actual redox potential at a specified pH. E_h,7 is the redox potential at pH 7 (V or mV).
E_m	Midpoint potential at a specified pH; E_m,7 refers to the E_m at pH 7 (V or mV). This term is used in the bioenergetics literature in place of E_0'.
eV	Electron volt. The energy acquired by an electron when it moves across a potential gradient of one volt.
F	Faraday's constant (96,500 coulombs/mol, 23.1 kcal/V·mol, or 96.5 kJ/V·mol).
F	Force (g·cm/s^2).
G	Gibbs free energy (kcal/mol or kJ/mol).
H	Enthalpy, or heat content (kcal/mol or kJ/mol).
$h\nu$	Energy content of a photon, the product of Planck's constant and the frequency (kJ).
J	Joule.
J	Flux of a substance across a membrane (mol/g or mol/cm^2 surface area).
K	Absolute equilibrium constant.
K'	Apparent equilibrium constant at pH 7.
K_m	Michaelis constant. That substrate concentration at which the rate of an enzyme-catalyzed reaction is half maximum.
k	Boltzmann's constant ($R/N = 1.38 \times 10^{-23}$ J/K).
kcal	Kilocalorie.
kdal	Kilodalton (units of molecular mass × 1000).
kJ	Kilojoule.
l	Length.
m	Mass.
N	Avogadro's number (6.023×10^{23} particles/mol).
P	Pressure.

P	Permeability constant (cm/s).
p	Probability.
Q	Heat.
q	Mass-action ratio.
R	Gas constant (2.0 cal/mol·K or 8.3 J/mol·K).
S	Entropy.
[S]	Concentration of substance S (mol/L).
T	Absolute temperature (K).
t	Time.
U	Internal energy.
V	Volt. When used as the unit of electromotive force, it is the difference in potential required to make a current of one ampere flow through a resistance of one ohm. It is also that potential difference against which one joule of work is done in the transfer of one coulomb of charge.
V	Volume.
W	Work (kcal/mol or kJ/mol).
w	Number of possible ways in which a system may be arranged.
λ	Wavelength of light (nm).
π	Osmotic pressure (atm or bar; 1 bar = 0.99 atm.)
$\tilde{\mu}_{ion}$	Electrochemical potential. The sum of the chemical and the electric potentials of the ion (named in the subscript) in a given phase. In practice, only the *difference* in electrochemical potential between two phases can be measured.
Ψ	Electric potential, the work required to move a test charge within an electric field from infinity (zero potential) to the point in question. In practice, only the *difference* in electric potential can be measured.
$\Delta E'_0$	Difference in standard redox potential at pH 7 (V or mV).
$\Delta G°$	Gibbs free-energy change under standard conditions; based on absolute equilibrium constant (kcal/mol or kJ/mol).
$\Delta G°'$	Gibbs free-energy change under standard conditions except that pH = 7. Based on apparent equilibrium constant (kcal/mol or kJ/mol).

ΔG_p	"Phosphorylation potential": ΔG of ATP synthesis under the particular experimental conditions (kcal/mol or kJ/mol).
ΔH	Enthalpy change.
ΔS	Entropy change.
Δp	Proton-motive force. The electrochemical potential gradient for protons between two bulk phases separated by a membrane; defined as $\Delta \tilde{\mu}_{H^+}/F$ (mV).
ΔpH	Difference in pH between two sides of a membrane, usually $pH_i - pH_o$.
$\Delta \tilde{\mu}_{ion}$	The electrochemical potential difference for the ion named in the subscript between two bulk phases separated by a membrane (kcal/mol or kJ/mol).
$\Delta \tilde{\mu}_{H^+}$	The electrochemical potential difference for protons between two bulk phases separated by a membrane (kcal/mol) or kJ/mol). The proton-motive force, $\Delta \tilde{\mu}_{H^+}/F$, is expressed in mV.
$\Delta \Psi$	Membrane potential. The difference in electric potential between two phases separated by a membrane (V or mV).

Conversions

$2.3RT = 1.37$ kcal/mol $= 5.72$ kJ/mol.

$$\frac{2.3RT}{F} = 0.059 \text{ V} = 59 \text{ mV at } 25°C.$$

To convert mV into kJ/mol, multiply by 0.0965.

To convert mV into kcal/mol, multiply by 0.0231.

A tenfold change in concentration is equivalent to $\Delta \tilde{\mu}$ of 59 mV, 1.37 kcal/mol, or 5.72 kJ/mol.

Abbreviations

ADP	Adenosine-5′-diphosphate.
AMP	Adenosine-5′-monophosphate.
ATP	Adenosine-5′-triphosphate.
cAMP	Cyclic AMP, cyclic adenosine–3′,5′-monophosphate.

CCCP	Carbonylcyanide *m*-chlorophenylhydrazone (a proton conductor).
CTP	Cytidine-5′-triphosphate.
DCCD	*N,N*′-dicyclohexylcarbodiimide (inhibitor of ATP synthase).
DNA	Deoxyribonucleic acid.
DCMU	3-(3,4-dichlorophenyl)-1,1-dimethylurea (inhibitor of photosystem II).
FAD, FADH$_2$	Flavin adenine dinucleotide (oxidized, reduced).
FCCP	Carbonylcyanide *p*-trifluoromethoxyphenylhydrazone (a proton conductor).
Fe·S center	Prosthetic group of a class of redox proteins that contain nonheme iron and acid-labile sulfur.
FMN, FMNH$_2$	Flavin mononucleotide (oxidized, reduced).
F$_1$F$_0$ ATPase	The proton-translocating ATPase, or ATP synthase, of bacterial, mitochondrial, and chloroplast membranes. F$_1$ is the catalytic headpiece, F$_0$ the membrane sector or proton channel.
GDP	Guanosine-5′-diphosphate.
GTP	Guanosine-5′-triphosphate.
HOQNO	Hydroxyquinoline-N-oxide (inhibitor of respiration).
LHC I, LHC II	Light-harvesting complexes I and II.
NAD$^+$, NADH	Nicotinamide adenine dinucleotide (oxidized, reduced).
NADP$^+$, NADPH	Nicotinamide adenine dinucleotide phosphate (oxidized, reduced).
P$_i$	Inorganic orthophosphate.
PQ, PQH$_2$	Plastoquinone (oxidized, reduced).
PS I, PS II	Photosystems I and II of chloroplasts.
RNA	Ribonucleic acid..
UQ, UQH$_2$	Ubiquinone (oxidized, reduced).
UTP	Uridine-5′-triphosphate.

The Vital Force:
A STUDY OF BIOENERGETICS

1

Energy, Work, and Order

*The trouble with simple things is that one must
understand them very well.*

Anonymous

• • •

The flow of matter through individual organisms and biological communities is part of everyday experience; the flow of energy is not, even though it is central to the very existence of living things. What makes concepts such as energy, work, and order so elusive is their insubstantial nature: we find it far easier to visualize the dance of atoms and molecules than the forces and fluxes that determine the direction and extent of natural processes. The branch of physical science that deals with such matters is thermodynamics, an abstract and demanding discipline that most biologists are content to skim over lightly. Yet bioenergetics is so shot through with concepts and quantitative relationships derived from thermodynamics that it is scarcely possible to discuss the subject without frequent reference to free energy, potential, entropy, and the second law. This chapter is not intended as a substitute for the effort and time required to master thermodynamic principles and computations and does not pretend to rigor; my purpose is merely to collect and explain, as simply as may be, the fundamental concepts and relationships that recur throughout this book. Readers who prefer a more extensive treatment of the subject should consult either the elementary introduction by Klotz (1967) or the advanced texts by Morowitz (1978) and by Edsall and Gutfreund (1983).

Thermodynamics evolved during the nineteenth century out of efforts to understand how a steam engine works and why heat is produced when one bores cannon. The very name thermodynamics, and much of its language, recalls these historical roots, but it would be more appropriate to speak of energetics, for the principles involved are universal. It is just a little disconcerting that occupations "of such unblushing economic tinge" (as P. W. Bridgman put it) should have given birth to our most profound insights into what is possible in this world and what is not. Living things, like all other natural phenomena, are constrained by the laws of thermodynamics. By the same token, thermodynamics supplies an indispensable framework for the quantitative description of biological vitality.

Energy Conservation

Let us begin with the meaning of "energy" and "work." Energy is defined in elementary physics, as in daily life, as the capacity to do work. The meaning of work is harder to come by, and rather more narrow. Work, in the mechanical sense, is the displacement of any body against an opposing force. The work done is the product of the force and the distance displaced, or

$$W = f \Delta l \qquad \text{(Eq. 1.1)}$$

(We may note in passing that the dimensions of work are complex, ml^2/t^2, where m denotes mass, l distance, and t time, and that work is a scalar quantity that is the product of two vectorial terms. This is an aspect to which we shall return when we come to vectorial metabolic reactions.) Mechanical work appears in chemistry because, whenever the final volume of a reaction mixture exceeds the

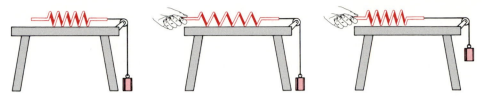

FIGURE 1.1   Energy and work in a mechanical system.

initial one, work must be done against the pressure of the atmosphere; conversely, the atmosphere performs work when a system contracts. This work is given by $P\Delta V$ (where P stands for pressure and V for volume), a term that appears frequently in thermodynamic formulas. In biology, work is employed in a broader sense to describe displacement against any of the forces that living things encounter or generate: mechanical, electrical, osmotic, or even chemical potential.

A familiar mechanical illustration may help to clarify the relationship of energy to work. The spring in Figure 1.1 can be extended by applying to it a force over some particular distance, that is, by doing work on the spring. This work can be recovered by an appropriate arrangement of pulleys, so as to lift a weight onto the table. The extended spring can thus be said to possess energy that is numerically equal to the work it can do on the weight (neglecting friction). The weight on the tabletop, in turn, can be said to possess energy by virtue of its position in the gravitational field of the earth, which can be utilized to do some other work, such as the turning of a crank. The weight thus illustrates the concept of potential energy, a capacity to do work that arises from the position of an object in a field of force, and the sequence as a whole illustrates the conversion of one kind of energy into another, or energy transduction.

It is common experience that such mechanical devices involve both the performance of work and the production or absorption of heat. We are at liberty to vary the amount of work done by the spring, up to some maximum, by using various weights, and the amount of heat produced will also vary. But much experimental work has shown that, under ideal circumstances, the sum of the work done and of the heat evolved is constant, and depends only on the initial and final extensions of the spring. We can thus envisage a property, the internal energy of the spring, with the characteristic that

$$\Delta U = \Delta Q + \Delta W \qquad\qquad \text{(Eq. 1.2)}$$

Here Q is the amount of heat absorbed by the system and W is the work done on the system. In the present illustration the work is mechanical, but it could just as well be electrical, chemical, or any other kind of work.[1] Thus ΔU is the net

---

[1] Equation 1.2 is more commonly encountered in the form $\Delta U = \Delta Q - \Delta W$, which results from the convention that Q is the amount of heat absorbed by the system from the surroundings and W is the amount of work done by the system on the surroundings. This convention affects the sign of W but does not alter the meaning of the equation.

amount of energy put into the system, either as heat or as work; conversely, both the performance of work and the evolution of heat entail a decrease in the internal energy. We cannot specify an absolute value for the energy content; only changes in internal energy can be measured. Note that Equation 1.2 assumes that heat and work are equivalent; its purpose is to stress that, under ideal circumstances, ΔU depends only on the initial and the final state of the system, while the partitioning between heat and work is variable.

Equation 1.2 is a statement of the first law of thermodynamics, which is the principle of energy conservation. If a particular system exchanges no energy with its surroundings, its energy content remains constant; if exchange occurs, the change in internal energy will be given by the difference between the energy gained from the surroundings and that lost to the surroundings. The change in internal energy depends only on the initial and the final states of the system, not on the pathway or mechanism of energy exchange. Energy and work are interconvertible; even heat is a measure of the kinetic energy of the molecular constituents of the system. To put it as simply as possible, Equation 1.2 states that no machine, including those chemical machines that we recognize as living, can do work without an energy source.

The equivalence of energy and work must be qualified by invoking "ideal conditions"; that is, by requiring that the process be carried out reversibly. The meaning of "reversible" in thermodynamics[2] is a special one: the term describes conditions under which the opposing forces are so nearly balanced that an infinitesimal change in one or the other would reverse the direction of the process. Under these circumstances the process yields the maximum possible amount of work. It will be obvious that reversibility in this sense does not often hold in nature: ideal conditions differ so little from a state of equilibrium that any process or reaction would require infinite time and would therefore not take place at all. Nonetheless, the concept of thermodynamic reversibility is a useful one: if we can measure the change in internal energy that a given process entails, we have an upper limit to the work that it can do; for any real process the maximum work will be less.

Several sources of energy are encountered in a study of biology, notably light and chemical transformations, as well as a variety of work functions. Pressure-volume work, which occupies so prominent a place in classical thermodynamics, is uncommon, but many examples of mechanical, osmotic, electrical, and chemical work will be discussed in later chapters. The meaning of the first law in biology stems from the certainty, painstakingly achieved by nineteenth-century physicists, that the various kinds of energy and work are measurable, equivalent, and within limits interconvertible. Energy is to biology what money is to economics: the means by which living things purchase useful goods and services.

---

[2] In biochemistry, reversibility has a different meaning. Usually the term refers to a reaction whose pathway can be reversed, often with an input of energy.

TABLE 1.1 Potential and Capacity Factors in Energetics

Type of energy	Potential factor	Capacity factor
Mechanical (PV)	Pressure	Volume
Electrical	Electric potential	Charge
Chemical	Chemical potential	Mass
Osmotic	Concentration	Mass
Thermal	Temperature	Entropy

It will be obvious that the amount of work that a system can do, be it mechanical or chemical, is a function of its size. Work can always be defined as the product of two factors, force and distance for example. One is a potential or intensity factor, which is independent of the size of the system; the other is a capacity factor and is directly proportional to the size (Table 1.1). In biochemistry, energy and work have traditionally been expressed in calories, the calorie being the amount of heat required to raise the temperature of one gram of water from 15.0 to 16.0°C. In principle one can carry out the same process by doing the work mechanically with a paddle; such experiments led to the establishment of the mechanical equivalent of heat as 4.186 joules per calorie (J/cal). In current standard usage based on the meter, kilogram, and second, the fundamental unit of energy is the joule (1 J = 0.24 cal) or the kilojoule (kJ), 1000 times larger. We shall often have occasion to use the equivalent electrical units, based on the volt: a volt is the potential difference between two points when one joule of work is involved in the transfer of a coulomb of charge from one point to the other. [A coulomb is the amount of charge carried by a current of one ampere (A) flowing for one second. Transfer of one mole (mol) of charge across a potential of one volt (V) involves 96,500 joules of energy or work.] The difference between energy and work is often a matter of the sign. Work must be done to bring a positive charge closer to another positive charge, but the charges thereby acquire potential energy, which in turn can do work. Finally, we shall be obliged to make reference to the much more elusive kind of work that is embodied in the concepts of order and information.

The Direction of Spontaneous Processes

Left to themselves, events in the real world take a predictable course. The apple falls from the branch. A mixture of hydrogen and oxygen gases will be converted into water. The fly trapped in a bottle is doomed to perish, the pyramids to crumble into sand; things fall apart. But there is nothing in the principle of energy conservation that forbids the apple to return to its branch with the absorption of heat from the surroundings or that prevents water from dissociating into its constituent elements in a like manner. The search for the reason why neither of these things ever happens led to profound philosophical insights and also generated useful quantitative statements about the energetics of chemical reactions and the amount of work that can be done thereby. Since living things are, in many respects, chemical machines, we must examine these matters in some detail.

From daily experience with falling weights and warm bodies growing cold, one might expect the direction of spontaneous processes to be that which lowers the internal energy, that is, those in which ΔU is negative. But there are too many exceptions for this to be a general rule. The melting of ice is one: an ice cube placed in water at $1°C$ will melt, yet measurements show that liquid water (at any temperature above 0 degrees) is in a state of higher energy than ice; evidently, some spontaneous processes are accompanied by an increase in internal energy. Our melting ice cube does not violate the first law, for heat is absorbed as it melts. This suggests that there is a relationship between the capacity for spontaneous heat absorption and the criterion determining the direction of spontaneous processes, and that is the case. The thermodynamic function we seek is called entropy, mathematically the capacity factor corresponding to temperature, Q/T. We may state the answer to our question, as well as the second law of thermodynamics, thus: The direction of all spontaneous processes is such as to increase the entropy of a system plus its surroundings.

Few concepts are so basic to a comprehension of the world we live in, yet so opaque, as is entropy—presumably because entropy does not intuitively relate to our sense perceptions, as mass and temperature do. The explanation given here follows the particularly lucid exposition by Atkinson (1977), who states the second law in a form bearing, at first sight, little resemblance to that given above:

> We shall take [the second law] as the concept that any system not at absolute zero has an irreducible minimum amount of energy that is an inevitable property of that system at that temperature. That is, a system requires a certain amount of energy just to be at any specified temperature.

The molecular constitution of matter supplies a ready explanation: some energy is stored in the thermal motions of the molecules and in the vibrations and oscillations of their constituent atoms. We can speak of it as "isothermally unavailable energy," since the system cannot give up any of it without a drop in temperature (assuming that there is no physical or chemical change). The isothermally unavailable energy of any system increases with temperature, since the energy of molecular and atomic motions increases with temperature. Quantitatively, the isothermally unavailable energy for a particular system is given by ST, where T is the absolute temperature and S is the entropy.

But what is this thing, entropy? Reflection on the nature of the isothermally unavailable energy suggests that, for any particular temperature, the amount of such energy will be greater the more atoms and molecules are free to move and to vibrate—that is, the more chaotic is the system. By contrast the orderly array of atoms in a crystal, with a place for each and each in its place, corresponds to a state of low entropy. Indeed, at absolute zero, when all motion ceases, the entropy of a pure substance is likewise zero; this is sometimes called the third law of thermodynamics.

A large molecule, a protein for example, within which many kinds of motion can take place, will have considerable amounts of energy stored in this fashion—more than would, say, an amino acid molecule. But the entropy of the protein molecule will be less than that of the constituent amino acids into which it can dissociate, because of the constraints placed on the motions of those amino acids so long as they are part of the larger structure. Any process leading to the release of these constraints increases freedom of movement, and hence entropy. This is the universal tendency of spontaneous processes as expressed in the second law; it is why the costly enzymes stored in the refrigerator tend to decay, and why ice melts into water. The increase in entropy as ice melts into water is paid for, as it were, by the absorption of heat from the surroundings. So long as the net change in entropy of the system plus its surroundings is positive, the process can take place spontaneously. That does not necessarily mean that the process will in fact take place: the rate is usually determined by kinetic factors quite separate from the entropy change. All the second law mandates is that the fate of the pyramids is to crumble into sand, while the sand will never reassemble itself into a pyramid; the law does not tell how quickly this must come about.

There is nothing mystical about entropy; it is a thermodynamic quantity like any other, measurable by experiment and expressed in entropy units. One method of quantifying it is through the heat capacity of a system, the amount of energy required to raise the temperature by one degree Celsius. In some cases the entropy can even be calculated from first principles, though only for simple molecules. For our purposes what matters is the sign of the entropy change, ΔS: A process can take place spontaneously when ΔS for the system and its surroundings is positive; a process for which ΔS is negative cannot take place spontaneously, but the opposite process can; and for a system at equilibrium the entropy of the system plus its surroundings is maximal and ΔS is zero.

Equilibrium is another of those familiar words that are easier to use than to define. Its everyday meaning implies that the forces acting on a system are equally balanced, so that there is no net tendency to change; this is the sense in which equilibrium will be used here. A mixture of chemicals may be in the midst of rapid interconversion, but if the rates of the forward and the back reaction are equal, there will be no net change in composition, and equilibrium obtains.

The second law has been stated in many versions over the years. One version forbids perpetual motion machines. That is because energy is, by the second law, perpetually degraded into heat and rendered isothermally unavailable ($\Delta S > 0$); continued motion therefore requires an input of energy from the outside. Lord Kelvin's formulation is, "It is impossible to devise any engine which, working in a cycle, shall produce no effect other than the extraction of heat from a reservoir and the performance of an equal amount of work." You cannot, in other words, run an ocean liner with the heat of the sea. To obtain useful work one needs both a source of heat and a sink, to absorb the heat generated by the engine; the performance of work requires a gradient. The most celebrated yet perplexing version of the second law was provided by R. J. Clausius: "The energy of the universe is constant; the entropy of the universe tends towards a maximum." How can entropy increase forever, created out of nothing? The root of the difficulty is verbal, as Klotz (1967) neatly explains. Had Clausius defined entropy with the opposite sign (corresponding to order rather than to chaos) its universal tendency would be to diminish; it would then be obvious that spontaneous changes proceed in the direction that decreases the capacity for further spontaneous change. Solutes diffuse from a region of higher concentration to one of lower, heat flows from a warm body to a cold one. Sometimes these changes can be reversed by an outside agency so as to reduce the entropy of the system under consideration, but if so, that external agency must change in such a way as to reduce its own capacity for further change. In sum, "Entropy is an index of exhaustion; the more a system has lost its capacity for spontaneous change, the more this capacity has been exhausted, the greater is the entropy" (Klotz, 1967). Conversely, the farther a system is from equilibrium, the greater is its capacity for change and the less its entropy. Obviously, living things fall into the latter category: a cell is the very epitome of a state remote from equilibrium.

Free Energy and Chemical Potential

Many energy transactions that take place in living things are chemical in nature; we therefore need a quantitative expression for the amount of work a chemical reaction can do. For this purpose, relationships that involve the entropy change in the system plus its surroundings are unsuitable. We need a function that does not depend on the surroundings but that, like ΔS, attains a minimum under conditions of equilibrium and so can serve both as a criterion of the feasibility of a reaction and as a measure of the energy available from it for the performance of work. The function universally employed for this purpose is the free energy, abbreviated G in honor of the nineteenth-century physical chemist J. Willard Gibbs, who first introduced it.

In the preceding section we spoke of the isothermally unavailable energy, ST. Free energy is defined as the energy that *is* available under isothermal conditions, and by the relationship:

$$\Delta H = \Delta G + T \, \Delta S \qquad \text{(Eq. 1.3)}$$

The term H, enthalpy or heat content, is not quite equivalent to U, the internal energy, used earlier. To be exact, ΔH is a measure of the total energy change, including work that may result from changes in volume during the reaction, whereas ΔU excludes this term. However, in the biological context we are usually concerned with reactions in solution, for which volume changes are negligible. For most purposes, then,

$$\Delta U \simeq \Delta G + T \, \Delta S \qquad \text{(Eq. 1.4)}$$

and

$$\Delta G \simeq \Delta U - T \, \Delta S \qquad \text{(Eq. 1.5)}$$

What makes this a useful relationship is the demonstration that for all spontaneous processes at constant temperature and pressure, ΔG is negative. The change in free energy is thus a criterion of feasibility. Any chemical reaction that proceeds with a negative ΔG can take place spontaneously; a process for which ΔG is positive cannot take place, but the reaction can go in the opposite direction; and a reaction for which ΔG is zero is at equilibrium and no net change will occur. For a given temperature and pressure, ΔG depends only on the composition of the reaction mixture; hence the alternative term "chemical potential" is particularly apt. Again, nothing is said about rate, only about direction. Whether a reaction having a given ΔG will proceed, and at what rate, is determined by kinetic rather than thermodynamic factors.

There is a close and simple relationship between the free-energy change of a chemical reaction and the work that it can do. Provided that the reaction is carried out reversibly,

$$\Delta G = - W_{\text{max}} \qquad \text{(Eq. 1.6)}$$

That is, given a reaction taking place at constant temperature and pressure, $-\Delta G$ is a measure of the maximum work that the process can perform. To put it more precisely, $-\Delta G$ is the maximum work possible exclusive of pressure-volume work, and thus a quantity of great importance in bioenergetics. Any process going toward equilibrium can, in principle, do work. We can therefore describe processes for which ΔG is negative as "energy-releasing," or exergonic. Conversely, for any process moving away from equilibrium ΔG is positive, and we speak of an "energy-consuming," or endergonic, reaction. Actually, of course, an endergonic reaction cannot occur: all real processes go toward equilibrium, with a negative ΔG. The concept of endergonic reactions is nevertheless a useful abstraction, for many biological reactions appear to move away from equilibrium. A prime example is the synthesis of ATP during oxidative phosphorylation, whose apparent ΔG is as high as 16 kcal/mol (67 kJ/mol). Clearly, the cell must do work so as to render the reaction exergonic overall. The occurrence of an endergonic process in nature thus implies that it is coupled to a second, exergonic process. Much of cellular and molecular bioenergetics is concerned with the mechanisms by which energy coupling is effected.

Free-energy changes can be measured experimentally by calorimetric methods. They have been tabulated in two forms: as the free energy of formation of a compound from its elements and as ΔG for a particular reaction. It is of the utmost importance to remember that by convention, the numerical values refer to a particular set of conditions. The standard free-energy change, $\Delta G°$, refers to conditions such that all reactants and products are present at a concentration of 1 M; in biochemistry it is more convenient to employ $\Delta G°'$, which is defined in the same way except that the pH is taken to be 7. The conditions that obtain in the real world are likely to be very different from these, particularly with respect to the concentrations of the participants. To take a familiar example, $\Delta G°'$ for the hydrolysis of ATP is about -8 kcal/mol (-33 kJ/mol). In the cytoplasm, the actual nucleotide concentrations are approximately 3 mM ATP, 1 mM ADP, and 10 mM P_i. Free-energy changes depend strongly on concentrations, and ΔG for ATP hydrolysis under physiological conditions is much more negative than $\Delta G°'$, about -12 to -15 kcal/mol (-50 to -65 kJ/mol). Thus, whereas values of $\Delta G°'$ for many reactions are easily accessible, they must not be used uncritically as guides to what happens in cells.

From the preceding discussion of the concept of free energy it is apparent that there must be a relationship between ΔG and the equilibrium constant of a reaction: at equilibrium ΔG is zero, and the farther a given reaction is from equilibrium, the larger is ΔG and the more work the reaction can do. The quantitative statement of this relationship is:

$$\Delta G° = -RT \ln K = -2.3RT \log K \qquad \text{(Eq. 1.7)}$$

where R is the gas constant, T the absolute temperature, and K the equilibrium constant of the reaction. This equation is one of the most useful links between thermodynamics and biochemistry, with a host of applications. For example, the equation is easily modified to allow computation of the free-energy change for concentrations other than the standard ones. For the reactions shown in Equation 1.8,

$$A + B \; \rightleftharpoons \; C + D \qquad \text{(Eq. 1.8)}$$

the actual free-energy change ΔG is given by

$$\Delta G = \Delta G° + RT \ln \frac{[C][D]}{[A][B]} \qquad \text{(Eq. 1.9)}$$

The terms in brackets refer to the concentrations at the time of the reaction. Strictly speaking, one should use activities, but these are usually not known for cellular conditions, and so concentrations must do.

Equation 1.9 can be rewritten to make its import a little plainer. Let q stand for the mass-action ratio $[C][D]/[A][B]$. Substitution of Equation 1.7 into

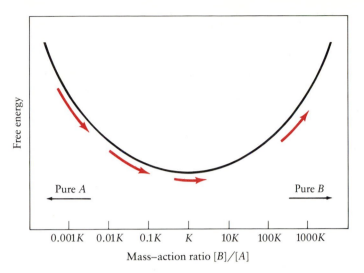

FIGURE 1.2   Free energy of a chemical reaction as a function of displacement from equilibrium. Imagine a closed system containing components A and B at concentrations [A] and [B]. The two components can be interconverted by reaction $A \rightleftharpoons B$, which is at equilibrium when the mass-action ratio B/A equals unity. The curve shows qualitatively how the free energy G of the system varies when the total [A] + [B] is held constant, but the mass-action ratio is displaced from equilibrium. The arrows represent schematically the free-energy change ΔG for a small conversion of [A] into [B] occurring at different mass-action ratios. (From Nicholls, 1982, with permission of Academic Press.)

Equation 1.9, followed by rearrangement, then yields Equation 1.10:

$$\Delta G = -2.3RT \, \log \frac{K}{q} \qquad \text{(Eq. 1.10)}$$

In other words, the value of ΔG is a function of the displacement of the reaction from equilibrium. In order to displace a system from equilibrium, work must be done on it and ΔG is positive. Conversely, a system displaced from equilibrium can do work on another system, provided the kinetic parameters allow the reaction to proceed and provided a mechanism exists that couples the two systems. Quantitatively, a reaction mixture at 25°C whose composition is one order of magnitude away from equilibrium (log $K/q = 1$) corresponds to a free-energy change of 1.36 kcal/mol (5.7 kJ/mol). The value of ΔG is negative if the actual mass-action ratio is less than the equilibrium ratio and positive if the mass-action ratio is greater.

   The point that ΔG is a function of the displacement of a reaction (indeed, of any thermodynamic system) from equilibrium is central to an understanding of biological energetics. Figure 1.2 illustrates this relationship diagrammatically for the chemical interconversion of substances A and B, and it will shortly reappear in other guises.

Redox Reactions

Oxidation and reduction refer to the transfer of one or more electrons from a donor to an acceptor, usually to another chemical species: for instance, the oxidation of ferrous iron by oxygen, with the formation of ferric iron and water. Reactions of this kind require special consideration, for they play a central role in both respiration and photosynthesis.

Redox reactions can be quite properly described in terms of their free-energy changes. However, the participation of electrons makes it convenient to follow the course of the reaction with electrical instrumentation and encourages the use of an electrochemical notation. It also permits one to dissect the chemical process into separate oxidative and reductive half-reactions. For the oxidation of iron, we can write

$$2Fe^{2+} \rightleftharpoons 2Fe^{3+} + 2e^- \qquad \text{(Eq. 1.11)}$$

$$\tfrac{1}{2}O_2 + 2H^+ + 2e^- \rightleftharpoons H_2O \qquad \text{(Eq. 1.12)}$$

$$2Fe^{2+} + \tfrac{1}{2}O_2 + 2H^+ \rightleftharpoons 2Fe^{3+} + H_2O \qquad \text{(Eq. 1.13)}$$

The tendency of a substance to donate electrons, its "electron pressure," is measured by its standard reduction (or redox) potential, E_0, with all components present at a concentration of 1 M. In biochemistry, it is more convenient to employ E'_0, which is defined in the same way except that the pH is 7. By definition, then, E'_0 is the electromotive force given by a half-cell in which the reduced and oxidized species are both present at 1.0 M, 25°C, and pH 7, in equilibrium with an electrode that can reversibly accept electrons from the reduced species. By convention the reaction is written as a reduction. The standard reduction potential of the hydrogen electrode[3] serves as reference: at pH 7, this equals -0.42 volt. The standard redox potential as defined above is often referred to in the bioenergetics literature as the midpoint potential: $E_{m,7}$ refers to the midpoint potential E'_0 at pH 7. A negative midpoint potential marks a good reducing agent; oxidants have positive midpoint potentials.

The redox potential for the reduction of oxygen to water is $+0.82$ volt, and that for the reduction of Fe^{3+} to Fe^{2+} (the direction opposite to Equation 1.11) is $+0.77$ volt. We can therefore predict that under standard conditions, the Fe^{2+}-Fe^{3+} couple will tend to reduce oxygen to water rather than the reverse. A mixture containing Fe^{2+}, Fe^{3+}, and oxygen will probably not be at equilibrium, and the extent of its displacement from equilibrium can be expressed either in terms of the free-energy change for Equation 1.13 or by the difference in redox

---

[3] The standard hydrogen electrode consists of platinum, over which hydrogen gas is bubbled at a pressure of one atmosphere. The electrode is immersed in a solution containing hydrogen ions. When the activity of hydrogen ions is 1, approximately 1 M H^+, the potential of the electrode is taken to be 0.

potential $\Delta E_0'$ between the oxidant and the reductant couples ($+0.05$ volt in the case of iron oxidation). In general,

$$\Delta G^{\circ\prime} = -nF\,\Delta E_0' \qquad \text{(Eq. 1.14)}$$

where n is the number of electrons transferred and F is Faraday's constant (23.06 kcal/V·mol). In other words, the standard redox potential is a measure, in electrochemical units, of the free-energy change of an oxidation-reduction process.

As with free-energy changes, the actual redox potential measured under conditions other than the standard ones depends on the concentration of the oxidized and reduced species according to Equation 1.15 (note the similarity in form to Equation 1.9):

$$E_h = E_0' + \frac{2.3RT}{nF}\,\log\frac{[\text{oxidant}]}{[\text{reductant}]} \qquad \text{(Eq. 1.15)}$$

Here E_h is the measured potential in volts and the other symbols have their usual meanings. It follows that the redox potential under biological conditions may differ substantially from the standard reduction potential.

The Electrochemical Potential

In the preceding section we introduced the concept that a mixture of substances whose composition diverges from the equilibrium state represents a potential source of free energy (Figure 1.3). Conversely, a like amount of work must be done on an equilibrium mixture in order to displace its composition from equilibrium. In this section, we shall examine the free-energy changes associated with another kind of displacement from equilibrium, namely, gradients of concentration and of electrical potential.

Consider a vessel divided by a membrane into two compartments that contain solutions of an uncharged solute at concentrations C_1 and C_2, respectively. The work required to transfer one mole of solute from the first compartment to the second is given by:

$$\Delta G = 2.3RT\,\log\frac{[C_2]}{[C_1]} \qquad \text{(Eq. 1.16)}$$

This expression is analogous to that for a chemical reaction (Equation 1.10) and has the same meaning. If C_2 is greater than C_1, ΔG is positive and work must be done to transfer the solute. The free-energy change for the transport of one mole of solute against a tenfold gradient of concentration is again 1.36 kcal, or 5.7 kJ. The reason that work must be done to move a substance from a region of lower concentration to one of higher is that the process entails a change to a less probable state and therefore a decrease in the entropy of the system. Conversely, diffusion of the solute from the region of higher concentration to that of lower concentration takes place in the direction of greater probability; it results in an increase in the entropy of the system and can proceed spontaneously. The sign of ΔG turns negative and the process can do the amount of work specified by Equation 1.16, provided a mechanism exists that couples the exergonic diffusion process to the work function.

Matters become a little more complex if the solute in question bears an electric charge. Transfer of positively charged solute from compartment 1 to compartment 2 will then cause a difference in charge to develop across the membrane, with the second compartment becoming electropositive relative to the first. Since like charges repel one another, the work done by the agent that moves the solute from compartment 1 to compartment 2 is a function of the charge difference; more precisely, it depends on the electric potential difference across the membrane. This term, called membrane potential for short, will recur constantly in these pages. The membrane potential is defined as the work that must be done by an agent to move a test charge from one side of the membrane to the other. When one joule of work must be done to move one coulomb of charge, the potential difference is said to be one volt. The absolute electric potential of any one phase cannot be measured, but the difference between two phases can be. By convention, the membrane potential is always given in reference to the movement of a positive charge. It states the intracellular potential relative to the extracellular one, which is defined as zero.

The work that must be done to move one mole of an ion against an electric potential of $\Delta\Psi$ volts is given by

$$\Delta G = mF\,\Delta\Psi \qquad \text{(Eq. 1.17)}$$

where m is the valence of the ion and F is Faraday's constant. The value of ΔG for the transfer of cations into a positive compartment is positive and so calls for work. Conversely, the value of ΔG is negative when cations move into the negative compartment, and so work can be done. The electric potential is negative across the plasma membrane of the great majority of cells; therefore cations tend to leak in but have to be "pumped" out.

In general, ions moving across a membrane are subject to gradients of both concentration and electric potential. Consider, for example, the situation depicted in Figure 1.3, which corresponds to a major event in energy transduction during photosynthesis. A cation of valence m moves from compartment 1 to compartment 2, against both a concentration gradient ($C_2 > C_1$) and a gradient of electric potential (compartment 2 is electropositive relative to compartment 1). The free-energy change involved in this transfer is given by Equation 1.18; ΔG is positive and the transfer can proceed only if coupled to a source of energy, in this instance the absorption of light:

$$\Delta G = mF\,\Delta\Psi + 2.3RT \log \frac{[C_2]}{[C_1]} \qquad \text{(Eq. 1.18)}$$

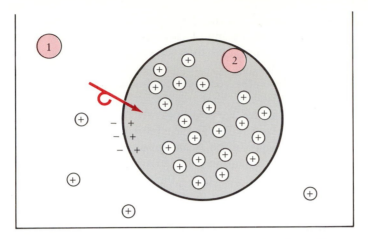

FIGURE 1.3   Transport against an electrochemical potential gradient. The agent that moves the charged solute $\oplus$ from compartment 1 to compartment 2 must do work to overcome both the electric and the concentration gradient. As a result, cations in compartment 2 have been raised to a higher electrochemical potential than those in compartment 1. Neutralizing anions have been omitted.

As a result of this transfer, cations in compartment 2 can be said to be at a higher electrochemical potential than the same ions in compartment 1. The electrochemical potential for a particular ion is designated $\tilde{\mu}_{ion}$. Ions tend to flow from a region of high electrochemical potential to one of low potential and can in principle do work thereby. The maximum amount of this work, neglecting friction, is given by the change in free energy of the ions that flow from compartment 2 to compartment 1 (Equation 1.6) and is numerically equal to the difference in electrochemical potential, $\Delta\tilde{\mu}_{ion}$. This principle underlies much of biological energy transduction.

The electrochemical potential difference $\Delta\tilde{\mu}_{ion}$ is properly expressed in kilocalories per mole or kilojoules per mole. However, it is frequently convenient to state the driving force for ion movement in electrical terms, with the dimensions of volts or millivolts. To convert $\Delta\tilde{\mu}_{ion}$ into millivolts (mV), all the terms in Equation 1.18 are divided by F:

$$\frac{\Delta\tilde{\mu}_{ion}}{F} = m\,\Delta\Psi + \frac{2.3RT}{F}\log\frac{[C_2]}{[C_1]} \qquad \text{(Eq. 1.19)}$$

An important case in point is the proton-motive force, or protonic potential, which will be considered at length in Chapter 3.

Equations 1.18 and 1.19 have proved to be of central importance in bioenergetics. First, they measure the amount of energy that must be expended on the active transport of ions and metabolites, a major function of biological membranes. Second, since the free energy of chemical reactions is often transduced into other forms via the intermediate generation of electrochemical potential gradients, they play a major role in descriptions of biological energy coupling. It should be emphasized that the electrical and concentration terms may either add, as in Equation 1.18, or subtract and that the application of the equations to particular cases requires careful attention to the sign of the gradients. We should also note that free-energy changes in chemical reactions (Equation 1.10) are scalar, whereas transport reactions have direction; this turns out to be a subtle but critical aspect of the biological role of ion gradients.

Ion distribution at equilibrium is an important special case of the general electrochemical equation (Equation 1.18). Figure 1.4 shows a membrane-bound vesicle (compartment 2) that contains a high concentration of the salt K_2SO_4, surrounded by a medium (compartment 1) containing a lower concentration of the same salt; it is stipulated that the membrane is impermeable to anions but freely passes cations. Potassium ions will therefore tend to diffuse out of the vesicle into the solution, whereas the anions are retained. Diffusion of the cation generates a membrane potential, vesicle interior negative, which restrains further diffusion. At equilibrium, ΔG and $\Delta \tilde{\mu}_{K^+}$ equal zero (by definition). Equation 1.18 can then be arranged to give

$$\Delta \Psi = \frac{-2.3RT}{mF} \log \frac{[C_2]}{[C_1]} \qquad \text{(Eq. 1.20)}$$

where C_2 and C_1 are the concentration of K^+ ions in the two compartments; m, the valence, is unity; and $\Delta \Psi$ is the membrane potential in equilibrium with the potassium concentration gradient.

This is one form of the celebrated Nernst equation. It states that at equilibrium, a permeant ion will be so distributed across the membrane that the chemical driving force (outward in this instance) will be balanced by the electric driving force (inward). For a univalent cation at 25°C, each tenfold increase in concentration factor corresponds to a membrane potential of 59 mV, 29.5 mV for a divalent ion. ·

The preceding discussion of the energetic and electrical consequences of ion translocation illustrates a point that must be clearly understood, namely, that an electric potential across a membrane may arise by two distinct mechanisms. The first mechanism, illustrated by Figure 1.4, is by the diffusion of charged particles down a preexisting concentration gradient, an exergonic process. A potential generated by such a process is described as a diffusion potential, or even as a Donnan potential. Many ions are unequally distributed across biological membranes and differ widely in their rates of diffusion across the barrier; therefore diffusion potentials always contribute to the observed membrane potential. But in most biological systems the measured electric potential differs from that which would be expected on the basis of passive ion diffusion. In these cases one must invoke electrogenic ion pumps, transport systems that carry out the exergonic process indicated in Figure 1.3, at the expense of an external energy source.

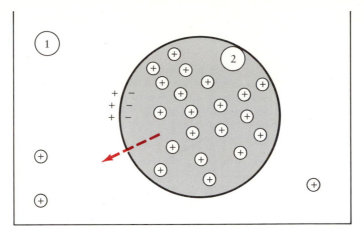

FIGURE 1.4 Generation of an electric potential by ion diffusion. Compartment 2 has a higher salt concentration than compartment 1 (anions are not shown). If the membrane is permeable to the cations but not to the anions, the cations will tend to diffuse into compartment 1, generating a membrane potential with compartment 2 negative.

Transport systems of this kind transduce the free energy of a chemical reaction into the electrochemical potential of an ion gradient and play a leading role in biological energy coupling.

Order, Information, and Maxwell's Demon

The concept of entropy was introduced early in the nineteenth century in the context of heat engines, but its essential nature became clear only when the molecular constitution of matter was understood. Molecules and atoms are not static but rather are subject to ceaseless thermal agitation: they vibrate, oscillate, wander, and collide. The entropy content of a sample of matter is, as we saw above, a measure of the energy associated with these motions. By the same token, entropy is an index of the degree of disorder that prevails in our sample. The molecules of a liquid are free to assume a very large number of arrangements having equivalent internal energy; this is a disordered system, high in entropy. Conversely, the same molecules in the crystalline state can be arranged in a much smaller number of configurations and possess less entropy. The relationship of entropy to order takes the form

$$S = k \ln w = 2.3k \log w \qquad \text{(Eq. 1.21)}$$

where w is a measure of the number of ways in which the system can be arranged and k is Boltzmann's constant. (Numerically k equals R/N, the ratio of the gas constant to Avogadro's number; k enters here because the average kinetic energy of a molecule is a function of k.) In much the same general sense we can speak of a cell as possessing an extraordinary degree of order and low entropy. The atoms of which cellular constituents are composed occupy quite definite positions in space, far fewer in number than would be available to those atoms if they were thrown together at random. Growth, reproduction and development of cells and organisms can be seen as the sum total of processes that impose order on the anarchic motions of free atoms and molecules. The gain in order must be paid for by doing work on the system (Equation 1.4), by increased disorder in the environment, or both. The problem of the nature and genesis of biological order is therefore an aspect of bioenergetics, albeit a peculiarly elusive one.

A relationship analogous to Equation 1.21 links entropy to a second everyday concept, information. This can be best explained with the aid of Maxwell's demon. In 1867, James Clerk Maxwell posed the following puzzle. Suppose a demon were stationed at a wall that separates two insulated chambers filled with gas at equal temperature and pressure (Figure 1.5). The demon operates a microscopic, frictionless door between the two chambers; whenever he sees a molecule approaching the door from the right but none from the left, he opens the door. In time a pressure will develop between the two chambers, which can be used to do work. The trouble is that the demon violates the second law. By virtue of nothing more than his cognitive abilities, and without himself doing any work, he has brought about a decrease in the entropy of the gas; the demon operates a *perpetuum mobile*. Evidently no such demon can exist, but why not? The problem is of more than passing biological interest, for living things have the demon's capacity to recognize and manipulate single molecules, establishing order in apparent contravention of the second law.

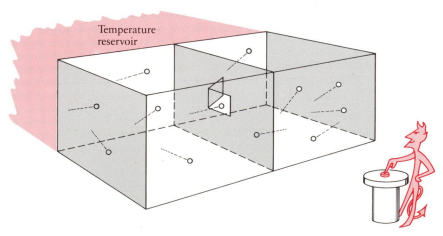

FIGURE 1.5   Maxwell's demon; for exorcism of same see text. (After Morowitz, 1978, with permission of Academic Press.)

The demon was exorcized by the efforts of Leo Szilard and L. Brillouin, who showed that in order to do his job the demon would require information about the molecules, such as their location and direction of travel. To obtain this information he would have to expend energy, and this energy cost would always be as great or greater than the gain resulting from the demon's cleverness. The relationship can be expressed quantitatively in terms of information theory. The amount of information gained when we know in which chamber a molecule is equals one bit. (The elementary unit of information is the binary unit, contracted to bit, which corresponds to a single choice between yes and no, on and off.) The energy cost that this knowledge entails is calculated to be $kT \ln 2$, which is also the maximum decrease in entropy per molecule that the demon can obtain from his device. The moral is, in Harold Morowitz's words, that you don't get something for nothing—not even information.

Thermodynamics, then, confirms the lesson of intuition and experience that information and order are closely allied and that neither is achieved without effort. In order to gather information from their surroundings, organisms must expend energy on the construction and operation of sensory devices. If order is to be transmitted from one generation to the next, work must be done to encode information in genetic messages, ensure their accuracy, transmit the messages, and ultimately translate them without error. Conversely, one can regard enzymic catalysis as a clever trick that uses information (in the form of specific ligand-binding and catalytic sites) to overcome the activation energy barrier, thereby converting thermal energy into work. I shall not pursue the thermodynamic analysis of these relationships (see Morowitz, 1968, 1978) but wish to penetrate a little deeper into the meaning of those common words, information and order. The argument follows that presented by Rupert Riedl in his demanding but trail-blazing book, *Order in Living Organisms* (1978).

Let us begin with messages. Suppose we are playing a game. I have a source capable of producing the numbers from 1 to 32 whenever I tap the corresponding keys, and if I tap out the winning number (say, 7), I receive a prize. Being uninformed about what the winning number is, I can produce it only by chance, and the probability of my doing so is $\frac{1}{32}$. Now, suppose a message puts me wise. I can use probability to assign an information content to the message: to pick any one number out of 32, I must make five binary choices (either yes or no), and so the message has an information content of five bits. If the winning combination were the sequence 1, 2, 3, 4, 5, 6, this would entail the selection of six numbers at five bits each, and the message would have an information content of 30 bits. In general, for a single event the information content, I, is

$$I = \log_2 \frac{1}{p} \qquad \text{(Eq. 1.22)}$$

where p is the probability of the event occurring by chance. The larger the number of possible choices, the greater the information content of the particular symbols that constitute the message; if the desired effect is probable to begin with, the message has little value. Information, as the word is used by information theorists, corresponds to surprise, unexpectedness, unlikelihood. Its biological equivalent is sensory perception.

Returning to my source of numbers, were I to learn that the message 1, 2, 3, 4, 5, 6 had been built into the wiring, so that it is produced without any choice on my part, all elements of surprise would vanish and the information content of the message would fall to zero. But this is precisely the situation that corresponds to biological order. What impresses us about living things is not the information conveyed by surprise, but rather the opposite: the regularity and predictability that assures me that the next lilac I see will be in essence the same as that in my own garden, from the pattern of branching and the shape of the leaves down to the amino acid sequences of thousands of proteins. We can obtain a quantitative measure of regularity or predictability by also expressing it in bits. To produce the message 1, 2, 3, 4, 5, 6, my source had to have 30 bits of "determinacy" wired into it. Let us say, then, that the total information content of a system is the sum of the indeterminate and the determinate information, $bits_I$ and $bits_D$. Biological order is a function of the extent of determinacy. In Riedl's words, "information surprises—determinacy informs."

One can obtain a numerical estimate of the degree of order that a given system represents by calculating how many binary decisions must be made to specify its structure. Selection of one amino acid out of 20 possible ones requires $D = \log_2 20$, or 4.3, bits. The total determinacy content of a protein 300 residues long is $300 \log_2 20$, or 1300, bits, whereas the corresponding length of DNA, with 900 nucleotides, holds $900 \log_2 4$, or 1800, bits of information. As we approach the dimensions of a cell this procedure becomes increasingly tenuous; nevertheless, several investigators, employing different procedures, arrived at a determinacy content as high as 10^{12} or 10^{13} bits per bacterial cell (Setlow and Pollard, 1964; Morowitz, 1978; Riedl, 1978). For present purposes it does not matter much if this figure is off by a few orders of magnitude, for it represents an extraordinary level of determinate order. Morowitz (personal communication) puts the determinacy content of a bacterial cell at near 10^9 $bits_D$, comparable to that of a volume of the *Encyclopaedia Britannica* (also 10^9 $bits_D$) or of the Bell Telephone system (for 100 million subscribers, $10^8 \log_2 10^8$, or about 4×10^9, $bits_D$). For an entire human being, the estimate of determinate order has been put at an astronomical 10^{28} $bits_D$.

We can get a better sense of what information and determinacy mean by reflecting on some of the paradoxes that amuse the experts. From the standpoint of information theory, "a theorem by Einstein or a random assemblage of letters contains the same information, provided the number of letters is the same" (Lwoff, 1968). True, but the perplexity vanishes when we see that shaking up Einstein's letters converts determinacy content into indeterminacy. Lwoff also posed a more biological conundrum. A mutation may replace one nucleotide in the genetic material with another and cause the death of the organism, yet substitution of a guanine for an adenine should not alter the gene's information content. What has happened here is that the original message has ceased to apply: its determinacy content has vanished. For a third puzzle, consider the informational enrichment that takes place when a fertilized ovum grows into an adult. The egg's DNA has an information content of 10^5 to 10^6 bits$_D$, the germ cell perhaps 10^{11}, but the adult may have 10^{28}. Where does the enormous increment come from? Quite a lot of it is represented simply by repeated instances of the same information content: multiple copies of DNA, RNA, proteins, mitochondria, liver cells, erythrocytes, hairs, and so on.

It would be instructive to pursue Riedl's analysis and consider the various kinds of order found in the living world in relation to morphology, development, and evolution, but this would take us beyond the scope of a book on bioenergetics. For now, what is important is that biological order can be described with several notations: as a state of high potential energy and low entropy, as a most improbable configuration of atoms and molecules, and as a state of matter that displays an extraordinary degree of regularity and predictability. Biological order is hard to quantify and the entire subject seems arid and abstruse, yet it conceals the most direct approach to the ultimate question, What is life? For it is in its extreme order, more than in any other aspect, that the organic world differs from the inorganic.

Energy Flow in the Biological World

Physicists inhabit a universe of abstract concepts such as systems, entropy, sources, and sinks. The biologist's world is populated by singular and wriggly objects such as fungi and butterflies, genes and enzymes, and it is not always obvious how the former world impinges upon the latter. At one level, this is the theme of my book as a whole, but at the risk of belaboring what may be obvious to many, an outline of the argument seems useful at this stage.

Let us begin with the naive question of whether living things in fact obey the first and second laws. Extensive experimental studies along this line have been carried out, beginning with the first ice calorimeter constructed by Antoine Lavoisier and Pierre Laplace in 1781 and culminating with the refined studies of W. A. Atwater and F. G. Benedict in 1903. The principle is to confine animals, or human beings, in a chamber that is kept at a constant temperature. The heat produced by the subject, the heats of combustion of foodstuffs and excreta, and the production of CO_2 and water are precisely monitored. With this procedure one can draw up an energy balance sheet and document that the law of energy conservation holds within very narrow limits of error: the sum total of work done and heat produced equals the net energy input determined by the combustion of nutrients and waste products.

The second law is more difficult, for cells and organisms that perform work, grow, reproduce, and evolve toward greater complexity clearly maintain states of low entropy. Does this not violate the second law? No, for living things are not closed systems but open ones, part of a vast network of thermodynamic interactions that extends clear out to the sun. The entropy of this large slice of the universe no doubt increases with time as the sun consumes its nuclear fuel, but there is nothing in the second law to forbid local decreases in entropy. It is, as Atkinson (1977) remarks, "as if a small droplet of spray, rising a few inches from the foot of Niagara Falls, were to cause an observer to doubt the validity of the second law as applied to water in a gravitational field." Nor do evolution and the biological production of vast amounts of organic substances such as coal and petroleum violate the second law, for they too draw on the torrent of energy from the sun to pay for the local decrease in entropy. One should add only that the second law itself remains beyond proof, an axiom consistent with all that we know about the universe, including life.

But the matter does not end here. Those droplets of spray were thrown up by the force of the waterfall, and a similar relationship exists between the living world and the great stream of energy that continually bathes the earth's surface and eventually radiates out again into the reaches of space. All the vital activities of living things that underlie growth, reproduction, development, and evolution are endergonic processes that can take place only by virtue of being coupled to a source of energy. Living things are not in a state of equilibrium. Quite the contrary, they represent steady states that are maintained far from equilibrium by a continuous stream of energy and matter. Any creature cut off from contact with the stream dies by asphyxiation or starvation. Structures and complex molecules come unglued, motion ceases, gradients dissipate, order decays; equilibrium is death. Life implies holding out against equilibrium, converting energy into organization, and these things are possible only at the price of ceaseless work.

Many years ago Erwin Schrödinger (1945) expressed this insight in the statement, "Life feeds on negentropy." It was not a happy choice of phrase, and I prefer to restate Schrödinger's point in the words of Harold Morowitz (1968):

> An isolated organism will be subject to a series of processes tending toward equilibrium. In statistical terms, it will tend to move from the very improbable state that it is in to one of the very probable states associated with the equilibrium ensemble. In order to prevent this drift toward equilibrium, it is constantly necessary to perform work to move the system back into the improbable state that it is drifting out of. An isolated system, however, cannot do steady work. The necessary condition for this is that the system be connected with a source and a sink and the work be associated with a flow of energy from source to sink.... When Schrödinger says that the organism feeds on negentropy, he means simply that its existence depends upon increasing the entropy of the rest of the universe.

A glance out the window shows that life has managed not only to exist but to flourish, multiply, and inherit the earth. That may not be a problem from the viewpoint of global energetic bookkeeping, but from any other it is a marvel beyond compare, and bioenergetics occupies a central position in its scientific exegesis.

The physical environment supplies many kinds of energy, but only two are of major importance as primary energy sources in biology: the free energy of a handful of inorganic transformations, and sunlight. A few organisms, chiefly bacteria, can utilize the oxidation of hydrogen sulfide or of ferrous iron or the reduction of CO_2 plus H_2 to methane. The great majority, however, are either themselves capable of photosynthesis (some bacteria, algae, higher plants) or else live by the metabolic degradation of organic matter produced by photosynthesis (most bacteria, fungi, protozoa, and animals of all kinds). Out of the roughly 5×10^{20} kcal (2×10^{21} kJ) of radiant energy that reach the earth's surface each year, only 1 to 2 percent on the average is captured by photosynthetic organisms. But this pittance is crucial, for the biosphere as we know it depends almost entirely on the mechanisms that harness the energy of sunlight. Directly or indirectly, this harvest supports the many kinds of work by which living things hold out against disorder and procures the goods and services required for growth. Biological energy coupling is the central motif of bioenergetics, and it is one of the grand themes of molecular biology in general.

Scientists take it for granted that the whole of the natural universe is a seamless web whose many strands are securely linked by a common obedience to the laws of physics and chemistry. The weakest point in this awesome intellectual construct is that ultimate riddle, the origin of life. For when all is said and done, it remains as mysterious as ever why life exists at all. "In the world there is nothing to explain the world," Loren Eiseley put it at the end of his life; "Nothing to explain the necessity of life, nothing to explain the hunger of the elements to become life, nothing to explain why the stolid realm of rock and soil and mineral should diversify itself into beauty, terror and uncertainty" (Eiseley, 1975). We may never be able to explain how life arose. Yet one cannot help but suspect that the great stream of energy that passes across the earth plays a larger role in biology than our current philosophy knows: that perhaps the flood of power not only permitted life to evolve, but called it into being.

The reason is that a flux of energy through a homogeneous system can elicit a considerable degree of chemical and structural organization, even including chemical cycles reminiscent of those found in living systems (Morowitz, 1968). A simple example is the irradiation of a mixture of CO_2 and water vapor with short-wavelength ultraviolet light. Photodissociation generates CO, O, and also H and OH radicals, which react to produce formaldehyde and higher aldehydes together with O_2; aldehydes and oxygen, in turn, undergo a chain reaction that regenerates CO_2, CO, and water among other products. Thus the flux of radiative energy induces a cyclic sequence:

$$CO_2 + H_2O \rightleftharpoons \text{aldehydes} + O_2 \qquad \text{(Eq. 1.23)}$$

Prolonged irradiation with trapping of the intermediates leads to more complex compounds, particularly when the system is composed of more than one phase. A celebrated case in point is the experiment of S. L. Miller and H. C. Urey (Miller, 1953) in which mixtures of methane, hydrogen, ammonia, and water were subjected to prolonged electric discharges under conditions so arranged that the products were allowed to accumulate in the water reservoir. These products, after a period of weeks, came to include a variety of amino acids as well as polymeric compounds. Subsequent variations on this experiment have yielded purines, pyrimidines, sugars, porphyrins—a cornucopia of "prebiotic" chemistry (Miller and Orgel, 1974). From another point of view, Prigogine and his associates have analyzed the instabilities and inhomogeneities of physical structure induced by energy flow (Nicolis and Prigogine, 1977), which they believe ultimately underlie cellular organization, growth, and development. As matters presently stand, these discoveries from physics and chemistry cannot be unambiguously connected to the origin of life as we know it: the gap in organizational level, or degree of order, between the reaction vessel and even the simplest living cell is too wide for anything but faith. Here lies the highest challenge to bioenergetics and its noblest promise: that we may come to understand the genesis of life, not as an event improbable beyond belief that did, all the same, happen once on this earth, but as the probable, even inevitable, result of the flux of energy from the sun impinging on our sort of planet.

Does this mean that it should be our object to reduce biology to chemistry and thermal physics? I believe that the answer is an emphatic no, however popular and successful reductionism may be as a research strategy, for such a goal bespeaks a profound misunderstanding of the nature of complex systems. The laws of physics and chemistry constitute boundary conditions that living things cannot transgress, but they do not usually determine the mechanisms actually employed. The structures and functions that biologists dissect are emergent properties of complex organizations, consistent with the laws that govern the behavior of quanta and molecules but not reducible to those laws (Polanyi, 1968). To understand the workings of cells and organisms we must come to terms with the fact that their intricate organization is the daughter of time, the product of millions of years of history. Moreover, living things (unlike nonliving ones) are clearly endowed with purpose, directed toward survival and the reproduction of their own kind. This property, called teleonomy (Pittendrigh, 1958; Monod, 1971), distinguishes living things (and their artifacts, such as our machines) from all other objects and systems in the universe and tells us plainly that we must look beyond physics and chemistry for understanding. You will not find "teleonomy" in the dictionary, but the term is useful all the same to designate the purposeful character of living things without the entanglements of the more familiar term, teleology. The machinery of life owes its teleonomic character to evolutionary design; it reflects at all levels the interplay of physical laws with the contingencies of environment and history, the tension of chance and necessity (Monod, 1971).

A moment more before we leave the windy uplands of generalization for the reassuring (if often illusory) solidities of experiment and observation. In the chapters that follow, we shall primarily seek to understand, as best we presently can, the cellular and molecular mechanisms that have evolved for the teleonomic purpose of capturing energy and performing work. The constant theme is energy coupling—the variety of means by which exergonic processes, moving downhill toward equilibrium, are linked to endergonic ones and make the latter move uphill, away from equilibrium. It is helpful to distinguish two kinds of energy coupling. Energy transduction refers to the conversion of one form of energy into another, such that potential energy is (largely) conserved and may be recovered in a later step. Work, or useful work, refers to the utilization of energy to effect some physiologically desirable change without necessarily conserving the potential for future use.[4] The difference is not always sharp and may even be largely semantic, but it is of practical value when one is discussing cellular events. For instance, the conversion of the energy of light, first into the electrochemical potential of a proton gradient and then into the free energy of a high ATP/ADP ratio, is an instance of energy transduction. The coupling of the ΔG of ATP hydrolysis to the extrusion of Ca^{2+} ions from the cytoplasm or to the contraction of muscle would be described as work, even though in both cases things could be so arranged that free energy is largely conserved: in the normal course of events calcium transport and muscle contraction represent energy expenditure for a purpose rather than energy conservation.

The dynamic nature of living organisms raises a question concerning the applicability of such functions as free-energy changes and electrochemical potentials to biology. These concepts are rooted in classical thermodynamics and, strictly speaking, apply only to reversible processes: those operating so close to equilibrium that a small change in the driving force can alter the direction of the reaction. This is not as a rule the situation in biology: as noted above, living organisms correspond to a complex steady state, maintained far from equilibrium by the flux of energy and of matter. Consequently, as in all real processes, the entropy of the system comprised by the organism plus its environment increases continually, and this flow of energy and entropy cannot be satisfactorily described by the formalism of equilibrium thermodynamics. A treatment appropriate to open systems, which exchange both energy and matter with their environment, is becoming available through the development of nonequilibrium (or "irreversible") thermodynamics; one of its most important general insights is the demonstration that the steady state (in which such parameters as mass, composition, and function remain invariant with time) corresponds to minimum entropy production. The application of nonequilibrium thermodynamics to bioenergetics has been explored by Katchalsky and Curran (1965), Morowitz (1968, 1978), and Caplan (1971), but it remains an abstruse and mathematically demanding approach. I have therefore elected to adhere throughout this book to the language of equilibrium thermodynamics. This means that we can make

---

[4] I am indebted to Ian West for drawing my attention to this distinction.

quantitative statements only about the maximum capacities of biological processes, under ideal circumstances. However, for the purposes of a volume that is primarily concerned with cellular and molecular mechanisms, this familiar classical framework remains adequate.

References

ATKINSON, D. E. (1977). *Cellular Energy Metabolism and Its Regulation*. Academic, New York.

BRIDGMAN, P. W. (1961). *The Nature of Thermodynamics*. Harper & Row, New York.

BRILLOUIN, L. (1962). *Science and Information Theory*. Academic, New York.

CAPLAN, S. R. (1971). Nonequilibrium thermodynamics and its application to bioenergetics. *Current Topics in Bioenergetics* 4:1–79.

EDSALL, J. T., AND GUTFREUND, H. (1983). *Biothermodynamics: The Study of Biochemical Processes at Equilibrium*. Wiley, New York.

EISELEY, L. (1975). *All the Strange Hours: The Excavation of a Life*. Scribner, New York.

KATCHALSKY, A., AND CURRAN P. F. (1965). *Nonequilibrium Thermodynamics in Biophysics*. Harvard University Press, Cambridge.

KLOTZ, I. M. (1967). *Energy Changes in Biochemical Reactions*. Academic, New York.

LWOFF, A. (1968). *Biological Order*, 2d ed. M.I.T. Press, Cambridge.

MILLER, S. L. (1953). A production of amino acids under possible primitive earth conditions. *Science* 117:528–529.

MILLER, S. L., AND ORGEL, L. E. (1974). *The Origins of Life on Earth*. Prentice-Hall, Englewood Cliffs, N.J.

MONOD, J. (1971). *Chance and Necessity*. Knopf, New York.

MOROWITZ, H. J. (1968). *Energy Flow in Biology: Biological Organization as a Problem in Thermal Physics*. Ox Bow Press, Wood Bridge, Conn.; (reprinted 1979).

MOROWITZ, H. J. (1978). *Foundations of Bioenergetics*. Academic, New York.

NICHOLLS, D. G. (1982). *Bioenergetics: An Introduction to the Chemiosmotic Theory*. Academic, New York.

NICOLIS, G., AND PRIGOGINE, I. (1977). *Self-Organization in Non-Equilibrium Systems*. Wiley, New York.

PITTENDRIGH, C. S. (1958). Adaptation, natural selection and behavior, in *Behavior and Evolution*, A. Roe and A. A. Simpson, eds. Yale University Press, New Haven, Conn., pp. 391–416.

POLANYI, M. (1968). Life's irreducible structure. *Science* 160:1308–1312.

RIEDL, R. (1978). *Order in Living Organisms*. Wiley, New York.

SCHRÖDINGER, E. (1945). *What Is Life? The Physical Aspect of the Living Cell*. Macmillan, New York.

SETLOW, R. B. AND POLLARD, E. C. (1964). *Molecular Biophysics*. Addison-Wesley, Reading, Mass.

2

The Metabolic Web

But surely this is an old tale you tell, they say;
But surely this is a new tale you tell, others say.
Tell it once again, they say;
Or, do not tell it yet again, others say.
But I have heard all this before, say some;
Or, but this is not how it was before, say the rest.

Naqshbandi recital, from *The Way of the Sufi*, by Idries Shah

• • •

The Logic of Metabolism
What Enzymes Do
ATP and Energy Coupling
Patterns of Energy Generation
Reducing Power
Functional Organization of Metabolism

The Logic of Metabolism

Metabolism can be succinctly defined as the sum total of all the chemical reactions that take place in a cell or in an organism. Its complexity is graphically illustrated by the familiar wall charts that display upward of 500 molecular structures linked by a tangle of arrows. The nature of metabolic order, however, is better perceived from a teleonomic viewpoint: metabolism is the ensemble of reactions by which cells utilize resources to obtain useful energy and chemical building blocks and to provide the goods and services required for their continued existence, growth, and reproduction. The object of this chapter is to consider the meaning of energy coupling and work in the context of metabolism.

This subject is covered to a greater or lesser degree in courses and textbooks of biochemistry (for example, Lehninger, 1982; Metzler, 1977; and Stryer, 1981) and is likely to be generally familiar to most readers. A brief recapitulation of the main findings of metabolic bioenergetics is nevertheless desirable here, for membrane bioenergetics and other recent developments are historically and conceptually rooted in classical biochemistry. In the preparation of this chapter I have drawn particularly on D. E. Atkinson's thoughtful book, *Energy Metabolism and Its Regulation* (1977), whose influence will be plain to the reader.

For present purposes, a convenient point of departure is the block diagram of cellular metabolism shown in Figure 2.1, which emphasizes the functional relationships at the expense of all detail and allows one to visualize the flow of matter from organic nutrients to cell constituents. Figure 2.1 is particularly applicable to organisms, from *Escherichia coli* to humans, that live by the oxidative degradation of preformed organic substances to CO_2 and water and synthesize most of their cell constituents de novo. The first, or catabolic, block consists of the degradative pathways by which carbon, energy, and sometimes nitrogen, sulfur, and phosphorus are made available. Bacteria in particular differ greatly with respect to their catabolic aptitudes: some degrade dozens of organic substances, others are largely restricted to glucose. Yet in most bacteria, indeed in organisms generally, the special pathways that initiate substrate degradation soon converge onto a few great metabolic highways: glycolysis, the citric acid cycle, and the pentose phosphate cycle (Figure 2.2). The almost universal distribution of these sequences throughout the biological world is evidence of their evolutionary antiquity; it also exemplifies the biochemical unity that underlies the stunning diversity of living things and permits a reasonably coherent treatment of cellular physiology.

The second, or anabolic, block amalgamates hundreds of biosynthetic reaction sequences that generate small molecules, some common to all cells and others confined to but a few: amino acids, nucleotides, lipids, sugars, steroids, pigments, antibiotics, and pheromones. Biochemical unity, less evident than in the first block, still prevails in the sense that for each of the basic molecules the number of known pathways is remarkably small, usually just one or two. Finally, the third block represents the synthesis of the macromolecules required for growth: proteins and nucleic acids, cell wall polymers, and complex membrane

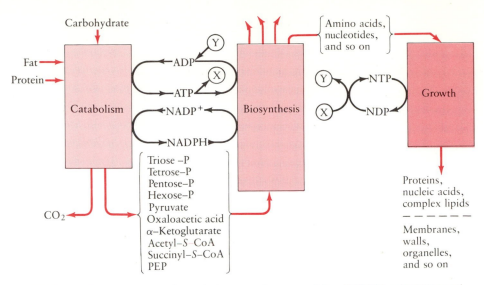

FIGURE 2.1   Block diagram of metabolism in a heterotrophic cell. NTP and NDP stand for nucleoside triphosphates and diphosphates; X and Y are the "terminals" that connect growth to the catabolic block. (From Atkinson, 1975, with permission of Plenum Press.)

lipids. In growing cells there is a net flux of matter through the metabolic web as substrate is converted into new protoplasm, but much metabolic activity continues in nongrowing cells, especially in the first two blocks.

Each block conceals much chemical complexity, but we should note the small number of molecules that connect the blocks, being produced in one and utilized in the next. Fewer than a dozen metabolites generated in the catabolic block constitute the raw materials for most of biosynthesis: four sugar phosphates (triose, tetrose, pentose, and hexose phosphates), three α-keto acids (pyruvate, oxaloacetate, and α-ketoglutarate), two activated carboxylic acids (acetyl-S-coenzyme A and succinyl-S-coenzyme A), and phosphoenolpyruvate. By the same token, a relatively small number of amino acids and nucleotides are the building blocks from which the multitudinous proteins and nucleic acids are constructed. To these we must add two special molecules, set apart by their unique and central biological role: ATP and NADPH or, more precisely, the ATP-ADP and NADPH-NADP$^+$ couples. Most biosynthetic sequences are reductive, and in the majority of cases the electron donor is NADPH, which is concurrently oxidized to NADP$^+$. Likewise, almost all biosynthetic pathways involve ATP or one of its congeners, for reasons generally described as energy input; in the process ATP is degraded either to ADP or to AMP. However, unlike the metabolic precursors that are incorporated into later products and must be made afresh every time, NADPH and ATP act in a cyclic manner and can be regenerated or recharged: they are the basic coupling agents of cellular metabolism. We may, then, without gross distortion summarize the functional relationships by stating that the catabolic pathways supply building blocks, reducing

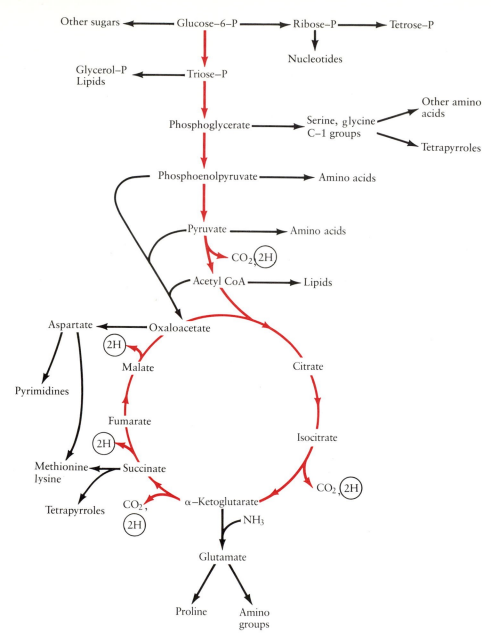

FIGURE 2.2   Core reactions of metabolism. The colored arrows identify the chief amphibolic routes that supply precursors and ATP; the black arrows identify selected biosynthetic pathways. The symbols 2H designate sites at which reducing power is produced in the form of NADH (succinate dehydrogenase reduces FAD instead).

power, and useful energy for the synthesis of molecules large and small and for various work functions as well (Lipmann, 1941; Krebs and Kornberg, 1957).

Before we turn to a closer examination of the meaning and mechanism of energy coupling in metabolism, a word is in order concerning some important features omitted from Figure 2.1. First, the distinction between catabolic and anabolic pathways is not rigid, particularly in the core area of metabolism. Glycolysis, the citric acid cycle, and other pathways (Figure 2.2) serve both to degrade substrates and to produce intermediates for biosynthesis and are better described as amphibolic pathways. Second, many reductive reactions depend on NADH and $FADH_2$ as electron donors, even a few biosynthetic ones. But these two electron carriers function primarily in glycolysis and the citric acid cycle and as links between these pathways and membrane-bound electron transport chains. We can thus say that NADH and $FADH_2$ serve as electron carriers within the catabolic block rather than as coupling agents between catabolism and anabolism. Finally, cellular growth entails the net consumption of ATP and other nucleoside triphosphates for nucleic acid synthesis; only in nongrowing cells would it be literally true that ATP is regenerated rather than synthesized afresh. Furthermore, there are many reactions in which other nucleotides take the place of ATP: UTP in the synthesis of polysaccharides, CTP in that of lipids, GTP in protein synthesis. However, the various nucleoside triphosphates are thermodynamically equivalent and chemically similar; their identification with particular functions is evidently related to the regulation of metabolism, not to its driving force.

Again, while Figure 2.1 is a fair representation of the metabolic pattern of a heterotrophic organism, autotrophic organisms arrange matters somewhat differently. In a green plant, for instance, catabolic pathways generate precursors for biosynthesis but make only a minor contribution to the production of ATP and NADPH. However, two new blocks appear: one for the photochemical reactions that harvest light and capture part of the energy in the form of ATP and NADPH, and another for the "dark reactions" of photosynthesis by which CO_2 is reduced to the carbohydrate level (Chapter 8). But none of these refinements affects the principal conclusion of this section, to wit, that the ATP-ADP and NADPH-$NADP^+$ couples perform a unique metabolic function in linking together the segments of the metabolic web.

What Enzymes Do

From the thermodynamic point of view, the conversion of nutrients into cell constituents can be regarded as a grand chemical reaction that proceeds with an overall decrease of free energy. This is the warp of the metabolic web. Its woof is the work of a multitude of enzymes that select the particular strands out of many that are thermodynamically possible in order to carry the flux of matter through the web. Enzymes, like all catalysts, accelerate reactions but cannot alter the position of equilibrium. They direct the course of metabolic events because their kinetic parameters favor one pathway over another, while their specificity ensures the absence of by-products. This is a striking feature that distinguishes metabolic chemistry from the organic chemist's. Equally important, the cell's ability to regulate enzymatic activity makes functional sense of the metabolic economy. As Monod (1971) put it, enzymes and proteins in general are the executors of biological purposes. Like Maxwell's troublesome demon, enzymes are able to recognize molecules with extreme precision, to the point of distinguishing one stereoisomer from the other. This cognitive function is the basis for the generation of chemical order and determinacy.

Enzymatic catalysis itself illustrates how energetic principles pervade every level of the metabolic economy. The driving force for each metabolic sequence is the difference in free energy, ΔG, between the starting materials and the end products. This parameter measures how far a mixture is from its equilibrium composition, but it tells us nothing about the rate of reaction. Gasoline, for example, is thermodynamically unstable in the presence of air yet kinetically so stable that it can be handled quite safely. The reason is that before a molecule can undergo a chemical reaction, it must first acquire enough internal energy to reach an activated condition in which there is a high probability of an existing bond being broken or a new one formed. This transition state corresponds, then, to a structure of lower stability (that is, of higher free energy) than that of the bulk of the population. The rate of any chemical reaction is proportional to the number of activated molecules and therefore depends strongly on the activation energy, defined as the free energy required to raise one mole of the substance to the level of the transition state (Figure 2.3). The higher the energy barrier, the fewer the molecules that can surmount it spontaneously and the slower the reaction. When we apply a match to gasoline, we raise the kinetic energy of a few molecules sufficiently to pass the barrier, and the heat released by their combustion carries the rest along. Living things, being isothermal, must employ another strategy: enzymes, like other catalysts, accelerate reactions by reducing the activation energy. As a rough measure, lowering the free energy of activation by 1.4 kcal/ mol (5.8 kJ/mol) is worth a tenfold increase in rate, and enzymes accelerate reactions by factors ranging from 10^8 to 10^{20}, far more than nonprotein catalysts can.

How is this reduction in activation energy accomplished? The critical step in enzymatic catalysis turns out to be the formation of the particular transition complex that can go on to decompose into the final product. There is no need to invoke special reaction mechanisms in order to explain the tremendous catalytic powers of enzymes. To be sure, the chemical versatility of some enzymes is enhanced by special cofactors, and in some cases particular amino acids enter into covalent reaction with the substrate to form more reactive structures than were originally present. But by and large the chemistry of enzyme-catalyzed reactions is quite conventional, depending on general acid or base catalysis and on the propensity of many amino acid side chains to donate electrons (Walsh, 1979; Fersht, 1985). What makes enzymes such effective catalysts is in general not

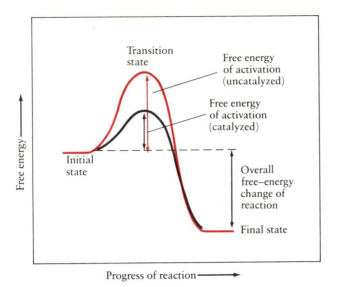

FIGURE 2.3   Free-energy diagram of a chemical reaction. Black line, uncatalyzed reaction; colored line, catalyzed reaction. (After Lehninger, 1982, with permission of Worth Publishers.)

a matter of chemical mechanism but of geometric complementarity between the substrate and the catalytic site. More precisely, the catalytic site is thought to be complementary to the configuration of the transition state. Binding of the substrate to the enzyme raises its effective concentration at the site of the reaction. Beyond that, the substrate molecule is presented in a particular configuration that may strain it, distort it, or place it in close proximity and in the correct orientation to a reactive group on the protein or on a second substrate molecule. Thus congruence of the substrate and the catalytic site enhances both the rate of the reaction and its specificity.

The pioneers of enzymology likened the relationship between enzyme and substrate to that between lock and key. Today's conception, described by Daniel Koshland's term "induced fit," resembles rather the relationship between hand and glove (Koshland, 1973). The point is that so long as substrate and enzyme are free in solution, neither has the configuration necessary for reaction. A close match only develops at the time binding takes place, often in a crevice or cavity of the protein molecule. This view explicitly recognizes that proteins are not rigid structures but undergo substantial changes in molecular configuration before, during, and after catalytic activity. The enzyme carboxypeptidase supplies a striking illustration. Formation of the enzyme-substrate complex apparently entails interaction between the terminal carboxylate residue and a positive group on the enzyme; this in turn frees an arm of the enzyme to swing over the substrate, creating the catalytic site and at the same time secluding it from the external medium (Hartsuck and Lipscomb, 1971). The conformational mobility of enzyme proteins is not an accidental feature but an essential element in their design. We shall return to this matter in subsequent chapters, for substrate-induced changes in protein conformation also underlie the transport of metabolites across membranes, the movement of cells and internal structures, and many examples of biological signaling and regulation.

Twisting protein side chains out of their favored position, straining and distorting substrate molecules, or transferring them from aqueous solution into a hydrophobic pocket all represent the performance of work that must be included in the activation energy. How is that work paid for? It seems increasingly likely that what ultimately pushes the reaction over the hill is the intrinsic energy of the substrate binding to the catalytic site. We can describe the formation of the enzyme-substrate complex by the equilibrium reaction shown in Equation 2.1, with a characteristic formation constant K_f that measures the strength of binding:

$$E + S \rightleftharpoons ES \qquad \text{(Eq. 2.1)}$$

$$K_f = \frac{[ES]}{[E][S]} \qquad \text{(Eq. 2.2)}$$

The formation constant is directly related to the binding energy by

$$\Delta G_f^\circ = -RT \ln K_f \qquad \text{(Eq. 2.3)}$$

For example, if $K_f = 10$, the binding energy equals -5.7 kJ, only a little in excess of the thermal energy of molecules in solution. A formation constant of 10^4, however, corresponds to a binding energy of -22.8 kJ, which can do a significant amount of useful work.

Jencks has picturesquely dubbed this the Circe effect: "The utilization of strong attractive forces to lure a substrate into a site in which it undergoes an extraordinary transformation of form and structure." Specificity and rate acceleration are not entirely independent variables of enzymatic catalysis, for both arise from the free energy made available by binding:

> The manifestation of specificity in the maximum velocity of the covalent step of enzymic reactions appears to require the utilization of the free energy that is made available from binding interactions with specific substrates. The observed free energy of binding ordinarily represents what is left over after this utilization. . . . The principal difference between enzymic and ordinary chemical catalysis is that enzymes can utilize noncovalent binding interactions with substrates to cause catalysis, in addition to the chemical mechanisms utilized by ordinary catalysis. (Jencks, 1975).

In other words, the precision of complementary fit between substrate and catalytic site provides the thermodynamic basis for the cognitive as well as the catalytic functions of enzymes.

The preceding discussion emphasized the specificity of enzymes and their catalytic prowess. To round off the subject, at least passing mention must be made of the regulation of enzymatic activities. In the living cell, most enzymes do not perform solo but as members of some functional ensemble that exchanges products, precursors, and coupling agents with other ensembles. For the metabolic economy to perform efficiently, many enzymes must be subject to regulation. This is particularly true of enzymes located at branch points, where an intermediate must be partitioned between two sequences according to need. The metabolic stream is therefore divided into interlocking cybernetic loops that allow signals from within and without to modulate the rates of critical reactions. Control is exercised at more than one level. We distinguish first between the regulation of enzyme synthesis and that of enzyme activity. The latter includes modulation by covalent modification of the enzyme or by noncovalent association with small modifying ligands, and the ligands in turn may compete with the substrate at the active site or exert remote control by binding to a distinct regulatory site. This is not the place to explore the diversity of regulatory mechanisms and their kinetic consequences, matters that are covered in textbooks of biochemistry and will crop up again in Chapter 13. Their existence, however, must be kept in mind when we consider how the functional organization of the metabolic web is achieved.

ATP and Energy Coupling

Few biochemical generalizations have proved as fruitful and illuminating as Fritz Lipmann's (1941) formulation of the special role of "energy-rich" phosphate compounds in cellular economics. By the time Lipmann wrote his celebrated article, ATP had already been recognized as a phosphate donor in the fermentative metabolism of yeast and muscle. H. A. Krebs had proposed the citric acid cycle in 1938, and H. Kalckar and V. Belitser (building on earlier work by V. A. Engelhardt) had discovered oxidative phosphorylation (1940). Even earlier, E. Lundsgaard's finding in 1931 that muscle poisoned with iodoacetate can contract anaerobically at the expense of stored creatine phosphate had established a connection between phosphate bond energy and mechanical work. Lipmann's contribution was nothing less than the construction of a conceptual framework for bioenergetics and, indeed, for much of biochemistry. Its essense is the thesis that ATP and other compounds containing "energy-rich" phosphate bonds function as a kind of energy currency, not unlike the role of money in our own economy. Lipmann himself (1941) spoke of it as the phosphate cycle: he saw cellular metabolism as a dynamo that picks up P_i and generates a current of energy-rich phosphate bonds (symbolized with a squiggle, $\sim P$); these are distributed throughout the cell and utilized to do work, being in the process reconverted to P_i.

The catchy term "energy-rich" phosphate was to prove unfortunate, misleading generations of beginning students (including this author) who understood it to imply some special kind of bond that releases energy when broken. In fact, energy must be expended to break a bond, and phosphate compounds are no exception. The special character of the two terminal phosphate groups of ATP is better expressed by the term "group transfer potential," also coined by Lipmann. This is a measure of the tendency of the group in question to migrate to a suitable acceptor: ATP is a better donor of phosphate than glucose-6-phosphate, and it can be said to have a higher phosphoryl transfer potential. When water is chosen as the standard acceptor, the group transfer potential is numerically equal to the free energy of hydrolysis. But hydrolysis, being a forbidden or at least useless reaction, is not of interest in itself; what matters is the "pressure" to transfer phosphoryl groups to metabolites other than water.

The metabolic role of the terminal phosphoryl groups of ATP is central but by no means unique. GTP, UTP, CTP, and other nucleotides share to some extent the role of ATP as a phosphoryl donor. In addition, there are compounds of still higher phosphoryl group transfer potential, such as phosphocreatine and phosphoenolpyruvate, that perform more specialized metabolic functions. Analogous compounds mediate the transfer of methyl groups, of acetyl and sulfate groups, and even of amino acids during protein synthesis (Table 2.1). In each case the group to be transferred is chemically activated, and it is possible to recognize structural features that render the molecule as a whole a good donor. In the case of ATP, for example, several factors combine to make the terminal phosphoryl unit a better leaving group than the phosphate of glucose-6-phosphate. Among these are the mutual repulsion of the negatively charged phosphoryl groups, the difference in resonance energy between ATP and its hydrolysis products, and the free energy of neutralization of the extra proton. But none of this quite answers the question of why ATP came to play so prominent a role in the energy economy of all living things: we can see features that make it suitable, but none that makes

TABLE 2.1 Some Metabolic Group Transfer Agents

Compound	Group activated	$\Delta G^{o\prime}$ (kJ/mol)
ATP	Phosphate	−33.5
1,3-Diphosphoglycerate	Phosphate	−54.5
Phosphoenolpyruvate	Phosphate	−62
Creatine phosphate	Phosphate	−43
Acetyl phosphate	Acetate and phosphate	−47.7
Acetyl-S-coenzyme A	Acetate	−35
UDP glucose	Glucose	−30.5
Valyl tRNA	Valine	−35
N^{10}-formyltetrahydrofolate	Formate	−26

it unique. Presumably the answer is that the position of ATP in the metabolic web reflects evolutionary design rather than chemical necessity; the reasons are at bottom historical and will not be apparent unless and until we solve the mystery of biological origins.

That living things use ATP as a kind of energy currency is a statement so familiar it has become a cliché, but it is not so obvious just how ATP performs that function or indeed just what is meant by energy coupling in the metabolic context. Fundamentally the answer is to be found in the thermodynamic coupling that arises between any two reactions sharing a common intermediate, but the principle is elusive enough to require closer scrutiny. Let us first consider a simple reaction: the role of ATP in the formation of glucose-6-phosphate. This compound will be formed to a very limited extent if we incubate glucose and phosphate (with an appropriate catalyst to speed things up), according to Equation 2.4:

$$\text{glucose} + \text{P}_i \rightleftharpoons \text{glucose-6-P} + \text{H}_2\text{O} \qquad K = 7 \times 10^{-3}$$

$$\Delta G^{\circ\prime} = 12.5 \text{ kJ/mol} \qquad \text{(Eq. 2.4)}$$

To put it another way, if we wished to establish a 1:1 ratio of glucose-6-phosphate to glucose, the concentration of P_i would have to be $1/7 \times 10^{-3} \, M$, or 143 M, an unattainable concentration. (By convention, water is assigned an activity of unity in this and subsequent calculations.) Generation of a biologically useful concentration ratio is, however, possible by an alternative route such as the reaction catalyzed by hexokinase:

$$\text{glucose} + \text{ATP} \xrightarrow{\text{hexokinase}} \text{glucose-6-P} + \text{ADP} \qquad K = 3600$$

$$\Delta G^{\circ\prime} = -21 \text{ kJ/mol} \qquad \text{(Eq. 2.5)}$$

The ratio of glucose-6-phosphate to glucose at equilibrium will now be a function of the ratio of ATP to ADP. If a cell were to maintain the latter at 5:1, well within the physiological range, the ratio of sugar phosphate to the parent sugar must be:

$$\frac{[\text{glucose-6-P}]}{[\text{glucose}]} = K \frac{[\text{ATP}]}{[\text{ADP}]} = 18,000 \qquad \text{(Eq. 2.6)}$$

What ATP does in this process is serve as an activated donor of phosphate, a function analogous to that of the activated reagents often employed in organic syntheses to enhance both the rate of a reaction and its yield. By using activated intermediates as stoichiometric participants, a cell can attain a useful ratio of products to reactants while keeping the concentration of all the reagents sufficiently low to be compatible with other functions.

It is sometimes said that a particular reaction or process is driven by the free energy of ATP hydrolysis. This misleading phrase stems from the following calculation. We can consider the hexokinase reaction, Equation 2.5, to be the sum of two separate reactions:

$$\text{ATP} + \text{H}_2\text{O} \rightleftharpoons \text{ADP} + \text{P}_i \qquad \Delta G^{\circ\prime} = -33.5 \text{ kJ/mol} \quad \text{(Eq. 2.7)}$$

$$\underline{\text{glucose} + \text{P}_i \rightleftharpoons \text{glucose-6-P} + \text{H}_2\text{O} \quad \Delta G^{\circ\prime} = 12.5 \text{ kJ/mol} \quad \text{(Eq. 2.4)}}$$

$$\text{glucose} + \text{ATP} \rightleftharpoons \text{glucose-6-P} + \text{ADP} \quad \Delta G^{\circ\prime} = -21 \text{ kJ/mol} \quad \text{(Eq. 2.5)}$$

This procedure is legitimate because the free-energy change of a reaction depends only on the initial and the final state, not on the path of the reaction. It is therefore permissible to write any feasible sequence of reactions, whether or not they actually occur, and add up their ΔG's algebraically. But this is no more than a useful computational device, for to make glucose-6-phosphate in significant yield by the sum of reactions 2.7 and 2.4 would be possible only at the price of having most of the phosphorus accumulate as P_i. The direct hexokinase reaction evades this difficulty. So it is not the free energy of ATP hydrolysis that drives the synthesis of glucose-6-phosphate but the stoichiometric participation of ATP in the reaction. By making the ATP-ADP couple part of the reaction pathway, the ratio of glucose-6-phosphate to glucose is shifted to an extent determined by the ATP/ADP ratio.

Examples of this principle are encountered throughout the metabolic web and are not restricted to phosphorylated compounds. Sucrose, the usual transport form of sugar in green plants, can be made from its constituents by the reaction:

$$\text{glucose} + \text{fructose} \rightleftharpoons \text{sucrose} + \text{H}_2\text{O} \qquad K = 1 \times 10^{-4}$$

$$\Delta G^{\circ\prime} = 23 \text{ kJ/mol} \quad \text{(Eq. 2.8)}$$

Once again, no more than negligible amounts of sucrose will be present at equilibrium at any reasonable levels of glucose and fructose, but then this is not how plants produce the disaccharide. They rely instead on a sequence of phosphorylative reactions that involve both ATP and UTP:

$$\text{ATP} + \text{glucose} \longrightarrow \text{glucose-6-P} + \text{ADP}$$

$$\text{glucose-6-P} \rightleftharpoons \text{glucose-1-P}$$

$$\text{glucose-1-P} + \text{UTP} \longrightarrow \text{UDP glucose} + \text{PP}_i$$

$$\text{ATP} + \text{fructose} \longrightarrow \text{fructose-6-P} + \text{ADP}$$

$$\text{UDP glucose} + \text{fructose-6-P} \longrightarrow \text{sucrose-6-P} + \text{UDP}$$

$$\text{PP}_i + \text{H}_2\text{O} \longrightarrow 2\text{P}_i$$

$$\underline{\text{sucrose-6-P} + \text{H}_2\text{O} \longrightarrow \text{sucrose} + \text{P}_i}$$

$$\text{glucose} + \text{fructose} + 2\text{ATP} + \text{UTP} \longrightarrow \text{sucrose} + 2\text{ADP} + \text{UDP} + 3\text{P}_i$$

$$\text{(Eq. 2.9)}$$

Overall, the standard free-energy change is about -67 kJ/mol, which, by Equation 1.7, corresponds to an equilibrium constant greater than 10^{11}. Plants thus carry out efficient and virtually unidirectional synthesis of sucrose at the expense of two molecules of ATP and one of UTP. This conclusion holds even though none of the participants is present in the plant at standard concentration. To put the matter in more general terms, by having sucrose formation stoichiometrically linked to the conversion of ATP into ADP (and of UTP into UDP), its synthesis is favored to an extent that depends on how far from equilibrium the cell maintains the poise of the ATP-ADP and UTP-UDP couples.

The magnitude of this effect is far greater than is suggested by the physiological ratio of ATP to ADP, which is normally around 10. If we take $\Delta G^{\circ\prime}$ for ATP hydrolysis (Equation 2.7) to be around -33.5 kJ/mol (-8 kcal/mol) and assume physiological concentrations of the adenylate nucleotides and of P_i, then the concentration ratio [ATP]/[ADP] at equilibrium would be about 4×10^{-8}. In fact, cells maintain this ratio above unity, or higher than the equilibrium value by a factor of 10^8. It can be shown that any reaction involving the conversion of one ATP into ADP will thereby be displaced from equilibrium by a like factor. The utilization of ATP in a biosynthetic reaction sequence thus makes any reasonable concentration of the product thermodynamically available (Atkinson, 1971, 1977).

Let us now turn to the mechanisms by which the cell maintains the [ATP]/[ADP] ratio so far from equilibrium. The chief processes involved, photosynthesis and oxidative phosphorylation, will be described shortly, but the thermodynamic principles can be conveniently discussed here in reference to ATP generation in the glycolytic pathway (Figure 2.2). The anaerobic fermentation of glucose to lactic acid can be written as

$$\text{glucose} \longrightarrow \text{lactic acid} \qquad \Delta G^{\circ\prime} = -197 \text{ kJ/mol} \qquad \text{(Eq. 2.10)}$$

The actual pathway involves a series of phosphorylated intermediates, with the overall stoichiometry

$$\text{glucose} + 2\text{ADP} + 2P_i \longrightarrow 2 \text{ lactic acid} + 2\text{ATP} + 2H_2O$$
$$\Delta G^{\circ\prime} = -135 \text{ kJ/mol} \qquad \text{(Eq. 2.11)}$$

Two reactions in the sequence involve the stoichiometric conversion of ADP into ATP. The first of these, Equation 2.12, involves the intermediate formation of 1,3-diphosphoglycerate, a highly activated phosphoryl donor (Table 2.1). The second reaction, Equation 2.13, includes the intermediate formation of another such compound, phosphoenolpyruvate:

$$\text{glyceraldehyde-3-P} + P_i + \text{ADP} + \text{NAD}^+ \rightleftharpoons$$
$$\text{3-phosphoglycerate} + \text{ATP} + \text{NADH} + H^+ \qquad \Delta G^{\circ\prime} = -12.5 \text{ kJ/mol}$$
$$\text{(Eq. 2.12)}$$

$$\text{2-phosphoglycerate} + \text{ADP} \rightleftharpoons \text{pyruvate} + \text{ATP} \qquad \Delta G^{\circ\prime} = -29 \text{ kJ/mol}$$
$$\text{(Eq. 2.13)}$$

It is the stoichiometric participation of ATP and ADP in the chemical reactions of glycolysis that allows entities as different as erythrocytes and fermentative bacteria to maintain their [ATP]/[ADP] ratio near 10, far above its equilibrium position. Since the free-energy change of a reaction is closely related to its equilibrium constant (Equation 1.7), we can also express this conclusion by the statement that part of the free energy of glucose degradation (Equation 2.10) has been captured in the form of ATP. More precisely, it has been captured by displacement of the [ATP]/[ADP] ratio from the value it would have at equilibrium.

The ATP/ADP ratio at equilibrium is a function of the free energy of ATP hydrolysis (Equation 2.7), and biochemists have expended much effort on measuring the numerical value of this parameter. The inherent difficulties of measuring a large ΔG are compounded by the fact that ATP and ADP are polyanionic molecules that give rise to multiple ionic species, all of which form complexes with divalent cations. The free energy of hydrolysis thus becomes a function of the pH and of the concentration of cations, particularly of Mg^{2+}. Recent measurements of the standard free energy of ATP hydrolysis $\Delta G^{\circ\prime}$ range from -6.8 to -8.7 kcal/mol (-28.4 to -36.6 kJ/mol), with -7.6 kcal/mol (-31.8 kJ/mol) being perhaps the best measurement available. The free energy required to support ATP synthesis in vivo depends on this value and also on the activities of ATP, ADP, P_i, and Mg^{2+} and on the cytoplasmic pH. These parameters vary, and the data required for accurate calculations are in general not available. However, -10 to -12 kcal/mol (-41.8 to -50.2 kJ/mol) is a reasonable estimate for the free energy of ATP hydrolysis ΔG under conditions prevailing in cytoplasm. In other words, to support ATP synthesis in vivo a reaction must deliver at least 10 to 12 kcal/mol at the actual concentrations of substrates and products.

The central role of ATP in the metabolic economy is universally accepted, but it has been seriously questioned whether this role is correctly described in thermodynamic terms. For example, Banks and Vernon (1970) argued in a provocative article that even if one discounts the confusion stemming from misuse of the term "high-energy phosphate," the concept is fundamentally flawed. Living things are open systems and therefore tend to a steady state in which the amount of matter entering equals that leaving, and the concentrations of intermediates become invariant with time. Since the concentration of ATP remains constant, the $\Delta G^{\circ\prime}$ of its hydrolysis becomes irrelevant to the search for an understanding of its actual role, which must be formulated in terms of kinetics and mechanism, not of thermodynamics. "One thing would appear certain: the standard free energy of hydrolysis of ATP is its least interesting property" (Banks and Vernon, 1970). This is an extreme position that beclouds the issue; the point is dealt with in detail by Atkinson (1977). The gist of the argument is that in the steady state, the flux of free energy through the system corresponds to the rate of conversion of ATP into ADP, and this is a real and meaningful reaction even though the concentrations of the two substances are kept constant by recycling ADP back into ATP. The conversion of ATP into ADP results from the stoichiometric participation of ATP in the reaction by which reactant A is transformed into product B, and it results in a displacement of the equilibrium ratio of B to A by a factor of as much as 10^8. It is true that the effect does not depend on the free energy of ATP hydrolysis as such, and that in principle another couple could take the place of ATP-ADP. However, in order to perform its function, any substitute for ATP-ADP would also have to be maintained far from equilibrium at concentrations that fell within the physiological range. In practice this means that the substitute would likewise have to have a large equilibrium constant (a large negative $\Delta G^{\circ\prime}$) for hydrolysis.

To summarize the preceding discussion, we can describe the role of ATP in the metabolic economy in two related yet distinct ways. The familiar formula states that part of the free energy released by glycolysis, a highly exergonic process (Equation 2.10), is captured and made to drive the endergonic process of ATP synthesis (Equation 2.7 in reverse). The free energy thus stored in the form of ATP is utilized in turn by coupling ATP breakdown to the synthesis of physiological products such as glucose-6-phosphate or sucrose. In much the same manner ATP supplies energy for muscular contraction, mediated by actomyosin ATPase. It is also the energy donor for the active transport of Na^+ and K^+ across animal cell membranes, mediated by the Na^+-K^+ ATPase, and for other ion translocations. All these processes, which are inherently endergonic, are rendered exergonic through linkage to the conversion of ATP into ADP (or sometimes into AMP) and so can go forward. There is nothing wrong with this formulation from the viewpoint of energetic bookkeeping, and I shall at times make use of it for brevity. Unfortunately the emphasis on the free-energy balance obscures both the biochemical mechanisms involved and their physiological meaning. It is preferable, albeit sometimes clumsy, to emphasize the role of the ATP-ADP couple as a carrier of useful free energy. Glycolysis, or respiration, maintains the [ATP]/[ADP] ratio at a value some 10^8-fold higher than its equilibrium position by means of such reactions as Equations 2.12 and 2.13. The free energy of glycolysis is thus conserved, not in energy-rich phosphate bonds, but in the form of an ATP/ADP ratio displaced far from equilibrium. The stoichiometric participation of ATP in biosynthetic reactions allows the latter to be displaced from their intrinsic equilibrium position by many orders of magnitude, thereby making the products available at physiologically useful concentrations. This viewpoint comes closer to the heat of the matter, for the teleonomic purpose of the metabolic web is to supply chemical products at relatively high concentrations, far higher than would be possible if their formation were not stoichiometrically bound up with the ATP-ADP couple. The meaning of energy coupling and work in the metabolic context is precisely the maintenance of concentration ratios far from equilibrium.

Patterns of Energy Generation

One of the primary tasks of the metabolic web is the production of useful energy for biosynthesis and work functions. In some circumstances, living organisms utilize environmental energy directly: examples are the butterfly basking on a leaf and the vulture or the dandelion seed soaring on currents of air. But these trivial exceptions must not obscure the general rule, that energy drawn from the environment is harnessed to the purposes of life through processing by the metabolic web. The output of useful energy takes several forms, chiefly high ratios of ATP/ADP, NADH/NAD$^+$, NADPH/NADP$^+$, and certain ionic gradients; but for our immediate purposes it is convenient to set out the patterns of biological energy production as though their sole object were the provision of ATP. This simplification is warranted because the various coupling agents are interconvertible and can thus be regarded as manifestations of a single energy pool. Besides, thanks to the central role of the ATP-ADP couple, only those energy sources can support growth that deliver enough free energy to significantly displace the ATP/ADP ratio from equilibrium, usually about 10 to 12 kcal/mol (40 to 50 kJ/mol).

The molecular mechanisms of ATP generation fall into just three classes, which correspond approximately to the time-honored concepts of fermentation, respiration, and photosynthesis.

1. Substrate-level phosphorylation. This term refers to an array of group transfer reactions, catalyzed by soluble enzymes, that generate various activated compounds with the ultimate production of ATP.

2. Electron-transport phosphorylation. This class includes numerous redox reactions that transfer electrons from a reduced substrate to a terminal electron acceptor, generally oxygen, coupled to the phosphorylation of ADP. Electron-transport phosphorylation is invariably associated with membranes and, as we shall see, its molecular mechanism relies on the translocation of protons across those membranes.

3. Photophosphorylation. Transduction of light energy into the free energy of the ATP-ADP couple is again dependent on the translocation of protons across a membrane. In the majority of organisms, light is absorbed by chlorophyll and subsequent steps involve an electron-transport chain.

These three modes of energy transduction will now be examined in a little more detail.

Substrate-Level Phosphorylation.   In the glycolytic fermentation of sugars into alcohol or lactic acid, the free energy of the reaction is partly conserved through the formation of 1,3-diphosphoglyceric acid and phosphoenolpyruvic acid (Equations 2.12 and 2.13); the activated phosphoryl group is subsequently transferred to ADP. These are widely distributed reactions, common to all anaerobic and aerobic organisms that use carbohydrates or glycerol as energy sources and familiar to any student of elementary biochemistry; they served for years as the paradigm of biological energy conservation in general.

Glycolysis is the main anaerobic mode of ATP production available to animal cells and to other eukaryotes. Bacteria are more versatile in this respect:

many anaerobic bacteria live by the fermentation of amino acids, carboxylic acids, purines, or pyrimidines, with the aid of phosphorylative reactions that are sometimes quite bizarre (see Figure 4.2). Though chemically diverse, all substrate-level phosphorylations share two important features exemplified by those of the glycolytic pathway. The first is that ATP formation is intimately and stoichiometrically coupled to the catabolic reaction, with the intermediate formation of an activated phosphoryl donor. Second, the enzymes are soluble, or at least can carry out the phosphorylative reaction after having been put in solution; association with membranes is not obligatory. Largely for that reason the molecular mechanisms of most substrate-level phosphorylations are quite well worked out.

Electron-Transport Phosphorylation.   Aerobic organisms from bacteria to humans generate the bulk of their ATP by the oxidation of reduced substrates coupled to the phosphorylation of ADP to ATP, with oxygen serving as the terminal electron acceptor. From the biochemical standpoint, the chief function of respiration is ATP production. Oxidative phosphorylation was recognized as a distinctive phenomenon in 1940 by H. Kalckar and by V. Belitser; for the past 40 years its mechanism has been one of the grand themes of biochemical research, with contributions from many of the founders of modern bioenergetics.

A major landmark along the way was David Keilin's discovery of the respiratory electron-transport chain. This is a particulate complex of redox carriers including flavoproteins, quinones, and cytochromes; it catalyzes the oxidation of NADH, succinate, and a few other reduced substrates by passing electrons sequentially from one carrier to the next (Figure 2.4). The reduced

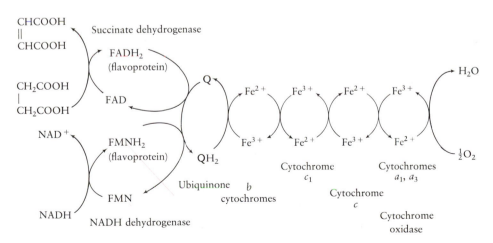

FIGURE 2.4   Pathway of electron transport in the respiratory chain of mitochondria. Coupling to ATP synthesis via the electrochemical potential of protons will be discussed in Chapter 3. (After Prebble, 1981, with permission of Longman Group, Ltd.)

substrates that serve as electron donors may be present in the external environment but are more commonly produced by metabolic reactions within the cell, particularly by the citric acid cycle (Figure 2.2). ATP synthesis proper is catalyzed by a second particulate enzyme complex, the ATP synthase.[1] By 1950 it was widely recognized that electron-transport particles do not exist in the cell as such but arise by the fragmentation of mitochondria, and that the mitochondrion serves as the powerhouse of eukaryotic cells. In bacteria the machinery of oxidative phosphorylation is housed in the plasma membrane, but its molecular basis is once again the collaboration of a respiratory chain with an ATP synthase.

The thermodynamic driving force for ATP generation during oxidative phosphorylation is clearly the difference in redox potential between the two half-reactions linked by the electron-transport cascade (Table 2.2). For example, the oxidation of NADH to NAD^+, with oxygen as the electron acceptor, spans a potential difference of 1140 mV under standard conditions (the span is even greater under physiological conditions). From Equation 1.14 we can calculate that the redox span for a pair of electrons corresponds to a free-energy difference of $\Delta G^{\circ\prime} = nF \, \Delta E^{\circ\prime} = -52.6$ kcal/mol $= -220$ kJ/mol. The phosphorylation of ADP to ATP requires an input of 10 to 12 kcal/mol. In principle, then, the oxidation of NADH to NAD^+ can support the production of several moles of ATP, in actuality probably three.

Thermodynamics illuminates the driving force but does not tell us how the exergonic and endergonic limbs are coupled, and that was for two decades the most contentious issue in bioenergetics. Historically, it was the failure of biochemists to devise a satisfying molecular explanation for energy coupling in oxidative phosphorylation that provoked Peter Mitchell to formulate his chemiosmotic theory and impelled the rise of membrane bioenergetics. The role of membranes in biological energy transduction will be explored in detail in later chapters. Suffice it here to state magisterially that the respiratory chain is an elaborate device to translocate protons across the membrane in which it is housed, and that the electrochemical potential of protons across the membrane is the driving force for ATP synthesis. Association of respiratory chains with membranes is not fortuitous; it is crucial to the mechanism of electron-transport phosphorylation.

The free energy available from oxidative reactions, and especially those that terminate with oxygen, is generally far greater than that released by fermentation reactions, and the yield of ATP per mole of substrate is higher. From a knowledge of intermediary metabolism one can calculate that the complete oxidation of glucose to CO_2 plus water yields as many as 36 moles of ATP per mole of glucose. By contrast, fermentation of glucose to lactic acid has a net yield of only two moles of ATP.

---

[1] The membrane-bound enzyme that produces ATP from ADP and P_i is designated ATP synthase. The same enzyme is often described as an ATPase because it is generally assayed in the hydrolytic direction. The factors that determine the direction of this reaction are considered in Chapter 3.

The amounts of ATP produced during metabolism are prodigious. Broda (1975) has calculated that one gram dry weight of the fermentative bacterium *Lactobacillus* produces and decomposes some 180 grams of ATP per day. The cells' ATP pool turns over some 300,000 times per day, and the mean lifetime of an ATP molecule is only one-third of a second. *Azotobacter*, an actively respiring bacterium, generates more than 7000 grams of ATP per gram dry cells per day. The daily ATP production of a human being comes to 75 kilograms, but the average turnover rate of the ATP pool is a modest 1000 times per day. Even so, the rate of energy production by a human being per unit weight is 10^4 times greater than that of the sun![2]

Photosynthesis and Photophosphorylation.   Photosynthetic organisms have historically been defined as those that can utilize light energy to convert CO_2 into biomass. This way of life is characteristic not only of the eukaryotic algae and green plants but also of a number of bacterial genera. Photosynthesis is by far the major route of energy transformation in the biosphere. The great majority of organisms today are either themselves capable of photosynthesis or live by the metabolism of organic substances originally produced by photosynthetic organisms. The only exceptions are to be found among the chemolithotrophic bacteria, which draw energy from a handful of inorganic reactions such as the oxidation of H_2S.

The amount of energy that light supplies is a function of its wavelength, or frequency, according to the relationship

$$U = Nhv = \frac{Nhc}{\lambda} \qquad\qquad \text{(Eq. 2.14)}$$

where U is the energy per mole of photons (a mole of photons is referred to as an einstein), N is Avogadro's number, v is the frequency, h is Planck's constant, c is the velocity of light, and λ is the wavelength. The shorter the wavelength, the higher the energy content. Photosynthesis spans the spectrum from 400 to 700 nanometers (nm), which corresponds to an energy content of 40 to 71 kcal/einstein (171 to 300 kJ). Each photon absorbed can thus in principle support the production of several molecules of ATP.

The assimilation of CO_2 is a strongly endergonic process: $\Delta G^{\circ\prime}$ for the conversion of CO_2 into $C_6H_{12}O_6$ plus free oxygen is 686 kcal/mol (2880 kJ/mol), which must be supplied by the concerted action of several quanta of light. It is also a reductive process, and the electron donors used vary considerably. Many photosynthetic bacteria are strict anaerobes and require either

[2] Implausible but nevertheless true. The sun's mass is given as 2×10^{33} grams and its total energy emission as 3.8×10^{33} ergs per second (erg/s), which comes to a mere 16 J/kg·day. The great bulk of the sun's mass is energetically inert. By contrast, a 70-kg human consuming 300 grams of carbohydrates per day (worth 17 kJ/g) turns over 7×10^4 J/kg·day.

organic or inorganic reducing agents: H_2, H_2S, lactate, or succinate. The overall stoichiometry of bacterial photosynthesis is given by Equation 2.15. The cyanobacteria (formerly known as blue-green algae but now recognized as prokaryotes), as well as the eukaryotic algae and all plants, rely on a second mode of photosynthesis in which water serves as the ultimate reductant and free oxygen is produced (Equation 2.16). These two modes of photosynthesis differ substantially but, as C. B. Van Niel recognized in the 1940s, their stoichiometry bespeaks a deeper unity: both the bacterial and the plant version can be described by Equation 2.17, suggesting a common underlying mechanism. In the following equations, (CH_2O) refers to biomass at the overall oxidation level of carbohydrates, while H_2A stands for all reducing agents.

$$CO_2 + 2H_2S \longrightarrow (CH_2O) + 2S + H_2O \qquad \text{(Eq. 2.15)}$$

$$CO_2 + 2H_2O \longrightarrow (CH_2O) + O_2 + H_2O \qquad \text{(Eq. 2.16)}$$

$$CO_2 + 2H_2A \longrightarrow (CH_2O) + 2A + H_2O \qquad \text{(Eq. 2.17)}$$

The metabolic reactions by which CO_2 is assimilated were largely worked out during the 1950s. The Calvin cycle, by which CO_2 is converted into triose phosphate, is a "dark reaction," and the same is true for the alternative pathways of CO_2 assimilation that were discovered subsequently. Dark reactions do not require light, only ATP and reducing power, and they are widely distributed among nonphotosynthetic organisms. For instance, the hydrogen bacteria and other chemolithotrophic bacteria assimilate CO_2 in this manner. The light reactions, intimately bound up with the membranes of chloroplasts and of photosynthetic bacteria, are exclusively concerned with the generation of ATP and of reducing power in the form of NADH and NADPH. These discoveries sharpened the focus of inquiry and altered the very meaning of the term photosynthesis. The heart of the matter is the mechanism by which photosynthetic organisms harness light energy to the production of ATP and reducing agents.

The molecular mechanisms of photosynthetic energy transduction have much in common with those of respiration. The unique feature is the presence of chlorophyll pigments. In the evocative words of Albert Szent-Györgyi (1961):

> When a photon interacts with a [chlorophyll molecule], it lifts one electron from an electron pair to a higher level. This excited state as a rule has but a short lifetime and the electron drops back within 10^{-7} to 10^{-8} seconds to the ground state, giving off its excess energy in one way or another. Life has learned to catch the electron in the excited state, uncouple it from its partner, and let it drop to the ground state through its biological machinery, utilizing its excess energy for life processes.

That machinery includes an electron-transport chain, organized within and across a membrane and composed of nonheme iron centers, cytochromes, and quinones. The function of that electron-transport chain is once again to pass protons across the membrane. The final step is mediated by an ATP synthase that belongs to the same molecular family as do those of mitochondria and nonphotosynthetic bacteria. The two versions of photosynthesis differ with respect to their light-harvesting chlorophyll centers and the organization of their electron-transport chains, but both rely on a chemiosmotic mechanism of energy transduction.

Reducing Power

The biological role of the nicotinamide adenine nucleotides is somewhat analogous to that of ATP: we can regard them as carriers of activated electrons that cycle between the oxidized and the reduced forms. Just as there is a range of phosphoryl donors differing in group transfer potential (Table 2.1), there is also a series of biological redox carriers (Table 2.2), with the NADH-NAD$^+$ and NADPH-NADP$^+$ couples near the reducing end of the scale. Furthermore, generation of NADH and NADPH constitutes a form of energy conservation that is no different in principle from the production of ATP. More precisely, displacement of the [NADH]/[NAD]$^+$ and [NADPH]/[NADP]$^+$ ratios from their expected equilibrium positions can in turn displace equilibria to which these nucleotides are stoichiometrically coupled and thus make reduced substances thermodynamically available.

TABLE 2.2   Reduction Potentials of Some Biologically Important Redox Couples

Half-reaction	E'_0 (V)	$\Delta G^{\circ\prime}$ (kJ/mol for oxidation by O_2, per two electrons)
$O_2 + 4H^+ + 4e^- \longrightarrow 2H_2O$	+0.82	0.0
$Fe^{3+} + e^- \longrightarrow Fe^{2+}$	+0.77	−9
Cytochrome a (Fe^{3+}) + $e^- \longrightarrow$ cytochrome a (Fe^{2+})	+0.29	−102
Cytochrome c (Fe^{3+}) + $e^- \longrightarrow$ cytochrome c (Fe^{2+})	+0.25	−110
Cytochrome b (Fe^{3+}) + $e^- \longrightarrow$ cytochrome b (Fe^{2+})	+0.08	−141
Ubiquinone + $2H^+ + 2e^- \longrightarrow$ ubiquinol	+0.04	−150
Fumarate + $2H^+ + 2e^- \longrightarrow$ succinate	+0.031	−153
Oxaloacetate + $2H^+ + 2e^- \longrightarrow$ malate	−0.17	−191
Pyruvate + $2H^+ + 2e^- \longrightarrow$ lactate	−0.19	−195
$NAD^+ + H^+ + 2e^- \longrightarrow$ NADH	−0.32	−220
$NADP^+ + H^+ + 2e^- \longrightarrow$ NADPH	−0.32	−220
Ferredoxin (Fe^{3+}) + $e^- \longrightarrow$ ferredoxin (Fe^{2+})	−0.41	−238
$2H^+ + 2e^- \longrightarrow H_2$	−0.41	−238

Data from Metzler, 1977.

The standard redox potentials of the two niotinamide adenine dinucleotides are the same but, as mentioned earlier, their biological functions are quite different. The NADH-NAD$^+$ couple is the main electron carrier in glycolysis, the citric acid cycle, and catabolic reactions in general. NADH is the main electron donor for the respiratory chain and is continually regenerated by various dehydrogenases, some linked to particular substrates and others associated with the metabolic highways. NADPH, by contrast, is the usual reductant in biosynthetic processes: the synthesis of fatty acids and steroids, the reduction of ribo- to deoxyribonucleotides, and others. NADPH is produced by the pentose phosphate cycle and by a number of other cytoplasmic reactions.

In general the enzymes that employ NADH or NADPH are specific for the one nucleotide or the other; the extra phosphate group in NADPH serves as a recognition code. But it is not sufficient for the enzymes of reductive biosynthesis to be able to recognize NADPH. To serve as an effective reductant, the NADPH-NADP$^+$ couple must be kept in a relatively reduced state, and that is indeed what one observes. In liver cytoplasm, for example, the ratio [NADH]/[NAD$^+$] is 0.001, but the ratio [NADPH]/[NADP$^+$] is 71, greater by a factor of 10^5 (Krebs and Veech, 1969). The disparity is much less marked in bacteria, but the two couples still differ in poise by a factor of 10 (Andersen and von Meyenburg, 1977). It is not entirely clear how this difference is maintained. On the basis of a very careful study of nucleotide ratios in liver, Krebs and Veech (1969) concluded that the difference between the ratios [NADPH]/[NADP$^+$] and [NADH]/[NAD$^+$] is the result of thermodynamic equilibria: each ratio is determined by the equilibrium constants of a few highly active, reversible enzymes (lactate dehydrogenase, glucose-6-phosphate dehydrogenase, glyceraldehyde phosphate dehydrogenase, malic enzyme) and by the concentrations of the substrates and products of these enzymes. The crucial feature is that certain substrates interact with both couples; the nucleotide ratios thus become linked and can be shown to be in good accord with the ratios calculated from the concentration of metabolites in liver cytosol.

In addition, both mitochondria and bacteria contain a membrane-bound enzyme that reversibly transfers hydrogen from NADH to NADP$^+$, poising the ratios [NADPH]/[NADP$^+$] and [NADH]/[NAD$^+$] at a value determined by the electrochemical potential of protons across the membrane. It seems intuitively likely that this "energy-linked transhydrogenase" (which will be more fully discussed in Chapter 5) serves to make NADPH available for biosynthetic reactions, but convincing evidence to this effect has not been forthcoming. The information at hand suggests that cytosolic and membrane-linked processes collaborate in maintaining the differential in reduction poise, but the details are uncertain.

The preceding summary applies to organisms that live by the oxidation of reduced organic substances, such as glucose. Special mechanisms appear in organisms that lack such obvious sources of chemical reducing power. The generation of NADH in certain bacteria by reversed electron transport will be discussed in Chapter 5; the manner in which green plants utilize water plus light to reduce $NADP^+$ to NADPH will be a major topic of Chapter 8.

Functional Organization of Metabolism

The metabolic web is designed to supply both useful energy and building blocks for cell constituents, and the chemical processes by which these two objectives are attained are intertwined. The usefulness of the web therefore depends on mechanisms that allow cells to monitor their metabolic status and to direct the flux of carbon and energy according to need. Metabolic control is a work function, on which cells expend a portion of the energy generated in metabolism. The adenine and nicotinamide nucleotides function both as carriers of metabolic energy and as regulators of energy flux.

The Direction of Metabolic Pathways.   Inspection of the metabolic web reveals many reaction sequences that run in opposite directions: oxidation of fatty acids and their synthesis, storage of glycogen and its consumption, synthesis of proteins and their breakdown. Such opposing sequences serve different physiological functions and generally utilize different enzymes and reaction mechanisms. But it is a remarkable feature that each individual sequence, regardless of its direction, proceeds with a large negative $\Delta G^{\circ\prime}$ and hence with a large equilibrium constant: each pathway is effectively unidirectional. Whenever necessary, this is accomplished by the stoichiometric participation of the coupling agents ATP-ADP or $NADPH\text{-}NADP^+$. We may describe this as an input of metabolic energy, but it is more precise to say that one of the chief physiological functions of coupling agents is the establishment of metabolic pathways that have large overall equilibrium constants. The course of metabolism is then determined not by the position of equilibrium but by regulatory signals that reflect physiological needs. These usually take the form of molecules that do not participate in the reaction but control the activities of key enzymes.

A case in point is the glycolytic pathway, which has been described so far only as a route for ATP production. When energy supplies are ample, many bacteria accumulate glycogen as a reserve material. Moreover, many bacteria grow happily on three-carbon compounds such as lactate, provided they can carry out oxidative phosphorylation. Growth on such substrates again requires the synthesis of hexoses as precursors for cell wall polymers and for nucleic acid pentoses. The pathway of glucose and glycogen biosynthesis shares many enzymes with that of glycolysis, yet proceeds in the opposite direction. This is possible because glucose breakdown and glucose synthesis differ at several steps that proceed with large equilibrium constants, so that each is effectively unidirectional.

Four such reactions are shown in Figure 2.5. During glycolysis, phosphoenolpyruvate is converted into pyruvate by a highly exergonic reaction that generates

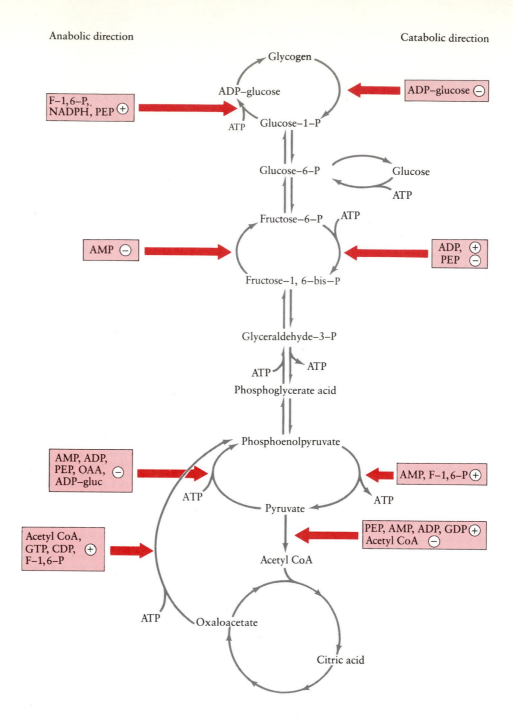

FIGURE 2.5 Direction and control of sugar metabolism in *E. coli*. Reactions involved in the utilization of glycogen and glucose are on the right, those of biosynthesis on the left. Compounds in boxes are modifying ligands, either stimulatory (+) or inhibitory (−). (After Mandelstam and McQuillen, 1973; Sanwal, 1970.)

ATP. This reaction is bypassed in gluconeogenesis in favor of one of several alternative reactions that produce phosphoenolpyruvate with the consumption of multiple ATP equivalents, chiefly the reaction catalyzed by the enzyme phosphoenolpyruvate synthase:

$$\text{pyruvate} + \text{ATP} \xrightleftharpoons{\text{PEP synthase}} \text{phosphoenolpyruvate} + \text{AMP} + 2P_i \quad \text{(Eq. 2.18)}$$

Two other phosphorylative reactions are bypassed by specific phosphatases, one for fructose-1,6-bisphosphate and another for glucose-6-phosphate. Finally, glycogen synthesis in *E. coli* relies on the formation of the activated glucosyl donor ADP glucose with the consumption of yet another molecule of ATP. Overall, for each molecule of glucose formed from lactate six ATP equivalents are consumed and two molecules of NADH are required as reductant.

   Pairs of opposing reactions, such as those catalyzed by phosphofructokinase and fructose bisphosphatase, exemplify what biochemists call "futile cycles." If both reactions were to proceed simultaneously at high rates, the resulting ATP dissipation would be catastrophic. In fact, the reactions are normally under kinetic control, which ensures that, even though both are thermodynamically favored, carbon flows predominantly either up or down (Figure 2.5). The adenine nucleotides themselves are among the chief regulators of metabolite flux.

Adenine Nucleotides as Metabolic Regulators.   Since the central role of the ATP-ADP couple stems from the stoichiometric participation of the adenine nucleotides in metabolic reactions, cells must endeavor to keep the [ATP]/[ADP] ratio high and relatively constant. Actually the matter is more complex, for the adenine nucleotide pool consists of ATP, ADP, and AMP in a dynamic equilibrium dominated by the enzyme adenylate kinase:

$$2\text{ADP} \xrightleftharpoons{\text{adenylate kinase}} \text{ATP} + \text{AMP} \quad \text{(Eq. 2.19)}$$

The equilibrium constant of adenylate kinase is not far from unity, and therefore the energy status of the adenine nucleotide pool is a function of all three nucleotides, not just of ATP and ADP. Moreover, the adenine nucleotide pool is small (on the order of 5 mM in growing bacteria) and turns over several times per second. Any imbalance between the rates of ATP synthesis and consumption will therefore cause sharp fluctuations in the cell's energy status, surely an undesirable state of affairs. One may therefore expect to find sensitive and powerful mechanisms to monitor the state of the adenine nucleotide pool and to ensure an appropriate response. For example, if ATP production falls behind momentarily, cells should increase the rate of ATP generation and also reduce its consumption.

   Such mechanisms take the form of allosteric inhibition or activation of key enzymes by ATP, ADP, or AMP; Figure 2.5 lists several examples. AMP, whose level fluctuates most dramatically, is a particularly effective regulator. But as examples multiplied, it became apparent that in many cases the regulatory parameter is not the concentration of any one nucleotide but the ratio of concentrations, [ATP]/[AMP] or [ATP]/[ADP]. (This may seem puzzling—how can binding sites sense a ratio? The answer is that the sites have high affinity for both nucleotides, but only one causes the response. An enzyme that is activated by AMP will therefore respond to some function of the mole fraction of AMP.) These regulatory effects show a clear relationship to cellular function. Enzymes whose activity brings about the synthesis of ATP tend to be activated as the proportion of ADP and AMP in the pool rises. Examples of such R regulation (for regeneration) include phosphofructokinase, pyruvate kinase, pyruvate dehydrogenase, citrate synthase, and isocitrate dehydrogenase. Conversely, enzymes whose function entails ATP consumption exhibit the opposite type of control, being stimulated by ATP but inhibited by ADP or AMP. Cases of U regulation (for use) are phosphoribosylphosphate synthase, aspartate transcarbamylase, nucleoside diphosphokinase, and the key enzyme in glycogen synthesis, adenosine diphosphate glucose synthase. At the same time these enzymes are also under feedback control by the end products of their activities. To put it anthropomorphically, cells try to make biosynthetic products only as needed and when they can afford the expense.

Reflection on the regulation of enzymes by adenine nucleotides led Atkinson (1970, 1977) to propose the adenylate energy charge as a useful expression to describe the energy status of the adenine nucleotide pool. The energy charge is defined by the relation

$$\text{energy charge} = \frac{[\text{ATP}] + \frac{1}{2}[\text{ADP}]}{[\text{ATP}] + [\text{ADP}] + [\text{AMP}]} \qquad \text{(Eq. 2.20)}$$

and can range in value from 1 (all ATP) to 0 (all AMP). It is an index of the fraction of the nucleotide pool that is charged with activated phosphoryl groups. The actual proportions of ATP, ADP, and AMP at any given value of energy charge are determined by the intervention of adenylate kinase (Equation 2.19). This enzyme performs the important function of catalyzing the rapid return of the adenine nucleotide pool to equilibrium following a change in any one of its constituents, but it does not alter the energy charge. Mutants lacking adenylate kinase are seriously impaired in growth, an indirect testimonial to the importance of its function.

We can conveniently express the regulatory effects of adenine nucleotides on R and U enzymes by plotting velocity against energy charge (Figure 2.6), without specifying which nucleotide is the actual effector. The velocity will be further modulated by other regulatory ligands, such as the end product of the pathway. In consequence the velocity can vary over a wide range, responding to both energy status and the demand for the end product.

The concept of adenylate energy charge has given rise to a certain amount of confusion. It was never meant to imply that enzymes that exhibit R or U

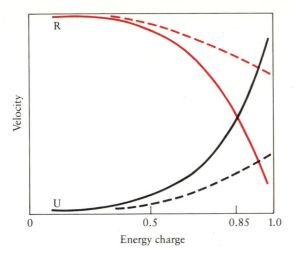

FIGURE 2.6   Theoretical curves showing the response of enzymes that catalyze ATP-utilizing reactions (U) and ATP-regenerating ones (R) to the adenylate energy charge. In each case the dashed line represents the effect of a regulatory ligand that counteracts the adenylate effect. (From Atkinson, 1977, with permission of Academic Press.)

regulation must sense and respond to the particular ratio that we define as the energy charge; most enzymes appear to establish meaningful relationships with two nucleotides at most and respond to their ratio. Neither was it intended to suggest that some central agency monitors the energy charge and adjusts the cellular economy accordingly. The energy charge concept is much more akin to the "invisible hand" that, according to Adam Smith, keeps the human economy in balance: the net effect of a large number of individual R and U responses, differing in kinetic detail, should be to stabilize the energy charge near the upper end of the scale, around 0.9. By the same token, the energy charge (or rather, the individual ATP/AMP and ATP/ADP ratios that enter into it) serves to regulate the individual R and U enzymes. The logic of regulatory systems is circular. In a constant-temperature water bath the temperature itself is the input that determines whether the heating coil will be on or off. In the same manner, the energy charge is at once the parameter to be stabilized and the signal to which the metabolic economy responds.

This is not to say that all enzymes should respond to the energy charge in precisely the same manner. On the contrary, one may expect a hierarchy of responses to help ensure that cells will deposit reserve substances (glycogen, poly-β-hydroxybutyrate) only when energy supplies are well in excess of the demands for growth, that macromolecule synthesis and growth will proceed as long as circumstances permit, and that in times of scarcity energy will be allocated to maintenance functions in preference to growth. The accumulating evidence

(Atkinson, 1977; Dawes and Senior, 1973; Knowles, 1977) is generally consistent with this pattern. For instance, the activities of nucleoside diphosphate kinase and of ADP-glucose synthase, key enzymes in growth and glycogen synthesis respectively, both exhibit steep activation at high values of energy charge. One should not, however, expect too simple a pattern, for each enzyme must correlate and integrate multiple inputs, both general and specific. It is likely, for instance, that cells also monitor the status of their nicotinamide adenine nucleotide pool and that the "reduction charge" (Andersen and von Meyenburg, 1977) is another general regulator of many enzymatic activities. The informational functions of the adenine and nicotinamide nucleotides complement their more familiar role in energy transfer.

Metabolic Work.   To conclude this chapter let us return to the question posed at the beginning: What is meant by energy coupling and work in the context of metabolism? The object of the metabolic enterprise is to convert simple molecules (CO_2, glucose) into complex ones (proteins, lipids, nucleic acids) at concentrations infinitely greater than would ever be present at equilibrium, to maintain elaborate ensembles composed of specific macromolecules in the face of the forces of decay, and to supply both the means and the materials for the enlargement and multiplication of such ensembles. This grand chemical synthesis has a positive change in free energy and can proceed only thanks to a corresponding decrease in the free energy of some primary energy source, ultimately the sun. Consequently, we must envisage a steady flux of free energy through the metabolic web, coupled to that of matter. Research is preoccupied with mechanistic questions: How do living organisms unlock the free energy potentially available from light and from chemical gradients, and how do they harness it to drive reactions away from their "intrinsic" equilibrium positions and do whatever else is necessary to support the synthesis of biomass?

But we can also regard the metabolic web as a device for generating dynamic order at the molecular level. Biological molecules, with their intricate yet entirely predictable structures, embody a remarkable degree of determinacy. As a result of metabolism, energy is somehow converted into molecular organization. What is it about all these reactions, which seem ordinary enough from the chemical viewpoint, that endows them with the mystical quality of generating order? The answer lies with the enzymes that create very special channels through which flows the stream of free energy, firmly and specifically linked to useful chemical transformations. Enzymes can accomplish this by virtue of their ability to recognize specific molecules, bind them, and facilitate particular reactions. This capacity ultimately stems from the enzymes' own highly determinate structure, refined by millennia of mutation and selection for a fitter organism. In the last analysis, the meaning of metabolic work lies in the production of a particular constellation of complex molecules, a unique order that is assigned by the historical forces that have shaped the organism as a whole.

References

ANDERSEN, K. B., and VON MEYENBURG, K. (1977). Changes of nicotinamide adenine nucleotides and adenylate energy charge as regulatory parameters for growth of *E. coli*. *Journal of Biological Chemistry* **252**:4151–4156.

ATKINSON, D. E. (1970). Enzymes as control elements in metabolic regulation, in *The enzymes*, P. D. Boyer, Ed., 3d ed., vol. 1. Academic, New York, pp. 461–489.

ATKINSON, D. E. (1971). Adenine nucleotides as stoichiometric coupling agents in metabolism and as regulatory modifiers: The adenylate energy charge, in *Metabolic Pathways*, H. J. Vogel, Ed., 3d ed. vol. 5. Academic, New York, pp. 1–21.

ATKINSON, D. E. (1975). Biochemical function and homeostasis: The payoff of the genetic program. In *Control Mechanisms in Development*, R. H. Meints and E. Davies, Eds. Plenum, New York, pp. 193–211.

ATKINSON, D. E. (1977). Op. cit., Chapter 1.

BANKS, B. E. C., and VERNON, C. A. (1970). Reassessment of the role of ATP *in vivo*. *Journal of Theoretical Biology* **29**:301–326.

BRODA, E. (1975). *The Evolution of the Bioenergetic Processes*. Pergamon, Oxford.

DAWES, E. A., and SENIOR, P. (1973). The role and regulation of energy reserve polymers in microorganisms. *Advances in Microbial Physiology* **10**:135–266.

FERSHT, A. (1985). *Enzyme Structure and Mechanism*, 2d ed. W. H. Freeman and Co., New York.

HARTSUCK, J. A., and LIPSCOMB, W. N. (1971). Carboxypeptidase A, in *The Enzymes*, P. D. Boyer, Ed., 3d ed., vol. 3. Academic, New York, pp. 1–56.

JENCKS, P. W. (1975). Binding energy, specificity and enzymic catalysis: The Circe effect. *Advances in Enzymology* **43**:219–410.

KNOWLES, C. J. (1977). Microbial metabolic regulation by adenine nucleotide pools. *Symposia of the Society for General Microbiology* **27**:241–283.

KOSHLAND, D. E. (1973). Protein shape and biological control. *Scientific American* **229**(4): 52–64.

KREBS, H. A., and KORNBERG, H. L. (1957). *Energy Transformations in Living Matter*. Springer-Verlag, Berlin.

KREBS, H. A., and VEECH, R. L. (1969). Equilibrium relations between pyridine nucleotides and adenine nucleotides and their roles in the regulation of metabolic processes. *Advances in Enzyme Regulation* **7**:397–412.

LEHNINGER, A. L. (1982). *Principles of Biochemistry*. Worth, New York.

LIPMANN, F. (1941). Metabolic generation and utilization of phosphate bond energy. *Advances in Enzymology* **1**:99–162.

MANDELSTAM, J., and MCQUILLEN, K., Eds. (1973). *Biochemistry of Bacterial Growth*, 2d ed. Halsted, New York.

METZLER, D. E. (1977). *Biochemistry: The Chemical Reactions of Living Things*. Academic, New York.

MONOD, J. (1971). Op. cit., Chapter 1.

PREBBLE, J. N. (1981). *Mitochondria, Chloroplasts and Bacterial Membranes*. Longman, London.

SANWAL, B. D. (1970). Allosteric controls of amphibolic pathways in bacteria. *Bacteriological Reviews* **34**:20–39.

STRYER, L. (1981). *Biochemistry*. 2d ed. W. H. Freeman and Co., New York.

SZENT-GYÖRGYI, A. (1961). Introductory comments, in *Light and Life,* W. D. McElroy and B. Glass, Eds. Johns Hopkins University Press, Baltimore, p. 7.

WALSH, C. (1979). Enzymes: Macromolecular biological catalysts of chemical change. *International Review of Biochemistry* 24:143–170.

3

Energy Coupling by Ion Currents

Every novel idea in science passes through three stages.
First people say it isn't true. Then they say it's true but
not important. And finally they say it's true and important,
but not new.

Anonymous

• • •

Membranes and Energy Transduction
The Chemiosmotic Theory
From Squiggle to Proton-Motive Force
Vectorial Metabolism and Ligand Conduction

Membranes and Energy Transduction

The electrical activities of living things were, until recently, of interest chiefly to neurophysiologists and to the handful of students concerned with such curiosities as electric eels and excitable plants. To them we owe the rigorous formulation of laws that govern the movement of ions across biological membranes and the potentials and currents that arise from this movement. As a consequence, electrophysiology became one of the most sophisticated branches of biological science but rather an arcane one, seemingly remote from the concerns of cell biologists, microbiologists, and biochemists. This comfortable apartheid has been eroded by developments in bioenergetics over the past 20 years. It is now clear that the circulation of ions across biological membranes, notably that of protons and of sodium ions, is one of the fundamental processes in cellular energetics. Ion currents play the title role in energy capture during respiration and photosynthesis; they mediate the interconversion of chemical, osmotic, and electrical forms of biological energy; and they support a range of physiological work functions, from the active transport of nutrients to motility, adaptive behavior, and perhaps the spatial localization of growth and development. Ion currents thus constitute an alternative kind of energy currency, complementing chemical coupling agents such as the ATP-ADP pair.

The recurrent theme is the "coupling ion," whose function is suggested in Figure 3.1a. A transport system "pumps" an ion across a membrane at the expense of some energy source, establishing an ion gradient whose electrochemical potential represents "stored" energy. Return of that ion across the membrane, down the electrochemical potential gradient, is mediated by a second transport system so designed as to link the downhill flux of the ion to the performance of some useful work. In Figure 3.1a this work is symbolized by the lifting of a weight to emphasize that the nature of the work does not depend on the chemical identity of the coupling ion but on the design of the transformer. A current of the coupling ion flows from the pump, or current source, via the aqueous medium and back across the membrane through the transformer. The rate of doing work depends on the rate of current flow, while its intensity is a function of the potential gradient established by the pump. Efficient coupling of the exergonic limb of the current circuit to the endergonic one requires an insulating membrane that is topologically closed and has a low ion permeability (or conductance) and further, that return of the coupling ion via the transducer be tightly linked to whatever work is to be effected. Any leak or slippage dissipates the current and entails a loss of energy. The linkage of fluxes, such that the coupling ion can only complete the circuit in conjunction with the performance of work, corresponds to the common intermediate of chemically coupled reactions.

Cellular energetics rings the changes upon this theme, varying the species of the coupling ion, the initial driving force, and the nature of the work done. Coupling ions usually carry a positive charge (H^+, Na^+), but anions serve in special cases. In principle, coupling could be effected by a flux of uncharged

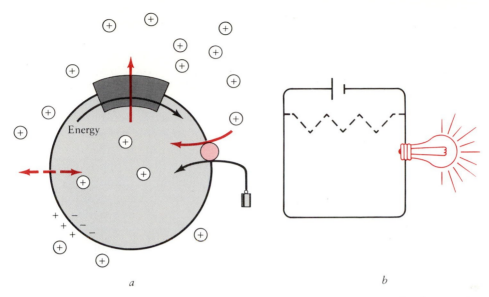

FIGURE 3.1   Coupling of work to metabolic energy by an ion current. (*a*) An electrogenic ion pump drives cations out of the vesicle, converting metabolic energy into a gradient of electrochemical potential across the membrane. The cations return by another transport system, designed to couple the ion flux downhill to the performance of some useful work. If ion leakage (dashed arrow) is excessive, the system cannot sustain an electrochemical potential and cannot perform work. (*b*) Performance of work by an electric circuit. This system can also be short-circuited by the insertion of a conductive shunt.

solute, but ions have the virtue that both electrical and concentration gradients contribute to the driving force. Besides, electrical imbalance is rapidly transmitted over relatively large distances along the surface of a membrane and is thus particularly well suited to both energy transduction and communication. Currents of protons and sodium ions dominate the coupling of metabolism to work, while calcium ions are preeminent in biological signaling.

Energy coupling by an ion current has many parallels to the performance of work by an electrical device (Figure 3.1*b*). There are, to be sure, important differences that stem from the composition of the biological machinery. Living things are predominantly made of water rather than of copper wiring; therefore ions, rather than electrons, carry the current and the path of current flow is delocalized. Moreover, biological work does not depend on the induction of secondary currents or magnetic fields but on molecular effects that are expressed in chemical terms. Nevertheless, one can legitimately think of an ion pump as a battery, generating a potential that is customarily expressed in volts or millivolts. (Recall that a tenfold difference in concentration is equivalent to a potential difference of 59 mV and to a free-energy change of 5.72 kJ/mol or 1.37 kcal/mol.) The potential across both kinds of circuit is maximal when no current is flowing and falls as the current drawn increases. Both kinds of circuit can be shorted out by the insertion of a highly conducting shunt, and the ion permeability of a membrane is quantitatively expressed by its conductance.

Historically, the realization that the electrochemical potential of an ion gradient can be drawn on for physiological work is rooted in the study of metabolite transport. Physiologists have traditionally classified the movements of molecules and ions across membranes into three categories: simple diffusion, facilitated diffusion by interaction with a carrier, and active transport. The last is defined (Rosenberg, 1954) as the net movement of an ion or molecule counter to its electrochemical potential gradient; active transport implies the performance of work and therefore requires that the transport process be coupled to a source of energy. The Na^+-K^+ ATPase of animal cell membranes (Skou, 1957, 1965) is the prototype of an active-transport system, but it soon became clear that not all active-transport systems are so directly linked to a metabolic reaction. In 1961, Robert K. Crane proposed that glucose accumulation by epithelial cells is effected by a carrier that mediates the simultaneous movements of both glucose and Na^+ (cotransport). By this coupling of fluxes, the movement of sodium ions down the gradient generated by the Na^+-K^+ ATPase provides the driving force for the movement of glucose uphill, against its concentration gradient (Crane et al., 1961; Crane, 1965, 1977). The sodium circulation is now recognized to be a central feature of animal cell physiology and will be discussed at some length in Chapter 9.

Energy coupling by an ion gradient is conventional enough from a biophysical standpoint, but it seemed to pose a serious challenge to biochemical orthodoxy when, also in 1961, Peter Mitchell postulated it in the context of oxidative and photosynthetic phosphorylation. As related in the preceding chapter, by the early 1960s it had been established that ATP synthesis during respiration is effected by the interaction of two elements, both firmly associated with the inner membrane of mitochondria (and with the cytoplasmic membrane of bacteria). One is the electron-transport chain, composed of a cascade of flavins, quinones, and cytochromes, which mediates the oxidation of NADH and of citric acid cycle intermediates; the other is the ATP synthase complex, which mediates the production of ATP at the expense of the free energy of oxidation. Likewise, in chloroplasts and in photosynthetic bacteria an electron-transport chain mediates between the pigments that harvest the light and ultimate production of ATP by an ATP synthase. The nature of the linkage between the exergonic and the endergonic limbs proved stubbornly elusive, but it became apparent that membranes were intimately involved. By 1965, students of oxidative phosphorylation and photosynthesis were generally agreed that free energy is first conserved in the form of an "energized state" of the mitochondrial or chloroplast membrane, sometimes designated simply by a squiggle ($\sim$). The energized state was considered to be the immediate energy donor for ATP synthesis, but it could also serve directly to support certain physiological functions such as ion accumulation or the transhydrogenation of nicotinamide adenine nucleotides (Ernster and Lee, 1964; Slater, 1971). Most investigators envisaged the link between electron transport and phosphorylation in terms of activated chemical intermediates, analogous to those that participate in glycolysis but more unstable (Slater, 1953). Membrane lipids were thought to provide a hydrophobic milieu for these labile bonds, and the barrier function of the membrane quite escaped notice in the frustrating and ultimately fruitless search for chemical intermediates.

Mitchell's chemiosmotic hypothesis was formulated in the spirit of a radical alternative (Mitchell, 1961). According to his provocative proposal (Figure 3.2), electron transport and phosphorylation are not chemically linked at all but are coupled only by a transmembrane current of protons. The electron-transport chain is a metabolic pathway arranged within and across the membrane to translocate protons across it. Since the membrane is conceived to have a low intrinsic conductance for protons (and for ions in general), a gradient of pH and of electric potential will develop across the membrane. The free energy released by the electron-transport chain is thus transduced into and stored as the electrochemical potential of protons. Finally, the ATP synthase is a second proton-translocating pathway, which utilizes that proton potential to drive the phosphorylation of ATP.

Some elements of this thesis had appeared in the literature before. About 1940, H. Lundegård, studying salt uptake and respiration in plants, realized that the cytochrome chain might be so arranged within a membrane that the passage of electrons would lead to the consumption of protons on one side and their release on the other. R. E. Davies, A. G. Ogston, E. J. Conway, and R. N. Robertson developed this idea as a possible basis for acid secretion by the

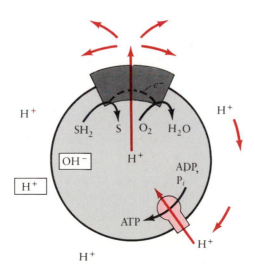

FIGURE 3.2   Chemiosmotic coupling of oxidation to phosphorylation. The respiratory chain extrudes protons across the membrane, generating an electrochemical potential gradient, interior alkaline and negative. Return of the protons through the ATP synthase is coupled to ATP formation.

stomach and for ion accumulation by plant and animal tissues (Robertson, 1968). What was novel was Mitchell's proposal that the ATP synthase, catalyzing an altogether different chemical reaction, was also a reversible proton-translocating device, and that these two transport systems were coupled together by the flux of their common substrate, the proton. The chemiosmotic hypothesis, although formulated specifically as a solution to the riddle of electron-transport phosphorylation, can now be seen as much broader in scope. For it was clear to Mitchell from the beginning that a proton current may power not only the phosphorylation of ADP but also a host of other work functions, in particular the accumulation of ions and metabolites (Mitchell, 1961, 1963, 1966, 1972). At bottom, the chemiosmotic hypothesis is not really about phosphorylation so much as about the role of ion currents in cellular energetics. It proved to be the most powerful unifying concept in bioenergetics since Lipmann's discovery of the "high-energy bond."

One of the major themes of research on biological membranes has been the elucidation of their structure, culminating in the celebrated model (Figure 3.3) developed by J. S. Singer and his colleagues (Singer, 1971; Singer and Nicolson, 1972). Happily, this model displays precisely the structural features that energy coupling by ion currents demands. The continuous matrix of almost all biological membranes consists of a phospholipid bilayer with a very low conductance for all ions and polar solutes, including H^+ and OH^-, and it can sustain an electric potential of about 500 mV. A few substances of biological importance diffuse passively across the lipid bilayer: O_2, CO_2, NH_3, and most notably, water itself. In general, however, passage of metabolites and ions across

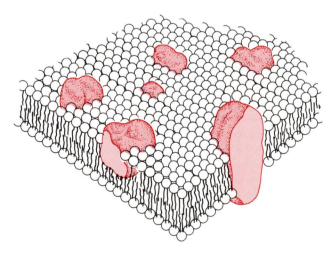

FIGURE 3.3 Fluid-mosaic model of membrane structure. The continuous, ion-impermeable matrix consists of a phospholipid bilayer. The ion pumps and other transport systems that transduce metabolic energy into work are integral proteins that span the bilayer. (From Singer and Nicolson, 1972; copyright 1972 by the American Association for the Advancement of Science.)

a membrane is not a matter of passive permeability but of transport, involving interaction with specific membrane proteins that span the bilayer. The proteins that make up the great variety of known transport systems confer specificity and control on translocation, just as enzymes both catalyze and regulate chemical transformations. Membranes differ with respect to their complement of lipids and proteins, the fluidity of the bilayer, and the mobility of their functional elements. Some approximate the "fluid mosaic" model better than others, and all serve some functions that do not come within the purview of bioenergetics. Nevertheless, it seems clear that the central role of biological membranes is the generation and utilization of ion currents; their general structure may well have evolved to fit this teleonomic objective.

The course of events during the decade and a half from the initial formulation of energy coupling by ion currents to its general acceptance by the scientific community (signaled by the award of the 1978 Nobel prize in chemistry to Peter Mitchell) has many of the hallmarks of a scientific revolution, in Thomas Kuhn's sense of the word (Kuhn, 1970). A historical treatment of this upheaval would be both entertaining and instructive, not least as an illustration of the human quality of this supposedly detached and objective enterprise called science. But that is not my purpose. The object of this and succeeding chapters is rather to display, as simply and clearly as can be, our present understanding of the genesis of transmembrane ion currents and their many roles in cellular physiology. Much of our knowledge developed directly out of the prolonged debate over the validity of chemiosmotic principles and practice, and some insight into this controversy is indispensable as a background to contemporary research at the frontiers of molecular bioenergetics. However, a strictly historical approach proved incompatible with the logical presentation of the material, and the conflicting claims of history and pedagogy were resolved in favor of clarity. The chemiosmotic theory will be presented in terms of bacteria, which illustrate most plainly the role of ion currents in general cell physiology. Chapters 7 and 8 will consider oxidative phosphorylation and photosynthesis, respectively, providing an opportunity to examine more closely the contentious issue of the molecular mechanics of proton translocation. Finally, in Chapter 9 we shall survey the manifold applications of ion currents across the plasma membrane to the workings of eukaryotic cells.

The Chemiosmotic Theory

For the past decade, most serious research on biological energy transduction has been conducted within the context of the chemiosmotic hypothesis. It should be recalled that when the hypothesis was first propounded, there were no experimental data with which its validity could be assessed. Mitchell's formulation, elegance and originality apart, had the special virtue of offering an abundance of specific predictions for critics to challenge and for proponents to verify. As a result, the chemiosmotic hypothesis itself was directly responsible for the creation of much of the present corpus of membrane bioenergetics. It is the breadth of Mitchell's conception as much as the extent to which particular elements have been confirmed by experiment that warrants its designation as a theory. Even today, with the central thesis all but universally accepted, research on unsolved problems in bioenergetics is strongly conditioned by ideas that were part of Mitchell's original proposal. Let us now examine what the chemiosmotic theory came to state by 1970, when its own internal evolution had largely run its course (Mitchell, 1966, 1968, 1970a, 1970b).

Proton Translocation and the Proton-Motive Force.   According to the chemiosmotic theory, the cytoplasmic membrane of bacteria and the inner membranes of mitochondria and chloroplasts form topologically closed structures that are essentially impermeable to H^+, to OH^-, and, indeed, to ions generally. These are the coupling membranes that effect energy conservation and the performance of useful work. The manner in which proton currents are generated and utilized is schematically illustrated in Figure 3.4. We shall focus here on the operations of bacteria, which in general consist of but a single compartment bounded by the cytoplasmic membrane. The same principles apply to mitochondria and chloroplasts, but some of the details are significantly different.

Several of the major energy-generating metabolic highways are so arranged within and across the cytoplasmic membrane as to transport protons across it. One example is the respiratory chain, which Mitchell envisaged as consisting of alternating carriers for hydrogen and for electrons arrayed into redox loops that span the membrane. Oxidation of a substrate thus results in the net extrusion of protons from the cytoplasm, two protons at each of the traditional "coupling sites" of oxidative phosphorylation (Chapters 4 and 7). In the chemiosmotic view, the sole function of the respiratory chain is to pump protons across the cytoplasmic membrane. The ATP synthase is a second, entirely independent proton-translocating system that operates reversibly. The ATP synthase is so oriented in the membrane that ATP hydrolysis brings about the extrusion of protons, while ATP synthesis is accompanied by proton uptake. Since the ATP synthase is conveniently assayed in the hydrolytic direction, we shall also refer to it as a proton-translocating ATPase. The photosynthetic apparatus is the third major electrogenic proton-translocating complex; in bacteria, at least, its polarity is the same as that of the respiratory chain and the ATP synthase (Figure 3.4).

FIGURE 3.4   The chemiosmotic theory in principle. The top row shows three of the known proton-extruding pathways: (*a*) the respiratory chain, (*b*) the proton-translocating ATPase, and (*c*) cyclic electron transport in photosynthesis. All three generate an electrochemical gradient of protons across the plasma membrane (*d*), which tends to pull protons back into the cytoplasm. The bottom row shows the utilization of the proton circulation to perform three kinds of useful work: (*e*) coupling of proton movements to transport carriers, (*f*) ATP synthesis driven by a proton flux, and (*g*) proton-linked transhydrogenation. For clarity the arrows point in only one direction, but all the reactions shown are in principle reversible. (After Harold, 1978, with permission of Academic Press.)

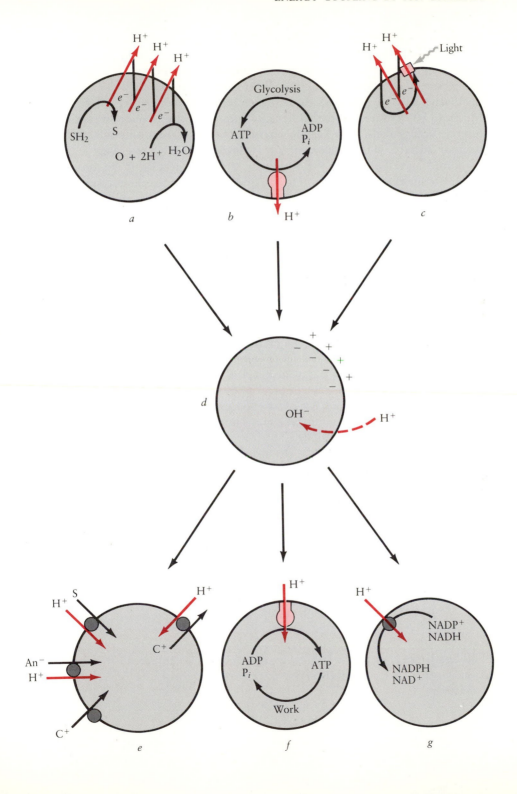

Several general features of metabolic proton transport should be noted at this point. First, Mitchell conceived of proton movements as resulting from the vectorial organization of catalysts in space, which would lend a spatial dimension to the scalar chemical reactions. Protons do not necessarily pass across the membrane as such. The respiratory chain, for example, was thought to be so organized within and across the plasma membranes that reduction of a flavin or quinone occurs at the cytoplasmic surface, while its oxidation takes place at the external surface. The movement of H (or $H^+ + e^-$) one way across the membrane and of electrons the other way effectively results in the movement of H^+ from one side to the other (Figure 3.11). Since protons do not move as such, translocation is a more appropriate term than transport. Second, it is generally not possible to distinguish the translocation of protons in one direction from the translocation of hydroxyl ions in the opposite direction. In the absence of information to the contrary, reactions are conventionally represented as proton translocations. Finally, all three pathways translocate electric charges and are therefore described as electrogenic. This usage is confusing at times, for it applies indiscriminately to all processes that translocate charge, whether their function is to generate an electric potential or to utilize one.

All three vectorial pathways transport electric charge outward across an ion-impermeable insulating membrane and therefore generate an electric potential across it, cytoplasm negative. Under circumstances that permit compensatory ion movements a pH gradient may also arise, interior alkaline. Both gradients contribute to the electrochemical potential of protons, $\Delta\tilde{\mu}_{H^+}$, which is the force that tends to pull protons back across the membrane into the cell. Metabolic energy is thus conserved as, or transduced into, the electrochemical potential of protons (Equations 1.18 and 1.19). The free-energy change as protons move back into the cell, down along both the electric and the chemical gradients, is negative:

$$\Delta G = \Delta\tilde{\mu}_{H^+} = F\,\Delta\Psi + 2.3RT \log \frac{[H^+]_i}{[H^+]_o} \qquad \text{(Eq. 3.1)}$$

In this equation $\Delta\tilde{\mu}_{H^+}$ is defined as the difference in electrochemical potential between protons on the inside and those on the outside. It is convenient and customary to express this relationship in units of electric potential:

$$\frac{\Delta G}{F} = \frac{\Delta\tilde{\mu}_{H^+}}{F} = \Delta\Psi + \frac{2.3RT}{F} \log \frac{[H^+]_i}{[H^+]_o} \qquad \text{(Eq. 3.2)}$$

The logarithmic factor simplifies to the pH difference across the membrane, $\Delta pH = pH_i - pH_o$. We thus arrive at the familiar form

$$\frac{\Delta\tilde{\mu}_{H^+}}{F} = \Delta p = \Delta\Psi - \frac{2.3RT}{F} \Delta pH \qquad \text{(Eq. 3.3)}$$

At 25°C, $2.3RT/F$ equals 59 mV, so that $\Delta\tilde{\mu}_{H^+}/F$ in millivolts becomes

$$\frac{\Delta\tilde{\mu}_{H^+}}{F} = \Delta p = \Delta\Psi - 59\Delta\text{pH} \qquad \text{(Eq. 3.4)}$$

Mitchell's term for the electrochemical driving force, the proton-motive force (Δp), has been widely adopted. He now prefers to speak of the protonic potential; in this book both terms will be used interchangeably. The electrochemical potential $\Delta\tilde{\mu}_{H^+}$ should be expressed in kilocalories or kilojoules and the protonic potential in millivolts, but both refer to the same driving force. The proton potential, or proton-motive force, is the immediate energy donor for work functions that are driven by the circulation of protons across the membrane; the "energized state of the membrane" is simply the proton-motive force across it.[1]

The membrane potential, $\Delta\Psi$, is related to the difference in charge between the inside and the outside of the compartments shown in Figure 3.4, with an excess of negative ions inside. Biological membranes have a low electric capacitance, and therefore the net transfer of just a few thousand protons across the plasma membrane of a bacterial cell suffices to generate a membrane potential of -60 mV. By contrast, the cytoplasm is well buffered and the number of protons that must be moved to generate a pH gradient of one unit is larger by some two orders of magnitude. These two components of the proton-motive force, albeit thermodynamically additive, therefore have different kinetic characteristics. The membrane potential rises quickly once proton translocation begins but stores little energy; a pH gradient develops slowly but represents a significant energy reservoir. This would seem to pose a problem: a bacterial cell has a volume on the order of one cubic micrometer (μm^3); at pH 7 it will contain just a few protons per cell and at pH 8 none at all! But the difficulty is only apparent. Protonation and deprotonation reactions are so rapid that protons associated with buffering groups are in fact freely available for translocation.

The proton potential that can be established by a given proton pump depends on the free-energy change of the driving reaction and on the number of protons transported per cycle (n). At equilibrium, the free-energy change of the driving reaction will just balance the electrochemical potential of protons:

$$\Delta G = nF\,\Delta p \qquad \text{(Eq. 3.5)}$$

---

[1] There is no uniform convention for the sign of Δp. Mitchell originally defined Δp as positive when protons tend to flow from the external medium into bacteria or mitochondria. In electrophysiology, the potential of a measuring electrode in the cytoplasm is stated relative to a reference electrode in the external medium, and ΔpH is defined as $\text{pH}_i - \text{pH}_o$. This should make Δp a negative quantity in mitochondria and bacteria, the convention employed in this book. Thylakoids, whose lumen is electropositive, can be considered to be effectively external to the cytoplasm; the side from which protons are pumped becomes electronegative (see Lowe and Jones, 1984).

Note that, contrary to one's intuitive feeling, a reaction that translocates a single proton will generate a steeper protonic potential than one that translocates two protons. As a practical example, the proton-translocating ATPase is known to generate a Δp on the order of -200 mV (-4.6 kcal/mol, -19.3 kJ/mol). Since the free energy of ATP hydrolysis in the cytoplasm is about -10 kcal/mol (-440 mV), the data are consistent with the transport of at least two protons per cycle. Both a membrane potential and a pH gradient may arise; their proportions depend on a variety of circumstances, which will not be discussed here.

Primary and Secondary Transport.   In the chemiosmotic view, the movements of solutes and chemical groups across membranes can be divided into two categories according to their relationship to enzymatic reactions. This classification, shown in Table 3.1, avoids the ambiguities inherent in the traditional terms "active transport" and "facilitated diffusion" and directs attention to the underlying molecular mechanisms. Proton translocation by respiratory chains, ATP synthase, and the photosynthetic apparatus is tightly and stoichiometrically linked to the chemical reaction: thanks to the vectorial organization of these pathways within the membrane, proton movements are an integral part of the reaction mechanism. These processes exemplify what Mitchell called *primary transport.* By definition, primary transport systems catalyze concurrently both a chemical reaction and a transport process. The category is a broad one. It includes the familiar ATP-linked transport systems of animal cells, such as the Na^+-K^+ ATPase and the Ca^{2+} ATPase, which are colloquially described as ion pumps. But it also embraces transport processes that do not fit the classical conception of active transport, such as cases where a metabolite is chemically modified in the course of transport, or the movements of electrons. In all primary transport processes, translocation is the direct result of a vectorial enzymatic reaction. These are the "osmoenzymes," or chemiosmotic reactions, that have both an osmotic and a chemical aspect and lend the chemiosmotic theory its name.

The primary proton-motive reactions generate a circulation of protons across the membrane as protons "seek" to return to the cytosol down the electrochemical potential gradient. One of the most important work functions supported by the proton current is the accumulation of nutrients and metabolites with the aid of specific carriers or "porters" (Table 3.1). These illustrate the concept of *secondary transport*: by definition, secondary transport systems (or porters for short) catalyze the movement of one or more solutes across a membrane but are not associated with any chemical reaction. Porters catalyze nothing more than facilitated diffusion but may perform osmotic work by coupling the flux of one solute to that of another—protons, for example. The carrier protein mediates the transport process and ensures specificity for both substrate and coupling ion; the driving force is the electrochemical potential gradient of the protons. Thus while primary transport systems are intrinsically vectorial, secondary ones mediate metabolite flux in either direction in accordance with the prevailing thermodynamic gradients.

TABLE 3.1   The varieties of transport

Primary Transport Processes
1. *Group translocation:* An enzymatic reaction vectorially arranged across a membrane so as to catalyze simultaneously the chemical transformation of the substrate and the transport of all or part of it.
   Example: Sugar uptake by vectorial phosphorylation.
2. *Primary solute transport:* A vectorial enzymatic reaction that catalyzes transport of a solute without altering it chemically. (This category corresponds to the traditional concept of active transport.)
   Examples: Na^+-K^+ ATPase, Ca^{2+} ATPase.
3. *Proton and electron translocation.* Pending clarification of their molecular mechanisms, these are best placed in a separate category; eventually they may be accommodated in one or both of the preceding categories.
   Examples: Various redox chains, proton-translocating ATPases.

Secondary Transport Processes
1. *Uniport:* Transport of a single solute by a carrier or channel.
   Examples: Glucose porter of erythrocytes; valinomycin-mediated K^+ transport; ion channels such as gramicidin also mediate uniport.
2. *Symport:* Coupled transport of two solutes in the same direction by a single carrier center, also called cotransport.
   Examples: Cases of H^+-linked or Na^+-linked systems for the uptake of metabolites by bacteria and animal cells.
3. *Antiport:* Coupled transport of two solutes in opposite directions by a single carrier center, also called exchange diffusion.
   Examples: Na^+-H^+ antiporters in bacteria; nigericin-mediated K^+-H^+ exchange.

To understand the energetics of secondary porters, consider the hypothetical examples shown in Figure 3.4*e*. (Note that these cases are intended to illustrate and clarify the concept of secondary transport; to what extent they describe actual transport processes is a separate question, which will be taken up in Chapter 5.) Porters for cationic substrates such as K^+ or lysine translocate positive charge; accumulation of the substrate in the cytoplasm occurs electrophoretically in response to the membrane potential. Carriers of this kind, which simply allow an ion or molecule to move in response to its electrochemical potential, are said to mediate "uniport." Anionic metabolites, such as P_i or succinate, cannot accumulate in this manner in the face of a negative membrane potential. Mitchell proposed that these would be transported together with a proton; this linkage is called "symport" since the coupled fluxes have the same direction in space. In the example shown in Figure 3.4*e*, the transport process is electroneutral overall; the membrane potential makes no contribution to the driving force, but a pH gradient (interior alkaline) can support accumulation of the metabolite. Uncharged metabolites such as sugars or amino acids will also be transported by symport with protons, but in this case the process involves the movement of both protons and electric charge; in consequence, both $\Delta\Psi$ and ΔpH contribute to the driving force. Finally, extrusion of Na^+ or Ca^{2+} ions can be attributed to "antiporters," carriers so articulated that the driving force on

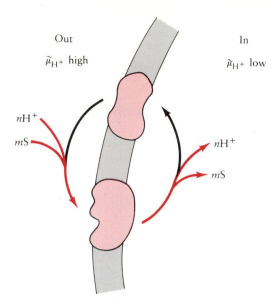

Out   In

$\tilde{\mu}_{H^+}$ high $\tilde{\mu}_{H^+}$ low

nH$^+$

mS nH$^+$

                                                              mS

FIGURE 3.5   Coupled translocation of n protons and m molecules of substrate (S) by a
symport carrier. The conformational change that effects translocation takes place only
when both H$^+$ and S bind to the carrier protein. The carrier operates reversibly,
transporting H$^+$ and S together either into the cell or outward. The arrows point in the
influx direction; the colored arrows designate the flows of S and H$^+$ and the black
the transformation of the protein. Accumulation of S depends on the cell's ability to
maintain a protonic potential gradient. At equilibrium the tendency of protons to flow into
the cell is balanced against the tendency of S to flow out.

the protons (directed inward) is translated into the movement of the metabolite
in the opposite direction. Symporters and antiporters are equivalent to the
cotransport and exchange carriers in the traditional nomenclature of membrane
transport.

How much work can a porter do? Specifically, how steep a concentration
gradient can it generate? A porter mediates a cyclic, reversible process. Figure 3.5
depicts the operation of a porter that couples the flux of m molecules of the
substrate S to that of n protons. The heart of the matter is that the porter is
mobile only when it is either unloaded or bears both S and H$^+$, but not when it
bears S or H$^+$ alone. At equilibrium, the tendency of S to diffuse downhill, out of
the cell, would be balanced by the porter against $\Delta\tilde{\mu}_{H^+}$, the inward driving force
exerted on the coupling ion H$^+$. The gradient of S therefore depends not only on
$\Delta\tilde{\mu}_{H^+}$ but also on the stoichiometry of the transport process: the coupling rules
determine the gradient that a particular porter can establish. For an uncharged
substrate S carried by symport with one proton, the relationship at equilibrium
between Δp and the concentration gradient of S should be

$$59 \log \frac{[S]_i}{[S]_o} = -\Delta p = \frac{-\Delta\tilde{\mu}_{H^+}}{F}$$

(Eq. 3.6)

TABLE 3.2 Thermodynamic driving forces on proton-linked porters

Substrate	Protons translocated	Symbol	Distribution at equilibrium*
Neutral	1	S^0, H^+ →	$Z \log \dfrac{[S^0]_i}{[S^0]_o} = -(\Delta\Psi - Z\,\Delta pH) = -\Delta p$
Neutral	2	S^0, $2H^+$ →	$Z \log \dfrac{[S^0]_i}{[S^0]_o} = -2(\Delta\Psi - Z\,\Delta pH) = -2\Delta p$
Cationic	0	S^+ →	$Z \log \dfrac{[S^+]_i}{[S^+]_o} = -\Delta\Psi$
Cationic	1	S^+, H^+ →	$Z \log \dfrac{[S^+]_i}{[S^+]_o} = -(2\Delta\Psi - Z\,\Delta pH)$
Anionic	1	S^-, H^+ →	$Z \log \dfrac{[S^-]_i}{[S^-]_o} = Z\,\Delta pH$
Anionic	2	S^-, $2H^+$ →	$Z \log \dfrac{[S^-]_i}{[S^-]_o} = -(\Delta\Psi - 2Z\,\Delta pH)$

* $Z = 2.3RT/F = 59$ mV.
After Rottenberg, 1976.

Some of the basic thermodynamic relationships are collected in Table 3.2, but their meaning is better conveyed by examples. A univalent cation will distribute itself across the membrane in accordance with the Nernst equation (Equation 1.20); a $\Delta\Psi$ of -120 mV may thus support a hundredfold concentration gradient. For a divalent cation the driving force is squared, and the concentration gradient may rise to 10^4. By the same token, suppose $\Delta\Psi$ were -120 mV and the pH gradient -60 mV (one unit), for a total proton-motive force of -180 mV. A neutral substrate carried by symport with one proton could attain a concentration gradient of 10^3, but if there were two protons, the same Δp could support a gradient of 10^6. The coupling rules for each porter cannot be predicted by theory but must be determined experimentally. In principle, at least, measurements of the substrate distribution can be used to infer the number of protons transported with each cycle.

Ionophores. Uniport, symport, and antiport are particularly well exemplified by a series of reagents, many of them antibiotics, that exert their effects by mediating the transport of particular cations. These ionophores (Harold, 1970; Pressman, 1976; Bakker, 1979) illustrate the meaning and thermodynamic consequences of coupled fluxes in a straightforward manner (Figure 3.6); they also supply a set of incisive tools that have allowed investigators to generate, modulate, and dissipate gradients of ion concentration and of electric potential almost at will.

As early as 1961, Mitchell recognized that lipid-soluble weak acids can traverse a membrane both in the protonated form (HAn) and as the anion (An^-), thereby effectively translocating protons across the membrane (Figure 3.6). This is in essence the mechanism by which 2,4-dinitrophenol and other "uncouplers" dissociate the linkage between processes that generate the proton potential and those that consume it. In the presence of proton-conducting uncouplers the membrane becomes specifically permeable to protons and can no longer sustain a proton potential; the proton current is effectively short-circuited. Other ionophores have more subtle effects. Valinomycin, for example, mediates the electrogenic uniport of K^+. When valinomycin is added to cells that have

Ionophores

(a) *Valinomycin*
Uniport, electrogenic, carrier mediated
Rb^+, $K^+ \gg Na^+$, Li^+, H^+

(b) *Nigericin*
Antiport, electroneutral, carrier mediated
K^+, $Rb^+ > Na^+$, Li^+

(c) *Monensin*
Antiport, electroneutral, carrier mediated
Na^+, $Li^+ > K^+$, Rb^+

(d) *Gramicidin*
Uniport, electrogenic, channel mediated
$H^+ > Rb^+$, K^+, Na^+

(e) *Dinitrophenol and other proton-conducting uncouplers*
Carrier-mediated symport of H^+ and An^- one way, uniport
of An^- the other way
$H^+ \gg K^+$, Rb^+, Na^+

(f) *A23187*
Antiport, electroneutral, carrier mediated
Ca^{2+}, $Mg^{2+} \gg K^+$, Na^+, Li^+

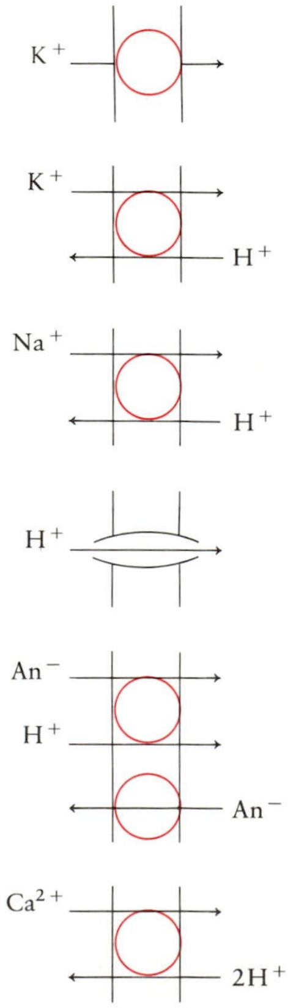

FIGURE 3.6 Some ionophore-catalyzed transport processes. For clarity, arrows point in only one direction, but transport is in fact reversible.

established a metabolic membrane potential, it permits the electrophoretic uptake of K^+ with dissipation of the electrical gradient. Conversely, valinomycin can be used to generate an artificial membrane potential by allowing K^+ to flow out of a resting cell, down the concentration gradient; this valuable procedure will be more fully discussed below (Figure 3.8). Nigericin and monensin catalyze antiport of K^+ for H^+ and of Na^+ for H^+, respectively; they can therefore be used to dissipate gradients of cations or of pH in an electroneutral manner. The formulation and testing of the chemiosmotic theory were immeasurably aided by natural as well as synthetic ionophores, and they are now proving equally valuable in the dissection of the roles of ion gradients in cellular physiology.

Proton-Transport Phosphorylation. The chemiosmotic interpretation of oxidative phosphorylation, like that of metabolite transport, invokes the coupling of two transport systems by a proton current. However, in this instance both the respiratory chain and the ATP synthase are primary transport systems and the output takes the form of chemical work. The most innovative element of Mitchell's chemiosmotic theory, and also the most controversial, was the proposal that the ATP synthase is a primary, reversible transport system so arranged that the hydrolysis of ATP is coupled to the translocation of protons outward, while the passage of protons through the enzyme, under the influence of a proton potential generated by respiration, is linked to the formation of ATP by reversal of the hydrolytic reaction (Figures 3.2 and 3.4). The mechanism of the proton-translocating ATPase is a particularly thorny issue, a discussion of which will be deferred for the present. What matters here is not the molecular mechanism of vectorial hydration and dehydration but the insight that, since the chemical events are integrally and stoichiometrically linked to the effective translocation of protons across the catalytic center, $\Delta\tilde{\mu}_{H^+}$ must poise the equilibrium of the catalytic reaction.

The thermodynamics of this mode of secondary coupling are analogous to those of carrier-mediated transport. Hydrolysis and synthesis of ATP are strictly linked to proton translocation, and the reaction is properly written

$$ATP + H_2O + nH_i^+ \rightleftharpoons ADP + P_i + nH_o^+ \qquad \text{(Eq. 3.7)}$$

where H_i^+ and H_o^+ designate protons on the inside and on the outside, respectively, and n is the number of protons traversing the membrane per cycle. The equilibrium constant of the reaction therefore includes not only the concentrations of ADP, P_i, and ATP but also the electrochemical activities of protons on the two sides of the membrane. It is important that, since protons cross the membrane during the reaction, the term that enters the equation is the electrochemical activity of protons, $\{H^+\}$, rather than their concentration:

$$K'_{eq} = \frac{[ADP][P_i]\{H^+\}_o^n}{[ATP][H_2O]\{H^+\}_i^n} \qquad \text{(Eq. 3.8)}$$

In the absence of net proton transport (for example, if the membrane is leaky), the ratio $[ATP]/[ADP][P_i]$ approaches 10^{-5}, and K'_{eq} (setting the activity of water at unity) is about 10^5. It can be seen, however, that when a difference in the electrochemical activity of protons exists across the membrane, the position of equilibrium will be greatly displaced:

$$\frac{\{H^+\}_o^n}{\{H^+\}_i^n} = K'_{eq} \frac{[ATP]}{[ADP][P_i]} \qquad \text{(Eq. 3.9)}$$

This relationship can be more usefully expressed in terms of the difference of protonic potential $\Delta\tilde{\mu}_{H^+}$:

$$n(-\Delta\tilde{\mu}_{H^+}) = \Delta G^{\circ\prime} + 2.3RT \log \frac{[ATP]}{[ADP][P_i]} \qquad \text{(Eq. 3.10)}$$

The right-hand term is the free-energy change of ATP synthesis, in kilojoules or kilocalories per mole, a quantity usually termed "phosphorylation potential," ΔG_p. Under the conditions prevailing in the cell this is a strongly endergonic reaction. In short,

$$n(-\Delta\tilde{\mu}_{H^+}) = \Delta G_p \qquad \text{(Eq. 3.11)}$$

When the electrochemical potential of protons is expressed in electrical units, the relationship is

$$\frac{n(-\Delta\tilde{\mu}_{H^+})}{F} = n(-\Delta p) = \frac{\Delta G_p}{F} \qquad \text{(Eq. 3.12)}$$

It may be noted again that the larger the number of protons translocated by the ATPase per cycle, the less is the $\Delta\tilde{\mu}_{H^+}$ required to obtain a particular poise of ΔG_p. In practice, values for n range from 2 or 3 to a maximum of 4. Assuming a physiological P_i concentration of 10 mM and also that the ATPase transports two protons per cycle, the ATP/ADP ratio would be unity at a proton-motive force of about -200 mV. It is extremely important that this hold whether the $\Delta\tilde{\mu}_{H^+}$ consists entirely of a membrane potential, a pH gradient, or some combination of the two.

Turning now to the respiratory chain, it likewise translocates protons and transduces the free-energy change of the redox reaction into the electrochemical potential of protons. By combining Equations 1.14 and 3.5 we obtain, for the passage of a pair of electrons over a particular redox span,

$$2\Delta E_h = n' \Delta p \qquad \text{(Eq. 3.13)}$$

where ΔE_h is the actual redox potential difference in millivolts and n' is the number of protons transported over that span per electron pair. Equations 3.12 and 3.13 state the quantitative relationships that should hold if a redox reaction is coupled to ATP synthesis via a proton current. For example, in respiring mitochondria the difference in reduction potentials between oxygen and NADH is about 1 V, while the measured Δp is close to -240 mV. By Equation 3.13 we infer that the oxidation of NADH can translocate as many as $2 \times 1000/240 \cong 8$ protons out of the matrix. If we assume that the ATP synthase translocates two protons per cycle, Equation 3.12 states that a Δp of -240 mV could support a $\Delta G_p/F$ of 480 mV ($\Delta G_p = 11.1$ kcal/mol or 46 kJ/mol). If n were 3, ΔG_p could reach 16.5 kcal/mol, well in the physiological range (see also Chapter 7).

Reduction and Transhydrogenation.   In principle the respiratory chain, like the ATPase, is reversible. One may therefore expect to find conditions under which the hydrolysis of ATP drives the reduction of a substrate, with the proton-motive force as energy donor. Such reversed electron transport, resulting in the reduction of fumarate to succinate or of NAD$^+$ to NADH, is well known from mitochondria and bacteria and is readily accommodated in the chemiosmotic theory. A more surprising case is the energy-linked nicotinamide adenine nucleotide transhydrogenase of mitochondrial and bacterial membranes, which may be involved in generating NADPH for reductive biosynthesis (Chapter 2). It was known from the work of L. Ernster and his colleagues that the equilibrium of the reaction

$$\text{NADH} + \text{NADP}^+ \rightleftharpoons \text{NAD}^+ + \text{NADPH} \qquad \text{(Eq. 3.14)}$$

is shifted sharply to the right when the membrane is energized; the equilibrium constant rises from unity to about 500. Mitchell proposed that this, too, is a proton-translocating reaction, which should be written

$$\text{NADH} + \text{NADP}^+ + n\text{H}_o^+ \rightleftharpoons \text{NAD}^+ + \text{NADPH} + n\text{H}_i^+ \quad \text{(Eq. 3.15)}$$

In this manner, the passage of protons down the electrochemical potential gradient would shift the reaction in the direction of NADPH production. The molecular mechanism underlying this mode of energy transduction will be considered later; for the present, we are concerned only with the principle that the equilibrium of any chemical reaction that proceeds with the concurrent translocation of protons will be poised by the proton-motive force.

The principles of energy coupling by an ion gradient are simple, but working out their quantitative ramifications is far otherwise and necessarily beyond the scope of this book. Readers who wish to delve deeper into these mysteries must consult the sacred writings, known to initiates as the "Little Gray Books" (Mitchell, 1966, 1968). Somewhat more accessible, but hardly light fare, are Mitchell's major reviews (Mitchell, 1966, 1970*a*, 1970*b*, 1976*a*, 1979*a*, 1979*b*). The "worm's-eye scrutiny" of the chemiosmotic hypothesis presented by Greville (1969) is still required reading, but novices may prefer to begin with

David Nicholls' admirable and contemporary book (Nicholls, 1982). It is a pleasure to acknowledge here my indebtedness to this lucid and concise work, which now provides the gentlest entry into the research literature on chemiosmotic energy transduction by mitochondria and chloroplasts.

From Squiggle to Proton-Motive Force

For at least a decade the central issue in bioenergetics was, What physical reality does the "energized state" of the membrane represent? Biochemists traditionally envisaged the coupling of, say, respiration to phosphorylation or to active transport in terms of activated chemical intermediates (Slater, 1953). When these proved elusive, more imaginative models began to proliferate, which drew inspiration from the orderly arrangement of energy-transducing proteins into more or less solid structures. Several investigators attributed energy coupling to energized conformational states of membrane proteins analogous to those found in muscle. Some called for phonons, packets of resonant conformational energy. In some quarters, charges were allowed to move separately, while others insisted on strict pairing. Williams (1962, 1978) and his followers attributed energy coupling to anhydrous protons localized *within* the hydrophobic phase of the membrane. And to chemiosmoticists it was obvious all along that the energized state is nothing but the proton-motive force *across* the membrane. Today it is all but universally accepted that the energized state of bacterial, mitochondrial, and chloroplast membranes corresponds to an electrochemical potential of protons. It is not practical here to review the evidence in detail; the arguments have been marshaled more than once, and readers may still catch the flavor of a debate that has not entirely subsided (Mitchell, 1966, 1970a, 1970b, 1976a; Skulachev, 1971, 1975; Slater, 1971; Harold, 1972, 1974, 1978; Racker, 1976; Boyer et al., 1977; Hinkle and McCarty, 1978; Green, 1981; Skulachev and Hinkle, 1981; Nicholls, 1982). But it may be useful to restate briefly the major points on which the present consensus rests.

1.   Metabolic reactions that energize the membrane always generate an electrochemical proton potential across it. Methods have been developed to estimate both membrane potentials and pH gradients from the distribution of permeant ions and acids or from the fluorescence characteristics of certain dyes that dissolve in the lipid phase and respond to the electric potential across it. The absolute values of the protonic potential during glycolysis, respiration, and photosynthesis vary considerably, but pH gradients of one to two units and membrane potentials of 120 to 220 mV are representative. Small cells and organelles necessitate the use of indirect methods, but recent technical refinements promise to bring even bacterial cells within the range of intracellular microelectrodes (Felle et al., 1980).

2.   Respiratory chains, photosynthetic centers, and ATP synthase all catalyze the electrogenic translocation of protons. This assertion is validated both by monitoring pH changes in the medium and by using ionophores and indicators to demonstrate concurrent movements of electric charge. Studies on purified systems reconstituted into artificial membranes document that proton translocation is an intrinsic activity, tightly coupled to the chemical reactions.

3.   The polarity of the protonic potential corresponds to that of the energized membrane. Bacterial cells, mitochondria, and also membrane vesicles prepared from them by gentle lysis all extrude protons; the lumen becomes alkaline and electrically negative. By contrast, chloroplasts, bacterial chromatophores, and membrane vesicles prepared with a French press are everted: when energized by light or by respiratory substrates, protons are taken up and the lumen turns acid and electrically positive (Figure 3.7). The dual polarity of membranous structures explained many observations that had long been puzzling.

4.   Energy coupling requires a topologically closed structure. This is a crucial assertion since, were it proved false, the chemiosmotic interpretation of oxidative and photosynthetic phosphorylation would have to be abandoned. Indeed, claims that oxidative phosphorylation persists in soluble, or at least nonvesicular, preparations continue to be made from time to time (most recently by Storey et al., 1980). The experience of the great majority of investigators is, however, that coupling between respiration or photosynthesis and such functions as ATP synthesis, transhydrogenation, and proton transport is seen only in vesicular structures.

5.   Reagents or mutations that dissipate the proton circulation also deenergize membranes. The first major success of the chemiosmotic theory was to provide a chemically plausible explanation for the "uncoupling" of oxidation from phosphorylation and other energy-linked functions by halogenated phenols: uncouplers are proton-conducting ionophores and thereby collapse the protonmotive force (Mitchell, 1963; Mitchell and Moyle, 1967a). Analogous effects are produced by mutations and by treatments that increase the ionic conductance of biological membranes.

6.   A membrane can be "energized" by imposing an artificial gradient of pH or of electric potential across it. It was first shown in the mid-1960s that ion gradients can elicit ATP synthesis by chloroplasts and mitochondria (Jagendorf and Uribe, 1966; Cockrell et al., 1967). Analogous experiments showed that the accumulation of metabolites by bacterial cells and membrane vesicles can likewise be coupled to gradients of pH or of $\Delta\Psi$. The technique illustrated in Figure 3.8 has been widely applied and will be referred to below more than once. Such nonmetabolic phosphorylation and active transport, like those supported by metabolism, are abolished by reagents that dissipate the electrochemical potential gradient.

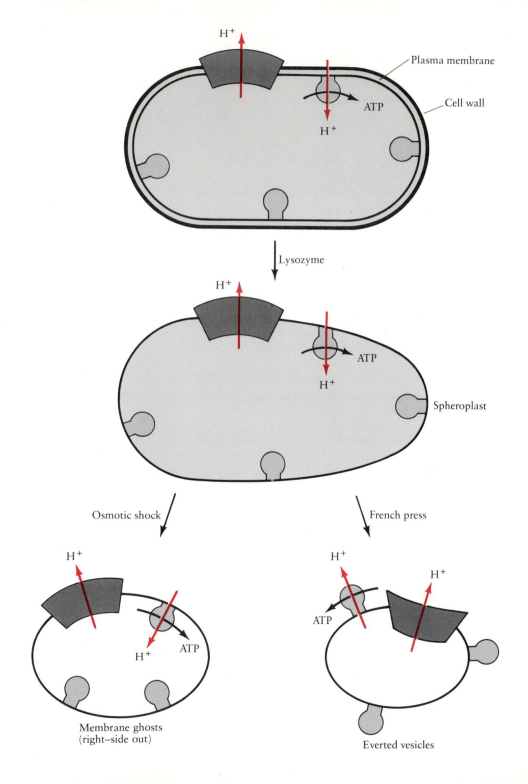

Plasma membrane

Cell wall

Lysozyme

Spheroplast

Osmotic shock

French press

Membrane ghosts
(right–side out)

Everted vesicles

7.   The accumulation of many metabolites is accompanied by proton movements of the expected direction and magnitude. The celebrated lactose "permease" of *E. coli* is a particularly well documented case of secondary proton
symport. The transhydrogenase and other work functions likewise display the
predicted proton fluxes (Chapter 5).

If all this still falls short of rigorous proof that the electrochemical potential
of protons is the obligatory intermediate in energy transduction, it is undoubtedly a very strong case. Nevertheless, it would be misleading to leave the
impression that all the doubts have been resolved and sweetness and light reign
over the field. Bioenergeticists are a contentious lot and it will be helpful to
indicate in a general way what the continuing pother is about. But it must first be
said that, while the great bulk of the experimental data is consistent with the
preceding theses, one can cite observations that are, or seem to be, inconsistent
with every one. Many such discrepancies have been resolved over the past two
decades and were found to be attributable to errors of fact or of interpretation.
Others persist and must not be glibly dismissed, for each stubborn loose end
holds the promise that it may eventually generate new and deeper insight. Yet it
is fair to say that today discrepant findings are likely to be seen not as challenges
to the chemiosmotic framework but as puzzles to be resolved within it.

For a recent case in point, consider picrate. Like other organic acids, picrate
transports protons across artificial lipid bilayers and would be expected to
uncouple oxidative phosphorylation. Hanstein and Hatefi (1974) made the
puzzling observation that picrate uncouples submitochondrial particles but not
intact mitochondria. Since the particles are topologically everted, the authors
inferred that picrate does not work by proton conduction but by reaction with a
coupling protein exposed at the inner surface of the mitochrondrial membrane.
Indeed, picrate binds to one of the subunits of the ATP synthase. However, its
failure to uncouple intact mitochondria is not inconsistent with chemiosmotic
principles. Picrate has a very low pK and exists chiefly as the anion, which is
exceedingly permeant. Submitochondrial particles, whose lumen is positive,
accumulate picrate and are uncoupled by the efflux of picric acid; intact
mitochondria, whose lumen is negative, repel the anion and remain coupled. An
analogous argument applies to bacterial vesicles (McLaughlin and Dilger, 1980;
Michels and Bakker, 1981). Incidentally, uncouplers also provide examples of
unresolved issues. If they exert their effects by conducting protons across the lipid
bilayer, one would not expect to find uncoupler-resistant mutants; such mutants
do exist, however (Guffanti et al., 1981), and their nature is presently quite
unclear.

FIGURE 3.7   Membrane vesicle preparations derived from bacterial cells may be either
right-side out or everted. The cell wall is removed by digestion with lysoyme. Lysis of the
spheroplasts by osmotic shock yields closed vesicles whose polarity is the same as that of
the parent cells. Passage of the spheroplasts through a French press generates everted
vesicles. (From Nicholls, 1982, with permission of Academic Press.)

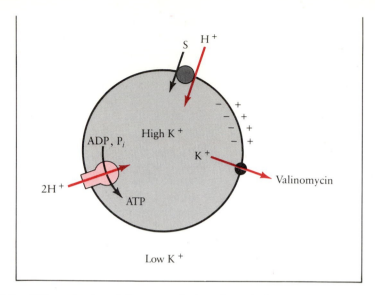

FIGURE 3.8   ATP synthesis and the accumulation of metabolites can be supported by an artificial membrane potential. Cells or rightside-out membrane vesicles are loaded with K$^+$ and suspended in buffer devoid of K$^+$ in the absence of an energy source. Addition of valinomycin elicits efflux of K$^+$ down its concentration gradient and generates an electric potential across the membrane, interior negative. This artificial proton-motive force can support diverse cellular work functions.

Let us now turn to a subtler issue: the physical nature of the proton current and its location in space. The chemiosmotic theory explicitly refers to the electrochemical potentials of ions in the bulk aqueous phases on either side of the membrane, such as the cytoplasm and the medium. The assumption that the membrane is merely a barrier that does not itself participate in protonic couplings permitted Mitchell to treat the subject in terms of equilibrium thermodynamics and encouraged experimental tests based on quantitative measurements of ΔpH and $\Delta\Psi$. However, there has long been a school of thought that attributes energy coupling to protons somehow localized within the membrane (Williams, 1962, 1978). In a thoughtful article Kell (1979) has argued that proton pumps will eject protons into the interface between the lipid bilayer and the bulk aqueous phase. Energy coupling will be effected largely by relatively localized proton currents confined to this region, whose electrochemical potential may be greater than that of protons in the bulk water phase. Localized proton currents would explain a variety of discrepant observations, which suggest that the $\Delta\tilde{\mu}_{H^+}$ as measured is either kinetically or thermodynamically insufficient to support the work assigned to it (Kell, 1979; Chapter 7). Unfortunately, the existence of localized proton currents is not easily demonstrated and there are good physicochemical reasons to reject the proposal that the electrochemical potential of interfacial protons is higher than that of protons in the bulk phase; in the steady state, the electrochemical potential of all protons in a given phase must be the same. There remains a simple and physically plausible kind of proton "localization": if proton sources and sinks are closely apposed, protons may travel from one to the other without coming into equilibrium with the bulk phases. In my opinion, interfacial protons and proximity effects would call for modifications in the thermodynamic formulas but would not shake the central tenet of energy coupling by a proton current. A significant conceptual challenge would arise only if proton sources and sinks were to be connected by genetically specified conduits. This, too, has been suggested, but the balance of the evidence is firmly against this belated resurrection of chemical intermediates in energy coupling.

Most of the ongoing disputes in membrane bioenergetics center on the molecular mechanisms that generate and consume ion currents. A recent collection of papers in Mitchell's honor (Skulachev and Hinkle, 1981) testifies to the general consensus that energy transmission is indeed effected by a proton current. By contrast, Mitchell's proposals on how the respiratory chain and the ATP synthase actually work have found little favor. The stoichiometry of proton transport, the existence of redox loops, and ATP synthesis by vectorial phosphorylation are at the center of the debate, which will be examined in some detail in Chapter 7. Let me conclude this chapter with a brief discussion of the concept of ligand conduction and the meaning of the term "energy coupling."

Vectorial Metabolism and Ligand Conduction

In the context of the metabolic web, energy conservation and work depend on the fact that when two reactions share a common intermediate, the equilibrium of each reaction is shifted or displaced. Glycolyzing cells accumulate ATP despite the unfavorable ATP/ADP ratio at equilibrium because ATP formation is stoichiometrically linked to the conversion of glucose into lactic acid. An analogous displacement of equilibrium takes place in membrane bioenergetics. The redox chain of respiration is so arranged that the flow of electrons to oxygen displaces the distribution of protons across the membrane, and secondary porters displace the equilibrium distribution of a solute between two compartments by linking its flux to that of a coupling ion. However, a new element enters here that was not present so long as we confined our attention to soluble enzymes: reactions associated with membranes have a direction in space. In fact, the great majority of biological processes are vectorial—transport, movement, growth, behavior, and development all have a marked spatial dimension. Where does the spatial orientation of chemical processes come from?

In the 1950s this question appeared to raise a serious conceptual issue. According to Curie's theorem, an asymmetric process must result from causes at least equally asymmetric. How, then, can the scalar reactions of metabolism elicit such vectorial effects as transport? The paradox vanished with the

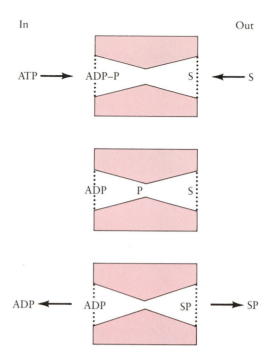

FIGURE 3.9   Vectorial transfer of a phosphoryl group from ATP to S via a channel or cavity in the enzyme. If this enzyme spans a membrane, the phosphoryl group is translocated from the inside to the outside. (After Mitchell and Moyle, 1958; reprinted by permission from *Nature*, Macmillan Journals, Ltd.)

recognition that at the molecular level, enzyme catalysis is not scalar at all but intrinsically vectorial (Mitchell and Moyle, 1958; Mitchell, 1963, 1970a). It is now common knowledge that enzymes are elaborately folded, with the catalytic site often located in a cleft or crevice. Catalysis depends on the precise positioning of substrates and products within the catalytic center, and it is often associated with changes in the conformation of the protein or in its tertiary structure (Chapter 2). In some reactions the entry of the substrate into the catalytic site and the exit of products may be strictly oriented with respect to the active center. Such molecular orientation is not of course apparent in solution, but it should become manifest when enzymes are built anisotropically into a suitable membrane.

The significant point is not merely that the occurrence of vectorial transport processes is consistent with Curie's principle. Rather, it is the recognition that soluble enzymes, no less than transport catalysts, often mediate an actual flow of ions or of chemical groups. Lipmann's group transfer potential is more than just a measure of the net tendency for chemical change; it corresponds to the field of action of a force in real space, and as the reaction proceeds, chemical particles

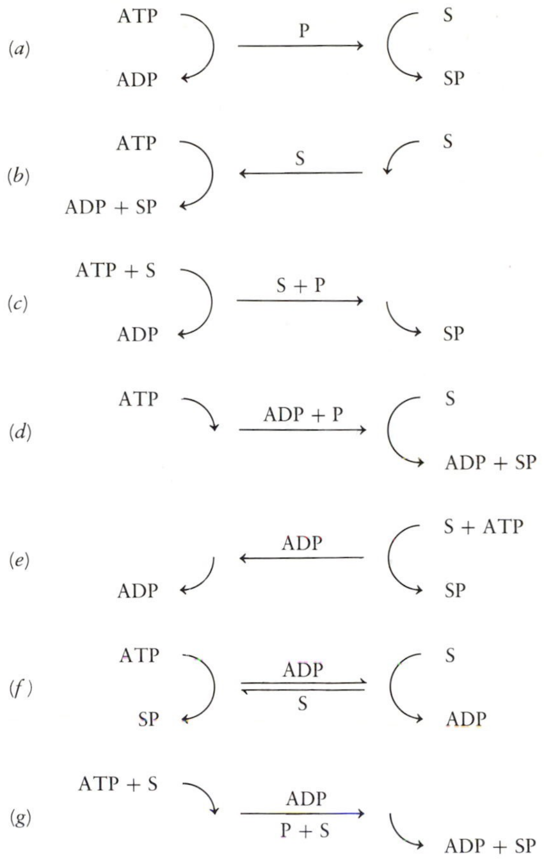

FIGURE 3.10 Chemiosmotic processes that correspond to phosphoryl transfer from ATP to S, assuming various translocational specificities. In this notation the chemical aspect of the reaction is represented as progressing up or down the page, while the osmotic (group translocation) aspect of the reaction progresses sideways across the catalytic complex. (From Mitchell, 1979b, with permission of Springer-Verlag.)

travel down that field. Figure 3.9 illustrates this spatial dimension of catalysis for phosphoryl transfer by a kinase endowed with overlapping binding sites for the acceptor and for ATP; if this enzyme were plugged across a membrane, it would catalyze the transfer of a phosphoryl group from one side to the other. It follows that in principle, all enzymes may have "chemicomotive" capacities, which can be exploited by the topological organization to produce osmoenzymes and chemiosmotic reactions; vectorial metabolism is the norm, not the exception.

The *extent* to which the substrate in Figure 3.9 will be phosphorylated is adequately predicted by scalar thermodynamic parameters, such as the equilibrium constant of the reaction or the group transfer potential. But the actual *course* of the reaction in space depends equally on the propensity of the catalyst to translocate chemical groups. Figure 3.10 displays an array of chemiosmotic processes, all of which correspond chemically to the transfer of a phosphoryl group from ATP to the substrate S. These processes reflect the various translocational specificities and mobilities with which the kinase may conceivably be endowed. Depending on the particular mechanism, a vectorial phosphorylation may result in the translocation from one side of the membrane to the other of the phosphate group, the substrate, an adenyl group, or of all of these together. In addition to the chemical groups that take part in the chemical transformation, accompanying ions may likewise by translocated, thanks to their association with ligand-binding groups located either on the substrate or on the catalyst. This mechanism of translocation is described as ligand conduction; Mitchell regards it as a general principle that unifies chemical catalysis by enzymes and osmotic catalysis by transport systems (osmoenzymes and porters). Both depend on the vectorial translocation of chemical particles down gradients of group transfer potential that represent real fields of chemical force directed in space. From this viewpoint, "it is easier to explain biochemistry in terms of transport than transport in terms of biochemistry" (Mitchell, 1979*b*).

The concept of vectorial metabolism preceded and inspired the formulation of the chemiosmotic hypothesis, and, under the guise of ligand conduction, it continues to chart Mitchell's approach to the molecular mechanism of energy transduction by primary and secondary transport processes. In practical terms, ligand conduction leads one to inquire: How direct are the connections between the forces and displacements of the particles involved in the chemical reaction, on the one hand, and on the other, the forces and displacements of the particles involved in the osmotic reaction? Molecular mechanisms vary, and it is certainly conceivable that in some cases any such connection will prove to be exceedingly indirect. Nevertheless, for heuristic reasons if for no others, it seems useful to begin with postulates that maximize this connection, for such hypotheses are most likely to be biochemically explicit and thus suggest handles for experimental attack.

Figure 3.11 outlines simplified ligand-conduction schemes for proton translocation by a respiratory redox loop and by the ATP synthase, intended only to illustrate the application of the principle to such complex chemiosmotic processes; the subject will be discussed more fully in Chapter 7. The redox loop (Figure 3.11*a*) consists of two limbs. In the upper limb, protons may be thought to move by symport with electrons, which themselves travel down the field of force represented by the redox potential along a spatially and conformationally defined channel within the catalyst. The lower limb allows the electrons to continue their journey alone, via appropriate electron-carrying centers, while the protons are liberated at the junction of the two limbs. The pathway of electron transfer is effectively looped across the osmotic barrier, and the protons are released at a site from which they travel (down a local gradient of $\Delta\tilde{\mu}_{H^+}$ in a

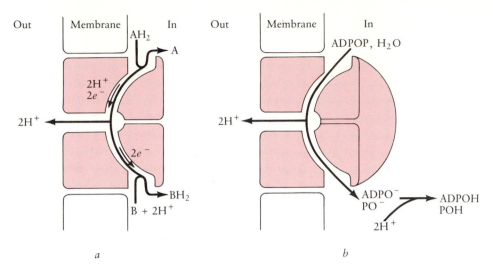

FIGURE 3.11 Ligand-conduction models for chemiosmotic proton translocation. (*a*) A redox loop. (*b*) ATPase. Both reaction pathways are drawn as loops that effectively cross the permeability barrier. Neither transports protons as such, but the chemical transformations are so arranged in space that protons would be consumed on one side of the barrier and released on the other. (After Nicholls, 1982, with permission of Academic Press.)

"proton well") into the bulk phase on the far side of the barrier. An analogous scheme can be written for proton translocation by the ATPase (Figure 3.11*b*). In this formulation, ATP and H_2O are conducted into the catalytic center by a pathway specific for a particular state of protonation that corresponds to ADPOP (OP designates the terminal phosphoryl group that will be split off by the enzyme). The products of hydrolysis are conducted back out by channels specific for states that may be written as $ADPO^-$ and PO^-, that is, states that bear two protons fewer than entered the catalytic site. The protons generated at the catalytic center are released toward the far side of the barrier via a proton well, and the net vectorial reaction consists of the effective translocation of two protons across the catalytic complex. Incidentally, it will be appreciated that schemes of this kind can be written for other cation-translocating ATPases, including the Na^+-K^+ ATPase and the Ca^{2+} ATPase of animal cell membranes (Mitchell, 1966, 1970*a*, 1970*b*, 1976*a*, 1979*b*, 1981*a*). It therefore seems highly significant that all the ion-transport ATPases known at present translocate cations.

Ligand-conduction mechanisms convey an almost tangible sense of what energy coupling means in the chemiosmotic context. Chemical energy ultimately depends on the sequestration of a chemical group in one or another covalent structure and is made available by unlatching the covalent bond and allowing the group to migrate (Chapter 2). Osmotic energy depends on the compartmentation

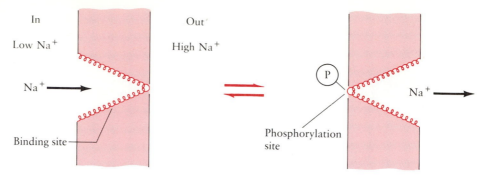

FIGURE 3.12   A conformational model for the Na$^+$-K$^+$ ATPase of animal cells. Na$^+$ ions bind to a site located in a cavity that faces the cytoplasm. Phosphorylation of the protein induces reorientation of the cavity and a change in the affinity of the binding site from Na$^+$ to K$^+$. (From Jardetzky, 1966; reprinted by permission from *Nature*, Macmillan Journals, Ltd.)

of a solute in alternative phases separated by an impermeable barrier, and it is released by allowing the solute to migrate across this barrier. Energy coupling is nothing more than the mutual dependence of the flows of "chemical" and "osmotic" particles when they are conducted jointly through the catalytic machinery. Energy exchanges are reversible because in principle there is not a driving reaction and a driven one: the gradients of thermodynamic potential determine which is cart and which is horse. Mitchell likes to go further and apply the principles of mechanics to the "thrusts [of] interacting trains of chemical particles." To this end he has developed a formal thermodynamic analysis that compares the "transmission of the products of the forces and displacements of the particles involved in the chemical process to those involved in the osmotic one" (Mitchell, 1977*a*, 1981*a*). But perhaps we have already dwelt sufficiently on what will seem to many readers no more than an elegant abstraction.

The principle of energy coupling by ligand conduction is not a hypothesis that can be falsified by experiment; it states a philosophical point of view. It is therefore instructive to contrast it with an alternative conception, no less philosophical in origin, that has much wider currency among students of membrane transport and among biochemists as well. The Na$^+$-K$^+$ ATPase, for example, has long been known to link the movements of Na$^+$ and K$^+$ across a membrane to the hydrolysis of ATP, with the intermediate formation of a phosphorylated enzyme. The mechanism by which the "sodium pump" transports ions is not yet known, but it is usually envisaged in terms of alternating conformational states much as Jardetzky depicted it long ago (Figure 3.12). Sodium ions bind tightly to sites within a cavity that faces the cytoplasm; phosphorylation of the enzyme induces an allosteric transition such that the cavity faces outward, releases Na$^+$, and picks up K$^+$ instead; finally, dephosphorylation restores the status quo ante, with discharge of K$^+$ to the cytoplasm. Note that this mechanism postulates no direct connection between the chemical hydrolysis of ATP and the transport of Na$^+$ and K$^+$; coupling between the two limbs would be effected exclusively through changes in the conformation and affinity of sites on the catalyst. Mechanisms of this kind can be, and have been, written for the proton-translocating ATPase and the proton-translocating redox chains as well, and we shall consider their implications in Chapters 7 and 10. Conversely, Mitchell has drawn up hypothetical ligand-conduction schemes for the Na$^+$-K$^+$ ATPase and allied systems (Mitchell 1979b, 1981a).

The contraposition of conformational (or "alternating access") mechanisms against those mediated by ligand conduction is artificial and probably misleading. These are two ends of a continuum, and transport by real osmoenzymes will display features of both. Neither "Occam's razor" nor evolutionary generalizations will justify a choice between the alternatives, only patient and open-minded reflection on whatever clues may be supplied by experiment. But I can think of no better basis for the design of such experiments than to ask for each system: How direct is the connection between the chemical transformation and the osmotic translocation?

References

BAKKER, E. P. (1979). Ionophore antibiotics, in *Antibiotics*, vol. V/1, *Mechanism of Action of Antibacterial Agents*, F. E. Hahn, Ed. Springer-Verlag, Berlin, pp. 67–97.

BOYER, P. D., CHANCE, B., ERNSTER, L., MITCHELL, P., RACKER, E., and SLATER, E. C. (1977). Oxidative phosphorylation and photophosphorylation. *Annual Review of Biochemistry* **46**:955–1026.

COCKRELL, R. S., HARRIS, E. J., and PRESSMAN, B. C. (1967). Synthesis of ATP driven by a potassium gradient in mitochondria. *Nature (London)* **215**:1487–1488.

CRANE, R. K. (1965). Na$^+$-dependent transport in the intestine and other animal tissues. *Federation Proceedings* **24**:1000–1006.

CRANE, R. K. (1977). The gradient hypothesis and other models of carrier-mediated active transport. *Reviews in Physiology and Biochemical Pharmacology* **73**:99–159.

CRANE, R. K., MILLER, D., and BIHLER, I. (1961). The restrictions on the possible mechanism of intestinal active transport of sugars, in *Membrane Transport and Metabolism*, A. Kotyk, Ed. Czechoslovak Academy of Sciences Press, Prague, pp. 439–449.

ERNSTER, L., and LEE, C.-P. (1964). Biological oxidoreductions. *Annual Review of Biochemistry* **33**:729–788.

FELLE, H., PORTER, J. S., SLAYMAN, C. L., and KABACK, H. R. (1980). Quantitative measurements of membrane potentials in *Escherichia coli*. *Biochemistry* **19**:3585–3590.

GREEN, D. E. (1981). A critique of the chemiosmotic model of energy coupling. *Proceedings of the National Academy of Sciences, USA.* **78**:2240–2243.

GREVILLE, G. D. (1969). A scrutiny of Mitchell's chemiosmotic hypothesis. *Current Topics in Bioenergetics* **3**:1–78.

GUFFANTI, A. A., BLUMENFELD, H., and KRULWICH, T. A. (1981). ATP synthesis by an uncoupler-resistant mutant of *Bacillus megaterium*. *Journal of Biological Chemistry* **256**: 8416–8421.

HANSTEIN, W. G., and HATEFI, Y. (1974). Trinitrophenol: A membrane-impermeable uncoupler of oxidative phosphorylation. *Proceedings of the National Academy of Sciences, USA* **71**:288–292.

HAROLD, F. M. (1970). Antimicrobial agents and membrane function. *Advances in Microbial Physiology* **4**:45–104.

HAROLD, F. M. (1972). Conservation and transformation of energy by bacterial membranes. *Bacteriological Reviews* **36**:172–230.

HAROLD, F. M. (1974). Chemiosmotic interpretation of active transport in bacteria. *Annals of the New York Academy of Sciences* **227**:297–311.

HAROLD, F. M. (1978). Vectorial metabolism, in *The Bacteria,* L. N. Ornston and J. R. Sokatch, Eds. Academic, New York, pp. 463–521.

HINKLE, P. C., and MCCARTY, R. E. (1978). How cells make ATP. *Scientific American* **238** (March):104–123.

JAGENDORF, A. T., and URIBE, E. (1966). ATP formation caused by acid-base transition of spinach chloroplasts. *Proceedings of the National Academy of Sciences, USA* **55**: 170–177.

JARDETZKY, O. (1966). Simple allosteric model for membrane pumps. *Nature (London)* **211**:969–970.

KELL, D. B. (1979). On the functional proton current pathway of electron transport phosphorylation: An electrodic view. *Biochimica et Biophysica Acta* **549**:55–79.

KUHN, T. (1970). *The Structure of Scientific Revolutions,* 2d ed. University of Chicago Press, Chicago, Ill.

LOWE, A. G., and JONES, M. N. (1984). Proton motive force—what price Δp? *Trends in Biochemical Sciences* **9**:11–12.

MCLAUGHLIN, S. G. A, and DILGER, J. P. (1980). Transport of protons across membranes by weak acids. *Physiological Reviews* **60**:825–863.

MICHELS, M., and BAKKER, E. P. (1981). The mechanism of uncoupling by picrate in *Escherichia coli* K12 membrane systems. *European Journal of Biochemistry* **116**: 513–519.

MITCHELL, P. (1961). Coupling of phosphorylation to electron and hydrogen transfer by a chemiosmotic type of mechanism. *Nature (London)* **191**:144–148.

MITCHELL, P. (1963). Molecule, group and electron transport through natural membranes. *Biochemical Society Symposia* **22**:142–168.

MITCHELL, P. (1966). Chemiosmotic coupling in oxidative and photosynthetic phosphorylation. *Biological Reviews of the Cambridge Philosophical Society* **41**:445–502.

MITCHELL, P. (1968). *Chemiosmotic Coupling and Energy Transduction.* Glynn Research, Ltd., Bodmin, Cornwall.

MITCHELL, P. (1970a). Reversible coupling between transport and chemical reactions, in *Membranes and Ion Transport,* E. E. Bittar, Ed., vol. 1. Wiley, New York, pp. 192–256.

MITCHELL, P. (1970b). Membranes of cells and organelles: Morphology, transport and metabolism. *Symposia of the Society for General Microbiology* **20**:121–166.

MITCHELL, P. (1972). Performance and conservation of osmotic work by proton-coupled solute porter systems. *Bioenergetics* **4**:265–293.

MITCHELL, P. (1976a). Vectorial chemistry and the molecular mechanics of chemiosmotic coupling. Power transmissions by proticity. *Biochemical Society Transactions* 4: 399–430.

MITCHELL, P. (1977a). From energetic abstraction to biochemical mechanism. *Symposia of the Society for General Microbiology* 27:383–423.

MITCHELL, P. (1979a). David Keilin's respiratory chain concept and its chemiosmotic consequences. *Science* 206:1148–1159.

MITCHELL, P. (1979b). Compartmentation and communication in living systems. Ligand conduction: A general catalytic principle in chemical, osmotic and chemiosmotic reaction systems. *European Journal of Biochemistry* 95:1–20.

MITCHELL, P. (1981a). Bioenergetic aspects of unity in biochemistry: Evolution of the concept of ligand conduction in chemical, osmotic and chemiosmotic reaction mechanisms, in *Of Oxygen, Fuels and Living Matter,* G. Semenza, Ed., vol. 1, part 1. Wiley, New York, pp. 1–56.

MITCHELL, P., and MOYLE, J. (1958). Group-translocation: A consequence of enzyme-catalyzed group transfer. *Nature (London)* 182:372–373.

MITCHELL, P., and MOYLE, J. (1967a). Acid-base titration across the membrane system of rat-liver mitochondria. *Biochemical Journal* 104:588–600.

NICHOLLS, D. G. (1982). Op. cit., Chapter 1.

PRESSMAN, B. C. (1976). Biological applications of ionophores. *Annual Reviews of Biochemistry* 45:501–530.

RACKER, E. (1976). *A New Look at Mechanisms in Bioenergetics.* Academic, New York.

ROBERTSON, R. N. (1968). *Protons, Electrons, Phosphorylation and Active Transport.* Cambridge University Press, London.

ROSENBERG, T. (1954). The concept and definition of active transport. *Society for Experimental Biology Symposia* 8:27–41.

ROTTENBERG, H. (1976). The driving force for proton(s) metabolite cotransport in bacterial cells. *FEBS Letters* 66:159–163.

SINGER, S. J. (1971). The molecular organization of biological membranes, in *Structure and Function of Biological Membranes,* L. I. Rothfield, Ed. Academic, New York, pp. 145–222.

SINGER, S. J., and NICOLSON, G. L. (1972). The fluid mosaic model of the structure of cell membranes. *Science* 175:720–731.

SKOU, J. C. (1957). The influence of some cations on an adenosine triphosphatase from peripheral nerves. *Biochimica et Biophysica Acta* 23:394–401.

SKOU, J. C. (1965). Enzymatic basis for active transport of Na^+ and K^+ across a cell membrane. *Physiological Reviews* 45:596–617.

SKULACHEV, V. P. (1971). Energy transformations in the respiratory chain. *Current Topics in Bioenergetics* 4:127–190.

SKULACHEV, V. P. (1975). Energy coupling in biological membranes: Current state and perspectives, in *Energy Transducing Mechanisms,* E. Racker, Ed. MTP International Reviews in Science 3:31–73.

SKULACHEV, V. P., and HINKLE, P. C., Eds. (1981). *Chemiosmotic Proton Circuits in Biological Membranes.* Addison-Wesley, Reading, Mass.

SLATER, E. C. (1953). Mechanism of phosphorylation in the respiratory chain. *Nature* 172: 975–978.

SLATER, E. C. (1971). The coupling between energy-yielding and energy-utilizing reactions in mitochondria. *Quarterly Review of Biophysics* 4:35–72.

STOREY, B. T., SCOTT, D. M., and LEE, C.-P (1980). Energy-linked quinacrine fluorescence changes in submitochondrial particles from skeletal muscle mitochondria. *Journal of Biological Chemistry* **255**:5224–5229.

WILLIAMS, R. J. P. (1962). Possible functions of chains of catalysts. *Journal of Theoretical Biology* **3**:209–229.

WILLIAMS, R. J. P. (1978). The multifarious couplings of energy transduction. *Biochimica et Biophysica Acta* **505**:1–44.

4

The Bacterial Paradigm: Energy Transduction

Our understanding of the world is built up of innumerable layers.
Each layer is worth exploring, as long as we do not forget that it
is one of many. Knowing all there is to know about one layer—a
most unlikely event—would not teach us much about the rest.

Erwin Chargaff, *Heraclitean Fire*

• • •

A Comprehensible Cell

Bacteria are the simplest form of life. Viruses do not qualify, for they are obligate intracellular parasites, dependent on one cell or another for growth and multiplication. Cells alone are free living, and the prokaryotic kind display the simplest organization compatible with the environments that now exist on earth. Moreover, if common sense and the meager fossil record can be trusted, bacteria come closest to the original mode of life from which all higher forms descend. I shall make no further apologies for devoting a substantial portion of this book to bacterial energetics: it is the proper foundation for the development of bioenergetics in general.

A bacterial cell typically consists of but a single compartment, a minute speck of protoplasm encased within its plasma membrane and shielded by an inert cell wall. The anatomical point of view allows one to appreciate why the chemiosmotic theory made so profound an impression on bacterial physiology. A bacterial cell corresponds in its entirety to the closed vesicle of chemiosmotic logic: all energy transformations and work functions take place either in the cytoplasmic metabolic web or at the membrane. Moreover, the chemiosmotic framework yields immediate insight into the directionality of cellular structure and functions. The eventual adoption of the chemiosmotic theory by the majority of biologists rests in considerable part on evidence obtained with bacteria. The object of these chapters, however, is not to argue once again the case for chemiosmosis. Rather, I propose to take the essential validity of the theory for granted and to ask how far we have come toward understanding how bacterial cells live and work.

Figure 4.1 depicts the fundamental patterns of energy transduction found in prokaryotic organisms in the form of energy-flow diagrams. These serve as a guide to this chapter and the next and make several important points. First, although bacteria differ widely with respect to the energy sources on which they draw, the primary mechanisms that harvest energy fall into just three classes:

1. Substrate-level reactions in the cytoplasm, leading to the production of activated compounds and ultimately of ATP;

2. Generation of a protonic potential, or proton-motive force, by a vectorial proton-translocating redox process that may be aerobic or anaerobic;

3. Transduction of light energy into a protonic potential, augmented in some organisms by the formation of NADPH.

These classes define more precisely the three traditional modes of energy generation described in Chapter 2 under the headings substrate-level phosphorylation, electron-transport phosphorylation, and photophosphorylation. They suffice to encompass not only the major routes of prokaryotic energy production but also those of the biosphere at large. Eukaryotic organisms have perfected several of them but invented no new ones.

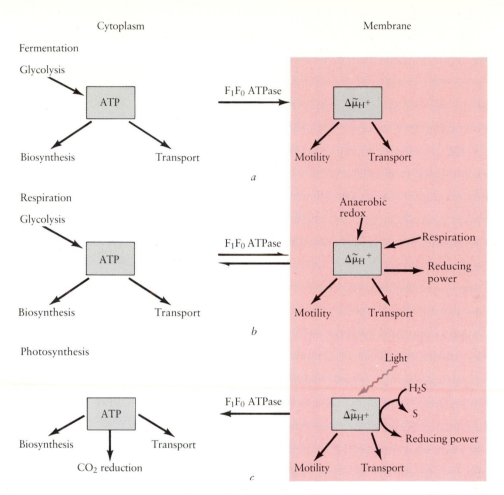

FIGURE 4.1   Patterns of energy flow in the bacterial world. (*a*) Fermentative metabolism in such organisms as streptococci or clostridia. (*b*) Metabolism of facultative organisms, which live either by fermentation or by respiration (e.g., *E. coli*). (*c*) Photolithotrophic metabolism in the green and purple sulfur bacteria. Direct generation of NADPH with the aid of light, characteristic of plant chloroplasts and of cyanobacteria, is not shown.

Second, irrespective of the primary energy source, the general organization of the bacterial economy is uniform. It depends on two complementary energy currencies: ATP in the cytoplasm and the electrochemical potential of protons across the plasma membrane. In fermentative organisms, the primary mode of energy transduction is substrate-level phosphorylation in the cytoplasm. Aerobes and photosynthetic bacteria capture energy initially in the form of a proton potential. Yet all bacteria require both phosphorylated intermediates in the cytoplasm and a protonic potential across the plasma membrane. The proton-translocating ATPase therefore occupies a central place in the metabolic economy, interconverting the two kinds of energy currency. In fermentative bacteria, its role is to "energize" the plasma membrane by translocating protons across it and generating $\Delta\tilde{\mu}_{H^+}$. During respiration and photosynthesis it consumes the proton potential set up by the primary energy donors with the production of ATP. And the two crucial potentials, the free energy of ATP hydrolysis in the cytoplasm and the proton-motive force across the plasma membrane, are kept by the ATPase in a steady state that is probably not far from equilibrium.

Third, the diagrams emphasize the division of labor between cytoplasm and membrane. Biosynthesis depends almost entirely on cytoplasmic energy donors, ATP and other activated compounds. Note particularly that the transcription, translation, and replication of the genome are entirely the business of the cytoplasm, and the membrane impinges on it only when the time comes for the cell to divide. By contrast, the membrane dominates energy transduction, the production of reducing power, and exchanges between the cytoplasm and the outer world. The membrane defines, and indeed creates, the functional unit denominated a cell.

Finally, some caveats are in order. Broad generalizations such as Figure 4.1 help one tell the wood from the trees, but they imply a degree of tidiness that does not usually obtain in biology. Indeed, to appreciate the nature of living things it is no less important to cherish their diversity than to proclaim the underlying unities. It is true, as Figure 4.1 emphasizes, that the proton is the central coupling ion of bacteria: even the archaebacteria (Chapter 6) employ the proton circulation for energy transduction and work. However, sodium ions do serve as auxiliary coupling ions in many bacteria, including E. coli, and have a major role in some, such as halobacteria and alkalophiles. Bacteria in which Na^+ is the central coupling ion, as it is in animal cells, are certainly conceivable and may have just been recognized. Few real organisms are restricted to a single metabolic pattern; most bacteria are sufficiently versatile to live either by fermentation or by respiration and some can function in all three modes as the occasion demands. Moreover, as will be emphasized again, each of the major pathways of energy transduction is multiform. In addition, we know of at least half a dozen other chemiosmotic processes that interconvert the free energy of some chemical reaction with that of an ion gradient: the proton-translocating pyrophosphatase of certain photosynthetic bacteria, for instance, or the ATP-driven sodium pump of streptococci. And ion gradients interact with each other and with chemical energy donors in a distracting variety of ways to convert energy into useful work. All these complications, omitted for the sake of clarity, serve as a warning that clarity is bought at a price.

The cellular economy, like that of nations and individuals, can be said to have a supply side and a demand side. The former refers to the processes that generate and interconvert the major energy currencies: the ATP-ADP couple and the protonic potential. The latter includes the many kinds of work that living things find necessary or useful. It is convenient to discuss these separately; the remainder of the present chapter will focus on the means of energy production.

Energy Sources for Prokaryotes

Microbiologists have traditionally based some of their most fundamental categories on the nature of the primary energy source for growth (Table 4.1). Perhaps the deepest division lies between the chemotrophic organisms, which depend on the energy released by the chemical transformation of either organic or inorganic compounds, and the phototrophic ones, which draw directly on sunlight. A second plane of cleavage arises from the source of reducing power. Cyanobacteria (formerly called blue-green algae) and green plants generate reducing power by the oxidation of water with the aid of light and are designated photoautotrophs. Other organisms, including many photosynthetic bacteria, require a chemical reductant, usually organic but in some cases an inorganic compound such as H_2S. The most self-sufficient organisms are the lithotrophic bacteria, often called chemoautotrophs; these generate both useful energy and reducing power from inorganic reactions, such as the oxidation of hydrogen gas, and also assimilate CO_2 as their source of carbon.

This nutritional and ecological diversity embroiders the plain pattern of microbial energy flow. Each of the three basic economies outlined in the preceding section (Figure 4.1) occurs in multiple versions that reflect the adaptation of organisms to particular environmental niches: The object of this section is to display a little of the variety of bioenergetics without losing sight of its essential unity.

ATP Generation by Cytoplasmic Enzymes. Bacteria are noted for their capacity to flourish in environments deficient in or devoid of oxygen (unlike eukaryotes, which are fundamentally aerobes). Anaerobic growth is often

TABLE 4.1   The Basic Types of Energy Metabolism

	Energy source	
Reductant	Light (phototrophs)	Chemical (chemotrophs)
Inorganic (lithotrophs)	Cyanobacteria; photosynthetic eukaryotes, algae, and plants (reductant H_2O) Photosynthetic sulfur bacteria (reductant mainly H_2S)	Prokaryotes only, e.g., nonphotosynthetic sulfur-oxidizing bacteria, iron bacteria, methanogens
Organic (organotrophs)	Prokaryotes only: photosynthetic bacteria utilizing organic reductants	Many prokaryotes and eukaryotes: most of the common saprophytic bacteria, fungi, symbionts; all animals

After Carlile, 1980.

sustained by catabolic reactions that generate ATP at the substrate level, without the intervention of chemiosmotic processes. The most familiar reactions of this kind form part of the glycolytic pathway, discussed in Chapter 2. Many anaerobic bacteria, however, ferment not only sugars but also amino acids, purines, pyrimidines, and other organic compounds. Figure 4.2 shows a selection of such pathways for the formation of activated phosphoryl compounds, taken from the excellent and detailed review by Thauer et al. (1977). A simple example is the fermentation of arginine with the formation of carbamylphosphate, which then donates its phosphoryl group to ADP; this pattern is characteristic of many streptococci and mycoplasmas. Purine fermentation by *Clostridium cylindro-sporum* is rather more recondite; the primary activated intermediate is formyl-tetrahydrofolate, which then reacts with ADP to yield ATP, formate, and tetrahydrofolate. Acetyl CoA is the primary energy-rich intermediate in the metabolism of pyruvate by a number of organisms, both aerobic and anaerobic. Aerobes oxidize it to CO_2 with concomitant ATP generation by oxidative phosphorylation, but anaerobes convert acetyl CoA to acetyl phosphate and that in turn to ATP (*Veillonella* and some species of clostridia). Lactobacilli have invented a variation in which acetyl phosphate is produced directly by the dehydrogenation of pentose phosphate.

An even more direct mode of ATP generation has recently come to light. According to Liu et al. (1982), certain anaerobes can take up inorganic pyrophosphate (PP_i) from the medium and utilize it as a donor of activated phosphoryl groups for ATP synthesis. This is an unexpected finding because $\Delta G^{\circ\prime}$ for the hydrolysis of PP_i is only -5.2 kcal/mol (-22 kJ/mol), not enough to sustain net ATP generation; how the organisms overcome the shortfall is not yet known. Since the utilization of PP_i bypasses the metabolic web, growth depends on a source of fixed carbon in addition to PP_i.

Cytoplasmic generation of ATP must be supplemented by a proton-trans-locating ATPase in the plasma membrane (Figure 4.1). The ATPases that have been examined so far, notably those of streptococci and clostridia, belong to the same molecular family as the ATPases of aerobic and photosynthetic bacteria (Harold, 1978; Thauer et al., 1977; Maloney, 1982). But it must not be taken for granted that all proton-translocating ATPases belong to that family; for instance, that of lactobacilli appears to be unusual (Biketov et al., 1982).

Electron Transport and Proton Translocation.   Most chemotrophic bacteria obtain the bulk of their energy not from substrate-level phosphorylation but from the oxidation of organic or inorganic substrates coupled to the generation of a protonic potential. This is accomplished with the aid of diverse electron-transport chains, which include flavoproteins, iron-sulfur proteins, quinones, and cytochromes embedded in the plasma membrane. The reduced substrates may occur as such in the medium or they may be products of metabolic reactions in the cytoplasm; the chains pass electrons either to oxygen or to an alternative electron acceptor (Table 4.1). Bacterial redox chains have been reviewed by

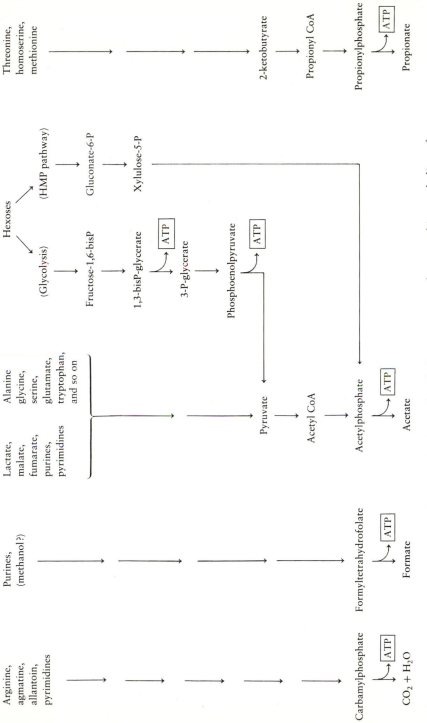

FIGURE 4.2 Substrate-level phosphorylation reactions in the anaerobic metabolism of carbohydrates, amino acids, purines, and pyrimidines. (After Thauer et al., 1977.)

Haddock and Jones (1977), Stouthamer (1978), Ingledew and Poole (1984), and in an extensive collection of specialized articles edited by Knowles (1980).

Figure 4.3 shows a selection of prokaryotic electron transport, or oxidation, chains. The order of the redox carriers, ultimately determined by their redox potentials, is everywhere more or less the same (and by no means as firmly established as the neat linear sequences imply), but bacterial respiratory chains differ in several respects from the mitochondrial chain. First, they are much more variable. The oxidation chain of *Paracoccus denitrificans* closely resembles that of mitochondria and probably includes three sites of proton translocation ("coupling sites"). *E. coli* lacks cytochrome *c* and is thought to have but two coupling sites. Some chains, whose function is the oxidation of substrates of high redox potential (e.g., Fe^{2+}), are truncated still further and probably have but one coupling site. Bacterial respiratory chains often contain unusual redox carriers, such as the menaquinones generally found in gram-positive bacteria in place of the more common ubiquinone, and the terminal electron acceptor may be nitrate, or even sulfate, in addition to oxygen.

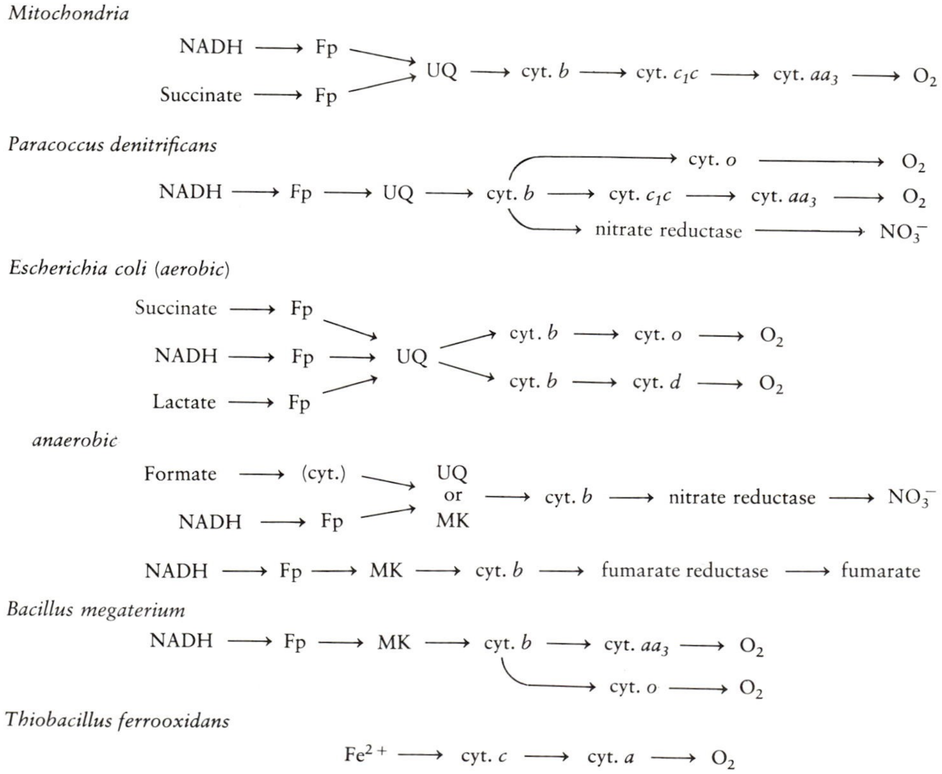

FIGURE 4.3   A selection of bacterial respiratory chains. UQ, ubiquinone; MK, menaquinone; Fp, flavoprotein. Multiple *b*-type cytochromes have been omitted.

Second, unlike the more familiar respiratory chain of mitochondria, bacterial oxidation chains are generally branched (Figure 4.3). Several alternative routes of electron transport may coexist in a single cell, and bacteria respond to environmental changes by elaborating an appropriate respiratory chain. For example, in *E. coli* growing aerobically under optimum conditions the terminal oxidase is a cytochrome of the *o* type; under conditions of low oxygen tension this is replaced by an oxidase of the *d* type and the number of coupling sites may fall from two to one. The branch of the chain leading to nitrate as terminal electron acceptor is produced only when nitrate is present in the environment and oxygen is absent.

Time was when respiration was held to require oxygen, and anaerobic bacteria were thought to generate all their ATP by substrate level phosphorylation. But this is not so: electron transport to acceptors other than oxygen, especially nitrate and fumarate, is an option available to many facultative bacteria. Even some strict anaerobes are now known to generate ATP by electron-transport phosphorylation (Thauer et al., 1977). Examples include the sulfate-reducing bacteria, strict anaerobes that live by the oxidation of hydrogen gas or organic reductants with sulfate, and perhaps the methanogens, which oxidize H_2 with CO_2 to produce methane. The electron-transport chains of these very ancient organisms are not yet well worked out and they may prove to be quite unlike those shown in Figure 4.3. But the data we have at present suggest that they too are proton-translocating pathways coupled to an ATP synthase.

Special mention should be made here of the chemolithotrophic bacteria, which obtain both energy and reducing power from inorganic reactions—the oxidation of H_2S or NH_3 by oxygen, the production of methane, and others shown in Table 4.2. This mode of life is confined to prokaryotes and has been attributed to unique metabolic capacities not shared by other organisms. Many chemolithotrophs inhabit restricted environments, and their energy metabolism is not as well known as that of their organotrophic cousins (Thauer et al., 1977; Kelly, 1978, 1981; Ingledew, 1982; Odom and Peck, 1984). Nevertheless, it appears that even obligate chemolithotrophs have quite conventional oxidation chains that translocate protons and that they utilize the proton potential to generate both ATP and reduced nicotinamide adenine nucleotides in a chemiosmotic manner (Table 4.2; Chapter 5). What makes these organisms unusual is their possession of special enzymes that give unconventional electron donors or acceptors access to these electron-transport chains. Of the organisms listed in Table 4.2, only the methanogenic bacteria are likely to prove fundamentally different; we shall return to these in Chapter 6.

Chemolithotrophs are sometimes dismissed as biological curiosities, denizens of such remote places as mine dumps and cow rumens. But in fact, habitats suited to chemolithotrophs are quite extensive. The most spectacular is surely the complex of deep-water volcanic vents and fissures, collectively called the midoceanic ridge, that extends for thousands of kilometers along the ocean floor. Here, at temperatures that may in places exceed 300°C, a little-known microbial community thrives on volcanic gases including H_2S, H_2, CO, CO_2, and methane and supports a strange fauna that includes clams and worms lacking both mouth

TABLE 4.2   Inorganic Reactions as Energy Sources for Bacterial Growth

Energy source	Reaction	$\Delta G^{\circ\prime}$ (kJ/mol)	ATP yield	Energy coupling	Representative organisms
Hydrogen oxidation	$H^+ + \frac{1}{2}O_2 \rightarrow H_2O$	−237	2–3	Cytochromes b, c, a; F_1F_0 ATPase, $\Delta\tilde{\mu}_{H^+}$	*Pseudomonas, Alcaligenes, E. coli*
Sulfur oxidation	$H_2S + \frac{1}{2}O_2 \rightarrow S^0 + H_2O$	−210	1?	Cytochromes b, c, a; F_1F_0 ATPase, $\Delta\tilde{\mu}_{H^+}$	*Thiobacillus*
	$S^0 + 1\frac{1}{2}O_2 + H_2O \rightarrow H_2SO_4$	−496	3?		
	$S_2O_3^{2-} + 2O_2 + H_2O \rightarrow 2SO_4^{2-} + 2H^+$	−436	6–7?		
Iron oxidation	$2Fe^{2+} + \frac{1}{2}O_2 + 2H^+ \rightarrow 2Fe^{3+} + H_2O$	−47	1	Cytochromes c, a; F_1F_0 ATPase, $\Delta\tilde{\mu}_{H^+}$	*Thiobacillus*
Ammonia oxidation	$NH_4^+ + 1\frac{1}{2}O_2 \rightarrow NO_2^- + 2H^+ + H_2O$	−272	2–4	Cytochromes b, c, a; F_1F_0 ATPase	*Nitrosomonas*
Nitrite oxidation	$NO_2^- + \frac{1}{2}O_2 \rightarrow NO_3^-$	−73	1	Cytochromes c, a; F_1F_0 ATPase	*Nitrobacter*
Methane production	$4H_2 + CO_2 \rightarrow CH_4 + 2H_2O$	−131	1?	No cytochromes; F_1F_0 ATPase, $\Delta\tilde{\mu}_{H^+}$?	*Methanobacterium, Methanosarcina*
Sulfate reduction	$SO_4^{2-} + 4H_2 + H^+ \rightarrow HS^- + 4H_2O$	−152	1?	Cytochromes b, c; menaquinone; F_1F_0 ATPase, $\Delta\tilde{\mu}_{H^+}$; also substrate level phosphorylation	*Desulfovibrio, Desulfotomaculum*

After Kelly, 1981, and elsewhere.

and anus but provided with masses of chemolithotrophic symbiotic bacteria (Cavanaugh et al., 1981; Felbeck, 1981; Arp and Childress, 1983). Environments such as this have a peculiar allure: they may prove to be closer to that of the primordial earth than the conventional dilute soup of organic compounds, and they may harbor relics of early evolution that survive nowhere else.

The Diversity of Bacterial Photosynthesis.   Bacteria have developed no fewer than three major variations on the general theme of transducing light energy into that of a proton potential: cyclic electron transport, noncyclic electron transport, and the unique constellation of mechanisms found in the halobacteria. In all cases ATP is produced by a proton-translocating ATPase.

Figures 4.4*a* and *b* depict the cyclic pathway characteristic of many bacteria, including the genera *Chromatium, Chlorobium, Rhodospirillum,* and *Rhodopseudomonas,* in which the electron ejected from chlorophyll in the photochemical act ultimately returns to its original place. The attendant translocation of protons across the plasma membrane generates a proton-motive force that supports both ATP synthesis and the reduction of nicotinamide adenine nucleotides. This is the process that will be analyzed below from the mechanistic point of view. Cyclic photosynthesis itself appears in several variations; the two shown in Figures 4.4*a* and *b* differ with respect to their chlorophyll pigments and electron carriers. In *Rhodopseudomonas* the primary electron acceptor has a relatively high midpoint potential and reduction of pyridine nucleotides requires reverse electron transport. *Chlorobium* operates over a more negative redox span and the primary electron acceptor can reduce NAD^+ directly. Moreover, the latter organisms contain specialized intracellular vesicles that bear accessory photosynthetic pigments; only the photochemical reaction centers are part of the plasma membrane proper. Another notable variation appears in the nonsulfur purple bacteria, *Rhodospirillum* for example, which can grow either anaerobically in the light or aerobically in the dark; a segment of the electron-transport chain does double duty, conducting electrons either to oxygen or back to chlorophyll.

Noncyclic electron transport, outlined in Figure 4.4*c*, is characteristic of the cyanobacteria and of all chloroplasts. This is a much more elaborate process, typically localized in intracellular membranous vesicles called thylakoids, which involves the absorption of two quanta of light by separate photochemical centers. It leads to the production of oxygen, and part of the incident light energy is conserved by the direct reduction of $NADP^+$ to NADPH. Cyanobacteria are prokaryotes, and their photosynthetic machinery logically forms part of the present chapter; considerations of space and clarity, however, dictate the deferral of this topic to Chapter 8.

In both the cyclic and the noncyclic mode of photosynthesis, light is captured by chlorophyll and its energy is transduced into that of a proton potential with the aid of an electron-transport chain. A fundamentally different mechanism is employed by the halophilic bacteria that inhabit salt lakes and brine ponds. Halobacteria capture light with the aid of a pigment called bacteriorhodopsin,

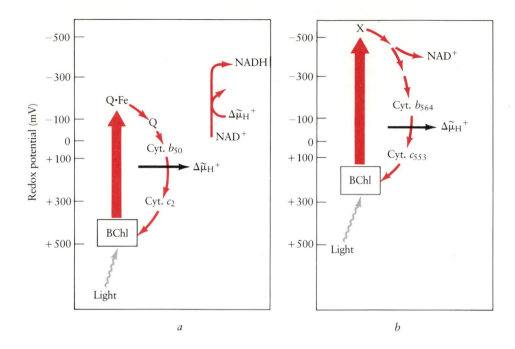

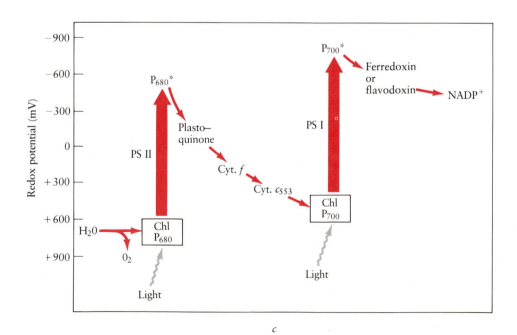

whose functional group is the carotenoid retinal (the same retinal that our own eyes use to register light). Retinal is the prosthetic group of a protein that is plugged through the plasma membrane and serves as a proton pump directly, without an ancillary electron-transport chain. The proton pump in turn supports ATP generation with the help of an ATP synthase, much like that of other bacteria. Halobacteria are fascinating creatures, unique in many respects, and will be more fully considered in Chapter 6; for present purposes we may regard them as a variation on the basic photosynthetic pattern.

The microbial world is vast, and its diversity overwhelming. It is quite possible that additional mechanisms of energy capture or transmission remain to be discovered among the manifold inhabitants of soil and sea; strict anaerobes and marine bacteria that employ sodium as the principal coupling ion underscore the point. But such eccentric organisms are likely to be confined to specialized niches. The major energy-transducing mechanisms, those that sustain the great majority of species and the bulk of living matter, are quite certainly known to us. And the wealth of microbial adaptations must not be allowed to obscure the prevalence of an energy economy based on the interplay of cytoplasmic phosphoryl donors with the electrochemical potential of protons (Figure 4.1).

Proton Translocation by Vectorial Enzymes: The F_1F_0 ATPase

The osmoenzymes that couple redox reactions, light absorption, and the synthesis (or hydrolysis) of ATP to the translocation of protons across a membrane are the prime movers of the cellular engine. From the molecular standpoint, their mechanism is the central issue in contemporary bioenergetics. Let it be said at once that this problem has not been satisfactorily solved and continues to generate a fury of research and controversy. However, we do possess an impressive body of knowledge about the structural basis of vectorial enzymology, which defines and circumscribes the issues that remain. The object of the present section is to summarize the main features of proton transport by the ubiquitous membrane-bound ATPase. Subsequent sections will deal with proton translocation by oxidation chains and the photosynthetic apparatus; bacteriorhodopsin will be discussed in Chapter 6.

FIGURE 4.4   Cyclic and noncyclic electron transport in bacterial photosyntheses. (a) Cyclic electron transport in *Rhodopseudomonas sphaeroides*. (b) Cyclic electron transport in *Chlorobium limicola*. These two organisms employ different redox carriers and operate over different redox spans. *Rps. sphaeroides* generates NADH by reversed electron transport, whereas *Chl. limicola* can reduce NAD^+ directly. (After Dutton and Prince, 1978, with permission of Academic Press.) (c) Noncyclic electron transport in cyanobacteria and chloroplasts. Two separate photosystems participate in the light-driven reduction of $NADP^+$ to NADPH; water serves as the reductant, with the liberation of oxygen. Chl, chlorophyll; BChl, bacteriochlorophyll; Q, quinone.

Bacterial membranes, like those of mitochondria and chloroplasts, are studded with knobs some 10 nanometers in diameter that project from the cytoplasmic surface and appear to be connected to the membrane proper by a short stalk. These have been identified as the ATP synthase, or ATPase, that mediates both the production of ATP during electron-transport phosphorylation and the generation of the protonic potential by ATP hydrolysis. Reviews by Mitchell (1976a), Haddock and Jones (1977), and Harold (1978) provide background knowledge, while current issues are the subject of excellent articles by Fillingame (1980a, 1980b), Maloney (1982), Futai and Kanazawa (1983), and Senior and Wise (1983).

In 1971, F. Gibson and his colleagues described a class of E. coli mutants that had lost the capacity to grow on oxidizable substrates (succinate, for example) but grew on fermentable ones; these were called *unc*, for uncoupled, and proved to lack the ATP synthase. In consequence, they were unable to produce ATP at the expense of oxidation, even though electron transport was unimpaired. When techniques became available for measuring the pH gradient and the membrane potential, it was found that wild-type cells can generate a proton potential at the expense of either respiration or ATP hydrolysis. *Unc* mutants could still utilize respiration for that purpose, but not the hydrolysis of ATP. Analogous experiments were done with everted membrane vesicles, produced by disrupting the cells with the aid of a French press. Everted vesicles incubated with ATP developed a membrane potential, interior positive, and their lumen became acidic relative to the medium by as much as two units. Removal of the ATPase (by mutation or by washing it off the membrane) prevented the establishment of both gradients, reinforcing the conclusion that the ATPase is an electrogenic proton pump.

The belief that the ATPase translocates protons is buttressed by extensive observations on streptococci, whose cytoplasmic membranes lack oxidation chains. Glycolyzing cells extrude protons, generating a membrane potential, cytoplasm negative, whose magnitude can be estimated from the accumulation of certain lipophilic cations with the aid of Equation 1.20. The potential is abolished by treating the cells with proton conductors or with DCCD (N,N'-dicyclohexylcarbodiimide), an inhibitor of the ATPase (Figure 4.5a). When everted membrane vesicles are incubated with ATP, protons accumulate in the lumen; enhanced fluorescence of ANS (8-anilino-1-naphthalene sulfonate) gives evidence that the lumen has become electrically positive, and quenching of quinacrine fluorescence indicates that it is acidic. These effects are again abolished by DCCD and by proton conductors (Figure 4.5b). Streptococci are fermentative bacteria that do not carry out oxidative phosphorylation. However, substantial ATP formation was seen when an artificial membrane potential (Figure 4.5c) or pH gradient of the correct polarity was imposed across the membrane; an influx of protons, mediated by the ATPase, accompanied ATP generation. The stoichiometry (the number of protons translocated per molecule of ATP hydrolyzed or formed) is still open to question. Earlier findings with bacterial systems converged on $2H^+$ per ATP, as Mitchell and Moyle had first reported. Recent measurements, however, increasingly favor $3H^+$ per ATP (Maloney, 1982, 1983; Kashket, 1982; Clark et al., 1983).

Whatever misgivings lingered concerning the capacity of the ATPase to translocate protons were dispelled by the reconstitution experiments of Y. Kagawa and his colleagues in Japan (Kagawa, 1980). Working with a thermophilic species of *Baccillus*, whose ATPase is particularly stable, they were able to purify the ATPase, reconstitute the purified enzyme into liposomes, and demonstrate that ATP hydrolysis is accompanied by proton translocation. Analogous experiments have now been done with several bacteria. They leave no doubt that the ubiquitous ATPase of bacterial plasma membranes is a proton pump that catalyzes the electrogenic transport of protons as envisaged by the chemiosmotic theory.

The ATPase is one of the most intricate enzymes known, almost rivaling the ribosome. The functional ensemble comprises two regions, labeled F_1 and F_0; henceforth we shall use the prefix F_1F_0 to designate enzymes that belong to this molecular family. (The terminology stems from Efraim Racker's pioneering work on the biochemical dissection of mitochondrial oxidative phosphorylation.) F_1 is a globular headpiece that is peripheral to the membrane and quite easily removed by washing with buffers of low ionic strength. The molecular mass of the headpiece is about 385 kilodaltons (kdal), and it is made up of five kinds of subunits whose proportions are probably $\alpha_3\beta_3\gamma\delta\epsilon$. It bears the catalytic site or sites as well as determinants for attachment to the membrane sector, F_0. The latter portion spans the cytoplasmic membrane. It consists of three kinds of subunits, in the proportions $a_1b_2c_{6-10}$. The smallest subunit, a proteolipid, is the one that reacts with DCCD, a powerful inhibitor of all ATPases of the F_1F_0 family. The primary sequences of all the subunits have been determined, but their arrangement in space is still quite uncertain.

The two regions of the F_1F_0 ATPase perform quite distinct functions. Removal of F_1 from the membrane, either by washing or by mutation, leaves the residual membrane highly permeable to protons but not to other ions. The inference is that F_0 functions as a proton channel across the membrane, as shown schematically in Figure 4.6. Reaction with DCCD blocks this channel. We arrive, then, at a general conception of how the F_1F_0 ATPase is constructed. The F_0 portion provides a passage for protons from the exterior to the catalytic site, or at least to the F_1 headpiece. But its true function is more subtle. As Mitchell recognized long ago (Mitchell, 1968, 1976a), a proton channel open to the exterior but blocked by the headpiece at its upper end is a "proton well." Protons will accumulate in the channel in response to the transmembrane electric potential; the effective pH at the blocked end will be lower than that at the open end. The F_0 basepiece thus transduces a membrane potential into a pH gradient, ensuring that both $\Delta\Psi$ and ΔpH contribute to whatever the protons do. F_0 is probably not a simple pore, but may consist of a chain of proton-binding groups. The headpiece is where the action is: F_1 catalyzes the interconversion of ATP and

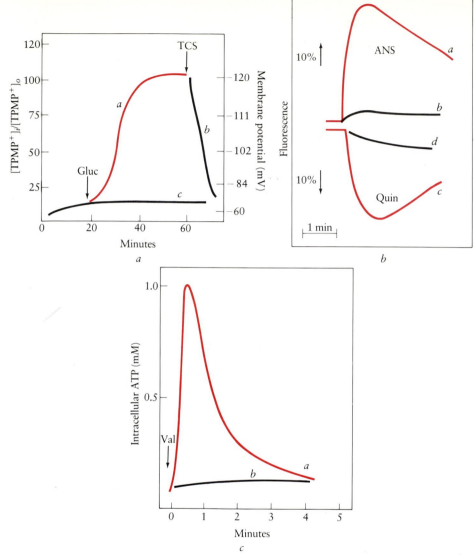

FIGURE 4.5   Indicators of proton translocation by the F_1F_0 ATPase of streptococci.
(*a*) Accumulation of triphenylmethyl phosphonium ion ($TPMP^+$) by glycolyzing cells serves
as a measure of the membrane potential (trace *a*). Accumulation is abolished by the
addition of tetrachlorosalicylanilide (TCS), a proton-conducting ionophore (trace *b*), and
also by preincubation of the cells with DCCD, an inhibitor of the F_1F_0 ATPase (trace *c*).
(*b*) Everted membrane vesicles, incubated with ATP and Mg^{2+} ions, generate a protonic
potential, lumen positive and acid. Enhanced fluorescence of 8-anilino-1-naphthalene
sulfonate (ANS) indicates that the vesicle lumen is positive (trace *a*); diminished
fluorescence of quinacrine indicates that it is acidic (trace *c*); both effects are abolished by
pretreatment of the vesicles with proton conductors or with DCCD (traces *b*, *d*).
(*c*) Synthesis of ATP in response to an imposed gradient of electric potential. Cells replete
with K^+ were suspended in buffer lacking K^+, in the absence of an energy source.
Addition of valinomycin at 0 minutes initiates transient ATP production (trace *a*). This is
abolished by pretreatment of the cells with DCCD or with proton conductors (trace *b*).
[Idealized from observation by F. M. Harold (*a*), H. Kobayashi (*b*), and P. C. Maloney and
T. H. Wilson (*c*).]

ADP + P_i, coupling it to the movement of two or three protons. The spatial orientation of the complex is such that ATP hydrolysis expels protons from the cytosol, while the passage of protons in the opposite direction induces the net formation of ATP.

One would now like to know just how F_1 and F_0 are put together, where the catalytic site, or sites, are situated, and how the chemiosmotic coupling of catalysis to proton transport is effected. But these are all bones of contention that must either be discussed in some detail or else set aside. Since most of the data and arguments derive from research with mitochondria, we shall defer this discussion to Chapter 7. Suffice it to state here that two contrasting views have emerged. Mitchell envisages a ligand conduction mechanism, such that the spatial pathway of proton movements corresponds closely to that of the chemical reaction. The catalytic center, situated at the interface of F_1 and F_0, is oriented with respect to the membrane, and the protons undergoing translocation participate directly in the chemical reactions of ATP hydrolysis or synthesis (Figure 3.11*b*). The alternative proposal, championed particularly by P. D. Boyer, denies any direct linkage between proton movements and the chemical transformations. In this view, binding sites for nucleotides are entirely separate from those for protons. ATP binding and hydrolysis induce a sequence of conformational changes that result in the uptake of protons on the cytoplasmic side and their discharge to the exterior, much as shown in Figure 3.12 for the Na-K ATPase. The experimental evidence relevant to these models will be summarized in Chapter 7, but it bears repeating here that the unresolved questions about the molecular mechanism of the F_1F_0 ATPase do not undermine its status as a physiological proton pump of central importance in biological energy transduction.

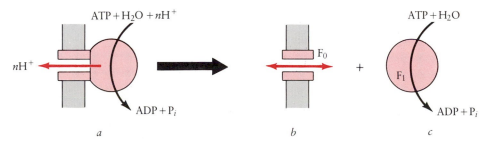

FIGURE 4.6   Topology of the proton-translocating ATPase. (*a*) In the intact complex, the membrane sector F_0 gives protons access to the catalytic site located on the headpiece F_1. Washing with buffers of low ionic strength releases F_1 from the membrane. F_0 remains behind, manifesting itself as a proton channel (*b*). Free F_1 is a soluble enzyme that catalyzes only the hydrolysis of ATP (*c*).

Proton-Translocating Oxidation Chains

It is now all but universally agreed that the passage of reducing equivalents from such substrates as NADH or lactate to the terminal electron acceptor is accompanied by the electrogenic translocation of protons outward across the plasma membrane. This belief rests primarily on experiments of the sort illustrated in Figure 4.7, in which proton extrusion in response to a small pulse of oxygen was monitored through the transient acidification of the external medium. Ordinarily the proton circulation is a cryptic process; protons extruded during respiration are immediately sucked back into the cell by the electric potential. In order to observe the extruded protons by means of pH changes, the proton circulation must be disrupted by allowing a permeant cation to flow into the cell and thus compensate for the protons; in Figure 4.7 both potassium plus valinomycin and the permeant thiocyanate anion were included to that end. Abolition of the proton pulse by the proton conductor CCCP shows that acidification is due to extruded protons, not to some acidic product of metabolism.

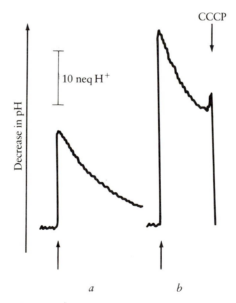

FIGURE 4.7   Proton translocation by anaerobic suspensions of *E. coli* on addition of an oxygen pulse. Cells were incubated anaerobically in the presence of substrate (malate), KSCN, 0.1 *M* KCl, and valinomycin. A small pulse of air-saturated KCl was added at the arrow, and the amount of H^+ translocated out of the cells was determined by the acidification of the medium. (*a*) From the amounts of O_2 added and of H^+ expelled, a stoichiometry of near 4H^+/O was calculated. (*b*) Addition of the proton conductor CCCP induced the protons to flow back into the cell, down the gradient of $\Delta\tilde{\mu}_{H^+}$ generated by respiration. (After observations by H. G. Lawford and B. A. Haddock, with permission of the Biochemical Society, London.)

Numerous variations on this experiment employing mutants, membrane vesicles, and inhibitors of respiratory enzymes supply ample evidence that proton transport is an intrinsic property of the redox cascade, requiring neither a functional ATPase nor any other ancillary device. The traditional "coupling sites" of oxidative phosphorylation correspond to the segments of the chain where protons are translocated. These are defined by redox complexes such as the NADH-ubiquinone oxidoreductase, the ubiquinone–cytochrome-*c* oxidoreductase, and cytochrome oxidase. It is likely that, as in mitochondria, each of these segments is a relatively independent unit effecting simultaneously both a redox reaction and proton transport. The diverse oxidation chains found in the bacterial world (Figure 4.3) reflect the combination of several such complexes. The burning issue is: How do the various segments link the movement of electrons down a scalar redox gradient to the vectorial movement of protons across the plasma membrane?

Figure 4.8 shows the core of the *E. coli* respiratory chain as it is presently understood, arranged in the form of two redox loops. The principle of the redox loop was an important element in Mitchell's original formulation of the chemiosmotic theory, presenting a clear example of vectorial metabolism. Mitchell envisaged oxidation chains as vectorial pathways so arranged that at each coupling site a hydrogen carrier (such as a flavin or a quinone) is reduced at the inner membrane surface. It then diffuses to the outer membrane surface, where protons and electrons separate: the electrons pass back across the membrane by an electron carrier (cytochromes, nonheme iron), while protons are released to the outside (Figure 4.8). Note that electrons traverse the membrane as such but protons do not. It is the consumption of protons by chemical reactions at the cytoplasmic surface, and their liberation by chemical reactions at the exterior surface, that brings about the effective translocation of protons across the membrane.

This arrangement of the respiratory chain is consistent with what is known of the topological positions of the redox carriers in the membrane and also with the finding that respiring cells and vesicles extrude two protons per electron pair when the substrate is lactate or succinate, four when the electron donor is NADH. But it must be added that the data on hand are not as extensive as one would wish and do not exclude alternative arrangements. In mitochondria, proton extrusion by the ubiquinone–cytochrome-*b* segment is now thought to involve a more sophisticated version of the redox loop called the Q cycle (Chapter 7), and this is likely to be true for bacterial oxidation chains as well. A more fundamental challenge to the entire concept of redox loops, and of vectorial chemistry in Mitchell's sense of the word, was raised by experiments suggesting that the cytochrome oxidase of mitochondria may catalyze not the transfer of electrons across the membrane but the active transport of protons (Chapter 7). In principle, at least, all redox-linked proton translocation might be attributed not to redox reactions looped across the membrane but to "conformational" proton pumps. This term designates the proposition that oxidation and reduction of the respiratory carriers are accompanied by changes in the conformation of mem-

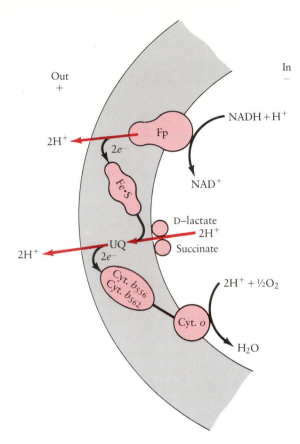

FIGURE 4.8   Redox loops in *E. coli*, a hypothetical arrangement of part of the main respiratory chain to form two proton-translocating loops. Oxidation of NADH involves both loops, oxidation of lactate or succinate only the second loop.

brane proteins, coupled to the uptake of protons at the inner surface and their release at the outer surface. Protons would be actively transported as such, but the redox pathway would not traverse the membrane. The fundamental issue, therefore, is whether transmembrane redox reactions as illustrated in Figure 4.8 actually exist at all.

Strong evidence that at least some bacterial redox chains are organized as redox loops comes from research on the anaerobic chain that allows *E. coli* to oxidize formate with nitrate as terminal electron acceptor. This is a simple and particularly well characterized pathway composed of two cytochromes *b*, quinone and nonheme iron centers (Figure 4.9). The chain translocates either two or four protons, depending on the substrate, and terminates with nitrate reductase. Peter Garland and his colleagues have carried out a thorough topological analysis of this chain, using electron donors that selectively react

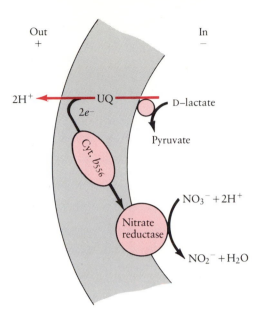

FIGURE 4.9   Nitrate reduction by *E. coli*, as a redox loop. Evidence that this process does in fact involve transfer of electrons across the membrane is cited in the text.

with either the external or the cytoplasmic face of the plasma membrane. They demonstrated that ubiquinone is oxidized and protons are released at the external face, while nitrate is reduced and protons are consumed at the cytoplasmic face (Jones et al., 1980). The final step in nitrate reduction is thus a redox reaction that traverses the membrane, and the physiological pathway must be depicted as a redox loop (Figure 4.9).

Several other bacterial redox chains have likewise been shown to involve the transfer of electrons across the membrane, thereby generating a protonic potential. The hydrogenase of *E. coli* ($H_2 \rightleftharpoons 2H^+ + 2e^-$) is one example, another is the oxidation of formate with fumarate as electron acceptor by the anaerobe *Vibrio succinogenes* (Kröger et al., 1980), and the oxidation of Fe^{2+} to Fe^{3+} by the chemolithotrophic thiobacilli is a third (Ingledew, 1982; Cobley and Cox, 1983). The pathway of electrons in cyclic photosynthesis likewise traverses the membrane. On the other hand, there is evidence that in *Paracoccus denitrificans,* whose respiratory chain is remarkably similar to that of mitochondria, the terminal cytochrome oxidase functions as a proton pump (Van Verseveld et al., 1981). Such findings have now been reported for several, but not all, cytochrome oxidases of bacterial origin. A conciliatory conclusion would be that both pumps and loops contribute to bacterial electron transport. It is tempting to suggest that redox loops may be the ancestral mode of proton translocation, which was buttressed by the addition of proton-pumping elements at a later stage of evolution; but this is rank speculation.

Light-Driven Proton Translocation

The utilization of light energy to drive the phosphorylation of ADP to ATP may be the epitome of biological energy transduction. Research in this field has made spectacular progress, thanks to technical developments that make it possible to elicit single turnovers of the photosynthetic apparatus and to monitor the sequence of events by ultrarapid spectrophotometry. I shall discuss photosynthesis in somewhat greater detail because its specialized methodology and language have hindered general appreciation of the insights obtained. A recent book by Clayton (1980) provides an excellent introduction to this subject; reviews of bacterial photosynthesis are provided by Dutton and Prince (1978), Crofts and Wood (1978), Baccarini-Melandri et al. (1981), Cramer and Crofts (1982), and Junge and Jackson (1982).

The Photochemical Reaction.   In 1952, H. K. Schachmann, A. B. Pardee, and R. Y. Stanier described the isolation of membranous preparations from the nonsulfur purple bacterium *Rhodospirillum rubrum* that were enriched in photosynthetic pigments. They thought that these might be the organelles of bacterial photosynthesis and called them chromatophores. Subsequent research revealed the presence of cytochromes and quinones and showed that chromatophores illuminated in the presence of ADP and P_i generate ATP. We now know that chromatophores do not exist as such in the living cell; the photosynthetic machinery is housed in invaginations of the plasma membrane that pinch off when the cells are broken to form closed vesicles 30 to 60 nanometers in diameter (Figure 4.10). The polarity of these vesicles is everted relative to that of the parent cell, and partly for that reason they proved far more convenient to work with than intact cells. The following account is based on work with *Rhodopseudomonas sphaeroides* and other nonsulfur purple bacteria, which have provided most of our knowledge.

The molecular dissection of the photochemical process began, also in the early 1950s, when L. N. M. Duysens discovered that illumination of a suspension of whole cells or of chromatophores induced a small reversible change in their absorption spectrum: the intensity of a band near 890 nanometers diminished and the peak of another near 800 nanometers shifted toward a lower wavelength. Duysens correctly ascribed this to the photo-oxidation of a special kind of bacteriochlorophyll that has a central role in photosynthesis. This pigment is now referred to as reaction center chlorophyll. It constitutes but a minor fraction of the total chlorophyll, one molecule in 80 or even less, intimately associated with an electron-transport chain and with other pigments. The reaction center is the focus of the photochemical act, where light energy is transduced into a more useful form. The bulk of the chlorophyll and also the carotenoid pigments play a less direct role. They serve as a collecting antenna that funnels the energy of incident light into the reaction center and greatly enhances the overall rate of energy capture and conversion. If light capture were restricted to the reaction center alone, then at normal light intensities each center would be activated only

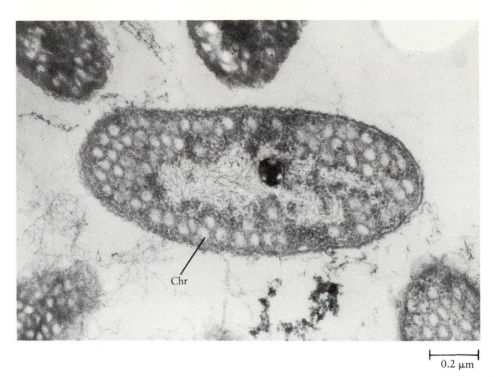

0.2 μm

FIGURE 4.10   Chromatophores. Section through a cell of *Rps. sphaeroides* containing relatively sparse photosynthetic membranes. Chromatophores (Chr) arise by invagination of the cytoplasmic membrane. When the cell is broken, these invaginations pinch off, yielding sealed vesicles whose polarity is everted with respect to the parent cell. (Electron micrograph courtesy of G. Cohen-Bazire.)

once every second; the actual rate of light processing is a thousand times greater, thanks to the increased cross section achieved by the cooperation of multiple chlorophylls.

Purified reaction centers were first isolated by R. K. Clayton in 1968 by using detergents to solubilize the chromatophore membrane. Highly purified reaction centers have a well-defined stoichiometry: four molecules of bacteriochlorophyll complexed with protein, two molecules of bacteriopheophytin (a chlorophyll derivative that contains two protons in place of Mg^{2+}), one atom of nonheme iron and two molecules of quinone (ubiquinone-10 in *Rps. sphaeroides*). Less rigorously purified preparations contain cytochromes b and c_2 as well. Remarkably, purified reaction centers respond to illumination much as do chromatophores or intact cells, with a reversible bleaching of the major absorption band of bacteriochlorophyll. The absorption peak varies from 860 to 890 nanometers in the various bacteria, but its essential chemistry is thought to be always the same. Reaction center chlorophyll is conventionally abbreviated P_{870} (for pigment) and consists of a dimer of the bacteriochlorophyll-protein complex.

Bleaching of P_{870} is complete within 10 picoseconds (ps), even at 80 degrees Kelvin, and requires one quantum of light per center. This is the primary photochemical event. In chemical terms, it corresponds to the photo-oxidation of the dimer with transfer of an electron from chlorophyll to a "primary acceptor," X, within the reaction center:

$$P_{870}X \xrightarrow{\text{light}} P_{870}^+ + X^- \qquad \text{(Eq. 4.1)}$$

The chemical identity of the primary acceptor has long been elusive. It now appears to be one of the bound molecules of quinone, closely associated with the atom of nonheme iron: the latter does not directly participate in the reaction and its function remains unknown.

Energy conservation within the reaction center can be seen as a light-driven redox reaction, whose pathway and thermodynamic features are shown in Figure 4.11. In the dark, P_{870} has a midpoint potential of $+500$ mV. When it absorbs a quantum of light, it undergoes excitation and becomes a powerful reducing agent, capable of donating one electron with an effective midpoint potential of

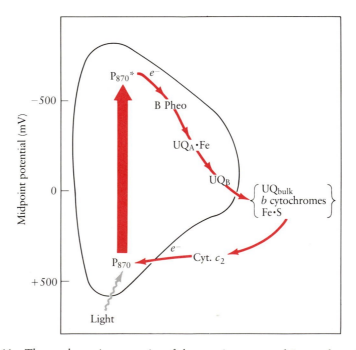

FIGURE 4.11   Thermodynamic properties of the reaction center of *Rps. sphaeroides*. The diagram shows the midpoint potentials of the major intermediates in electron transfer. The shape encloses constituents retained in purified reaction centers. (After Dutton and Prince, 1978, and Nicholls, 1982, with permission of Academic Press.)

−700 mV. The energy of the incident photon is thus transduced into and conserved as the free energy of a redox potential difference. The excited electron is transferred first to bacteriopheophytin and then, within picoseconds, to the UQ·Fe center, which corresponds to X and has an effective E_m of −180 mV. Subsequent transfers take place more slowly, in the microsecond range. The electron passes on to the second of the two quinones. This quinone (more precisely, the protein with which the quinone is associated) undergoes protonation; and that, as we shall see, initiates the conversion of redox potential into a transmembrane proton potential.

Meanwhile, back at the reaction center, oxidized bacteriochlorophyll does not long remain in this state. Within 100 microseconds it is re-reduced by accepting an electron from cytochrome c_2 (effective E_m about +300 mV). Therefore the overall result of the rapid reactions that follow the absorption of a quantum of light may be summarized by the equation

$$\text{cyt. } c_2 \cdot \text{P}_{870} \cdot \text{UQ} \cdot \text{Fe} \longrightarrow \text{cyt. } c_2^+ \cdot \text{P}_{870} \cdot \text{UQ}^- \cdot \text{Fe} \qquad \text{(Eq. 4.2)}$$

The difference in redox potential between oxidized cytochrome c_2 and reduced UQ·Fe is approximately 650 mV, equivalent to 63 kJ/mol. An essential feature of the primary photochemical events is that they are virtually irreversible. Due to the fall in potential from excited bacteriochlorophyll to reduce UQ·Fe, the reverse reaction occurs many orders of magnitude more slowly than the forward reaction, and a single photon results in the production of a single reducing electron.

Now, it is extremely important that the light-driven redox reaction catalyzed by the reaction center takes place vectorially across the membrane. Cytochrome c_2 is located on the periplasmic surface of the plasma membrane and is trapped in the lumen of chromatophores as they pinch off. Conversely, protonation of the second quinone involves the binding of protons on the outer aspect of the chromatophore membrane. Equation 4.2 therefore describes the separation of electric charge across the membrane such that the inner surface of the chromatophore vesicle becomes electrically positive. The generation of an electric field across the chromatophore membrane is the first step in the transduction of light energy into a gradient of protonic potential. The process is completed by the electron-transport chain associated with the reaction center.

Electrons and Protons.   In 1966, L. V. von Stedingk and H. Baltscheffsky observed reversible uptake of protons by illuminated chromatophores. They noted that proton movements were inhibited both by uncouplers and by antimycin, a reagent known to block mitochondrial electron transport in the vicinity of cytochrome b. The pathway by which electrons, deposited on the secondary quinone in the photochemical act, reduce the oxidized cytochrome c_2 with concurrent translocation of protons has not been fully worked out. However, it is clear that their route traverses the membrane and is best described as a sophisticated kind of redox loop.

The overall stoichiometry seems to be established: for each electron that passes over the cycle, two protons are translocated electrogenically into the chromatophore. Kinetic analysis of single and multiple turnovers has identified several charge-translocation steps, but the protons themselves move in an electroneutral manner, effectively as H. These findings are consistent with Mitchell's original proposal that the electron-transport chain is bent into two loops; in the first loop, quinone functions as the hydrogen carrier and cytochrome b conducts electrons, while the second loop requires a novel hydrogen carrier designated Z.

With the accumulation of further data, the concept of two sequential redox loops became unsatisfactory and has now been abandoned, chiefly because it did not account for the existence of multiple b-type cytochromes engaged in a complex redox interplay with multiple quinone pools. Analogous findings with mitochondria prompted Mitchell to propose a novel pathway called the Q cycle, a subtle idea whose development must be deferred to Chapter 7. For the present, Figure 4.12 illustrates the kind of pathway that best explains electron transport in the purple bacteria. It consists in effect of two interwoven loops; one is the light-driven circulation of one electron for each photon absorbed by the reaction center, while the other recycles one electron through cytochrome b_{50}. The net result is the translocation of two protons across the membrane for each photochemically excited electron.

It will be recalled that *Rhodopseudomonas* and its relatives can grow either anaerobically in the light or aerobically in the dark. Under the latter circumstances they produce a conventional respiratory chain that includes the very same cytochrome b_{50} that mediates light-driven electron flow. The *modus operandi* of this cytochrome must be the same whether it participates in respiratory or in photosynthetic flow. Most likely, then, respiratory electron flow in this region also takes the form of a Q cycle rather than of the traditional redox loops suggested earlier in Figure 4.8.

For present purposes, the mechanistic details are less important than the overall result. Each photon absorbed by bacteriochlorophyll in the reaction center elicits a vectorial redox reaction; in consequence, a single electron is transferred to the external surface of the chromatophore, while a positive charge (a "hole," as it were) remains on the cytochrome c_2 at the luminal surface. The subsequent passage of the electron back across the membrane, via the network of quinones and b- and c-type cytochromes, is stoichiometrically coupled to the uptake of two protons. The electric field generated in the primary photochemical act is thereby transduced into a gradient of protonic potential across the chromatophore membrane.

Phosphorylation.   Soon after the discovery of light-induced proton uptake, J. B. Jackson and A. R. Crofts interpreted it in chemiosmotic terms and began to accumulate evidence in support of this view. They as well as others demonstrated the development of a membrane potential (on the order of 200 mV, interior

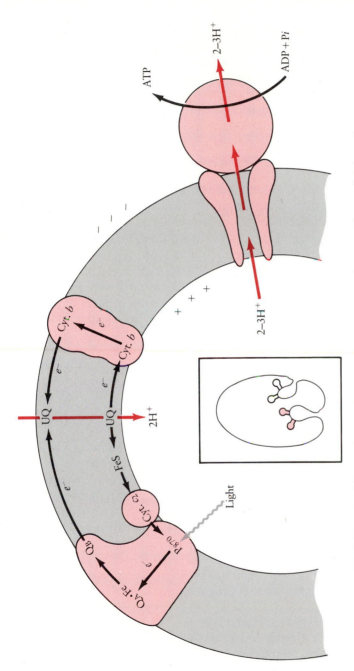

FIGURE 4.12   Electron flow and proton translocation in chromatophores of the purple bacteria. The electron-transport chain, arranged in the form of a Q cycle, translocates protons into the lumen and thereby transduces light energy into a protonic potential, lumen positive and acidic. The protons return to the exterior via the F_1F_0 ATP synthase, which mediates ATP generation. Note that the chromatophores are everted with respect to the intact cell (inset). (After Junge and Jackson, 1982, with permission of Academic Press.)

positive) and a pH gradient (one to two units, interior acid). Reagents and treatments that dissipate the proton-motive force or prevent its development also block photosynthetic phosphorylation; conversely, ATP production can be elicited in the dark by the imposition of a protonic potential of the appropriate magnitude and polarity. The prolonged argument over the thermodynamic and kinetic adequacy of the protonic potential as the intermediate between the light reactions and phosphorylation is now all but settled. The final step is mediated by an F_1F_0 ATPase of the conventional kind, which links the translocation of two or three protons out of the chromatophore to the production of one molecule of ATP (Figure 4.12). Incidentally, in *Rhodopseudomonas* and its relatives the same F_1F_0 ATPase can mediate both photophosphorylation and oxidative phosphorylation. Overall, one photon is thought to elicit the transport of two protons and eventually the synthesis of 0.66 to one molecule of ATP.

What remains unsettled, apart from the stoichiometry and molecular mechanics of proton translocation by the photosynthetic apparatus and the F_1F_0 ATPase, is the functional path of the protons from the one to the other. Figure 4.12 shows the source of the proton current and its sink as independent entities, linked only by the diffusion of protons through the bulk water phase. There is strong qualitative evidence in favor of this loose relationship, particularly the observation that it takes only a few molecules of gramicidin to uncouple an entire photosynthetic vesicle. Other data, however, point to a more intimate mode of coupling via a localized proton current (Baccarini-Melandri et al., 1981). The precise meaning of "localization" remains to be defined.

The Magnitude of the Proton Potential

By the central dogma of chemiosmotic logic, the primary pathways of energy transduction are linked to phosphorylation and to the performance of useful work by the current of protons across the plasma membrane. The magnitude of the protonic potential, or proton-motive force, is thus a central parameter in the operation of the cellular machinery. As explained in Chapter 3, $\Delta\tilde{\mu}_{H^+}$ at equilibrium is directly related to the free-energy change of the reactions that translocate protons. A living cell, however, is not at equilibrium; it is an open system whose activities are sustained by a continual flux of energy that is ultimately dissipated as heat. The protonic potential that exists at any time depends, then, both on the reactions that generate the potential and on those that consume it.

It is not difficult to measure ΔpH and $\Delta\Psi$ across bacterial membranes and then to compute $\Delta\tilde{\mu}_{H^+}$ by Equation 3.4; the problem is to decide what the numbers mean. Bacteria are too small for conventional intracellular microelectrodes; routine published measurements all rely on indirect methods that occasion misgivings (Rottenberg, 1979; Ferguson and Sorgato, 1982).

The pH gradient is usually determined from the distribution of permeant acids or bases across the plasma membrane and, more recently, from ^{31}P nuclear magnetic resonance. These are entirely independent procedures, and the substantive agreement between the results obtained is very gratifying. The measurement of $\Delta\Psi$ is rather more tricky. The membrane potential is usually calculated from the uptake of lipophilic cations or anions, which penetrate the plasma membrane and are assumed to be distributed electrophoretically according to the electric potential. The uptake of $^{86}Rb^+$ in the presence of valinomycin is also often used. In practice, data such as those obtained for Figure 4.5 are inserted into the Nernst equation (Equation 1.20). The calculation depends on the premise that the ions are not transported actively, that they are free in solution on both sides of the membrane, and that binding to the membrane itself is either negligible or open to correction. Alternatively, $\Delta\Psi$ has been estimated from the quenching of certain fluorescent dyes that are permeant and accumulate in the cells by electrophoresis in response to the membrane potential (Waggoner, 1979). In the photosynthetic bacteria, an electrochromic shift in the absorption spectrum of membrane carotenoids has been used as a built-in voltmeter that responds within nanoseconds and is suitable for monitoring electrical events during a single turnover of the photosynthetic machinery. However, what the electrochromic shift reports is probably the electric field across a local domain within the membrane phase, and the relationship of this local field to that across the membrane as a whole is not entirely clear. Given the indirect nature of the measurements, qualms inevitably persist; it seems unwise to rely too heavily on the resulting numbers for the calculation of stoichiometries or of absolute thermodynamic parameters.

Such reservations notwithstanding, much is now being learned concerning the effects of external conditions and physiological activities on the protonic potential. Figure 4.13 illustrates the dependence of $\Delta\tilde{\mu}_{H^+}$ generated by respiring cells of *E. coli* on the external pH. The pH gradient is maximum at acidic external pH, amounting in the present instance to nearly two units (-120 mV) at pH 6.0, but falls to zero at pH 7.8; $\Delta\Psi$ increases slightly over this range, from -100 to -150 mV; and the total proton potential declines from -200 to -130 mV. Note that the emphasis on the elements of the proton-motive force obscures the remarkable fact that the cytoplasmic pH remains nearly constant at pH 7.8 over the entire range; we shall return to this in the following chapter. This pattern of declining $\Delta\tilde{\mu}_{H^+}$ with constant internal pH has now been reported many times for aerobic and anaerobic organisms and for membrane vesicles as well as intact cells (Padan et al., 1981). Clearly, the protonic potential is not a fixed quantity like the line voltage in our homes, but a variable one.

The proton potential also reflects cellular activities. In a study with the fermentative bacterium *S. lactis,* Kashket et al. (1981) found that cells glycolyzing in buffer at pH 6 maintained a proton-motive force of -160 mV, while that of growing organisms was only -140 mV. The difference presumably results from the partial consumption of the proton potential by various transport processes that go on in growing cells but not in resting ones. It is also of general

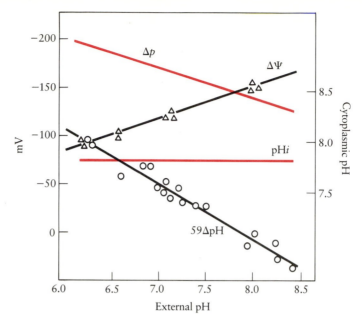

FIGURE 4.13   The proton potential of growing cells of *E. coli* varies with the pH of the medium. The membrane potential was estimated from the uptake of triphenylphosphonium ion and ΔpH from the distribution of benzoate. The values of $\Delta\tilde{\mu}_{H^+}$ and pH$_i$ were calculated from these data. (After Kashket, 1982, copyright American Chemical Society.)

significance that the $\Delta\tilde{\mu}_{H^+}$ of the fermentative streptococci is distinctly lower than that of the respiring cells shown in Figure 4.13. When growing cells of *Staphylococcus* were compared under aerobic and anaerobic conditions, $\Delta\tilde{\mu}_{H^+}/F$ at pH 6 was -250 mV and -160 mV, respectively. The difference presumably reflects the fact that aerobic cells extrude protons by redox reactions, while anaerobic ones rely primarily on the hydrolysis of ATP by the F_1F_0 ATPase. The physiological consequences are likely to be considerable: Δp is a logarithmic function, such that 59 mV corresponds to a gradient of tenfold. Any process that depends on the proton potential will therefore be carried on much more effectively by respiring than by fermenting organisms.

Is the proton-motive force sufficient to support the work functions assigned to it? This question is not as straightforward as it seems, because the answer depends on the stoichiometries of the coupled reactions and is further bedeviled by uncertainties regarding the magnitude of $\Delta\tilde{\mu}_{H^+}$ and the possible "localization" of the proton current. To a first approximation, the answer is clearly yes. With regard to the uptake of metabolites, if we assume symport of a metabolite with one proton, a $\Delta\tilde{\mu}_{H^+}$ of -180 mV would suffice to sustain a concentration gradient of 10^3, while -240 mV would support a gradient of 10^4; physiological gradients fall within this range. Likewise, in oxidative phosphorylation, a $\Delta\tilde{\mu}_{H^+}$

of -200 mV could support a phosphorylation potential, or ΔG_p, of 9 kcal/mol (38 kJ/mol), assuming that the F_1F_0 ATPase translocates two protons per cycle. If H^+/ATP is taken to be 3, a protonic potential of -200 mV would support a ΔG_p of up to 57 kJ/mol. Recent measurements of the relevant parameters (ΔpH, $\Delta \Psi$, and the cytoplasmic concentrations of ATP, ADP, and P_i) favor the latter stoichiometry (Kashket, 1982; Maloney, 1983; Clark et al., 1983).

However, ΔG_p values as high as 15 to 16 kcal/mol (63 to 68 kJ/mol) have been measured for phosphorylation by some membrane vesicle preparations, notably those from photosynthetic bacteria. These would call for proton potentials of at least -230 mV and here is the rub, for Δp as large as this is not usually seen. Only when $\Delta \Psi$ is measured by the electrochromic shift of membrane carotenoids, a procedure that consistently gives higher values than other methods, does $\Delta \tilde{\mu}_{H^+}$ come into the requisite range. It is likely that the carotenoids report the potential across some localized domain rather than across the plasma membrane as a whole; this, together with a number of other discrepancies, has suggested to some investigators that the linkage between the photosynthetic apparatus and the F_1F_0 ATPase may be closer (more localized) than is implied by the global proton-motive force between the aqueous phases on the two sides of the plasma membrane.

An even greater discrepancy has been reported in the alkalophilic bacteria that live in alkaline mud flats at pH 10 and above. Intact cells and membrane vesicles from alkalophilic bacilli carry out what appears to be conventional proton-transport phosphorylation with a conventional respiratory chain and F_1F_0 ATPase. They maintain a phosphorylation potential of 11 to 12 kcal/mol (46 to 50 kJ/mol); yet the proton potential across the vesicle membrane is only -40 mV (Guffanti et al., 1981, 1984). It is not clear what the explanation may be, but a strong possibility is that oxidation and phosphorylation are coupled through localized interactions between respiratory complexes and an adjacent F_1F_0 ATPase. Students of mitochondria have debated this issue for years, and we shall return to it in Chapter 7. It would be ironic (and, to this microbiologist, quite gratifying) if definitive data were at last to come from the specialized and obscure alkalophiles.

Energy Coupling by a Sodium Circulation

In bacterial energetics the proton circulation holds center stage, but a sodium circulation often plays a supporting role and sometimes a principal one. As a rule, the expulsion of Na^+ ions to generate an electrochemical potential gradient, $\Delta \tilde{\mu}_{Na^+}$, is itself driven by the flux of protons and the sodium circulation is ancillary to that of protons. However, we now recognize several primary sodium-translocating vectorial pathways. Bacterial sodium circulations are therefore of interest both to students of vectorial enzymology and to comparative physiologists.

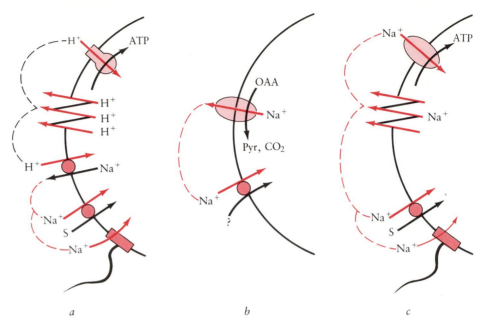

FIGURE 4.14   Energy coupling by a sodium current. (*a*) Alkalophilic bacilli. The respiratory chain extrudes protons and drives ATP production. An electrochemical gradient of Na^+ ions is generated by Na^+-H^+ antiport; this drives the accumulation of metabolites and the flagellar motor. (*b*) A primary sodium pump in *Klebsiella*. Sodium ions are extruded by the oxaloacetate decarboxylase. Whether the sodium current supports useful work is not yet known. (*c*) Sodium circulation in *Vibrio alginolyticus*. Speculative scheme showing a sodium-translocating respiratory chain, ATP synthesis by a hypothetical sodium-transport ATPase, and sodium-linked motility and metabolite uptake. Stoichiometries are uncertain and have been omitted throughout.

The obligate alkalophilic bacilli are one group in which both the genesis and the functions of the sodium circulation are well documented (Krulwich, 1983; Krulwich and Guffanti, 1983). These organisms grow at an external pH as high as 11, yet maintain their cytoplasmic pH at 9 to 9.5 by exchanging cytoplasmic Na^+ for protons with the aid of a Na^+-H^+ antiporter (Figure 4.14*a*). They are aerobes, endowed with a powerful respiratory chain that expels protons and maintains the membrane potential at about $-150\,mV$. However, since the cytoplasm is acidic relative to the medium by as much as 1.5 to two units, the concentration gradient for protons (ΔpH) is directed outward; and the protonic potential may be as low as $-30\,mV$ ($\Delta p = -150 + 120 = -30\,mV$). As mentioned in the preceding section, this scanty proton potential is no obstacle to proton-linked ATP synthesis by an apparently conventional F_1F_0 ATP synthase, though it is far from clear how the bacteria overcome the shortfall in driving force.

Significantly, the alkalophilic bacilli assign other work functions to a circulation of sodium ions across the plasma membrane. Thanks to the Na^+-H^+ antiporter (which, like the F_1F_0 ATPase, may be intimately associated with the respiratory chain), there is an electrochemical potential gradient for Na^+ ions, $\Delta\tilde{\mu}_{Na^+}$, directed inward. If it is assumed that the antiporter maintains the cytoplasmic sodium level at one-tenth that in the medium, the "sodium-motive force" is given by

$$\frac{\Delta\tilde{\mu}_{Na^+}}{F} = \Delta\Psi + \frac{2.3RT}{F}\log\frac{[Na^+]_i}{[Na^+]_o} = -150 - 60 = -210 \text{ mV} \qquad \text{(Eq. 4.3)}$$

The return of sodium ions down this gradient into the cytoplasm is coupled to a number of work functions, including the accumulation of metabolites and motility (Chapter 5). In these organisms the sodium circulation appears to be truly ancillary to that of protons.

Organisms that generate an electrochemical sodium gradient by means of a primary sodium pump represent a larger departure from the bacterial norm. This trail of research began two decades ago with the discovery that *Klebsiella* requires the presence of Na^+ for growth on citrate. The search for a reason led P. Dimroth (1982) to the discovery that one of the enzymes of citrate catabolism, oxaloacetate decarboxylase, requires Na^+ ions and transports them outward across the membrane. Part of the free energy of the decarboxylation is conserved in the form of $\Delta\tilde{\mu}_{Na^+}$ (Figure 4.14*b*). Sodium-translocating decarboxylases appear to be quite widely distributed among anaerobes, where the sodium circulation that they generate is clearly of physiological significance (Hilpert and Dimroth, 1983). In fact, in the anaerobic bacterium *Propionigenicum modestum*, Na^+ appears to be the sole coupling ion. According to Hilpert et al. (1984), these organisms extrude Na^+ by means of succinate decarboxylase; the resulting $\Delta\tilde{\mu}_{Na^+}$ supports ATP synthesis by a sodium-coupled ATPase and protons are not involved at all.

Another variation on the theme of energy coupling by a sodium circulation is *Vibrio alginolyticus*, a marine aerobe in which a sodium circulation is required for the uptake of metabolites and also for motility. When the organisms are grown at neutral pH, sodium is extruded by a Na^+-H^+ antiporter that is ancillary to a conventional proton-translocating respiratory chain. At alkaline pH, however, sodium extrusion appears to be more intimately linked to the respiratory chain. Tokuda and Unemoto (1982, 1984) have proposed that the respiratory chain itself transports Na^+ ions under these conditions, but alternatives to this provocative notion have not yet been rigorously excluded. A key question is whether ATP synthesis involves a sodium-transporting ATPase (Figure 4.14*c*) (Chernyak et al., 1983). Whatever the answer proves to be, these peculiar bacteria underscore the importance of a sodium circulation to organisms living in unusual habitats, and they prompt one to reconsider one's perception of the norm.

References

ARP, A. J., and CHILDRESS, J. J. (1983). Sulfide binding by the blood of the hydrothermal vent tube worm *Riftia pachyptila*. *Science* **219**:295–297.

BACCARINI-MELANDRI, A., CASADIO, R., and MELANDRI, B. A. (1981). Electron transfer, proton translocation and ATP synthesis in bacterial chromatophores. *Current Topics in Bioenergetics* **12**:197–258.

BIKETOV, S. F., KASHO, V. N., KOZLOV, I. A., MILEYKOVSKAYA, Y. I., OSTROVSKY, D. N., SKULACHEV, V. P., TIKHONOVA, G. V., and TSUPRUN, V. L. (1982). F_1-like ATPase from the anaerobic bacterium *Lactobacillus casei* contains six similar subunits. *European Journal of Biochemistry* **129**:241–250.

CARLILE, M. J. (1980). From prokaryote to eukaryote: Gains and losses. *Symposia of the Society for General Microbiology* **38**:1–40.

CAVANAUGH, C. M., GARDINER, S. L., JONES, M. L., JANNASCH, H. W., and WATERBURY, J. B. (1981). Prokaryotic cells in the hydrothermal vent tube worm *Riftia pachyptila* Jones: Possible chemoautotrophic symbionts. *Science* **213**:340–342.

CHERNYAK, B. V., DIBROV, P. A., GLAGOLEV, A. N., SHERMAN, M. YU., and SKULACHEV, V. P. (1983). A novel type of energetics in a marine alkali-tolerant bacterium. *FEBS Letters* **164**:38–42.

CLARK, A. J., COTTON, N. P. J., and JACKSON, J. B. (1983). The relation between membrane ionic current and ATP synthesis in chromatophores from *Rhodopseudomonas capsulata*. *Biochimica et Biophysica Acta* **723**:440–453.

CLAYTON, R. K. (1980). *Photosynthesis: Physical Mechanisms and Chemical Patterns*. Cambridge University Press, London.

COBLEY, J. G., and COX, J. C. (1983). Energy conservation in acidophilic bacteria. *Microbiological Reviews* **47**:579–585.

CRAMER, W. A., and CROFTS, A. R. (1982). Electron and proton transport, in *Photosynthesis*, vol. I: *Energy Conversion by Plants and Bacteria*, Govindjee, Ed., Academic, New York, pp. 387–467.

CROFTS, A. R., and WOOD, P. M. (1978). Photosynthetic electron transport chains of plants and bacteria and their role as proton pumps. *Current Topics in Bioenergetics* **7**:175–244.

DIMROTH, P. (1982). The generation of an electrochemical gradient of sodium ions upon decarboxylation of oxaloacetate by the membrane-bound and Na^+-activated oxaloacetate decarboxylase from *Klebsiella aerogenes*. *European Journal of Biochemistry* **12**:443–449.

DUTTON, L. P., and PRINCE, R. C. (1978). Energy conversion processes in bacterial photosynthesis, in *The Bacteria*, L. N. ORNSTON and J. R. SOKATCH, Eds., vol. VI. Academic, New York, pp. 523–584.

FELBECK, H. (1981). Chemoautotrophic potential of the hydrothermal vent tube worm *Riftia pachyptila* Jones (Vestimentifera). *Science* **213**:336–338.

FERGUSON, S. J., and SORGATO, M. C. (1982). Proton electrochemical gradients and energy transduction processes. *Annual Review of Biochemistry* **51**:185–217.

FILLINGAME, R. H. (1980*a*). The proton-translocating pumps of oxidative phosphorylation. *Annual Review of Biochemistry* **49**:1079–1113.

FILLINGAME, R. H. (1980*b*). Biochemistry and genetics of bacterial H^+-translocating ATPases. *Current Topics in Bioenergetics* **11**:35–106.

FUTAI, M., and KANAZAWA, H. (1983). Structure and function of proton-translocating adenosine triphosphatase (F_0F_1): Biochemical and molecular biological approaches. *Microbiological Reviews* 47:285–312.

GUFFANTI, A. A., BORNSTEIN, R. F., and KRULWICH, T. A. (1981). Oxidative phosphorylation by membrane vesicles of *Bacillus akalophilus*. *Biochimica et Biophysica Acta* 635:619–630.

GUFFANTI, A. A., FUCHS, R. T., SCHNEIER, M., CHIN, E., and KRULWICH, T. A. (1984). A transmembrane electrical potential generated by respiration is not equivalent to a diffusion potential of the same magnitude for ATP synthesis by *Bacillus firmus* RAB. *Journal of Biological Chemistry* 259:2971–2975.

HADDOCK, B. A., and JONES, C. W. (1977). Bacterial respiration. *Bacteriological Reviews* 41:47–99.

HAROLD, F. M. (1978). Op. cit., Chapter 3.

HILPERT, W., and DIMROTH, P. (1983). Purification and characterization of a new sodium-transport decarboxylase. *European Journal of Biochemistry* 132:579–587.

HILPERT, W., SCHINK, B., and DIMROTH, P. (1984). Life by a new decarboxylation-dependent energy conservation with Na^+ as coupling ion. *EMBO Journal* 3:1665–1670.

INGLEDEW, W. J. (1982). *Thiobacillus ferrooxidans*: The bioenergetics of an acidophilic chemolithotroph. *Biochimica et Biophysica Acta* 683:89–117.

INGLEDEW, W. J., and POOLE, R. K. (1984). The respiratory chains of *Escherichia coli*. *Microbiological Reviews* 48:222–271.

JONES, R. W., LAMONT, A., and GARLAND, P. B. (1980). The mechanism of proton translocation driven by the respiratory nitrate reductase of *Escherichia coli*. *Biochemical Journal* 190:79–94.

JUNGE, W., and JACKSON, J. B. (1982). The development of electrochemical potential gradients across photosynthetic membranes, in *Photosynthesis*, vol. I: *Energy Conversion by Plants and Bacteria*, Govindjee, Ed. Academic, New York, pp. 589–646.

KAGAWA, Y. (1980). Energy-transducing proteins in thermophilic biomembranes. *Journal of Membrane Biology* 55:1–8.

KASHKET, E. R. (1981). Protonmotive force in growing *Streptococcus lactis* and *Staphylococcus aureus* under aerobic and anaerobic conditions. *Journal of Bacteriology* 146:369–376.

KASHKET, E. R. (1982). Stoichiometry of the H^+-ATPase of growing and resting, aerobic *Escherichia coli*. *Biochemistry* 21:5534–5538.

KELLY, D. P. (1978). Bioenergetics of chemolithotrophic bacteria, in *Companion to Microbiology*, A. T. Bull and P. M. Meadow, Eds. Longman, London, pp. 363–386.

KELLY, D. P. (1981). Introduction to the chemolithotrophic bacteria, in *The Prokaryotes: A Handbook on Habitats, Isolation and Identification of Bacteria*, M. P. Starr, H. Stolp, H. G. Trüper, A. Balows, and M. G. Schlegel, Eds., vol. I. Springer-Verlag, Berlin, pp. 997–1004.

KNOWLES, C. J., Ed. (1980). *Diversity of Bacterial Respiratory Systems*, vols. I and II. CRC Press, Boca Raton.

KRÖGER, A., DORRER, E., and WINKLER, E. (1980). The orientation of the substrate sites of formate dehydrogenase and fumarate reductase in the membrane of *Vibrio succinogenes*. *Biochimica et Biophysica Acta* 589:118–136.

KRULWICH, T. A. (1983). Na^+/H^+ antiporters. *Biochimica et Biophysica Acta* 726:245–264.

KRULWICH, T. A., and GUFFANTI, A. A. (1983). Physiology of acidophilic and alkalophilic bacteria. *Advances in Microbial Physiology* **24**:173–214.

LIU, C.-H., HART, N., and PECK, H. D., JR. (1982). Inorganic pyrophosphate: Energy source for sulfate-reducing bacteria of the genus *Desulfotomaculum. Science* **217**:363–364.

MALONEY, P. C. (1982). Energy coupling to ATP synthesis by the proton-translocating ATPase. *Journal of Membrane Biology* **67**:1–12.

MALONEY, P. C. (1983). Relationship between phosphorylation potential and electrochemical H^+ gradient during glycolysis in *Streptococcus lactis. Journal of Bacteriology* **153**:1461–1470.

MITCHELL, P. (1968). Op. cit., Chapter 3.

MITCHELL, P. (1976a). Op. cit., Chapter 3.

NICHOLLS, D. G. (1982). Op. cit., Chapter 1.

ODOM, J. M., and PECK, H. D. (1984). Hydrogenase, electron-transfer proteins, and energy coupling in the sulfate-reducing bacteria *Desulfovibrio. Annual Review of Microbiology* **38**:551–592.

PADAN, E., ZILBERSTEIN, D., and SCHULDINER, S. (1981). pH homeostasis in bacteria. *Biochimica et Biophysica Acta* **650**:151–166.

ROTTENBERG, H. (1979). The measurement of membrane potential and pH in cells, organelles and vesicles. *Methods in Enzymology* **55**:547–569.

SENIOR, E., and WISE, J. G. (1983). The proton-ATPase of bacteria and mitochondria. *Journal of Membrane Biology* **73**:105–124.

STOUTHAMER, A. H. (1978). Energy-yielding pathways, in *The Bacteria*, L. N. Ornston and J. R. Sokatch, Eds. vol. 6, Academic, New York, pp. 389–462.

THAUER, R. K., JUNGERMANN, K., and DECKER, K. (1977). Energy conservation in chemotrophic anaerobic bacteria. *Bacteriological Reviews* **41**:100–180.

TOKUDA, H., and UNEMOTO, T. (1982). Characterization of the respiration-dependent Na^+ pump in the marine bacterium *Vibrio alginolyticus. Journal of Biological Chemistry* **257**:10007–10014.

TOKUDA, H., and UNEMOTO, T. (1984). Na^+ is translocated at NADH: Quinone oxidoreductase segment in the respiratory chain of *Vibrio alginolyticus. Journal of Biological Chemistry* **259**:7785–7790.

VAN VERSEVELD, H. W., KRAB, K., and STOUTHAMER, A. H. (1981). Proton pump coupled to cytochrome *c* oxidase in *Paracoccus denitrificans. Biochimica et Biophysica Acta* **635**:525–534.

WAGGONER, A. S. (1979). Dye indicators of membrane potential. *Annual Review of Biophysics and Bioengineering* **8**:47–68.

5

The Bacterial Paradigm: Useful Work

The ways of bacteria are odd
They're understood only by God
Any man who proclaims
That he's fathomed their aims
Is either a fool or a fraud.

With apologies to I. N. Dubin and
Perspectives in Biology and Medicine

• • •

The Varieties of Bacterial Transport
Movement and Behavior
Homeostasis
Costs, Profits, and Yield

In this chapter we turn from the supply side of bacterial economics to the demand side. The object of the biotic enterprise is not to conserve and transduce energy, but to produce cells. During exponential growth in a rich medium, *Escherichia coli* and *Streptococcus faecalis* may double every half-hour and a single cell could produce a mass of descendants exceeding that of the earth in about two days. One of the chief tasks of bioenergetics is to describe how the energy harvested in the course of metabolism supports the many kinds of integrated work required to make two cells grow where one grew before.

The production of ATP during oxidative and photosynthetic phosphorylation can itself be thought of as a kind of work, but since energy conserved in the ATP/ADP ratio is recoverable, it is better described as energy transduction. We shall return to this topic in Chapters 7 and 8. Of the various forms of energy available to microbial cells the lion's share, often two-thirds or more, goes for the biosynthesis of macromolecules, proteins in particular. This expenditure takes the form of ATP and other activated intermediates in the cytoplasm. The reactions involved make up the corpus of biochemistry, and despite their role as the major sink for biological energy, they are beyond the scope of this book. Our concern here is with work functions that involve the plasma membrane and with the integration of the bacterial economy into a functional whole.

The Varieties of Bacterial Transport

Molecular traffic across the cytoplasmic membrane is not a matter of selective diffusion but of transport. With a few exceptions, such as water, oxygen, and ammonia, the passage of metabolites is mediated by specific transport systems whose substrates range from trace metals and vitamins through the major ions and nutrients to transforming DNA and the precursors of extracellular cell walls. *E. coli*, for example, is thought to produce more than a hundred genetically distinct transport systems, which fall into several classes with respect to their structure and mechanism. Some bacterial transport systems need only equalize the concentration of substrate on both sides of the membrane (those for glycerol and lactic acid, for example). The majority, however, are geared to the performance of osmotic work: substrates are either accumulated or expelled. Bacteria provide many spectacular examples of active transport. In *E. coli* the cytoplasmic concentration of β-galactosides and amino acids may exceed that of the medium by a factor of 10^3 and galactose by 10^5, and K^+ ions may accumulate in the cells to a gradient of 10^6. Sodium and calcium are expelled from the cytoplasm such that the internal concentration is lower than that of the medium by at least two orders of magnitude.

Can all the diverse transport systems be accommodated in the chemiosmotic theory? It will be recalled that the basic scheme called for transport carriers to be linked to the metabolic machinery in a secondary manner, via the proton circulation (Figure 3.4). By definition, porters are substrate-specific but have no intrinsic orientation and merely catalyze transport in either direction in accord with the electrochemical potential of protons (Table 3.1). The proton-motive force does the work and determines the maximum concentration gradient that can be attained, while the vectorial nature of the proton-translocating pathways ultimately accounts for the direction of transport. The upshot of the past two decades of research, begun in Mitchell's laboratory and continued in many others, is that many bacterial transport systems do indeed function as secondary porters, with protons or sodium ions serving as the coupling ion. However, we have also come to recognize a surprising number of primary transport systems energized by phosphoenolpyruvate, ATP, and other phosphoryl donors that remain to be identified (Figure 5.1).

The field has burgeoned enormously. Reviews that deal with the transport of ions and metabolites by bacteria have been published by Postma and Roseman (1976), Saier (1977), Eddy (1978), Harold (1978, 1982), Lanyi (1979), Dills et al. (1980), West (1980), Padan et al. (1981), Hengge and Boos (1983), Kaback (1983), and Krulwich (1983). Major articles by Crane (1977), Eddy (1982), and Sanders et al. (1984), though primarily concerned with eukaryotic cells, provide valuable analyses of secondary transport in general. Rosen (1978) has edited a most useful compendium of bacterial transport, and many articles in the recent collection edited by Martonosi (1982) deal with this subject.

Porters.   The β-galactoside transport system of *E. coli* (or "lac permease") was first characterized from the genetic and kinetic standpoints 30 years ago by H. V. Rickenberg, G. N. Cohen, G. Buttin, and J. Monod, but recognition of its true nature had to await the discovery of energy coupling by proton currents. Dissection of the coupling mechanism was immensely facilitated by the use of plasma membrane vesicles that are devoid of cytoplasmic proteins and substrates but retain a functional respiratory chain, F_1F_0 ATPase, and a variety of proton-linked porters (Kaback, 1974, 1983). The assertion that this transport system mediates symport of lactose (and of other β-galactosides) with protons in both directions, in the classic Mitchellian manner, rests on five lines of argument. The evidence has been reviewed in detail by Kaback (1983), Hengge and Boos (1983), West (1980), and Overath and Wright (1983), where references to the original papers will be found. It should be noted that experiments with membrane vesicles and intact cells reinforce each other. This is of the utmost importance, since the intact cell is the standard by which the validity of work with resolved systems must be judged.

1.   Accumulation of galactosides depends on the existence of a proton-motive force of the requisite magnitude and polarity, but movement of the substrate across the membrane does not require $\Delta\tilde{\mu}_{H^+}$. Reagents and treatments that dissipate the proton-motive force dissociate transport from respiration or glycolysis; net accumulation is blocked, but movement of the sugar across the membrane continues to equilibrium (Figure 5.2*a*).

The extent of β-galactoside accumulation is proportional to $\Delta\tilde{\mu}_{H^+}$ (Figure 5.2*b*). Either a pH gradient or a membrane potential can support the accumula-

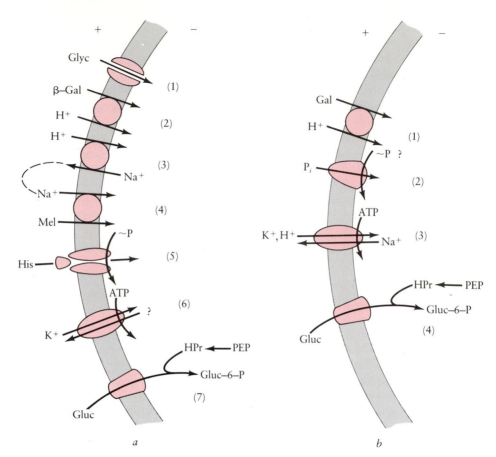

a

b

FIGURE 5.1   A sampler of primary and secondary systems for the transport of metabolites
across bacterial plasma membranes. (*a*) *E. coli.* Porters: (1) glycerol uniporter, probably
channel type; (2) lactose-proton symport; (3) sodium-proton antiport; (4) sodium-melibiose
symport. Primary systems: (5) histidine uptake by a shock-sensitive system with
unidentified energy donor and periplasmic binder; (6) Kdp-K^+ ATPase, counterion
uncertain; (7) glucose uptake by vectorial phosphorylation from PEP. The major proton
pumps, F_1F_0 ATPase and redox chains, have been omitted. (*b*) *S. faecalis* and *S. lactis.*
(1) galactose-proton symport; (2) uptake of P_i by a primary transport system that does not
include a periplasmic binder and whose energy donor is unidentified; (3) sodium ATPase,
exchanging Na^+ for H^+ or K^+; (4) glucose uptake by vectorial phosphorylation. The
only major electrogenic proton pump, the F_1F_0 ATPase, has been omitted.

tion, and the two parameters are additive. By Equation 3.6, if lactose moves by
symport with one proton, a proton potential of -180 mV would suffice to
establish a concentration gradient of 10^3. Most of the available data are
consistent with this stoichiometry, but some investigators favor two protons per
cycle under certain conditions.

2.   Accumulation can be supported by an artificially imposed proton-motive force. For example, vesicles in the absence of an energy source do not accumulate lactose. In order to impose an artificial membrane potential, vesicles containing as much as 0.4 M K^+ were suspended in buffer lacking K^+, so as to ensure a large outward gradient of K^+. Addition of valinomycin then elicited efflux of K^+, generating an electric potential, interior negative, and allowing a transient accumulation of lactose (Figures 3.7 and 5.2c). An artificial pH gradient, interior alkaline, also supported lactose accumulation. Vesicles treated with proton-conducting ionophores, and those obtained from mutants lacking the lactose porter, failed to accumulate the sugar.

3.   Movement of galactosides is tightly linked to a stoichiometric flux of protons in the same direction. For instance, when lactose is added to a dense suspension of vesicles in the absence of an energy source, equilibration of the sugar is accompanied by alkalinization of the medium; enhancement of this effect by valinomycin suggests that electric charge is transported into the vesicles together with protons (Figure 5.2d). By the same token, efflux of lactose via the porter is accompanied by the exit of a proton and can lead to the generation of an electrical potential, lumen negative. From the change in pH and concurrent movement of other ions, it can be calculated that one proton and a single positive charge travel with each molecule of sugar.

   Mutants have been isolated in which the porter is defective, such that β-galactosides are still translocated across the membrane but do not accumulate against a concentration gradient. These mutants no longer transport protons together with the sugar and therefore cannot respond to the proton-motive force.

4.   Net movement of sugar occurs in either direction, subject to the polarity of the proton-motive force. For example, everted membrane vesicles normally generate a proton-motive force such that the lumen is acid and electrically positive, and therefore they exclude β-galactosides. However, imposition of a proton-motive force of polarity such that the lumen is alkaline and negative allows these vesicles to accumulate the sugar. In other words, at least qualitatively the carrier is symmetrical.

5.   Very recently, purified lactose porter has been reconstituted into liposomes and shown to accumulate lactose in response to a gradient of $\Delta\tilde{\mu}_{H^+}$ generated either artificially or by purified cytochrome oxidase inlaid into the same vesicle. Only a single protein, specified by the Y gene of the lactose operon, is required to effect coupled transport of protons and β-galactosides. The characteristics of the transport process are essentially the same as those seen in plasma membrane vesicles and in intact cells.

   The preceding outline describes the thermodynamic basis of galactoside accumulation, but it conveys no more than the bare bones of the transport

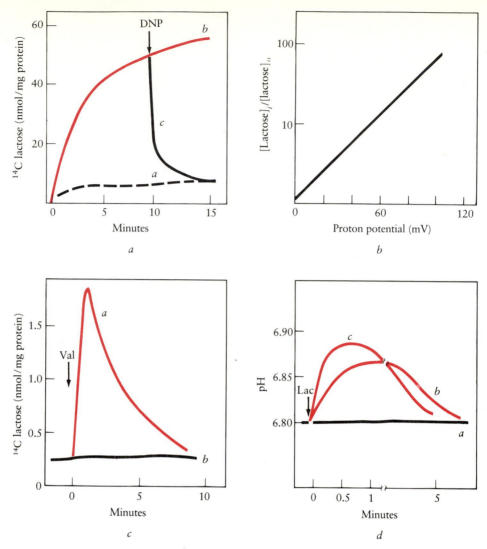

FIGURE 5.2   Coupling of lactose and proton transport by membrane vesicles. (*a*) Accumulation of lactose requires a proton potential. In the absence of an energy source, ^{14}C lactose equilibrated until the internal concentration equaled the external one (trace *a*). Respiring vesicles accumulated lactose (trace *b*). The lactose gradient was dissipated by 2,4-dinitrophenol (DNP), a proton-conducting uncoupler (trace *c*). (*b*) The lactose concentration gradient is a logarithmic function of the proton potential, as predicted by Equation 3.6. (*c*) Lactose accumulation in response to an artificial proton potential. Potassium-loaded vesicles were suspended in buffer lacking K^+. Addition of valinomycin induced electrogenic K^+ efflux and lactose uptake (trace *a*). Vesicles from a mutant lacking the lactose porter did not accumulate lactose (trace *b*). (*d*) Lactose uptake is accompanied by the uptake of protons. A dense vesicle suspension in the absence of an energy source maintained a constant pH (trace *a*). When lactose was added, protons moved transiently into the vesicles, raising the pH (trace *b*). Proton uptake was enhanced by valinomycin, suggesting charge transport (trace *c*). (Idealized after experiments from the laboratories of H. R. Kaback and F. M. Harold.)

mechanism. Granted that the lactose permease is a proton-sugar symporter, just how does this (or any other) porter operate as a molecular entity that converts $\Delta\tilde{\mu}_{H^+}$ into a substrate gradient? Let it first be said that few students of transport still think in terms of carriers that shuttle, rotate, or circulate. Porters are proteins that span the membrane, and vigorous gyrations are precluded by the polar segments that protrude into the aqueous phase like the heads of a rivet. Instead, one envisages a transmembrane channel within an oligomeric protein complex (or even within a single monomer). A specific substrate-binding site may be exposed alternately on the one surface or the other, thanks to quite minor changes in the three-dimensional structure involving movements of just a few angstrom units, and the electrochemical driving force exerted on the coupling ion may affect the binding of the organic substrate, the mobility of its binding site or sites, or even the tertiary structure of the porter. Figure 5.3 illustrates the general concept of transport through a regulated, substrate-specific pore. The difficulties arise when one tries to work out the molecular specifications of any particular porter.

The lactose porter protein was first isolated by E. P. Kennedy and his colleagues some 15 years ago and has been intensely studied ever since. Recent genetic and biochemical data indicate that the monomeric porter protein, 46.5 kdal, bears a single sugar-binding site; a proton-binding site resides on the same polypeptide chain. The protein has been shown to span the membrane; its amino acid sequence is known and indicates a secondary structure composed of as many as 12 α-helical segments that traverse the membrane in a zigzag fashion. The tertiary structure of the functional porter is not known. There is good evidence that the monomer is sufficient to catalyze transport, but recent work by H. R. Kaback and his colleagues suggests that the form that responds to the proton potential is the dimer (Kaback, 1983).

As mentioned above, reconstitution of all the manifestations of lactose transport by the insertion of the purified transport protein into liposomes gives evidence that the product of the Y gene is both necessary and sufficient. However, certain genetic lesions that map outside the lactose operon result in defects in lactose transport, suggesting that optimal performance of the porter may require additional polypeptides.

Reconstitution of the porter and even the determination of its molecular structure will not by themselves tell us how the porter works; they must be complemented by an account of the mechanism by which the flows of lactose and of protons are coupled. To this end it is useful to regard the porter as being analogous to an enzyme that has two substrates (H^+ and sugar), with reorientation of the carrier center corresponding to the catalytic step. The diagram shown in Figure 5.4 points out that in either orientation, the carrier center may be unoccupied or else complexed with either sugar, a proton, or both; eight possible states in all. Efficient energy coupling demands that in general reorientation of the carrier center be allowed only when it is either unloaded or fully loaded. However, "slip reactions," such as reorientation of the center with

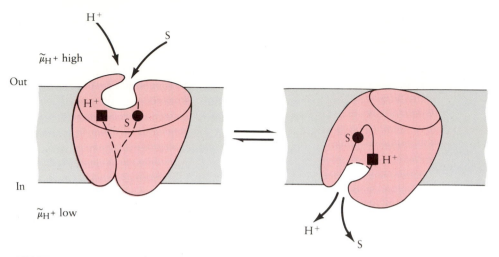

FIGURE 5.3 Transport by a regulated pore. The transport system consists of a protein that spans the membrane and bears binding sites for both the substrate S and a proton. When both are bound simultaneously, a conformational change exposes the binding sites to the opposite surface of the membrane. The proton-motive force may control the mobility of the sites, their affinity, or both.

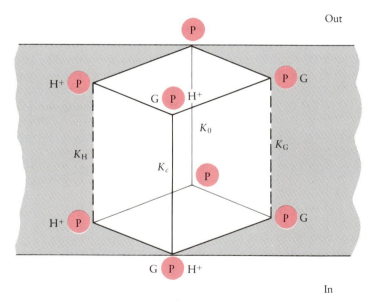

FIGURE 5.4 Kinetic model of galactoside-proton symport. P, porter, binding either a galactoside (G), a proton, or both; K_0, rate constant for reorientation of the unloaded porter; K_c, rate constant for the ternary complex; K_H and K_G, rate constants for the binary complexes. Dashed lines represent pathways that must be minimized for maximum energy coupling. (After Wright et al., 1981, with permission of the American Chemical Society.)

sugar alone, appear to play a significant role in efflux and must be taken into account. In general, the level of galactosides in the cytoplasm will almost always be less than that predicted from equilibrium considerations (Equation 3.6), thanks to efflux via the porter or by other pathways (Booth et al., 1979; Eddy, 1982). Slippage is not a failure of the energy-coupling mechanism but a safety valve; if porters necessarily attained equilibrium with $\Delta\tilde{\mu}_{H^+}$, cells incubated with modest concentrations of substrates would accumulate them to the point of bursting.

Accumulation of lactose can be understood intuitively as a consequence of coupling rules stating that net proton entry via the porter is possible if and only if net lactose flow can also take place. Conversely, net flow of lactose is allowed only when net proton movement can take place simultaneously (Figures 3.6 and 5.4). But sophisticated kinetic analysis is required to describe the pathway, that is, to define the set of intermediates involved in binding the sugar and proton(s) to the carrier center at one surface and their release on the other together with the pertinent stoichiometries, rate constants, affinities, and binding energies. This has not been fully accomplished. The literature records numerous measurements of such parameters for various substrates, in cells and membrane vesicles, and in either the presence or the absence of a proton-motive force. But it does not now provide a coherent and generally acceptable account of the transport process, and it seems better to set the matter aside than to present an annotated catalog of the findings. Readers who wish to delve deeper must consult the original literature (Ahmed and Booth, 1981; Wright et al., 1981; Page and West, 1982; Lancaster, 1982; Overath and Wright, 1983; Kaback, 1983; Sanders et al., 1984; Wright and Seckler, 1985). A circle with two arrows may be a ludicrously oversimplified representation of symport, but it does make the essential point forcefully and economically.

I have considered the lactose porter in some detail because it is the best-studied example of its genre, but it is by no means the only one. Other proton-linked symporters in *E. coli* include those for ribose, arabinose, proline and at least half a dozen other amino acids, glucose-6-phosphate, succinate, and gluconate. Evidence for secondary antiport of Na^+ and Ca^{2+} by exchange for protons is also strong, if not quite conclusive. Analogous systems occur in streptococci, bacilli, and other gram-positive bacteria. Curiously, while the uptake of most metabolites by bacteria is coupled to the flux of protons, some are accumulated by symport with Na^+. The requisite sodium electrochemical potential $\Delta\tilde{\mu}_{Na^+}$ (cytoplasm low in Na^+ and electrically negative) is generated by the extrusion of Na^+. Bacteria extrude Na^+ ions in a variety of ways (see below). *E. coli* employs a secondary antiport of Na^+ for H^+ (Figure 5.1a) so that the ultimate driving force for substrate accumulation is still the proton-motive force. It is not at all clear what advantage *E. coli* obtains from transporting melibiose or glutamate with Na^+ while most other metabolites travel with H^+. However, both exemplify the use of an ion current to effect energy coupling to transport.

Primary Transport Systems in Bacteria.   A significant departure from chemiosmotic expectations (Figure 3.4) came with the discovery that a large class of
bacterial transport systems relies for energy coupling not on an ion gradient but
on a donor of activated phosphoryl groups. The tale begins with the discovery by
L. A. Heppel two decades ago that the uptake of a number of metabolites by *E.
coli* is impaired by osmotic shock. This was traced to the loss of small proteins,
localized in the periplasmic space, that bind these metabolites avidly and
specifically. The molecular weights of the periplasmic binding proteins range
from 20 to 40 kdal; they have no enzymatic activities but bind one molecule of
the substrate per monomer. Binding proteins are known for a whole gamut of
organic and inorganic metabolites: histidine, glutamine, arginine, maltose,
ribose, arabinose, folic acid, sulfate, phosphate, and many others.

The complete transport system often includes three additional proteins that
are firmly associated with the plasma membrane and cannot be released by
osmotic shock (Higgins et al., 1982; Hengge and Boos, 1983). Just how this
quartet functions is not particularly clear. A common view holds that the binding
proteins recognize the metabolite and determine the selectivity of the transport
system but do not ferry the metabolite across the membrane. Instead, the
complex of metabolite and binding protein interacts with the trio of membrane-
bound proteins. These are thought to make up a channel across the membrane,
regulated by conformational changes elicited by the association of the binder-
metabolite complex with the channel complex (Figure 5.5).

We thus distinguish two classes of transport systems in gram-negative bacteria,
the "shock-sensitive" systems, which include a periplasmic binding protein, and
the "shock-resistant" ones, which depend on carrier proteins tightly associated
with the plasma membrane. In a perceptive contribution, E. A. Berger and L. A.
Heppel (1974) showed that these two classes also differ with respect to energy
coupling. The shock-resistant systems, those that are retained intact when one
prepares membrane ghosts by the gentle lysis of spheroplasts (Kaback, 1974), are
energized by the proton circulation. The shock-sensitive ones were found to
depend on the generation of ATP but did not require a proton potential. Some of
the experimental criteria on which this conclusion was based are summarized in
Figure 5.6; they have been applied by many subsequent investigators in order to
distinguish secondary, proton-linked systems from primary ones. Unfortunately,
the identity of the energy donor for shock-sensitive transport systems has
remained obstinately mysterious. It need not be ATP itself, but could be some
other phosphoryl donor metabolically derivable from ATP. For a few years
acetyl phosphate appeared to be a strong candidate (Hong et al., 1979), but more
recent findings have led the same laboratory to question this hypothesis and leave
the issue wide open (Hunt and Hong, 1983).

Transport systems dependent on periplasmic binding proteins appear to be
characteristic of gram-negative bacteria, but gram-positive ones also possess
some transport systems that are energized by phosphoryl donors rather than by
the protonic potential. Here again, the identity of the phosphoryl donor remains

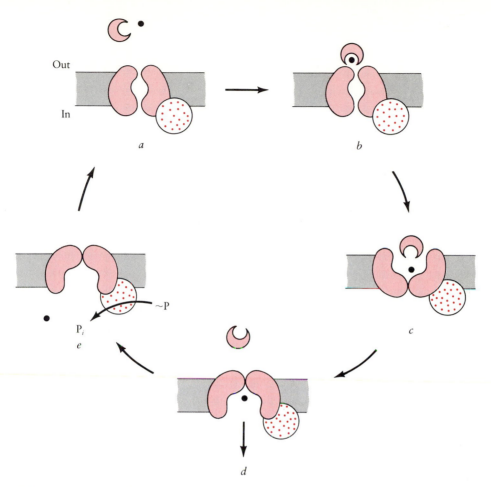

FIGURE 5.5   A hypothetical sequence of events in the uptake of a substrate molecule (black circle) by a transport system that involves a periplasmic binding protein. The energy donor remains unidentified. (After Hengge and Boos, 1983, with permission of Elsevier Biomedical Press.)

unknown. It should be noted that this mode of transport has no known parallel outside the prokaryotic world.

Cation Transport by ATPases and Porters.   Bacteria, like other cells, accumulate K^+ ions and extrude Na^+. In animal cells this is the function of the familiar Na^+-K^+ ATPase, which mediates a strictly coupled exchange of cytoplasmic Na^+ for external K^+ (Chapter 9). This enzyme is absent from bacteria. Instead, K^+ and Na^+ ions are transported by separate systems that are functionally connected with the proton circulation. The movements of K^+ and Na^+ ions are

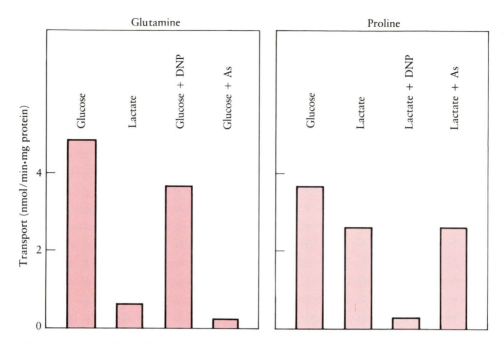

FIGURE 5.6   Evidence that glutamine uptake requires ATP whereas proline uptake requires a proton potential. The experiments were done with an *unc* mutant lacking the F_1F_0 ATPase in order to dissociate the ATP-ADP couple from the proton potential. Glutamine uptake was supported by the fermentable energy source glucose but not by lactate, which is oxidized by the respiratory chain. It was not inhibited by the proton conductor 2,4-dinitrophenol (DNP) but was blocked by arsenate (As), which interferes with ATP production. By contrast, proline uptake was supported by both glucose and lactate; dinitrophenol blocked it but arsenate had no effect. (After data by Berger and Heppel, 1974.)

heavily involved in the regulation of turgor and of the cytoplasmic pH and in other facets of homeostasis, matters that will be discussed in a later section of this chapter. Our concern here is with the transport systems per se. Although these mechanisms are not well worked out, it is clear that cation transport is more sophisticated than the elementary chemiosmotic scheme (Figure 3.4) predicts. It involves both novel ion-transport ATPases and unusual proton-linked porters, and the various kinds of bacteria differ significantly with respect to these pathways.

The uptake of K^+ ions by *E. coli* has been extensively studied from the genetic and biochemical standpoints, most notably by W. Epstein and his colleagues (Laimins et al., 1978, 1981; Epstein and Laimins, 1980; Helmer et al., 1982). Briefly, there appear to be at least two distinct potassium transport systems (Figure 5.7*a*), each composed of several polypeptides. The major one, designated TrK, is constitutive and has a modest affinity (K_m about 1 m*M*) but a very high rate, as much as 500 μmol/min·g cells. This is normally the major route

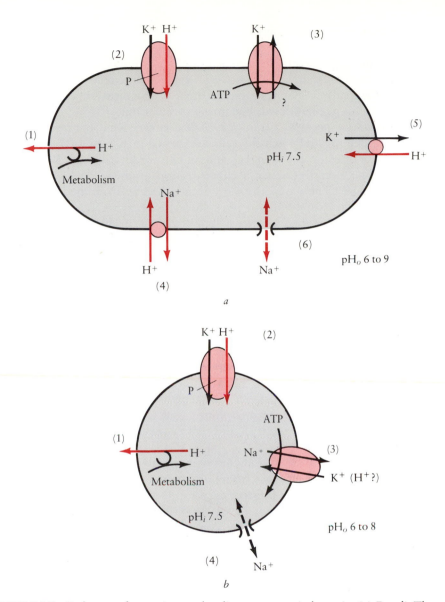

FIGURE 5.7   Pathways of potassium and sodium transport in bacteria. (*a*) *E. coli.* These organisms extrude protons electrogenically (1) by means of oxidation chains and of the F_1F_0 ATPase. Two systems for K^+ accumulation have been identified: TrK (2) may be a K^+-H^+ symporter regulated by phosphorylation; Kdp (3) is a K^+-translocating ATPase. Sodium is extruded by antiport for protons (4). A K^+-H^+ antiporter (5) and a pathway for Na^+ influx (6) have been proposed. (*b*) *S. faecalis.* Protons are extruded electrogenically by the F_1F_0 ATPase (1). The major system for K^+ accumulation, KtrI (2), resembles the TrK of *E. coli.* Sodium is extruded by a sodium-translocating ATPase (3), which under certain conditions appears to exchange Na^+ for K^+. A pathway for sodium entry (4) has been identified. All stoichiometries are uncertain.

of K^+ uptake. It requires the cells to generate both ATP and a proton potential, and there is evidence that it translocates positive charge into the cells. The molecular meaning of the dual requirement for $\Delta\tilde{\mu}_{H^+}$ and ATP is unknown; the best guess at present is that TrK is a K^+-H^+ symporter, effectively transporting K^+ as a divalent cation. The ATP requirement may indicate that the porter is active only when phosphorylated, but this is speculation.

By contrast, the minor K^+ transport system is very well characterized. The Kdp system is derepressed when the cells are grown in K^+-limited medium. It has a lower rate of transport than TrK but an extraordinarily high K^+ affinity, about 2 μM, and it can establish a concentration gradient between cells and medium of over 10^6. The requisite energy comes directly from the hydrolysis of ATP, and the mechanism involves a phosphorylated enzyme intermediate, as in the eukaryotic ATPases. In sum, the Kdp system is a primary K^+ pump; the counterion, if any, is still unknown. Originally the Kdp system was regarded as a scavenging device, intended to mop up the last traces of K^+ from a deficient medium. Recent results, however, suggest a primary involvement in osmoregulation since an increase in the external osmolarity induces expression of the Kdp system.

Streptococcus faecalis likewise possesses two potassium transport systems (Figure 5.7b). The major one, KtrI, resembles the TrK system of *E. coli*: it requires the cells to generate both ATP and $\Delta\tilde{\mu}_{H^+}$, is strongly electrogenic, and may again be a regulated K^+-H^+ symporter (Bakker and Harold, 1980). The minor one, KtrII, has just recently been discovered (Kobayashi, 1982). This is produced only by cells whose capacity to generate a proton-motive force has been impaired by mutation or by ionophores, and when Na^+ ions are present. KtrII probably mediates primary, ATP-linked exchange of Na^+ for K^+ (Kakinuma and Harold, 1985).

The two organisms differ even more markedly with respect to sodium extrusion. There is strong evidence from both membrane vesicles and intact cells that *E. coli* extrudes Na^+ ions by secondary antiport for protons, probably in an electrogenic manner (Figure 5.7a; Lanyi, 1979; Padan et al., 1981; Krulwich, 1983; Borbolla and Rosen, 1984). By contrast, streptococci have a primary

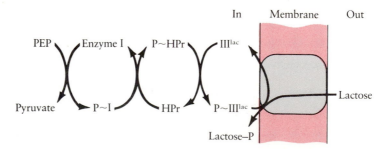

FIGURE 5.8   Schematic representation of the phosphotransferase system for lactose uptake by *Staphylococcus aureus*. (After studies by R. Simoni and S. Roseman.)

sodium pump, an ATPase thought to mediate electroneutral exchange of Na^+ for H^+ or K^+ (Heefner and Harold, 1982); the second potassium transport system, KtrII, is probably an expression of this ATPase (Figure 5.7b).

The variety of primary and secondary processes that bacteria employ to expel sodium ions is noteworthy, and somewhat mystifying. As mentioned above E. coli and many other bacteria extrude Na^+ by secondary antiport for protons. S. faecalis does it with a sodium-stimulating ATPase. A quite unexpected novelty, noted in the preceding chapter, is the electrogenic expulsion of Na^+ ions coupled to the decarboxylation of oxaloacetic acid and of methylmalonyl coenzyme A (Dimroth, 1981; Hilpert and Dimroth, 1983). And the extrusion of sodium ions by way of the respiratory chain may involve yet another primary sodium pump. The functional and adaptive significance of this diversity will be considered below.

Group Translocation. In addition to the various kinds of primary and secondary transport systems listed above, many bacteria absorb sugars by a vectorial process better described as group translocation, that is, a metabolic pathway so oriented across the plasma membrane as to catalyze at the same time the enzymatic transformation of the substrate and its translocation across the membrane. Several examples of this genre have been proposed over the years, but the only one that is solidly established is the phosphotransferase complex (Figure 5.8).

In the 1960s, S. Roseman and his associates clearly demonstrated that many bacteria effect the uptake of sugars by means of a vectorial cascade of enzymes that transfers a phosphoryl group from phosphoenolpyruvate to the incoming sugar, such that phosphorylation and transport are concurrent. Well-studied examples include the phosphorylation of glucose by E. coli and S. faecalis, also that of lactose by *Staphylococcus aureus*. Let me emphasize that this is not active transport in the thermodynamic sense: the phosphorylated sugar in the cytoplasm is a different chemical species from that in the medium, and the process cannot be described as movement against the electrochemical potential gradient. Nevertheless, it will be obvious that group translocation serves the same physiological function, namely, the acquisition of useful metabolites at the cost of metabolic energy. Experimentally, the critical issue is to distinguish concurrent transport and phosphorylation from transport followed by phosphorylation in the cytoplasm. Key points of evidence are that the sugar first appears in the cytoplasm as the phosphorylated derivative; that mutants deficient in phosphorylation are also defective in uptake, failing even to equilibrate the sugar concentration inside and out; and that in membrane vesicles internal glucose-6-phosphate was entirely and necessarily derived from external glucose.

The phosphotransferase systems found in various bacteria are not identical but share the essential features; uptake of lactose by staphylococci (Figure 5.8) is representative. A series of small cytoplasmic proteins carries the phosphoryl group from phosphoenolpyruvate to the cytoplasmic protein known as factor III.

Phosphorylated factor III forms a ternary complex with the incoming lactose molecule and with the lactose-specific enzyme II that spans the membrane. The vectorial reaction is that in which the phosphoryl group is transferred to form lactose-6-phosphate; a large decrease in free energy during this reaction denotes the performance of both chemical and osmotic work. Under certain conditions enzyme II catalyzes exchange transphosphorylation, a process in which extracellular sugar is phosphorylated and transported by exchange for an intracellular sugar-phosphate group. This may be regarded as a partial reaction that allows one to study the translocation of the sugar moiety separately from the overall pathway. It must be added that the phosphotransferase system has proved to be functionally much more sophisticated than Figure 5.8 suggests: it is subject to modulation both by the proton-motive force and by cyclic AMP, and it interacts with proton-coupled porters such that the phosphotransferase takes precedence. These fascinating interactions cannot be discussed here but have been described in reviews by Postma and Roseman (1976), Saier (1977), and Dills et al. (1980).

Finally, I would point out that the rubric of group translocation properly includes the transport and vectorial biosynthesis of macromolecules. A substantial fraction of the bacterial mass consists of macromolecules located in the cell envelope, external to the plasma membrane. Peptidoglycan, teichoic acids, and lipopolysaccharides are major structural components of the cell wall. In addition there is an abundance of proteins: the lipoprotein of murein, periplasmic binding proteins, exoenzymes, flagellin, porins, and others. Assembly of these external structures must include not only vectorial biosynthesis or some other mode of macromolecule transport but often their orderly incorporation into an elaborate supramolecular ensemble. Exclusion of such processes from this chapter is quite arbitrary, excused only by the need to set some limits to its scope.

On the Diversity of Coupling Mechanisms.   We began this section with the clear, if somewhat naive, postulation of a small number of primary pathways that translocate protons, coupled to a large array of secondary porters (Figure 3.4). In actuality, we find a bewildering wealth of devices that includes, in addition to the proton-coupled carriers, a sodium circulation with Na^+-linked symporters, a large number of primary transport systems energized by an unidentified phosphoryl donor (possibly acetyl phosphate), a growing list of cation-transporting ATPases, and pathways for the vectorial metabolism of sugars and perhaps some other metabolites. More than likely, additional mechanisms will be found among the transport systems for vitamins and trace metals that have been identified but not fully characterized. I would emphasize that this profusion has little to do with the diversity of bacterial habitats and evolutionary divergence: almost all the examples cited in the preceding sections were drawn from the small number of organisms commonly found on the microbiologist's slants—*E. coli* and *Salmonella*, streptococci, and staphylococci.

There is no quick answer to the puzzle of why bacteria are not as content with proton-linked porters as chemiosmoticists had expected, but it is helpful to reflect on the costs and benefits of the various modes of transport. Proton-linked systems have the virtues of versatility and genetic economy; they are small and structurally simple, and can be accommodated in large numbers in a membrane already crowded with bulky respiratory assemblies. On the debit side, the concentration gradients that can be attained are limited by the proton-motive force; moreover, since porters are freely reversible, a temporary power shortage may be aggravated by the leakage of critical metabolites. By contrast, primary transport systems powered by ATP or phosphoenolpyruvate are intrinsically unidirectional and are capable of attaining higher concentration gradients: hydrolysis of ATP under normal cellular conditions provides about 10 kcal/mol (42 kJ/mol) and could support a concentration gradient of more than 10^6-fold. Systems that involve binding proteins characteristically have higher affinities for their particular substrate and are often produced by nutrient-limited organisms, presumably to scavenge any traces of such metabolites as phosphate, sulfate, or amino acids. In general, gram-negative bacteria seem to favor the proton-linked porters for rapid growth when nutrients are plentiful, but fall back on the $\sim$P-driven systems in times of scarcity.

In some bacteria Na^+, rather than H^+, serves as the chief coupling ion for nutrient uptake. This preference makes sense for halobacteria and also for alkalophilic organisms, which cannot maintain a large proton-motive force across the plasma membrane (Chapter 4. Recall that in alkalophiles the cytoplasmic pH is more acidic than that of the medium; in Equation 3.4, the ΔpH term subtracts from the proton-motive force.) It is less easy to rationalize the existence in *E. coli* of Na^+-coupled secondary porters for certain metabolites, including melibiose and glutamate; the selective advantage, if any, is not obvious.

Streptococci possess ATP-driven primary pumps for Na^+, Ca^{2+}, and possibly also for P_i and glutamate; *E. coli* handles all these metabolites with secondary systems of one kind or another. The reason may be found in the primary energy-transducing pathways of the two organisms. *E. coli*, endowed with a range of proton-translocating redox pathways, can generate a proton-motive force of about -200 mV under most conditions. Streptococci, cramped by a fermentative economy with the F_1F_0 ATPase as their sole proton pump, achieve only -150 mV and often much less (Kashket, 1981). Is this perhaps why the latter organisms employ ATP-driven pumps for purposes that *E. coli* can entrust to the proton circulation? But if so, what adaptive significance should we assign to *Klebsiella*'s capacity for expelling sodium ions at the expense of the free energy of oxaloacetate decarboxylation?

The phosphotransferase system for sugar uptake is widely distributed in the bacterial world, but not universal. It is characteristic of anaerobes and facultative aerobes, including the genera *Escherichia*, *Streptococcus*, and *Staphylococcus*, but it is absent from such strict aerobes as *Pseudomonas* and *Azotobacter*. The latter organisms accumulate glucose by symport with protons. This distribution makes sense, but it is not obvious why lactose transport should be proton linked in *E. coli* yet phosphotransferase mediated in *Staphylococcus*.

Questions of this kind may be of greater interest to microbiologists than to students of energetics, but they are far from trivial. The diversity of bacterial transport systems is the equivalent, at the prokaryotic level, of the variety of form and pattern that delights the naturalist's eye; they constitute alternative solutions to common physiological problems. Are all of them necessarily adaptive, conferring a selective advantage? Do some hold clues to evolutionary lineage, or recall the successful transfer of a block of genes from one genus to another? Be this as it may, it is clear that students of bacterial transport can no longer take the universality of bacterial energetics for granted; we must acknowledge the exuberance of bacterial diversification, while keeping in mind the essential unity of the design.

Movement and Behavior

Many kinds of bacteria swim actively and purposefully in response to environmental stimuli. They sense gradients of light intensity, pH and oxygen tension and of a variety of sugars, amino acids, and inorganic ions, regarding some as attractants and others as repellents. Bacterial swimming was noted by microscopists in the late seventeenth century and chemotaxis in the nineteenth, but the wealth of information that we now possess was largely acquired in the past two decades. Bacterial motility is presently an exceedingly lively field, spurred by the fascination of one of nature's few rotary motors and by the prospect of coming to understand a sophisticated pattern of behavior at the molecular level. Progress can be traced in reviews by Adler (1975), Berg (1975a, 1975b), Macnab (1978), Glagolev (1981), and Macnab and Aizawa (1984). The present section considers bacterial motility from the viewpoint of energy transduction into mechanical work; information processing will be treated separately in Chapter 13.

The organelles of bacterial motility are called flagella (from the Latin word for whip), but structurally and functionally they are quite unlike the eponymous organelles of eukaryotes. Bacterial flagella (Figure 5.9), only 20 nanometers in diameter but several micrometers in length, are built of a single protein called flagellin. Each flagellum contains thousands of the monomers, packed helically into a hollow cylinder twisted into a regular helix of higher order, so that it appears wavy under the microscope. The flagellum is inserted into the plasma membrane via a bent segment called the hook and terminates in a complex basal body. Peritrichously flagellated cells, such as *E. coli* or *Salmonella*, may bear a dozen flagella randomly placed on the surface. When the cell swims, the individual flagella coalesce into a single helical propulsive bundle, working in unison to generate thrust (Figure 5.10a). Speeds of 10 to 20 micrometers per second, that is, 10 body lengths per second, are standard.

Until recently bacterial flagella were thought to propagate wave motions along the filament, as eukaryotic ones do, but it is now established that bacterial flagella are in fact fairly rigid and rotate about their point of insertion. One of the many experiments that support this conclusion takes advantage of mutants that produce straight rather than helical flagella. Such organisms cannot swim, but when the tip of a straight flagellum is tethered to a microscope slide coated with

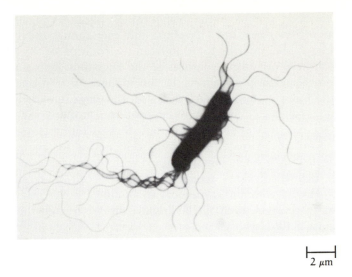

2 µm

FIGURE 5.9   Electron micrograph of a flagellated cell of *Salmonella typhimurium*. The individual flagella originate randomly around the cell body. (Courtesy of S. I. Aizawa and R. M. Macnab.)

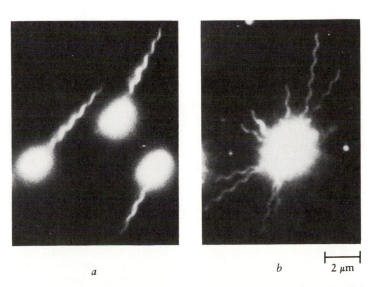

a   *b*   2 µm

FIGURE 5.10   The flagellar bundle of *Salmonella*. (*a*) Swimming cells, with all the flagella coalesced into a propulsive bundle; the individual filaments cannot be resolved by light microscopy. (*b*) Cells that have been rendered nonmotile. The flagellar bundle has drifted apart, revealing the random origins of the filaments. This state corresponds to an episode of tumbling. (Courtesy of R. M. Macnab.)

antiflagellin antibody, the cell is seen to spin rapidly about its flagellum. One can also attach minute latex beads to a flagellum and, even though the flagellum itself is too fine to be visible, infer its rotation from the rotation of the beads (Figure 5.11). In the normal organism, rotation of the flagellum generates thrust that pushes the cell through the medium in the manner of a propeller. But the mechanics are more complex than this, for the flagellum can rotate both clockwise and counterclockwise and the resulting motions are dramatically different. Flagella are helices of uniform handedness. Counterclockwise rotation (as seen by an observer situated behind the cell) makes for coalescence of the individual flagella into a single bundle aligned with the long axis of the cell, even though each filament rotates independently; the cell therefore progresses smoothly. Clockwise rotation is inimical to bundle formation; whenever the sense of rotation reverses, the flagellar bundle flies apart and the cell tumbles aimlessly (Figure 5.10*b*).

It is just this alternation of smooth swimming with tumbling that underlies bacterial behavior. Using an automatic tracking microscope, Howard Berg showed that the motions of an individual cell consist of straight runs interrupted by brief periods of tumbling; after each tumbling episode the cell sets off in a new, random direction. Tumbling results from a brief reversal of the sense of flagellar rotation; this is a spontaneous event, but its frequency is modulated by environmental signals. Chemotactic attractants, for example, temporarily suppress tumbling. A cell whose path brings it into a region of higher attractant concentration will therefore tend to make longer runs than one that encounters lower attractant levels. The consequence is that although the cell moves at random, runs and tumbles are biased so as to make the cell drift up an attractant

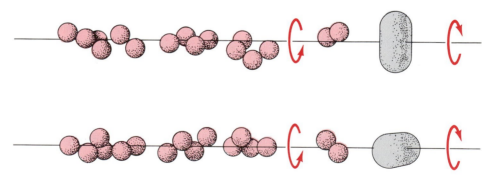

FIGURE 5.11   An experiment to visualize the rotation of bacterial flagella. Microscopic beads coated with antiflagellin antibody were adsorbed onto the flagella of a mutant in which the flagella are straight rather than helical. The filament cannot be resolved in the light microscope and the beads seem to mark an invisible line projecting from the cell. The beads revolve around the line in one direction while the cell body rotates in the other direction, as is shown here in two successive views. From time to time the sense of rotation reverses. (From "How Bacteria Swim," by H. C. Berg. Copyright © 1975 Scientific American, Inc. All rights reserved.)

gradient. By the same token, repellents increase the frequency of tumbling, making it more likely that the cell will travel away from the source of repellent. Chemotactic signals are first perceived by specific surface proteins, including the periplasmic binding proteins that participate in membrane transport. Just how the signal is transmitted to the flagellum is a mystery, to which we shall return below.

We must now ask how a bacterial cell makes its flagellum rotate. Work is obviously being done at the expense of metabolic energy, but the cost is not exorbitant: even though water is, from the bacterial point of view, a viscous fluid, *E. coli* probably expends less than 1 percent of its energy budget on movement (Macnab, 1978). Remarkably, the immediate energy donor proved to be not ATP but the proton-motive force. This conclusion was first drawn from studies with *unc* mutants lacking the F_1F_0 ATPase (Adler, 1975). For example, wild-type *E. coli* swam under both aerobic and anaerobic conditions; the mutants swam only if allowed to respire. Since *unc* mutants can generate $\Delta\tilde{\mu}_{H^+}$ by respiration but not by the ATPase, this observation implicates the proton circulation in motility. Further, movement was blocked by low concentrations of proton-conducting uncouplers but was unaffected by treatments that drastically lowered the cells' ATP pool. Direct evidence that the proton circulation supplies the driving force for flagellar rotation came from experiments in which an artificial proton-motive force was shown to support movement in the absence of metabolic energy. One of several such experiments is illustrated in Figure 5.12; it relies once again on a valinomycin-induced potassium flux to generate an electric potential across the plasma membrane. Starved cells of *S. faecalis* suspended in sodium phosphate buffer were nonmotile. When an electric potential, interior negative, was induced by addition of valinomycin, the cells thrashed about vigorously for approximately one minute; addition of valinomycin together with an attractant, leucine, induced a brief period of smooth swimming (Manson et al., 1977; Glagolev, 1981).

With the recognition that the flagellar motor is powered by the proton circulation, the focus of interest shifts to the molecular mechanics of energy transduction. The main clues available until recently came from electron microscopy. A decade ago DePamphilis and Adler showed that the flagellar basal body is embedded in the cell envelope by means of a series of rings: two in gram-positive bacteria, four in the gram-negatives (Figure 5.13). Berg (1975*a*) then proposed that the inner pair of rings, connected to the basal hook by a short rod, constitute part of the motor. The M ring rotates freely in the cytoplasmic membrane, but it is rigidly mounted on the rod so that rotation of the ring turns the flagellum. The rod passes freely through the S ring, which is rigidly attached to the wall, and torque is generated between the M and S rings. (If the torque is to be applied, some part of the motor must be static. Readers who have as much trouble grasping this as I did may imagine trying to turn the flagellum while floating in the fluid membrane; you'll find your foot twitching for a place to stand!) The outer set of rings are bushings that pass the rod through the outer lipoprotein membrane, and the hook serves as a swivel that links the rod to the

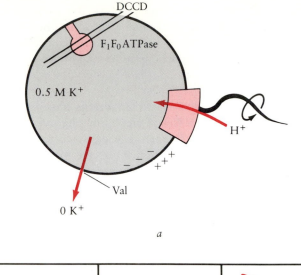

a

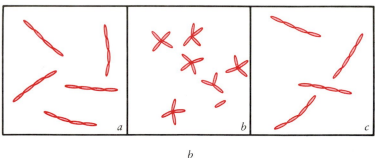

b

FIGURE 5.12   An artificial proton potential powers the flagellar motor of *S. faecalis*.
(*a*) The principle. Starved cells suspended in K$^+$-free medium are nonmotile. Addition of
valinomycin elicits efflux of K$^+$ and generates a membrane potential, interior negative;
protons flow into the cell and drive the rotation of the flagellar motor. In order to make
sure that ATP is not involved the F_1F_0 ATPase is blocked with DCCD. (*b*) The
observations. The diagram shows motility tracks of *S. faecalis* obtained by making four
successive photographs of the same field, 0.7 second apart. In panel *a*, with glucose present,
the cells swim in straight lines. In panel *b*, starved cells do not swim but valinomycin
induces tumbling. In panel *c*, valinomycin plus the chemoattractant leucine induces a brief
period of smooth swimming. (After Manson et al., 1977.)

flagellar filament proper. We can, then, envisage transduction of the proton-
motive force into mechanical work by the passage of protons across the M ring,
via a pathway that allows the protons to interact with fixed charges suitably
disposed on the surface of the S ring (Figure 5.14).

In order to obtain a closer view of energy transduction by the flagellar motor,
Berg and his associates undertook detailed studies on *S. faecalis* (Manson et al.,
1980; Berg et al., 1982; Block and Berg, 1984), from which they drew the
following conclusions.

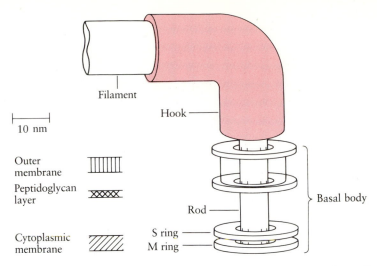

FIGURE 5.13   A model of the basal body and hook of *E. coli* flagella. (After Adler, 1975; reproduced with permission of the *Annual Review of Biochemistry* **44**, 1975, copyright by Annual Reviews.)

1.   The rotation rate of a metabolizing cell is inversely proportional to the viscosity of the medium; it follows that when $\Delta\tilde{\mu}_{H^+}$ is constant, the motor runs at a constant torque.

2.   A starved cell neither spins nor rotates randomly; when $\Delta\tilde{\mu}_{H^+}$ is zero, the motor must be rigidly engaged most of the time.

3.   Rotation of a starving cell can be driven by an artificial gradient of pH, of electric potential, or of both, and the rotation rate is a linear function of $\Delta\tilde{\mu}_{H^+}$. One of the implications is that a fixed number of protons carry the flagellum through each revolution; from the amount of work done, this number was estimated to be about 1000.

4.   The motor can be driven by a proton flux directed either inward or outward across the plasma membrane. Setting aside complications due to changes in cytoplasmic pH, it appears that the flagellar motor spins counterclockwise when protons are driven into the cell and clockwise when they are driven out.

Very recently Khan and Berg (1983) added two more observations, negative but highly pertinent. Replacement of protons in the external medium by deuterons has no effect: when $\Delta\tilde{\mu}_{H^+}$ and $\Delta\tilde{\mu}_{D^+}$ were the same, the motor generated the same torque. This suggests that the entry of protons into the motor and their exit are fast compared to the mechanical events. If rotation involved the formation and breakage of chemical bonds or a major structural rearrangement, its rate should be enhanced at higher temperatures. Surprisingly this was not the case: the torque remained nearly constant between 4 and 38°C. Khan and Berg

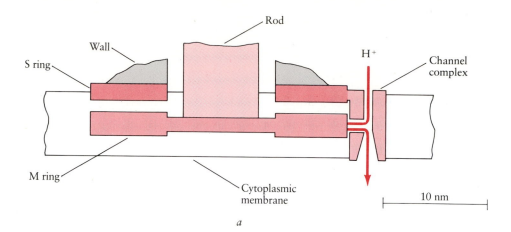

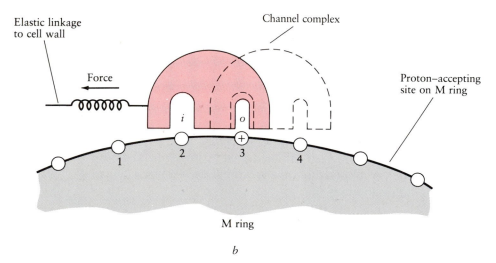

FIGURE 5.14   Model of the flagellar rotary motor of a gram-positive bacterium. (*a*) Cross section, drawn to scale. Torque develops between the mobile M ring and the static S ring. (*b*) A hypothesis to explain the generation of torque. Part of the motor is shown in a section cut parallel to the plane of the plasma membrane, through the middle of the bilayer. The channel complex is the means of torque generation, as explained in the text. Given that the channel complex occupies the position shown by the solid outline, the coupling rules permit it to move to the right when site 3 is protonated (+) while site 2 is empty. When it moves, taking up the position indicated by the dashed outline, a new proton can move from the outside onto site 4, while the proton on site 3 can move into the cytoplasm. The channel complex can continue to drift to the right, but not to the left. Movement of the complex to the right stretches the spring, which exerts an equal but opposite force on the M ring. (After Khan and Berg, 1983; copyright Massachusetts Institute of Technology.)

therefore concluded that the flagellar motor works as a reversible engine by simple acid-base dissociation. The proton-motive force generates a difference in the degree of protonation of two sets of proton-accepting groups, one set in contact with the external medium and the other set with the cytoplasm. This difference in occupancy provides the immediate driving force for rotation.

It remains to devise a model that explains how the flux of protons down their electrochemical potential gradient and across the M and S rings can generate torque between these rings. The model proposed by Khan and Berg (1983), illustrated in Figure 5.14, accounts for the observations in a most satisfactory way but is not easily apprehended by intuition. In addition to the M ring in the plasma membrane and the S ring attached to the wall, the model calls for a set of channel complexes distributed around the periphery of the S ring. Only one is shown in the illustration, but their actual number may be 16. Each channel complex contains a dual set of channels: "outer" channels communicate between the exterior and a set of proton-binding sites on the M ring; "inner" channels link these proton-binding sites to the cytoplasm. The channel complexes are attached to the S ring by linkers, symbolized by the spring, that are capable of a certain amount of elastic extension, so that each channel complex can jiggle circumferentially with respect to the M ring by a distance greater than d, the spacing of the proton-binding sites. However, movements of the channel complex along the periphery of the M ring are subject to two constraints, or coupling rules. First, a site cannot move past the center of the channel complex unless that site is protonated. Second, a site cannot move away from the channel complex into the surrounding hydrophobic region of the membrane unless that site is unprotonated. Suppose now that a proton potential exists across the membrane. In that case, whether a given site is protonated or not depends on whether it communicates with the exterior or with the cytoplasm. If the M ring is held in a fixed position, the channel complex in Figure 5.14b will "drift" to the right as far as it can, somewhat like a dog on a leash, and the spring will exert torque on the M ring. It can be shown that at equilibrium the torque is equal to $\Delta\tilde{\mu}_{H^+}$. Let us now allow the M ring to move in response to the torque. Protons will then move from the medium through the outer channels, onto the M ring, and through the inner channels into the cytoplasm. Provided the flux of protons is not too great, the torque can be maintained and the M ring will rotate at a rate proportional to $\Delta\tilde{\mu}_{H^+}$.

We cannot here enter into the thermodynamic calculations that flesh out this model, nor into the refinements required to explain how the sense of rotation can reverse even though protons continue to flow inward. Among the questions that the model does not address are the nature of the tumble generator and the nature of the signals that link receptors to the rotor so as to modulate the frequency of tumbling; we shall return to the latter question in Chapter 13. But Berg's model easily accommodates the recent discovery that in the alkalophilic bacteria it is a flux of sodium ions, not of protons, that drives the flagellar motor: the mechanism of rotation, like that of a porter, can be adapted to any ion flowing down its electrochemical potential gradient.

Homeostasis

The internal environment of bacteria, like that of other living things, generally differs substantially from the external one in chemical composition and in physical properties; it also remains remarkably constant in the face of external fluctuations. Bacteria, unlike higher organisms, can often survive gross changes in composition, but they reestablish the normal one before growth resumes. The reasons are generally plain enough. The cytoplasmic pH, for example, must not depart too far from neutrality, which is optimal for most enzymes. Turgor pressure supplies the driving force for expansion; it must always be sufficiently high to keep the membrane appressed to the cell wall, yet not rise to the point of bursting. Other examples are not so easily understood: why, for instance, is K^+ always the predominant cytoplasmic cation while the more abundant Na^+ is extruded? And how about such central metabolic parameters as the proton-motive force, the phosphorylation potential of ATP, the adenine nucleotide energy charge, and the redox potential: are these also subject to regulatory mechanisms that counter perturbations and endeavor to maintain optimal values? Homeostatic control circuits have been a central concern of animal physiologists for a century and more. Microbiologists have but recently begun to explore the maintenance of constancy, and the mechanisms are generally far from clear. But homeostasis is obviously an important use of biological energy, and it illustrates again the interplay of the proton circulation with other modes of energy generation.

The Constancy of Cytoplasmic pH.   From the chemiosmotic point of view, the pH gradient between cytoplasm and medium is part of the "line voltage" that powers physiological work functions. One may therefore naively expect bacteria to maximize ΔpH. Instead, as experimental data began to accumulate it became clear that the cytoplasmic pH remains quite constant despite large changes in the external pH. Figure 4.13 shows a typical set of data, obtained with *E. coli* growing in media ranging in pH from 6 to 8.5. Note that over this range ΔpH declines markedly, the total proton-motive force declines somewhat less due to the rise in $\Delta\Psi$, but the cytoplasmic pH holds steady near 7.8. Similar results have been obtained with other bacteria, not only those that favor neutral environments but also with acidophilic and alkalophilic ones. *Thiobacillus ferro-oxidans*, for example, grows best at pH 2 but is capable of maintaining a cytoplasmic pH near 6.5 over an external pH range between 2 and 8; *Bacillus alkalophilus*, whose optimal external pH is 10.5, maintains its cytoplasmic pH near 9 even when the external pH rises to 11. This degree of stability far exceeds the capacity of cytoplasmic buffers. It is the more remarkable when one considers the central role of the proton circulation in energy transduction. Evidently bacteria must possess the means to monitor their cytoplasmic pH and to respond to any fluctuations by stabilizing an optimum "set point" (Padan et al., 1981). Just how they do this is under intense investigation; the following account is intended to light the way rather than proclaim the answer.

Many bacteria produce copious amounts of acidic metabolites, particularly when growing fermentatively. Glycolyzing streptococci, for instance, excrete their own weight in lactic acid every hour. That they nevertheless maintain their cytoplasmic pH as much as one unit more alkaline than the medium is a tribute to the effectiveness of an electroneutral porter that mediates the efflux of lactic acid,[1] together with the F_1F_0 ATPase and that mystifying potassium transport system KtrI (Figure 5.7b). Between them they effectively carry out the accumulation of K^+ ions by exchange for protons and thus allow the cytoplasmic pH to rise. The increase in ΔpH compensates, at least in part, for the collapse of $\Delta\Psi$ due to the electrogenic influx of K^+ ions. But the object of the exercise is probably not so much to maximize the proton potential as to raise the cytoplasmic pH. In fact, in streptococci the maintenance of the cytoplasmic pH may be considered the chief function of the F_1F_0 ATPase (Harold and Van Brunt, 1977; Kobayashi et al., 1982).

Maintenance of a constant internal pH requires that there be mechanisms to acidify the cytoplasm as well as to alkalinize it. The production of metabolic acid is one way to lower the internal pH, but it is often unavailable to aerobic organisms (oxidation of succinate to fumarate, for example, generates no metabolic acid). Such circumstances call for devices that import protons without, however, uncoupling the proton current. Two hypothetical mechanisms that have been invoked to this end are shown in Figure 5.7a. The first calls for electroneutral exchange of K^+ for protons; there is some evidence that such a porter exists in E. coli, but the case is not yet compelling. Exchange of K^+ for H^+ would have to be strictly regulated lest the interplay of K^+ influx with K^+-H^+ antiport give rise to a dissipative net flux of protons. The second mechanism invokes exchange of Na^+ for protons, and this has received substantial experimental support. The most persuasive observation is that mutants of E. coli and Bacillus alkalophilus that lack the Na^+-H^+ antiporter concurrently lose the capacity to acidify their cytoplasm and can no longer grow in alkaline media above pH 8 (Padan et al., 1981; Krulwich, 1983). It appears, then, that the Na^+-H^+ antiporter serves not so much to export sodium ions as to import protons. Here again questions about the stoichiometry and control of the exchange need to be answered: a priori, an electroneutral process would seem to be best suited to pH regulation, but the data suggest that Na^+-H^+ antiporters are electrogenic.

If it is true that exchange of cytoplasmic Na^+ for protons is required in order to acidify the cytoplasm, then there must also be pathways to let Na^+ pass into the cell—preferably in an electroneutral manner. Oddly, bacteria do not usually require sodium ions for growth; perhaps the small amounts of Na^+ that invariably contaminate laboratory reagents suffice for their needs. It also seems

---

1. Can any statement about biology be left unqualified? In some streptococci the lactate porter appears to be electrogenic, with a stoichiometry of $2H^+$ per lactate. Net efflux of lactate plus protons contributes to the protonic potential and thus participates in the conversion of metabolic energy into that of an ion gradient (tenBrink and Konings, 1982).

necessary to postulate some molecular mechanism to monitor the cytoplasmic pH and to report the need for corrective action; evidence for the existence of such sensors has recently been published, but virtually everything else remains to be discovered. There is probably truth in Figure 5.7, but it is probably neither the whole truth nor nothing but the truth.

Cation Gradients, Turgor, and Energy Storage.   The universal asymmetry of cation distribution, such that K^+ is the predominant cation of the cytoplasm while Na^+ is excluded, has often been remarked but never satisfactorily explained. It is true that some metabolic enzymes require K^+ and that ribosomal protein synthesis demands K^+ concentrations of 0.1 M or more, but these facts do not explain why bacteria cease to grow as soon as they have exhausted the external K^+. Regulation of the cytoplasmic pH is almost certainly one function of potassium transport, the control of turgor pressure another, and buffering of the cell's energy supply a third. The large gradients of K^+ and Na^+ ions that almost all bacteria establish reflect the roles of these ions in several interwoven homeostatic mechanisms.

Bacterial cell walls are subject to osmotic pressures that range from about five atmospheres in gram-negative bacteria to 10 atmospheres in gram-positive ones. Turgor, the difference in osmotic pressure between cytoplasm and medium, must be positive for the cells to expand. When an impermeant solute is added to a growing culture in amounts sufficient to reverse the normal gradient, growth ceases while the organisms adjust the composition of their cytoplasm and resumes only when turgor is once again positive. How is this done?

The osmotic pressure of the cytoplasm reflects the sum total of its constituents, particularly the small molecules ("osmolites"). Potassium ions make a major contribution: *E. coli* achieves a cytoplasmic K^+ level of 0.25 M, *S. faecalis* 0.5 M, and halobacteria a prodigious 4 M. Osmolite concentrations vary in response to changes in the external osmolarity: *E. coli* uses potassium glutamate to maintain turgor, but gram-positive bacteria seem to prefer proline or γ-aminobutyric acid for this purpose (Measures, 1975). Little is known concerning the sequence of signals and responses by which a decrease in turgor brings about a compensatory increase in the cytoplasmic concentration of proline or glutamate. However, thanks chiefly to the work of W. Epstein and his colleagues, we are increasingly well informed about the accumulation of K^+ ions that is induced by such manipulations (Epstein and Laimins, 1980; Laimins et al., 1981; Helmer et al., 1982). Both the TrK and the Kdp systems (Figure 5.7a) are turned on by an increase in the external osmolarity. One wonders how *E. coli* measures turgor and reports the result: perhaps the *kdp-D* locus, known to code for a positive regulator of that transport system, specifies a receptor protein that senses pressure across the membrane.

In the preceding paragraphs, K^+ accumulation and Na^+ extrusion were discussed as forms of cellular work. However, the large pools and steep gradients of K^+ and Na^+ ions also represent stored energy that could in principle stabilize the protonic potential or support additional work functions. Recent experiments indicate that starving bacteria can draw on cation gradients to prolong motility and to maintain their protonic potential; the mechanism is uncertain (Skulachev, 1978; Brown et al., 1983). This is probably not the chief function of K^+ and Na^+ transport but an ancillary one of importance to certain organisms such as the halobacteria.

Reducing Power.   Biosynthesis of cell constituents generally requires the provision of reducing equivalents as well as energy. This is no great matter for anaerobic bacteria, which generate electrons at very negative redox potential during fermentation. Aerobes, however, as well as photosynthetic bacteria had to invent alternative strategies to generate reduced nicotinamide adenine nucleotides with energy input from the proton circulation.

One of these is the reduction of NAD^+ to NADH by reversed electron transport, a pathway particularly prominent in many photosynthetic and chemolithotrophic bacteria. Membrane preparations from *Rhodospirillum rubrum,* for instance, catalyze the light-dependent reduction of NAD^+ with H_2S or succinate as reductant. Reduction also proceeds in the dark provided ATP is supplied, and it is sensitive to proton-conducting uncouplers under both conditions. These and other observations point to the proton potential as the force that reverses the normal direction of electron flow from NADH to quinones. We can write the oxidative reaction catalyzed by the NADH-ubiquinone oxidoreductase as

$$NADH + H^+ + UQ + 2H_i^+ \rightleftharpoons NAD^+ + UQH_2 + 2H_o^+ \quad \text{(Eq. 5.1)}$$

In the forward direction this reaction generates a protonic potential across the membrane; in the reverse direction, an applied protonic potential (derived either from ATP hydrolysis or from photosynthetic proton translocation) can shift the position of equilibrium toward the formation of NADH (Figure 5.15a).

In many biosynthetic reactions, the reductant is NADPH rather than NADH, and bacteria often maintain their $NADPH/NADP^+$ ratio near unity, substantially more reduced than the $NADH/NAD^+$ ratio, which is about 0.1 (Chapter 2). It is likely that this differential redox ratio is at least in part the work of the "energy-linked" transhydrogenase, an enzyme first discovered in mitochondria but now known to be widely distributed among bacteria as well. The overall reaction catalyzed by the native, membrane-bound complex appears to be

$$NADH + NADP^+ + 2H_o^+ \rightleftharpoons NAD^+ + NADPH + 2H_i^+ \quad \text{(Eq. 5.2)}$$

Thus the proton-motive force enters into the reaction even though hydrogen is transferred from the reduced to the oxidized nucleotide without equilibrating with water, and there is no obvious transmembrane reaction at all (Mitchell, 1976a; Rydström, 1977). Figure 5.15b illustrates how a proton-motive force generated by respiration or photosynthesis can shift the equilibrium toward the reduction of $NADP^+$.

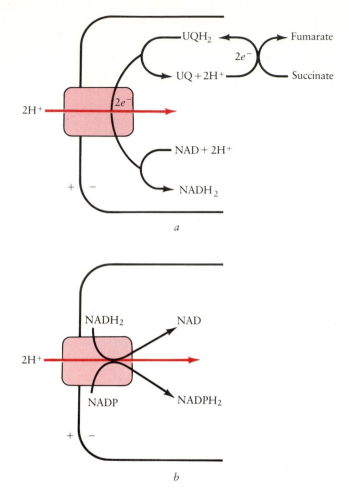

FIGURE 5.15 Chemiosmotic generation of reduced nicotinamide adenine nucleotides. (a) Reduction of NAD^+ by reverse electron transport, mediated by the NADH-ubiquinone oxidoreductase of the respiratory chain. (b) Reduction of $NADPH^+$ by NADH with the aid of a proton-translocating nicotinamide adenine nucleotide transhydrogenase. In both diagrams the protonic potential, interior negative, is generated by other proton-transporting pathways, such as the photosynthetic apparatus. Arrows are pointed in only one direction, but all the reactions are reversible.

The transhydrogenase from mitochondria has been purified, reconstituted into membrane vesicles, and shown to catalyze proton translocation. Its molecular mechanism, however, remains as uncertain as that of other proton-translocating pathways. In candor, it must be added that bacterial mutants lacking the transhydrogenase grow perfectly well (Zahl et al., 1978), suggesting that it is but one of several routes that contribute to the maintenance of a pool of reduced nicotinamide adenine nucleotides.

The Poise of Adenine Nucleotides and Proton Potential.   The ATP-ADP couple is the central energy currency of the cytoplasm. One would therefore expect bacteria to monitor the status of this pool and to respond to any decrease in the free energy available from ATP hydrolysis with remedial action; given the universal involvement of ATP and ADP in biosynthetic reactions, homeostatic control of their absolute concentrations might also be anticipated. Curiously, it is quite uncertain at present whether such regulatory mechanisms exist and, if so, what signals they monitor.

Research in this area is technically difficult. The ATP pool of respiring bacteria turns over several times per second. In order to obtain meaningful data about the concentrations of ATP, ADP, and AMP, it is necessary to stop metabolism instantly at the moment of sampling. The major effort in this direction has been made by the laboratory of D. E. Atkinson, who concluded that bacteria behave as though they monitored the energy charge (Chapter 2). In practice, this means that individual enzymes of both catabolic and anabolic sequences respond to the ATP/ADP or the ATP/AMP ratio; there need be no centralized supervision of the nucleotide pool as a whole. There is good evidence that glycolysis is controlled in this manner and that bacteria cease to grow if their energy charge falls below 0.8 (Atkinson, 1977; Knowles, 1977). The essential role of adenylate kinase, attested by the failure of mutants with a defect in this enzyme to grow, reinforces Atkinson's views. However, the parameter that is mechanistically tied to energy transduction is the phosphorylation potential ΔG_p, equivalent to the free energy of hydrolysis of cytoplasmic ATP. In mitochondria the rate of respiration appears to be controlled by ΔG_p. Comparable data are not available for bacteria, nor has it been established whether these cells seek to stabilize ΔG_p.

Information about the proton-motive force of growing bacteria is now becoming available, but it is not yet extensive enough to justify sweeping conclusions. It appears, however, that while the internal pH remains constant during growth, $\Delta\Psi$ and $\Delta\tilde{\mu}_{H^+}$ change with external conditions. It is also known that unlike mitochondria, bacteria do not in general exhibit respiratory control. All this produces the impression (and that is all it is!) that bacteria regulate neither ΔG_p nor $\Delta\tilde{\mu}_{H^+}$ but rather seek to maximize them by catabolizing substrates as rapidly as their enzymatic complement permits. Energy supply in excess of the demand for growth leads to spillage, while a deficiency is countered by adjusting the cells' enzyme composition to make additional substrates available. But we remain largely in the dark regarding the signals that notify the cell of deficiencies in the energy supply.

The topics discussed above must suffice to document the importance of the proton circulation in bacterial energy transduction and work. They do not exhaust the subject, however: recent research implicates the proton potential in nitrogen fixation, gliding motility, the regulation of adenylate cyclase, cellulose production, and the secretion of proteins across the plasma membranes. Not surprisingly, reagents that dissipate the protonic potential inhibit growth. A case in point are the bacteriocins, protein antibiotics that form ion-conducting channels across the plasma membrane and disrupt energy coupling.

It therefore comes as something of a surprise to discover that bacteria can nevertheless grow in the absence of any protonic potential (Harold and Van Brunt, 1977). *S. faecalis* grew perfectly well in the presence of the powerful ionophore gramicidin D, under conditions such that ΔpH, $\Delta \Psi$, and $\Delta \tilde{\mu}_{H^+}$ were all set to zero, provided the growth medium was adjusted to meet their requirements. These included a high concentration of external K^+, the absence of excessive Na^+, a pH above 7.5, and a sufficient concentration of amino acids and vitamins. The cells are obviously crippled with respect to their ability to maintain the normal cytoplasmic pH and ionic composition, to accumulate organic metabolites, and to cope with adversity; they are also nonmotile. However, the fact that they grow and divide normally, producing offspring morphologically indistinguishable from the controls, sets an upper limit to the biological importance of the proton circulation: the fabric of a bacterial cell can be constructed without it.

Costs, Profits, and Yield

The preceding sections of this chapter, together with Chapter 4, surveyed the means by which bacteria generate useful energy and the uses to which it is put. To conclude this chapter, let us glance at the quantitative relation between the input of energy and the output of living matter.

A routine procedure in any microbiological laboratory is to inoculate a few bacteria into a flask that contains a solution of inorganic salts and an organic substrate, perhaps glucose. The flask is then placed in an incubator overnight; the next morning the solution is turbid with cells, millions per milliliter, and the substrate has been degraded to waste products. In the 1940s, Jacques Monod's classic studies on bacterial growth in defined media showed that when limiting amounts of carbohydrate were supplied as the sole energy source, the cell yield was proportional to the amount of carbohydrate added. However, two complications were recognized right away. First, an organism growing in a minimal medium draws on the organic substrate for both energy and carbon skeletons. Second, catabolism need not be tightly coupled to growth, since cells suspended in buffer and unable to grow often degrade the substrate as rapidly as growing cells do.

Clear-cut results bearing on the energy requirements for growth were first obtained in studies with streptococci (Bauchop and Elsden, 1960). These are nutritionally exacting organisms that obtain practically all their assimilated carbon from amino acids and other components of the growth medium. Their typical energy source is glucose, which is fermented almost quantitatively to lactic acid by the glycolytic pathway, with a net yield of two moles of ATP per mole of glucose; respiratory metabolism is generally absent. Bauchop and Elsden found that the growth yield was directly proportional to the amount of glucose added and expressed this relationship as the yield of cells per mole of ATP produced, Y_{ATP}. The numerical value of Y_{ATP}, about 10.5 g dry weight/mol ATP, was soon found to hold for other bacteria growing anaerobically and even for yeast, and it appeared to be an approximate biological constant.

The following decade saw many analogous studies, most of them done under continuous culture conditions, which are better suited to quantitative analysis than the traditional batch cultures (Payne, 1970; Forrest and Walker, 1971). Most of these studies confirmed the impression that Y_{ATP} is at least approximately constant for organisms growing anaerobically (Table 5.1). Calculation of the amount of cells obtained per mole of ATP requires, of course, prior knowledge of the ATP yield of the catabolic pathway. For example, the yield of streptococci appears to be much lower when they are growing on arginine than on glucose, but taking into account that arginine metabolism generates only one mole of ATP, Y_{ATP} is not far from 10.5. Conversely, growth yields have been used to determine the ATP yield of unknown metabolic pathways, as Thauer et al. (1977) did in their studies on energy metabolism in clostridia. Growth yields made a particularly important contribution to our understanding of aerobic metabolism in microorganisms. Early work with bacterial membranes had suggested that bacterial oxidative phosphorylation was poorly coupled. This possibility could now be dismissed as the result of damage during cell breakage: assuming that Y_{ATP} for aerobic growth is again 10.5, the efficiency of bacterial oxidative phosphorylation turns out to be comparable to that of mitochondria, between

TABLE 5.1 Growth Yields of Selected Bacteria Growing Anaerobically

Organism	Substrate	ATP from substrate-level phosphorylation*	Y_{ATP}[†] (g/mol)
Clostridium perfringens	Glucose	3.1	14.6
Escherichia coli	Glucose	3	11.2
Klebsiella aerogenes	Glucose	3	10.2
	Gluconate	2.5	11.0
	Mannitol	2.5	10.8
Lactobacillus casei	Glucose	2	21–24
	Mannitol	2.2	18.2
	Citrate	1	19
Streptococcus faecalis	Glucose	2.0–3.0	10.9
	Ribose	1.7	12.6
	Pyruvate	1	10.4
	Arginine	1	10.2
Zymomonas anaerobia	Glucose	1	5.9

Data from Stouthamer, 1978.
* Fractional values of ATP yield are due to interacting minor catabolic routes.
[†] Y_{ATP} is expressed as grams of cells per mole of ATP generated.

two and three moles of ATP per gram-atom of oxygen (Payne, 1970; Forrest and Walker, 1971; Stouthamer and Bettenhaussen, 1973).

Nevertheless, as data from both aerobes and anaerobes mounted, it became unmistakable that Y_{ATP} can vary considerably from one organism to another and also varies for any single organism as a function of growth conditions. For example, Table 5.1 records Y_{ATP} values that range from a low of 5.9 for *Zymomonas anaerobia* to a high of 24 for *Lactobacillus casei*, both anaerobes. Extensive studies on the aerobic growth of *Klebsiella aerogenes* documented that Y_{ATP} is strongly influenced by the rate of growth, thus precluding a straightforward calculation of P/O ratios from growth yields. Organisms whose growth rate is constrained by the absence of some essential ion or nutrient have strikingly low growth yields. The result was to change our perception of the meaning of Y_{ATP}. It is not a biological constant but an index of the metabolic cost of living: Y_{ATP} reflects the way cells apportion their energy budget.

Energy requirements for growth can be divided into two categories: biomass synthesis and the "energy of maintenance"; that is, the cost of keeping the cell alive and working. The interplay of these two categories will determine the growth yield obtained under any given set of conditions. The great bulk of cell dry weight is due to biopolymers: proteins, nucleic acids, polysaccharides, and also lipids. Now, since the role of ATP in biosynthesis is a stoichiometric one, we can draw on our extensive knowledge of intermediary metabolism to estimate the cost, in ATP equivalents, of making tryptophan or NADPH, DNA or protein. Table 5.2 lists estimates of metabolic prices under two common conditions: with all basic monomers supplied and with glucose as the sole source of both carbon and energy. By taking into account the composition of cells and the price of each bond and metabolite, one arrives at the conclusion that the cost of biosynthesis for cells growing in a complex medium that supplies all the requisite small precursors is about 0.026 mole of ATP per gram of cells (Forrest and Walker, 1971; Stouthamer, 1978). To establish the best-possible growth yield, Stouthamer made an allowance for the cost of transporting metabolites into the cell (in ATP equivalents, even though the process itself may rely on the proton circulation) and concluded that the maximum yield attainable in complete medium, Y_{ATP}^{max}, should be around 32 grams of cells per mole of ATP (g/mol). It is noteworthy that some of the values listed in Table 5.1 approach this theoretical maximum. The bacterial parasite *Bdellovibrio* comes even closer: it grows in the host's periplasm, consumes the host's cytoplasmic constituents, and boasts a Y_{ATP} of 26 g/mol (Rittenberg and Hespell, 1975). Under more demanding conditions, the theoretical Y_{ATP}^{max} will be reduced. Stouthamer estimates 28.8 g/mol for growth on glucose plus inorganic salts, but only 10 g/mol for growth on acetate (chiefly due to the larger ATP requirement for monomer formation and transport). For autotrophic growth on CO_2, the theoretical Y_{ATP}^{max} is as low as 6.5 g/mol. Nitrogen fixation is another expensive process, requiring as much as 12 moles of ATP per mole of ammonia produced; organisms growing with glucose as carbon source and molecular nitrogen as sole nitrogen source have a theoretical Y_{ATP}^{max} of only 7.9.

TABLE 5.2   ATP Requirement for the Formation of Microbial Cells

Macromolecule	Amount (g/g cells)	ATP required per monomer* (mol/mol)	ATP required per gram of cells (mol × 10⁴)	
			From preformed monomers[†]	From glucose and inorganic salts
Polysaccharide	0.17	2	21	21
Protein	0.52			
Amino acid formation				14
Polymerization		4	191	191
Lipid	0.09	1	1	1
RNA	0.16			
Nucleotide monophosphate formation		3	15	34
Polymerization		2	9	9
DNA	0.03			
Deoxynucleoside monophosphate formation		4	4	9
Polymerization		2	2	2
Turnover of mRNA			14	14
Subtotal			257	295
ATP required for transport of:				
Amino acids			48	
Ammonium ions				42
Potassium ions			2	2
Phosphate			8	8
Total ATP requirement			315	347
Theoretical Y_{ATP}^{max}			32	29

Data from Stouthamer, 1978, rounded off.

* The ATP cost of nucleoside monophosphate formation is calculated from the nucleic acid bases.

[†] Preformed monomers refers to glucose, amino acid, and nucleic acid bases.

Actual growth yields always fall short of the theoretical maximum, often by as much as a factor of 2, and it is not entirely clear why this is the case. Part of the shortfall can be chalked up to energy expenditure for processes other than the net synthesis of macromolecules. Examples include the turnover and repair of membranes, cell walls, and nucleic acids; motility and behavioral responses; and particularly the transport of ions, molecules, and chemical groups from the medium to their ultimate locations in the growing cell. The cost can be substantial: under certain circumstances bacteria may have to devote a third of their energy budget to the transport of potassium ions and other metabolites across the plasma membrane. But there is something misleading about the fundamental assumption that the free energy of catabolism is fully conserved as ATP and expended necessarily either for biosynthesis or for useful work. Any departure from perfect coupling, either in the generation of ATP or in its utilization, will show up as a shortfall of the yield and exaggerate the apparent cost of cellular upkeep.

Reflection on the nature of the bacterial energy economy suggests many opportunities for imperfect coupling. ATP production by substrate-level phosphorylation is tightly and stoichiometrically linked to the catabolic reactions; by contrast, proton-transport phosphorylation is indirectly linked and the ATP yield will be affected by the manner and extent of branching in the oxidative chain and by alternative routes of proton flux. Not all transport processes can be regarded as immediately useful: the cycling of K^+, Na^+, and sugars (not to mention protons) inevitably dissipates some energy. Authorities disagree about how large a fraction of the energy potentially available is lost due to slippage (Forrest and Walker, 1971; Stouthamer and Bettenhaussen, 1973; Tempest, 1978; Tempest and Neijssel, 1984), but it may well be substantial. To put it another way, loose coupling means that the rate of substrate catabolism is not determined by the demand for ATP and $\Delta\tilde{\mu}_{H^+}$ but by the metabolic capacity of the cell. Conversely, even under conditions of energy limitation the rate of growth may not be determined entirely by the rate of generation of ATP and $\Delta\tilde{\mu}_{H^+}$.

Loose coupling between catabolism and growth is most evident when growth is restricted; in the extreme case of washed cell suspensions, all the free energy available must be dissipated. Surprisingly little effort has been made to identify specific mechanisms that dissipate the surplus, but in principle there is a wide choice. One way is the formation of reserve substances such as glycogen or poly-β-hydroxybutyrate, which characteristically accumulate during nitrogen starvation (Dawes and Senior, 1973). The extent of energy dissipation by this means will, of course, be limited by the cells' storage capacity. Another polymer that comes to mind is inorganic polyphosphate, an abundant and characteristic constituent of microorganisms (Dawes and Senior, 1973). There is little evidence that polyphosphate serves as a storage form of "energy-rich" phosphate, even though the free energy of hydrolysis of each phosphoryl bond is nearly equivalent to that of ATP. However, the formation of polyphosphate from ATP by phosphoryl transfer catalyzed by polyphosphate kinase, followed by hydrolysis of the polymer by polyphosphatases (Figure 5.16a), would assuredly dissipate ATP. Tempest (1978) mentions another "futile cycle," composed of glutamine synthase and glutaminase (Figure 5.16b), that may serve in the controlled dissipation of ATP by nitrogen-limited K. aerogenes. Even the F_1F_0 ATPase may perform this service, expelling protons and raising the electrochemical potential gradient to the point where proton backflow by leak pathways equals their rate of extrusion. Other chemiosmotic options include a controlled proton leak (Figure 5.16c), as in the case of heat production by adipose tissue mitochondria, or the production of modified respiratory chains that lack one or more coupling sites.

It must be concluded that the bacterial solution to energy surplus is in general to waste it. However reprehensible this attitude may seem to one who accepts the conservation ethic, it makes sense in the bacterial world. It is a general principle of ecology that when resources are scarce but dependable, their efficient use will confer selective advantage. In other circumstances, when resources are ample but only temporarily available, competition and reproductive advantage favor organisms that can rapidly acquire a large share of these resources, even if this involves waste. Most bacteria are plainly in the latter situation. Theirs is a feast-or-famine existence in which the race is to the swift, those who have a high metabolic capacity and can begin multiplying with little delay. It is the eukaryotes, rather than the prokaryotes, that came to dominate environments providing a meager but dependable livelihood.

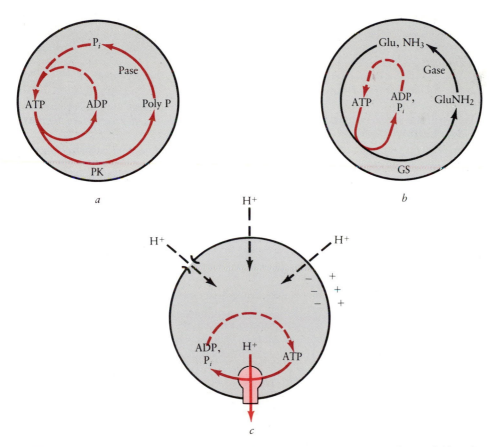

FIGURE 5.16 Possible mechanisms of ATP dissipation by nongrowing cells supplied with excess substrate. (*a*) Polyphosphate cycle, composed of polyphosphate kinase (PK) and polyphosphatase (Pase). (*b*) Glutamine cycle, composed of glutamine synthase (GS) and glutaminase (Gase). (*c*) A dissipative proton cycle, composed of the proton-translocating ATPase and proton leaks back into the cytoplasm. In each scheme, the dashed line stands for whatever reactions regenerate ATP.

References

ADLER, J. (1975). Chemotaxis in bacteria. *Annual Review of Biochemistry* **44**:341–356.

AHMED, S., and BOOTH, I. R. (1981). The effects of partial and selective reduction in the proton-motive force on lactose uptake in *Escherichia coli*. *Biochemical Journal* **200**:583–589.

ATKINSON, D. E. (1977). Op. cit., Chapter 1.

BAKKER, E. P., and HAROLD, F. M. (1980). Energy coupling to potassium transport in *Streptococcus faecalis*. *Journal of Biological Chemistry* **255**:433–440.

BAUCHOP, T., and ELSDEN, S. R. (1960). The growth of microorganisms in relation to their energy supply. *Journal of General Microbiology* **23**:457–469.

BERG, H. C. (1975a). Bacterial behavior. *Nature (London)* **254**:389–392.

BERG, H. C. (1975b). How bacteria swim. *Scientific American* **233**(3):36–44.

BERG, H. C., MANSON, M. D., and CONLEY, M. P. (1982). Dynamics and energetics of flagellar rotation in bacteria. *Society for Experimental Biology Symposia* **35**:1–31.

BERGER, E. A., and HEPPEL, L. A. (1974). Different mechanisms of energy coupling for shock-sensitive and shock-resistant amino acid permeases of *Escherichia coli*. *Journal of Biological Chemistry* **249**:7747–7755.

BLOCK, S. M., and BERG, H. C. (1984). Successive incorporation of force-generating units in the bacterial rotary motor. *Nature* **309**:470–472.

BOOTH, I. R., MITCHELL, W. J., and HAMILTON, W. A. (1979). Quantitative analysis of proton-linked transport systems: The lactose permease of *Escherichia coli*. *Biochemical Journal* **182**:687–696.

BORBOLLA, M. G., and ROSEN, B. P. (1984). Energetics of sodium efflux from *Escherichia coli*. *Archives of Biochemistry and Biophysics* **229**:98–103.

BROWN, I. I., GALPERIN, M. YU., GLAGOLEV, A. N., and SKULACHEV, V. P. (1983). Utilization of energy stored in the form of Na^+ and K^+ gradients by bacterial cells. *European Journal of Biochemistry* **134**:345–349.

CRANE, R. K. (1977). Op. cit., Chapter 3.

DAWES, E. A., and SENIOR, P. (1973). Op. cit., Chapter 2.

DILLS, S. S., APPERSON, A., SCHMIDT, M. R., and SAIER, M. H., JR. (1980). Carbohydrate transport in bacteria. *Microbiological Reviews* **44**:385–418.

DIMROTH, P. (1982). Op. cit., Chapter 4.

EDDY, A. A. (1978). Proton-dependent solute transport in microorganisms. *Current Topics in Membranes and Transport* **10**:279–360.

EDDY, A. A. (1982). Mechanisms of solute transport in selected eukaryotic microorganisms. *Advances in Microbial Physiology* **23**:1–78.

EPSTEIN, W., and LAIMINS, L. (1980). Potassium transport in *Escherichia coli*: Diverse systems with common control by osmotic forces. *Trends in Biochemical Sciences* **5**:21–23.

FORREST, W. W., and WALKER, D. J. (1971). The generation and utilization of energy during growth. *Advances in Microbial Physiology* **5**:213–274.

GLAGOLEV, A. N. (1981). Proton circuits of bacterial flagella, in *Chemiosmotic Proton Circuits in Biological Membranes*, V. P. Skulachev and P. C. Hinkle, Eds. Addison-Wesley, Reading, Mass., pp. 577–600.

HAROLD, F. M. (1978). Op. cit., Chapter 3.

HAROLD, F. M (1982). Pumps and currents, a biological perspective. *Current Topics in Membranes and Transport* **16**:485–516.

HAROLD, F. M., and VAN BRUNT, J. (1977). Circulation of H^+ and K^+ is not obligatory for bacterial growth. *Science* **197**:372–373.

HEEFNER, D. L., and HAROLD, F. M. (1982). ATP-driven sodium pump in *Streptococcus faecalis*. *Proceedings of the National Academy of Sciences USA* **79**:2798–2802.

HELMER, G. L., LAIMINS, L. A., and EPSTEIN, W. (1982). Mechanisms of potassium transport in bacteria, in *Membranes and Transport*, A. N. Martonosi, Ed., vol. II. Plenum, New York, pp. 123–128.

HENGGE, R., and BOOS, W. (1983). Maltose and lactose transport in *Escherichia coli*: Examples of two different types of concentrative transport systems. *Biochimica et Biophysica Acta* 737:443–478.

HIGGINS, C. F., HAAG, P. D., NIKAIDO, K., ARDESHIR, F., GARCIA, G., and FERRO-LUZZI AMES, G. (1982). Complete nucleotide sequence and identification of membrane components of the histidine transport operon of *S. typhimurium*. *Nature* 298:723–727.

HILPERT, W., and DIMROTH, P. (1983). Op. cit., Chapter 4.

HONG, J.-S., HUNT, A. G., MASTERS, P. S., and LIEBERMAN, M. A. (1979). Requirement of acetyl phosphate for the binding-protein dependent transport systems in *Escherichia coli*. *Proceedings of the National Academy of Sciences USA* 76:1213–1217.

HUNT, A. G., and HONG, J.-S. (1983). Properties and characterization of binding protein dependent active transport of glutamine in isolated membrane vesicles of *Escherichia coli*. *Biochemistry* 22:844–850.

KABACK, H. R. (1974). Transport studies in bacterial membrane vesicles. *Science* 186:882–892.

KABACK, H. R. (1983). The *Lac* carrier protein in *Escherichia coli*. *Journal of Membrane Biology* 76:95–112.

KAKINUMA, Y., and HAROLD, F. M. (1985). ATP-linked exchange of Na^+ for K^+ ions by *Streptococcus faecalis*. *Journal of Biological Chemistry* 260:2086–2091.

KASHKET, E. R. (1981). Op. cit., Chapter 4.

KHAN, S., and BERG, H. C. (1983). Isotope and thermal effects in chemiosmotic coupling to the flagellar motor of streptococcus. *Cell* 32:913–919.

KNOWLES, C. J. (1977). Op. cit., Chapter 2.

KOBAYASHI, H. (1982). Second system for potassium transport in *Streptococcus faecalis*. *Journal of Bacteriology* 150:506–511.

KOBAYASHI, H., MURAKAMI, N., and UNEMOTO, T. (1982). Regulation of the cytoplasmic pH in *Streptococcus faecalis*. *Journal of Biological Chemistry* 257:13246–13252.

KRULWICH, T. A. (1983). Op. cit., Chapter 4.

LAIMINS, L. A., RHOADS, D. B., ALTENDORF, K., and EPSTEIN, W. (1978). Identification of the structural proteins of an ATP-driven potassium transport system in *Escherichia coli*. *Proceedings of the National Academy of Sciences USA* 75:3216–3219.

LAIMINS, L., RHOADS, D. B., and EPSTEIN, W. (1981). Osmotic control of Kdp operon expression in *Escherichia coli*. *Proceedings of the National Academy of Sciences USA* 78:464–468.

LANCASTER, J. R. (1982). Mechanism of lactose-proton cotransport in *Escherichia coli*: Kinetic results in terms of the site exposure model. *FEBS Letters* 150:9–18.

LANYI, J. K. (1979). The role of Na^+ in transport processes of bacterial membranes. *Biochimica et Biophysica Acta* 559:377–397.

MACNAB, R. (1978). Bacterial motility and chemotaxis: The molecular biology of a behavioral system. *CRC Critical Reviews in Biochemistry* 5:291–341.

MACNAB, R., and AIZAWA, S. I. (1984). Bacterial motility and the bacterial flagellar motor. *Annual Review of Biophysics and Bioengineering* 13:51–83.

MANSON, M., TEDESCO, P., and BERG, H. C. (1980). Energetics of flagellar rotation in bacteria. *Journal of Molecular Biology* 138:541–561.

MANSON, M., TEDESCO, P., BERG, H. C., HAROLD, F. M., and VAN DER DRIFT, C. (1977). A protonmotive force drives bacterial flagella. *Proceedings of the National Academy of Sciences USA* 74:3060–3064.

MARTONOSI, A. N., Ed. (1982). *Membranes and Transport*, two vols. Plenum, New York.

MEASURES, J. C. (1975). Role of amino acids in osmoregulation of non-halophilic bacteria. *Nature (London)* **191**:144–148.

MITCHELL, P. (1976a). Op. cit., Chapter 3.

OVERATH, P., and WRIGHT, J. K. (1983). Lactose permease: A carrier on the move. *Trends in Biochemical Sciences* **8**:404–408.

PADAN, E., ZILBERSTEIN, D., and SCHULDINER, S. (1981). pH homeostasis in bacteria. *Biochimica et Biophysica Acta* **650**:151–166.

PAGE, M. G. P., and WEST, I. C. (1982). Alternative-substrate inhibition and the kinetic mechanism of the β-galactoside/proton symport of *Escherichia coli*. *Biochemical Journal* **204**:681–688.

PAYNE, W. J. (1970). Energy yields and growth of heterotrophs. *Annual Review of Microbiology* **24**:17–52.

POSTMA, P., and ROSEMAN, S. (1976). The bacterial phosphoenolpyruvate: Sugar phosphotransferase system. *Biochimica et Biophysica Acta* **457**:213–257.

RITTENBERG, S. C., and HESPELL, R. B. (1975). Energy efficiency of intraperiplasmic growth of *Bdellovibrio bacteriovorus*. *Journal of Bacteriology* **121**:1158–1165.

ROSEN, B. P., Ed. (1978). *Bacterial Transport*. Dekker, New York.

RYDSTRÖM, J. (1977). Energy-linked nicotinamide nucleotide transhydrogenase. *Biochimica et Biophysica Acta* **463**:155–184.

SAIER, M. H., JR. (1977). Bacterial phosphoenolpyruvate: Sugar phosphotransferase systems: structural, functional and evolutionary interrelationships. *Bacteriological Reviews* **41**:856–871.

SANDERS, D., HANSEN, U.-P., GRADMANN, D., and SLAYMAN, C. L. (1984). Generalized kinetic analysis of ion-driven cotransport systems: A unified interpretation of selective ionic effects on Michaelis parameters. *Journal of Membrane Biology* **77**:123–152.

SKULACHEV, V. P. (1978). Membrane-linked energy buffering as the biological function of the Na^+/K^+ gradients. *FEBS Letters* **87**:171–179.

STOUTHAMER, A. H. (1978). Op. cit., Chapter 4.

STOUTHAMER, A. H., and BETTENHAUSSEN, C. (1973). Utilization of energy for growth and maintenance in continuous and batch cultures of microorganisms. *Biochimica et Biophysica Acta* **303**:53–70.

TEMPEST, D. W. (1978). The biochemical significance of microbial growth yields: A reassessment. *Trends in Biochemical Sciences* **3**:180–184.

TEMPEST, D. W., and NEIJSSEL, O. M. (1984). The status of Y_{ATP} and maintenance energy as biologically interpretable phenomena. *Annual Review of Microbiology* **38**:459–486.

TENBRINK, B., and KONINGS, W. N. (1982). Electrochemical potential gradient and lactate concentration gradient in *Streptococcus cremoris* cells grown in batch culture. *Journal of Bacteriology* **152**:682–686.

THAUER, R. K., JUNGERMANN, K., and DECKER, K. (1977). Op. cit., Chapter 4.

WEST, I. C. (1980). Energy coupling in secondary active transport. *Biochimica et Biophysica Acta* **604**:91–126.

WRIGHT, J. K., RIEDE, I., and OVERATH, P. (1981). Lactose carrier protein of *Escherichia coli*: Interaction with galactosides and protons. *Biochemistry* **20**:6404–6415.

WRIGHT, J. K., and SECKLER, R. (1985). The lactose/H^+ carrier of *Escherichia coli*: *lac* Y^{un} mutation decreases the rate of active transport and mimics an energy-uncoupled phenotype. *Biochemical Journal* **227**:287–297.

ZAHL, K. J., ROSE, C., and HANSON, R. L. (1978). Isolation and partial characterization of a mutant of *Escherichia coli* lacking pyridine nucleotide transhydrogenase. *Archives of Biochemistry and Biophysics* **190**:598–602.

Vestiges of Evolution

The moving finger writes, and, having writ,
Moves on: nor all your piety nor wit
Shall lure it back to cancel half a line,
Nor all your tears wash out a word of it.

E. Fitzgerald, *The Rubaiyat of Omar Khayyam*

• • •

In the Beginning Was the Membrane
Oxygen, Respiration, and Photosynthesis
Molecular Phylogeny and Archaebacteria
The Rise of the Eukaryotes

Physics and chemistry seek to formulate universal laws that describe the properties of matter and energy wherever and whenever they occur. The generalizations of biology have a different status, for they are conditioned by historical factors that have no place in physical science. Every living thing descends from a chain of progenitors through modifications made by chance and passed through the filter of natural selection. Consequently, biological processes obey the laws of matter but, as a matter of principle, would not be deducible from these laws even if they were known to us in their entirety. In order to understand why things are as we find them we must trace the historical connections by which they arose.

Evolution has been likened to a tapestry, woven upon the loom of innovation and persistence. We tend to focus attention on the former, the continuous generation of novelty by mutation, duplication, and rearrangement of the genetic material. But the creativity of evolution, our sense that the living world unfolds in an orderly manner, depends as much on the stability of the basic patterns once they have been established. The retention of tetrapod organization by all animals since the first amphibians is a familiar instance of the principle that the generation of novelty is constrained, indeed canalized (Riedl, 1978), by what went before. The conservatism of evolution is exemplified still more dramatically in the persistence of basic cellular mechanisms, which have presumably undergone little change since they were set several billion years ago. It is not likely that ours is the only, or even the best, possible world. Rather, there are barriers that restrict the exploration of possible alternatives, and these grow higher as the millennia pass. Mutation and recombination can work only on the genetic information that lies to hand. Natural selection is exercised not on individual gene products but on the whole organism. Therefore only those genetic changes are acceptable that result in gene products judged superior by some functional criterion (or, at the least, selectively neutral) and also compatible with existing structures and functions. For any given process or molecule, the more deeply embedded it is in cellular operations, the more refractory it will be to modification by later mutation and selection.

For the purpose of unraveling the evolution of bioenergetic processes, little help is to be got from the fossil record. However, living organisms embody a far richer lode of genealogical information in the distribution and structure of their own constituents, macromolecules in particular. Speculation and analogy still play a large part in all reconstructions of the early phases of evolution, but these are now being supplemented with more objective information derived from the sequencing of proteins and nucleic acids. The fresh data lead one to question much of the received tradition concerning the origin of life and its history during the first three billion years.

In the Beginning Was the Membrane

In a seminal paper published nearly two decades ago, H. J. Morowitz (1967) attempted to define the minimum requirements for a self-replicating biological system and thus to gather some clues to the endowments of protobionts ancestral to present-day life. His reasoning was based in part on the principle that ubiquity implies antiquity. Universal constituents such as DNA, polypeptide enzymes, ribosomes, transfer RNA, ATP, NADH, and phospholipid membranes must have arisen early and set a lower limit to possible size and complexity. Based on what was known at the time, Morowitz suggested a minimum cell with 45 biochemical functions and a diameter on the order of 0.1 micrometers. Both numbers would be larger today because we can discern additional functions that are ubiquitous and logically required.

Morowitz was writing before the implications of a membrane-bound protocellular structure had been fully explored. Let us for the sake of argument adopt the conventional view that the earliest organisms were anaerobic chemotrophic prokaryotes, not unlike modern clostridia. They would have drawn both energy sources and precursors for biosynthesis from a primordial soup of carbon and nitrogen compounds generated by abiogenic processes; energy metabolism was fermentative, this being the simplest pattern, and they lacked cell walls. Now, most fermentation products are acidic (for example, lactic acid) and must be disposed of without acidifying the cytoplasm; passive efflux is insufficient, since phospholipid bilayer membranes are virtually impermeable to protons and even organic acids would be largely retained. The implication is that protobionts must have been equipped with porters to catalyze the efflux of metabolic wastes and with an outward-directed proton pump. Raven and Smith (1976, 1982) have argued persuasively that this was the original function of the F_1F_0 ATPase. It is certainly remarkable that this sophisticated enzyme or one much like it has been found in all prokaryotes examined so far and in mitochondria and chloroplasts as well; it may prove to be a universal cell constituent. Incidentally, F_1F_0 ATPases are electrogenic, and net expulsion of protons requires concurrent movement of other ions, usually an exchange of H^+ for another cation. If this was true from the beginning, selective cation permeability must have been another original endowment.

Cellular organization also entails the problem of osmotic stability. Macromolecules and metabolites enclosed within a relatively impermeable membrane give rise to an osmotic influx of water that swells the membrane and may burst it. The problem will be exacerbated by the presence of a proton pump that imposes a negative membrane potential, with consequent accumulation of cations. Modern bacteria cope by means of an external cell wall that restrains swelling. Wall-less protobionts, like animal cells today, may have achieved stability by effectively excluding a major medium constituent from the cytoplasm. This could only have been Na^+, as there is insufficient K^+ in natural environments. It seems quite possible that the universal asymmetry of intracellular K^+ and extracellular Na^+ has its roots in the maintenance of cellular volume. Mechanisms for the selective accumulation of K^+ and extrusion of Na^+ may even have antedated the fixation of the ribosomal mechanism of protein synthesis, which requires K^+. In any event, a primary or secondary sodium pump, in addition to a proton pump of some sort, seems to be a necessary condition for cellular stability

(Wilson and Lin, 1980). It appears, then, that the stage was set from the beginning for energy transduction by ion currents.

The sequestration of metabolism in closed vesicles entails a need for additional transport systems. Inorganic phosphate often proves to be a growth-limiting nutrient in nature, thanks to its propensity to form insoluble precipitates with ferric iron and other metals; transport systems that accumulate P_i may have been a prerequisite for metabolic energy conversion based on ATP. But P_i will also precipitate with calcium ions, particularly at alkaline pH. Calcium is abundant in natural waters and, however impermeant, must accumulate to inordinately high levels in any vesicle with a negative membrane potential. The capacity to expel Ca^{2+} is probably another sine qua non of cellular organization and a universal feature of both prokaryotes and eukaryotes. Finally, once the concentration of organic compounds in the primordial ocean fell to the micromolar level, pumps and porters would obviously have conferred great selective advantages upon their possessor.

The minimum requirements of any living system certainly include at least a rudimentary apparatus for the replication and expression of genetic information. These are prerequisites for evolution by mutation and natural selection, which many hold to be the very essence of the living state. Some sort of energy metabolism, together with a few anabolic enzymes, is likewise required, if only because the supply of activated precursors for the synthesis of proteins and nucleic acids obtainable from the primordial broth must surely have been limited. To make a believable protobiont, these must be encapsulated in a boundary membrane that is not merely "semipermeable" but capable of sustaining ion gradients and that is fortified with a basic kit of vectorial ion-translocating catalysts. Only when this stage of complexity had been reached could natural selection assess the quality, or fitness, of gene products; Darwinian evolution depends on cellular organization. It follows that the heart of the abiding mystery of the origin of life is not the abiogenic origin of genes and proteins; it is the spontaneous generation of cells.

This is hardly a resounding conclusion, but it is at variance with the impression one obtains from the literature of primordial evolution. With few exceptions, those who fish in Darwin's warm little pond hope to discover the manner in which self-replicating informational macromolecules may have arisen before there were cells. The conceptual framework was first formulated by J. B. S. Haldane and A. I. Oparin more than half a century ago and has since been extended and modified but not substantially altered (Oparin, 1938; Miller and Orgel, 1974; Fox and Dose, 1977; Crick, 1981). Briefly, it is held that a great variety of small molecules containing carbon, nitrogen, sulfur, and phosphorus were produced by the action of electric discharges or ultraviolet light from the sun impinging on the earth's primitive atmosphere. These accumulated in the primitive ocean, attaining concentrations in the micromolar or millimolar range. Complex structures, including macromolecules, arose either in the ocean or in locales that intermittently communicated with it, such as lagoons, volcanic vents, or the surface of clay particles. Now, polymerization reactions involve the removal of water and cannot be expected to occur spontaneously in aqueous solution. One must therefore envisage the local concentration of monomers increasing by freezing, drying, or heating or else assume a supply of activated precursors. Only later, after self-reproducing macromolecules made their appearance, would they somehow be compartmentalized into primitive cells, possibly by coacervation. A recent exemplar is the ambitious scheme advanced by Eigen and Schuster (Eigen et al., 1981; Eigen and Schuster, 1982), which derives the origin of genetic information from the rules that govern the chemistry and kinetics of RNA replication. This calls for a steady supply of nucleoside triphosphates and assumes that encapsulation of the evolving genetic machinery took place after coupled transcription and translation had arisen; the implication is that formation of cells, while not exactly trivial, is not the major issue.

I strongly suspect that this conception is flawed in principle: cellular organization, far from an afterthought, must have been from the beginning part and parcel of the origin of life. The vital force, that *vis vitae* which will not be exorcised without proper explanation, has its roots in the astonishing degree of organization that pervades the living world from the molecular level to the organismic and societal. Biological order must be maintained by a continuous flux of energy. Therefore a believable biopoietic scheme is one that creates mounting levels of biological order naturally, by providing the means to convert the flux of energy into the organization of matter. This seems to me inconceivable without compartments. Only with the aid of membranes can one rationalize, for example, the coupling of some exergonic reaction to the endergonic production of macromolecules. Primordial metabolism may conceivably have been based on the degradation of preformed organic compounds, at least for a little while until the supply was exhausted. However, like C. R. Woese (1979) in his trenchant critique of the Oparin-Haldane hypothesis, I prefer to believe that the first cells developed from sources, not sinks, of prebiotic chemistry. Sunlight is today the ultimate driving force for virtually all life, and therefore an early version of photosynthesis remains the most plausible source of biological organization (chemolithotrophy is an alternative possibility). And it is significant in this context that a closed vesicle is not merely a compartment that sequesters metabolites and genomes, it is also an elementary unit of biological energy transduction.

In this paper Woese (1979) also attempted to develop an explicit alternative proposal for the primal sequence that is cellular *ab initio*. In a radical departure from convention, he rejects the global broth of precursor molecules. Cosmological arguments suggest that the primitive earth was too hot to hold an ocean of liquid water. Instead its atmosphere, like that of Venus today, may have been laden with great clouds consisting in part of droplets of salty water. It is in these droplets, or rather at their surfaces, that prebiotic chemistry would have taken place, including the formation of lipid membranes and macromolecules. Just how this happened is not spelled out, nor does Woese's iconoclastic notion help one to understand whence came genetic information; but it does represent a first attempt to conceive of the origin of life in a physical state that would naturally lead to cellular order.

In truth, despite the expenditure of much effort and ink, genesis has lost little of its mystery. In the absence of a realistic model that describes the generation of protobionts of substantial complexity, we have no plausible scenario for the origin of life. That life did spontaneously originate on earth is a necessary article of faith if the subject is to be explored experimentally, but how it came about passes understanding.

Oxygen, Respiration, and Photosynthesis

One of the few firm facts about the early evolution of life is that it was accompanied by drastic changes in the composition of the earth's atmosphere. The atmosphere of the early earth had virtually no free oxygen except for traces arising from the photodissociation of water vapor. It may have been composed predominantly of CO_2, along with smaller amounts of H_2, H_2S, CO, and N_2, and was mildly reducing (Cloud, 1983; Schopf, 1983). (The earlier belief in a strongly reducing atmosphere dominated by methane and hydrogen gas has been largely abandoned.) The first living organisms, whose fossil remains have been found in rocks some three billion years old, were almost certainly anaerobes that obtained energy from photosynthesis or from such geochemical processes as the reduction of CO_2 with H_2 to produce methane. The modern atmosphere, containing little free CO_2 but 20 percent oxygen and strongly oxidizing, was produced by the metabolic activities of living things, first cyanobacteria and other oxygen-generating prokaryotes and later algae and green plants (Berkner and Marshall, 1965). The transition from a reducing atmosphere to an oxidizing one must have been one of the potent forces that shaped the evolution of the metabolism of cells.

Figure 6.1 depicts in broadest outline the present conception of the early history of life in relation to that of the atmosphere (Cloud, 1974, 1983; Margulis, 1981; Schopf, 1983). The earth itself is now thought to be some 4.5 billion years old. Rock formations in the Canadian shield and elsewhere, reliably dated from 1.7 to 2.5 billion years ago, contain a profusion of well-preserved microfossils of generally prokaryotic structure. These are the most ancient definitive remains of living organisms, but it is very likely that a far earlier microflora is represented by the Pilbara formation in Australia, some 3.5 million years old. The recent claim that traces of life remain in 3.8-million-year-old rocks found in Greenland may be questionable, but even so the antiquity of microbial life is beginning to approach that of the habitable earth itself. Particular importance has been attached to structures termed stromatolites. These are the fossil remains of thick microbial mats, very similar in appearance to those formed by communities of cyanobacteria and other prokaryotes in tropical waters today. The most ancient

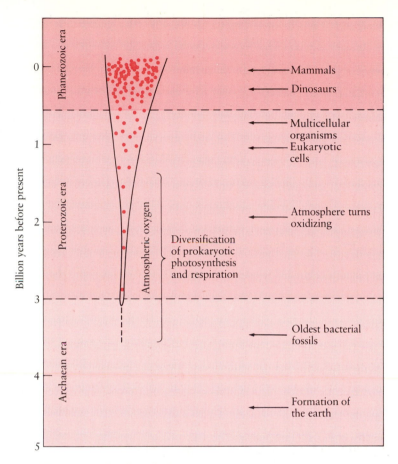

FIGURE 6.1   The time scale of evolution.

definitive stromatolites come from the Canadian cherts (2.5 billion years), but there are putative ones in the Pilbara assemblage. Also on morphological grounds many microfossils at least two billion years old have been identified as cyanobacteria. We may therefore, with due caution, take it that cyanobacteria arose 2.5 or even three billion years ago, and with them, presumably, arose the oxygen-generating mode of photosynthesis that was to exert so profound an influence on all subsequent forms of life. Finally, a number of deposits between 0.9 and 1.3 billion years old contain microfossils of cells 20 micrometers or more in diameter, some with traces of cell walls and dividing nuclei. These are the chief basis for placing the origin of eukaryotic cells at about 1.4 billion years ago. Metazoan traces appear late in the Proterozoic era, and with the opening of the Cambrian epoch a mere 600 million years ago, the fossil record burgeons with a profusion of forms, many endowed with hard parts, and the familiar succession of higher animals and plants begins. Note that the evolution of the simplest organisms, the prokaryotes and the early eukaryotes, took up the major part of evolutionary time; the subsequent proliferation of living forms was comparatively rapid.

The trouble with charts of this sort is that their time scale is so utterly remote from human experience as to be incomprehensible except by analogy. A graphic one has been provided by Crick (1981). Suppose that the length of this book corresponds to all 4500 million years of the earth's history. Each page, then, equals roughly eight million years, each line some 150,000 years, and each letter nearly 1500 years. The contents of all the fossil cabinets that grace our museums were written down in the last 80 pages, all of recorded history in the final four letters. By contrast, the evolution of life from hypothetical protobionts to the first eukaryotic cells occupies more than 400 pages. They are nearly blank, but traces of the gradual transformation of the atmosphere from reducing to oxidizing can be read both in the rocks and in the metabolic structure of contemporary organisms.

Geological evidence comes from datable mineral deposits, particularly those of iron. Much of the iron ore mined commercially is found in sedimentary deposits consisting of alternating iron-rich and iron-poor layers. These are called banded iron formations and are thought to have arisen from the local deposition of ferric iron at a time when the general oxygen level was very low. They may therefore chronicle the slow, fitful but relentless accumulation of biological oxygen in favorable locales. Banded iron formations extend into the upper Proterozoic but are replaced in younger strata by uniformly oxidized deposits of ferric iron, the red beds. The time of transition from a generally reducing atmosphere to an oxidizing one has been dated by Cloud (1983) to about two billion years ago (some authorities put it considerably earlier). Just how much free oxygen was present at this time is not known; a reasonable guess is 1 percent of the present oxygen level. Presumably both oxygen-generating photosynthesis and some aerobic respiratory pathways had developed well before the transition point.

Contemporary prokaryotes also retain marks of having evolved at a time when the environment was quite unlike the present one. Bacterial metabolism is fundamentally anaerobic, in the sense that oxygen is not involved in the formation of their basic constituents. Sterols, unsaturated fatty acids, and ascorbic acid, whose biosynthesis depends on oxygen, are absent, and biosynthetic pathways avoid oxygen-requiring hydroxylations. Exceptions, such as the phycocyanin pigments of cyanobacteria, which are produced by an oxygen-linked cleavage of chlorophyll, would have evolved only after oxygen became available. To be sure, many bacteria use oxygen as the terminal electron acceptor in respiration. But the very diversity of their respiratory chains (Figure 4.3) suggests that they evolved and diverged, probably from multiple origins, during the progressive buildup of atmospheric oxygen.

In marked contrast all eukaryotes are aerobes, the great majority obligate aerobes, and incorporate constituents whose biosynthesis requires oxygen: sterols, polyunsaturated fatty acids, ascorbic acid, chlorophyll *b*. Mitochondrial respiratory chains represent a small subset of the diversity found among the prokaryotes. It is not likely, as Berkner and Marshall (1965) thought, that eukaryotes evolved in response to the crisis engendered by the rise in atmospheric oxygen, but an oxidizing atmosphere was clearly a prerequisite for this momentous evolutionary development. By the time eukaryotes appear in the fossil record, a billion or so years ago, the major pathways of biological energy transduction had been established: they all belong to the first two billion years, the age of the prokaryotes.

Discussions of the origin and early development of energy-transducing mechanisms are inevitably speculative and burdened by assumptions about the metabolic patterns of the first protobionts. There is, however, one noteworthy feature of present-day systems that offers a measure of objective guidance. That is the marked similarity between the proton-translocating redox chains serving bacterial respiration and photosynthesis, particularly in the ubiquinone–cytochrome-*b* segment. In the extreme case, such as *Rhodopseudomonas* and its relatives, a given ubiquinone–cytochrome-*b* complex can participate in either respiration or photosynthesis, depending on how the cells were grown. *Chromatium* and *Chlorobium*, strict anaerobes and phylogenetically remote, still utilize a sequence of quinones and *b*-type cytochromes akin to that found in respiring bacteria. The implication that respiratory and photosynthetic electron transport share a common ancestry (rather than illustrating a remarkable degree of convergent evolution) is confirmed by the amino acid sequences of *c*-type cytochromes and of ferredoxins (Schwartz and Dayhoff, 1978; Dickerson, 1980). Since significant levels of oxygen only became available with the advent of oxygen-producing photosynthesis, it has been argued that photosynthesis preceded respiration and that aerobic organisms descend from photosynthetic ones with the loss of light-harvesting pigments (Uzzell and Spolsky, 1974; Broda, 1975).

There are genealogies that can be interpreted in this manner, for instance the descent of the nonphotosynthetic aerobe *Paracoccus* from what appear to be closely related members of the photosynthetic purple bacteria (Figure 6.2). But in general the thesis that respiration arose from photosynthesis by loss of functions is probably a simplistic one that would confine the many braided currents of evolution to a single swift channel. We are more cognizant now of the existence of a variety of anaerobic proton-translocating redox chains with sulfate or fumarate as acceptors (Chapter 4). Primitive "prerespiratory" pathways transferring electrons or reducing equivalents to extracellular oxidants may have developed quite early and served as the basis for an extensive filiation of both respiratory and photosynthetic redox chains (Broda and Peschek, 1979). A lengthy, convoluted history with multiple points of origin seems more probable than a single line of descent and is also in better accord with the newer findings of bacterial phylogeny, which will be discussed below.

Of the origins of biological energy transduction, no vestiges whatsoever have been identified. This has been no hindrance to speculation, starting from the conventional belief that the earliest organisms were chemotrophs immersed in a dilute soup of abiogenic organic compounds. On this premise it has been suggested that prerespiratory electron transport developed in response to selective pressures arising from the need to maintain cellular redox balance and later fused with light-harvesting pigments (Broda and Peschek, 1979; Gest, 1980; Wilson and Lin, 1980; Raven and Smith, 1981). As to the origins of chlorophyll itself, plausible lines of descent are not hard to imagine. Porphyrins are not universal among bacteria (they are absent, for instance, from most clostridia and methanogens), but the structurally related corrinoid pigments appear to be ubiquitous. The recent discovery that a nickel-containing tetrapyrrole plays the central role in the reduction of CO_2 to methane (see below) offers another attractive precursor. I shall refrain from pursuing the argument; what is needed is not more speculation, however closely reasoned, but pertinent data.

Molecular Phylogeny and Archaebacteria

Biological molecules and processes do not evolve as such; only organisms are subject to variation and selection. Just as we trace the evolution of the running foot in the descent of the horse, so we look to bacterial phylogeny for evidence about the origins of bioenergetic mechanisms. Unfortunately this has long been an intractable subject. A distinguished microbiologist once likened the bacterial world to a gigantic tree shrouded in dense fog that leaves only the tips of its branches exposed to view. In the absence of an informative fossil record, bacterial affinities have perforce been inferred solely from their shapes and biochemical characteristics. During the past decade, however, C. R. Woese and his colleagues have gone far toward working out a more objective phylogeny from comparisons of the nucleotide sequences in ribosomal RNA. This approach has produced a revolutionary insight. In place of a single kingdom Monera, containing all the prokaryotic organisms, there appear to be two entirely different lines of prokaryotic descent, designated eubacteria and archaebacteria. These share the universal genetic code but differ in many facets of their biochemistry and physiology, and they probably separated even before the apparatus for genetic transcription and translation had been fully fixed. Chloroplasts and mitochondria are both phylogenetic offshoots from the eubacterial stem. Eukaryotes, as judged by the RNA of their cytoplasmic ribosomes, did not evolve from either prokaryotic stem. They represent a third, independent line of descent from the universal common ancestor of all life, for which Woese has proposed the name "urkaryote" (Fox et al., 1980; Woese, 1981; Gray and Doolittle, 1982). These discoveries have profound implications for the early evolution of bioenergetics and for the subsequent origin of the eukaryotic cell.

Very briefly, to classify an organism purified ribosomal RNA (typically 16S and 18S RNA of prokaryotes and eukaryotes, respectively) is digested with T_1 ribonuclease. The resulting oligonucleotides are separated by electrophoresis, and the sequence of every fragment six or more nucleotides long is determined. We can think of each sequence as a word and of the entire oligonucleotide catalog as a dictionary of the RNA "language" of this particular organism. Related organisms will share many words, as Dutch and German do; Spanish would correspond to a more distant member of the family, Persian to one still more remote. A totally unrelated language (say, Chinese) would have no words in common with that of our test organism but would fall into an altogether separate cluster. The degree of relatedness can be expressed mathematically as an association coefficient, which bears some relationship to the time elapsed since two organisms diverged from their common ancestor.

It is rather remarkable that genealogical lines can be made out at all. What with divergence over billions of years plus gene transfer by viruses, episomes, and free DNA, one might have expected all traces of pedigree to have been obliterated. Evidently ribosomal RNAs, whose function has remained unchanged from the beginning, diverge slowly enough to clock the millennia. However, one cannot take it for granted that more peripheral functions (cytochrome oxidases, for instance) have not undergone redistribution by gene transfer. Unfortunately there is not yet any satisfactory way of translating association coefficients into years elapsed since the bifurcation; one of the fundamental difficulties is that the rates of RNA divergence are not everywhere the same.

The Primary Kingdoms.   Figure 6.2 provides an overview of the major lines of prokaryotic descent, with some emphasis on organisms that have received attention from students of bioenergetics. We will focus here on the two prokaryotic kingdoms, leaving the nature and origin of eukaryotes for a later section.

The eubacterial kingdom includes all the familiar organisms that have shaped our conception of the microbial world. It is divisible into at least eight major groupings, which appear to have separated over a relatively short period of time, so that their order of branching is uncertain. Several aspects are pertinent to our inquiry into the origin of energy-transducing mechanisms.

1.   Photosynthesis clearly holds an important place in the radiation of the eubacteria. Photosynthetic forms are intermingled with respiratory ones, particularly in the large assemblage of gram-negatives that includes both the purple bacteria and *E. coli*. A noteworthy instance is the close relationship between *Rhodopseudomonas* and *Paracoccus,* which had previously been inferred from the amino acid sequence of their cytochromes; we shall return to this in the context of mitochondrial origins. Taxonomy thus reinforces the long-held belief that some respiratory chains descended from photosynthetic ones by the loss of

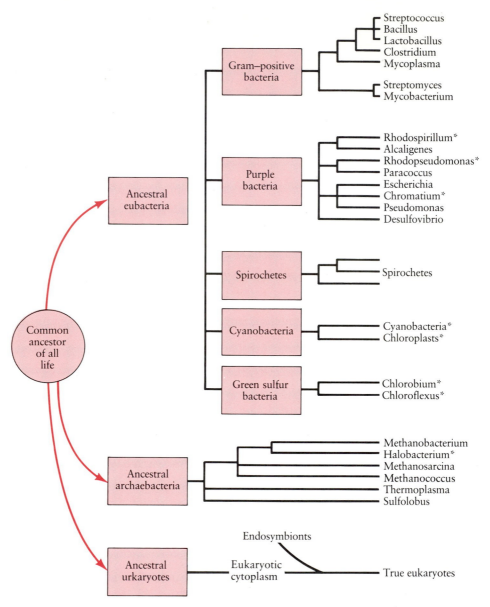

FIGURE 6.2   The major lines of prokaryotic evolution. The scheme shows three independent lines of descent from the hypothetical universal ancestor of all life. The eubacteria fall into at least eight major groups, of which five are shown. The archaebacteria comprise a second prokaryotic kingdom, also highly diversified. Asterisks mark photosynthetic organisms. The urkaryotic line is represented today by the cytoplasmic ribosomes of eukaryotes; fusion of these protoeukaryotes with endosymbionts gave rise to the true eukaryotes. (After Fox et al., 1980; copyright 1980 by the American Association for the Advancement of Science.)

the light-harvesting pigments. This must have happened many times; there are at least three independent instances in this one group.

2.   Photosynthetic bacteria are deeply split into two ancient lines of green sulfur bacteria and three of purple bacteria; their separation from each other and from the line that leads to cyanobacteria and chloroplasts goes back to the original common ancestor of all eubacteria. Photosynthesis, then, must be at least as ancient as the eubacteria are. Somewhat surprisingly, contemporary cyanobacteria form a tight cluster of significantly lesser antiquity, little greater than that of strict aerobes such as *Pseudomonas* and *Bacillus*. This conflicts with the prevailing belief that cyanobacteria akin to those we know today were responsible for generating the oxidizing atmosphere. Could it be that the microfossils and stromatolites that are so dominant a feature of deposits 2.5 billion years old were left by organisms other than cyanobacteria? Or that these remains testify to a line of photosynthetic organisms ancestral to cyanobacteria that survives today only in chloroplasts?

3.   Strictly anaerobic chemotrophic forms appear to have arisen independently in several lines. In particular, clostridia and their relatives are an ancient tribe, but there is no clear indication of exceptional seniority. This is again reason for some skepticism regarding the dogma that the earliest protobionts were chemotrophs of the clostridial kind.

The kingdom Archaebacteriae comprises a much smaller number of genera, all of which inhabit peculiar and extreme environments. Three major groups have been recognized: the methanogenic bacteria, strict anaerobes that obtain energy from the oxidation of H_2 with CO_2 to produce methane; the extreme halophiles; and thermoacidophiles, which flourish in environments of high temperature and low pH (Figure 6.2). All of these have long been known to be unusual in one respect or another, but it was taken for granted that their eccentricities reflected adaptation to their stressful environments. In 1977, when the revised classification of bacteria was first proposed, T_1-oligonucleotide catalogs seemed a flimsy foundation on which to erect a new primary kingdom.

The accumulation of new data has considerably solidified the belief that archaebacteria and eubacteria are indeed alternative kinds of prokaryotic organization. Archaebacteria feature a variety of cell walls, but muramic acid, the hallmark of eubacterial walls, is always absent. Plasma membranes are constructed of a novel class of lipids, phosphorylated diethers of glycerol and phytanols. (The latter are long-chain, branched alcohols structurally related to isoprene.) Eubacteria, by contrast, contain orthodox phospholipids made of the glycerol esters of straight-chain fatty acids. Archaebacterial transfer RNAs differ from those of eubacteria with respect to the location of certain modified bases; the subunit structure of their RNA polymerase is also distinctive. Finally, methanogenic bacteria boast an unusual spectrum of coenzymes; whether these occur in other archaebacteria as well is not yet known.

It is impossible to assign a date to the divergence of the archaebacteria from the line that was to generate today's profusion of eubacterial forms, but it must have happened early. Archaebacteria share the universal genetic code, but their idiosyncratic molecular biology suggests that the mechanisms of transcription and translation were still fluid. By the same token, their unusual habitats may bespeak a time when the typical environmental niche was very different from what would be so considered today. Archaebacteria may therefore hold clues to the patterns of energy metabolism that prevailed when the earth was young.

Methanogenic Bacteria.   Methanogens are strict anaerobes found in such places as the rumen of ungulates and the sediments of stagnant swamps, where they are responsible for the production of marsh gas. They are chemolithotrophs, producing all their organic constituents from CO_2 by uncommon pathways that are only partially understood (Fuchs and Stupperich, 1982). The requisite metabolic energy comes from the reduction of CO_2 by molecular hydrogen to methane:

$$CO_2 + 4H_2 \rightleftharpoons CH_4 + 2H_2O \qquad \Delta G^{\circ\prime} = -31 \text{ kcal/mol} \, (-131 \text{ kJ/mol})$$

(Eq. 6.1)

Under physiological conditions, with H_2 concentrations in the micromolar range, the free-energy change is considerably less, suggesting that no more than one mole of ATP can be formed per mole of CH_4 produced (Thauer et al., 1977).

Methanogens are a deeply diversified group and come in a variety of forms. Their metabolic processes are of the utmost interest, for two reasons. First, methane production involves coenzymes that have not so far been encountered in any other organism. Second, there is good reason to believe that CO_2 and H_2, produced by volcanic outgassing, were major constituents of the earth's metabolic economy at a very early stage of biological history, conceivably the primordial one that supported the rise of ancestral protobionts (Woese, 1979).

The molecular mechanism of methane production has been extensively studied in the laboratories of R. S. Wolfe, R. K. Thauer, and others but remains uncertain because almost every step is unprecedented (see the reviews by Thauer et al., 1977; Wolfe, 1979; Vogels et al., 1982; Daniels et al., 1984). The reduction of CO_2 to methane is mediated by an electron-transport system that involves dehydrogenases, electron carriers, and probably four reductases, as suggested in Figure 6.3. Many methanogens contain neither quinones nor *b*- or *c*-type cytochromes. The intermediate stages in CO_2 reduction take place as derivatives of four new coenzymes: CDR, a CO_2-binding factor of unusual methanofuran structure; FAF, a pterin of unknown structure that carries activated formaldehyde; methanopterin, whose place in the scheme is uncertain; and coenzyme M (2-mercaptoethane sulfonate). The terminal step, the reduction of methyl-*S*-coenzyme M to methane, requires a protein that has hydrogenase activity, a cofactor of unknown structure, and the enzyme methyl coenzyme M reductase.

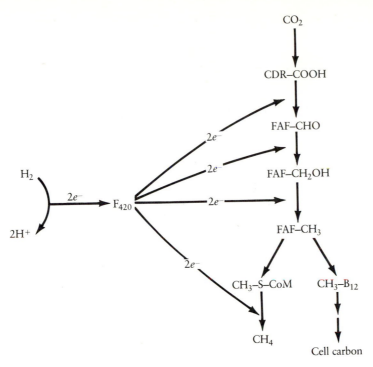

FIGURE 6.3   Methanogenesis: a tentative scheme for electron transport during the reduction of CO_2 by hydrogen gas. CDR, carbon dioxide reduction factor; FAF, formaldehyde activating factor; CoM, coenzyme M; F_{420}, a deazaflavin. (After Daniels et al., 1984.)

The latter has a yellow chromophore, F_{430}, recently shown to be a nickel-containing tetrapyrrole that is again unique.

How the free energy released in the reduction of CO_2 to methane is coupled to ATP generation is the subject of some debate. The enzymes involved have been described as soluble, and substrate-level phosphorylation is not entirely ruled out. Most investigators, however, favor the view that electron transport during methane production is organized vectorially across a membrane, effectively translocating protons across it. Interestingly, however "primitive" methanogens may be from the phylogenetic standpoint, some are structurally quite elaborate. The cytoplasm of *Methanobacterium autotrophicum* is packed with lamellae, and there is evidence that these rather than the plasma membrane may be the locus of both methane production and ATP synthesis. Apparently oxidation of H_2 leads to the generation of a proton potential that supports ATP production via an F_1F_0 ATPase. If the membrane in question is the lamellar one, the newly formed ATP must be exported to the cytoplasm, possibly with the aid of an ATP-ADP translocator as suggested in Figure 6.4 (Doddema et al., 1979;

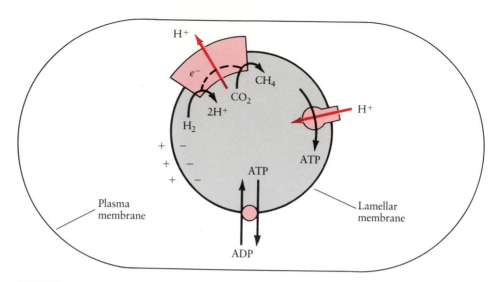

FIGURE 6.4   Proton transport and phosphorylation in methanogenic bacteria: a speculative scheme. The hypothesis proposes that the reduction of CO_2 to methane takes place in intracellular, vesicular lamellae that also house an F_1F_0 ATP synthase. Electron transport is accompanied by proton translocation out of the vesicles, generating a $\Delta\tilde{\mu}_{H^+}$ that drives ATP generation; ATP is exported to the cytoplasm by an ATP-ADP antiporter.

Kell et al., 1981). It must be added that the existence of "methanochondria" is far from being universally accepted; several alternative schemes have been evaluated by Daniels et al. (1984), who prefer a more conventional arrangement of electron-transport chain and ATP synthase within the cell's plasma membrane.

Halobacteria.   A purple sheen on the surface of a brine pond may betray the presence of *Halobacterium halobium* or its relatives, archaebacteria adapted to life in 5 *M* NaCl. Although their T_1-oligonucleotide catalogs suggest that halobacteria are related to the methanogens, the metabolic patterns of the halobacteria are entirely different and display a disconcerting blend of the commonplace and the unique.

Halobacteria are aerobes, and their plasma membranes contain a respiratory chain and an ATP synthase that collaborate in oxidative phosphorylation. (Recent findings suggest that the ATP synthase differs in some respects from the conventional F_1F_0 ATPase; will this prove to be another divergence characteristic of archaebacteria in general?) At low oxygen tension, in the light, halobacteria synthesize two unique light-transducing pigments: bacteriorhodopsin, an electrogenic proton pump; and halorhodopsin, originally reported to be an electrogenic sodium pump but now thought to transport chloride. ATP synthesis relies on the circulation of protons, but the accumulation of metabolites (and possibly

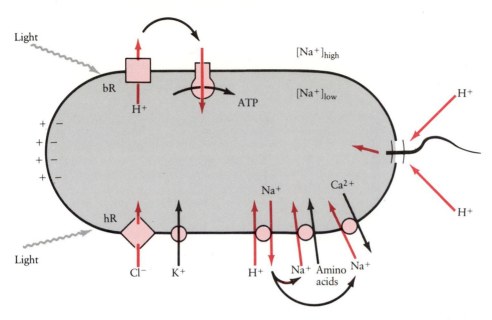

FIGURE 6.5   Ionic relations in *Halobacterium halobium*. Organisms growing under low oxygen tension synthesize two unique light-driven ion pumps: bacteriorhodopsin (bR), an electrogenic proton pump; and halorhodopsin (hR), an electrogenic chloride pump. ATP synthesis is effected by an F_1F_0 ATPase, with protons as the coupling ion; the flagellar motor is also driven by the proton current. Potassium ion accumulation has been attributed to a uniporter; other transport processes, including the accumulation of amino acids and the expulsion of Ca^{2+} ions, are driven by a sodium circulation. This begins with the expulsion of Na^+ from the cells by a secondary Na^+-H^+ antiporter. All stoichiometries are uncertain; the respiratory chain produced by well-aerated cultures has been omitted.

even motility) is effected by a sodium current (Figure 6.5). Thanks to these singular features halobacteria have become popular research organisms that deserve fuller treatment than the present context allows. Recent reviews by Henderson (1977), Lanyi (1978), Eisenbach and Caplan (1979), and by Stoeckenius and Bogomolni (1982) cover the field in detail.

Bacteriorhodopsin was discovered by W. Stoeckenius in the late 1960s in the course of an investigation on the structure of the halobacterial plasma membrane. This contains distinct purple patches that were easily isolated and found to consist of a single protein species with the carotenoid retinal as the chromophore. In 1973, D. Oesterhelt and W. Stoeckenius made the remarkable discovery that bacteriorhodopsin functions as a light-driven proton pump. The protein is situated within its patch of membrane in a vectorial manner: absorption of a photon by the retinal bleaches it, with simultaneous release of protons to the exterior; the original chromophore regenerates spontaneously with the uptake of protons from the cytoplasmic side. The cycle includes a

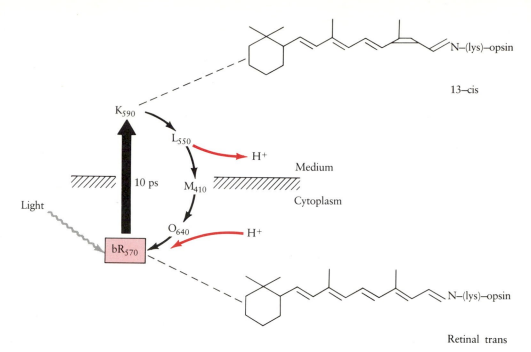

FIGURE 6.6   The photochemical cycle of bacteriorhodopsin. In bacteriorhodopsin, retinal is present in the trans configuration. On absorption of a quantum of light, bacteriorhodopsin is bleached to a pigment designated K (absorption peak at 590 nm), with photoisomerization of retinal about the 13,14 double bond. This is the step in which light energy is captured. Subsequent steps, which are slower and spontaneous, are accompanied by the release of protons to the exterior and their uptake from the cytoplasm. The photochemical cycle traverses the membrane, but the location of the retinal with respect to the membrane is not known.

number of intermediates that have been detected by flash spectroscopy. As depicted in Figure 6.6, the cycle translocates one proton per photon absorbed, but recent data suggest that two protons per cycle may be closer to the mark (Stoeckenius and Bogomolni, 1982). Let me emphasize that neither chlorophyll nor an electron-transport chain is involved in this novel mode of transducing light energy into a protonic potential.

Bacteriorhodopsin is the simplest known proton pump in terms of its protein composition and has attracted intense interest. It consists of but a single polypeptide chain that is bent into seven α-helical regions, each of which spans the membrane (Figure 6.7). A monomer can function as a proton pump, and the significance of the protein's aggregation into crystalline purple patches is not yet known. The mechanism of energy transduction is also still unclear, but there is widespread agreement that the central elements are the chromophoric group, the retinal, and the Schiff base by which the retinal is linked to a lysine residue on the

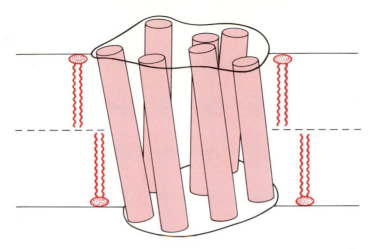

FIGURE 6.7   Drawing showing the arrangement of α helices in a monomer of bacteriorhodopsin in relation to the lipid bilayer. (From Henderson, 1977; reproduced with permission from the *Annual Review of Biophysics and Bioengineering*, vol. 6, by Annual Reviews.)

protein ("opsin"; Figure 6.6). The Schiff base is protonated in the original pigment, designated bR_{570} in Figure 6.6; on absorption of a quantum of light, this is converted successively into the intermediates designated K_{590}, L_{550}, and M_{410}; in the latter the Schiff base is deprotonated, presumably with concurrent expulsion of the proton from the cell. The chemical change that underlies the spectroscopic transitions is probably the light-induced isomerization of the 13,14 double bond, from the all-trans to the 13-cis configuration. The spatial configurations of these two isomers are very different, as indicated in Figure 6.6, with the result that the Schiff base finds itself in a new position that is spatially or chemically altered. If the photochemical cycle spans the membrane, as suggested in Figure 6.6, we can imagine that photoisomerization alters the exposure of the Schiff base from inside to outside or else affects its pK so as to facilitate proton release. Just how this proton gate, buried within the structure of the protein, communicates with the aqueous phases on either side of the plasma membrane is another matter that remains to be settled.

Proton translocation by the bacteriorhodopsin cycle generates a protonmotive force and supports ATP synthesis by the F_1F_0 ATPase. (As a historical aside, one of the most persuasive observations favoring the chemiosmotic hypothesis was made in this context, when Racker and Stoeckenius, 1974, showed that liposomes inlaid with purified bacteriorhodopsin and mitochondrial F_1F_0 ATPase produced ATP on illumination.) However, the transport of metabolites across the plasma membrane is linked not to a proton circulation in the standard bacterial fashion but to a circulation of sodium ions. The driving force for these and possibly other work functions is the electrochemical potential of sodium ions, $\Delta \tilde{\mu}_{Na^+}$ (Equation 4.3). Intact cells maintain a membrane potential of about $-150 \, mV$; they contain about $4 \, M$ KCl but only 0.5 to $1 \, M$ NaCl, so that there is a substantial gradient of sodium concentration between the brine ($5 \, M$ NaCl) and the cytoplasm. The high intracellular KCl concentration is required to keep the cells in osmotic equilibrium with their medium. Together the gradients of K^+ and Na^+ constitute an energy reservoir of high capacity that permits the cells to support some work functions anaerobically, even in the dark (Brown et al., 1983).

How do halobacteria generate $\Delta \tilde{\mu}_{Na^+}$, the sodium electrochemical potential gradient? There is much evidence from research with membrane vesicles that Na^+ is extruded by secondary electrogenic antiport for protons (Lanyi, 1978, 1979). To that extent, the proton circulation retains its status as the primary mode of energy conversion: from light energy to $\Delta \tilde{\mu}_{H^+}$ and thence to $\Delta \tilde{\mu}_{Na^+}$.

This pattern, comfortably familiar from research with other bacteria, has now been challenged by the discovery that halobacteria produce a second energy-transducing pigment called halorhodopsin, which serves as a light-driven primary electrogenic chloride pump (Figure 6.5; Schobert and Lanyi, 1982). The implications of this unprecedented primary transport system are yet to be worked out. Since the chief cytoplasmic osmolite is KCl, halorhodopsin may pump chloride into the cell while potassium ions enter by a uniporter in response to the membrane potential. Very recently a third retinal pigment has been reported; it has not been implicated in transport but may be the light receptor for phototaxis. In any event, it is clear that ionic relations in halobacteria differ even more drastically from those of ordinary bacteria than had been supposed.

So far, so good: if archaebacteria diverged from the ancestral stock so early that even the mechanisms of genetic transcription and translation were still fluid, an unconventional approach to energy coupling may not be out of line. Their possession of an F_1F_0 ATPase and of ion-linked porters would be consistent with the expectation that such ancient transport systems may have been present in the common ancestor of the archaebacteria and the eubacteria. But what is one to make of the fact that under aerobic conditions, halobacteria produce an orthodox respiratory chain complete with cytochrome b, cytochrome c, and an o-type cytochrome oxidase? These constituents have been characterized spectroscopically (for references see Lanyi, 1979), but nothing is known of their molecular structures. A similar puzzle, incidentally, arises with respect to another archaebacterium, the wall-less *Thermoplasma acidophilum,* which likewise contains what appear to be cytochromes b and c (Holländer, 1978). We may have here instances of lateral gene transfer or a strong hint that the divergence of the archaebacteria is not nearly as ancient as is now believed. But it remains possible that cytochromes predate the deep cleavage in bacterial phylogeny and gave rise to a number of quite independent types of respiration; amino acid sequences should eventually settle the issue.

The Rise of the Eukaryotes

Prokaryotes can sustain a biosphere by themselves, without help from any higher organism, and they appear to have done so for the first two billion years of the history of life. The metabolic patterns discussed in this and earlier chapters, in various combinations and permutations, produce organisms fit for any ecological niche—light or dark, aerobic or anaerobic, rich in organic matter or devoid of it. But this is not the world we know. In ours, most environments are dominated by eukaryotes, from single-celled algae and protozoa to beetles, tigers, and trees. Almost half of the world's primary biological productivity is due to marine phytoplankton alone, and most of that is composed of eukaryotic microscopic algae. On land also eukaryotes now constitute the great majority of species and the bulk of total biomass. We must at least conclude that the eukaryotic model is a successful one, and ask how it originated and what makes it superior in many respects to the older prokaryotic version.

The dichotomy between the prokaryotic and the eukaryotic modes of cellular organization was first recognized by E. Chatton in 1938, but its depth was not generally appreciated until the 1960s when R. Y. Stanier and C. B. van Niel collated the numerous differences that had by then become apparent. The divergence that separates the bacteria (including the cyanobacteria) from all other organisms remains, as Stanier said, the greatest single discontinuity to be found in the living world (Figure 6.8). Prokaryotes and eukaryotes share universal traits, such as the central metabolic highways and the genetic code. However, we now recognize that even the organization of their genetic machinery is not the same, and they seem to differ ever more markedly in cellular structure.

From the bioenergeticist's point of view, two differences stand out. In bacteria, the machinery for the production of useful energy is generally located in the plasma membrane (there are noteworthy exceptions, such as the thylakoids of cyanobacteria); eukaryotes assign this function to discrete intracellular organelles, the mitochondria and chloroplasts. The molecular mechanisms by which mitochondria and chloroplasts produce ATP are so similar to those of prokaryotes that a relatively direct descent of the former from the latter can hardly be doubted. Eukaryotes invented no major new means of energy production. What they did invent was a rich variety of molecules and mechanisms for the performance of useful work that have not (so far) been demonstrated in any prokaryote. Actin, myosin, microtubules, 9 + 2 flagella, mitosis, endocytosis and exocytosis, intracellular transport (not to mention true nuclei, the Golgi apparatus, and endoplasmic reticulum), all testify to a radically different cytoplasmic organization. Cells endowed with these capabilities were released from the restrictions on maximum size that prokaryotic structure engenders and could explore novel ways of harvesting resources and coping with the vicissitudes of the environment.

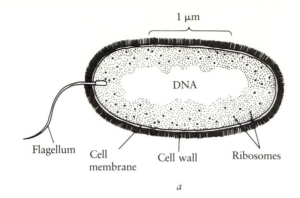

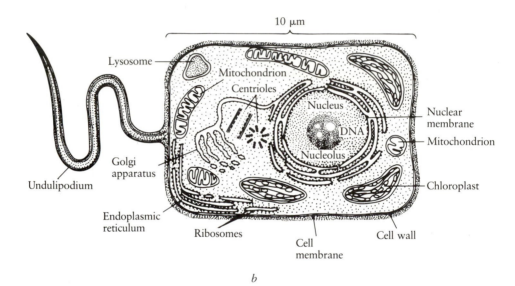

FIGURE 6.8   Major structural features of prokaryotic and eukaryotic cells. (After "Archaebacteria," by C. R. Woese. Copyright © 1981 by Scientific American, Inc. All rights reserved.)

The proposition that chloroplasts and mitochondria are in effect captive prokaryotes that long ago exchanged a free-living existence for the benefits of obligatory intracellular symbiosis has been on the table for nearly a century, but for most of this time few took it seriously. The contemporary version of the hypothesis was shaped by the writings of R. Y. Stanier (1970), P. H. Raven (1970), and especially by Lynn Margulis's scholarly and provocative books (Margulis 1970, 1981). She envisages the eukaryotic cell as a true chimera that arose from the successive fusion of three or even four separate genomic lineages: nucleus-cytoplasm, flagella, mitochondria, and chloroplasts. Margulis's "serial endosymbiosis model" of early evolution makes a convenient starting point for the present discussion.

Chloroplasts. That chloroplasts are the evolutionary descendants of oxygen-evolving prokaryotic endosymbionts now seems almost beyond question. Chloroplasts fall into several classes, distinguishable by their ultrastructure and their complement of light-harvesting pigments. These reflect the multiple origins of plastids: endosymbionts matured into organelles on at least three separate occasions. For two of these one can identify cousins, so to speak, among contemporary prokaryotes. The chloroplasts of red algae (*Rhodophytae*) resemble cyanobacteria in appearance and ultrastructure and also share the same pigments: chlorophyll *a* (but no chlorophyll *b*) together with the accessory light-harvesting phycobilin pigments that are found nowhere else. Compelling evidence now comes from sequence analysis of protein constituents and of ribosomal RNAs (Schwartz and Dayhoff, 1978; Gray and Doolittle, 1982). Phycobiliproteins, for instance, are all homologous and traceable to a single ancestral sequence. The ribosomal RNA of plastids from the red alga *Porphyridium* bears no relationship to that of ribosomes from eukaryotic cytoplasm but is clearly homologous to ribosomal RNA of bacteria in general and of cyanobacteria in particular. These findings go well beyond showing that chloroplasts and cyanobacteria are phylogenetically related: they are evidence that their last common ancestor was itself a free-living, oxygen-evolving prokaryote. The plastids of green algae (*Chlorophytae,* for example *Euglena*) and of higher plants, which feature chlorophyll *b* but lack phycobilins, cannot be traced back to the cyanobacteria. However, a plausible ancestry is represented by *Prochloron,* a recently discovered class of prokaryotes that produce oxygen during photosynthesis and are remarkably similar to algal chloroplasts with respect to both pigment composition and ultrastructure (Withers et al., 1978; Giddings et al., 1980).

In a few instances, endosymbionts have been caught in the very act of transmogrification into intracellular organelles. A remarkable case in point is the flagellated alga *Cyanophora paradoxa,* which contains cytoplasmic bodies called cyanelles. These look like cyanobacteria, possess pigments typical of these organisms, and are enclosed in a cell wall of the bacterial type, yet their genome is only a tenth of the size of that of free-living cyanobacteria and resembles that of chloroplasts in its low order of complexity (Stanier and Cohen-Bazire, 1977; Margulis, 1981). The inference is inescapable that cyanelles represent an intermediate evolutionary stage between free-living bacteria and full-fledged chloroplasts.

Mitochondria. It seems very likely that mitochondria also evolved from erstwhile bacterial endosymbionts, but one hastens to add that this is a matter of judgment upon which honest persons can and do differ. Many mitochondrial qualities point to a bacterial origin: their circular genomes, their mode of initiating protein synthesis, and the sensitivity of their ribosomes to antibiotic inhibitors fall into this category. The T_1-oligonucleotide catalog of mitochondrial rRNA shows it to be a distant offspring of the eubacterial stem and unrelated to the RNA of cytoplasmic ribosomes. Mitochondrial respiratory chains, all of which are much alike, closely resemble those of the purple bacteria, particularly *Paracoccus denitrificans* and *Rhodopseudomonas sphaeroides*. The degree of homology among their *c*-type cytochromes is so impressive that several authors have proposed that mitochondria may have arisen quite directly from an endosymbiont of this particular class, with the loss of some functions and the acquisition of others (John and Whatley, 1975; Schwartz and Dayhoff, 1978; Whatley et al., 1979; Dickerson, 1980). Among the changes that this proposal presupposes are the replacement of a simple cytochrome oxidase by a more complex one, the loss of the capacity to oxidize hydrogen, and the acquisition of an ATP-ADP antiporter, not to mention the massive transfer of genes from the genome of the endosymbiont to that of its host.

Despite these persuasive correlations, I must concur with the cautious judgment of Gray and Doolittle (1982) that "the evolutionary origin of mitochondria still defies ready explanation." Serious objections to the endosymbiotic scenario have been raised in several quarters (Raff and Mahler, 1972; Uzzell and Spolsky, 1974; Mahler and Raff, 1976; Gillham and Boynton, 1981). Why, if mitochondria sprang from the bacterial tree, do their DNAs vary so widely in size, form, base composition, and even genetic organization? What is one to make of the fact that the genetic code utilized by mitochondria deviates somewhat from that of both prokaryotes and eukaryotes? Even mitochondrial ribosomes are not truly bacterial, except in their response to antibiotics. It would be premature to reject an endosymbiotic origin, but the quick descent from a defective photosynthetic purple bacterium cuts too many corners. If mitochondria do derive from bacteria, they were transformed into organelles very long ago; only subsequently did some of the early eukaryotic lines embark on further partnerships that eventually produced chloroplasts. It is just conceivable that the origin of mitochondria precedes that of the contemporary eubacteria, but it then becomes hard to explain the congruence of cytochrome sequences that are a pillar of the endosymbiotic scenario.

There are no organisms that clearly represent an intermediate stage in the evolution of the mitochondrion, but there are some remarkable symbiotic associations to reflect on. A case in point is the giant amoeba, *Pelomyxa palustris*. It lacks mitochondria but harbors endosymbionts that are undoubtedly aerobic bacteria although they undergo regular distribution when the cell divides. It is a pity that *Pelomyxa* cannot be cultured and remains largely a curiosity.

Flagella and Undulipodia.   The most controversial facet of the serial endosymbiosis hypothesis is the proposition (Margulis, 1970, 1981) that eukaryotic cilia and flagella are the remnants of yet another bacterial endosymbiont. Margulis cogently emphasizes the profound difference in structure between the prokaryotic and the eukaryotic organs of movement and proposes that only the former be called flagella. The eukaryotic organs, which spring from a specialized basal body and are constructed of microtubules, are designated "undulipodia." In Margulis's view, these derive from association of the emerging eukaryotic chimera with bacteria akin to modern spirochetes.

The best argument in support of this provocative thesis is the existence of contemporary protozoa that do in fact move with the aid of domesticated spirochetes. *Mixotricha paradoxa*, an inhabitant of the hindgut of wood-eating termites, is an astonishing creature. Each cell is covered with thousands of hairlike projections whose coordinated undulations propel the organism; these were shown by L. R. Cleveland to be symbiotic spirochetes, some 500,000 of them, attached to specific sites on the external surface. There are in addition four undulipodia per cell, which serve as a rudder. *Mixotricha* lacks mitochondria but harbors what appear to be endosymbiotic bacteria. Several other examples of this kind have been described. They are impressive demonstrations of the lengths to which microbial symbioses go, but one must question whether they bear on the origin of undulipodia.

We do not know just how the spiral motions of spirochetes are generated, but it is likely that they are based on rotary flagella of the standard bacterial kind, powered by a proton circulation. There seems to be no mechanistic resemblance at all to the ATP-driven, dynein-mediated apparatus of undulipodial motility (Chapter 11). Proteins that resemble tubulin in size and immunological reactivity have been detected in certain spirochetes (Margulis et al., 1978), but nothing less than sequence homology will persuade the skeptics. As matters stand, I would favor the opposite thesis; there are no proteins in eubacteria that are homologous to actin, myosin, tubulin, dynein, nor perhaps to calmodulin. The entire machinery for intracellular motility developed in the lineage that produced the protoeukaryotic host or else arose after fusion with the mitochondrial genome. Needless to say, this thesis is also longer on logic than on factual evidence.

The Protoeukaryotic Host.   According to the hypothesis of serial endosymbiosis, the ancestors of mitochondria and chloroplasts fused with a rather shadowy host organism that is represented in living eukaryotes by the nucleus and the cytoplasm. Some of the characteristics of the host are implicit in this postulate. Since the Krebs cycle, like respiratory assemblies, is localized in the mitochondria but glycolysis takes place in the cytoplasm, the host is presumed to have been fermentative but aerotolerant. It would also have contributed histones and, perhaps, noncircular DNA. Stanier (1970) argued persuasively that the critical "invention" underlying the eukaryotic state was the capacity for invagination of the plasma membrane. This allowed a well-less bacterium, perhaps resembling the mycoplasmas, to harvest resources by phagocytosis and thus initiated the animal's way of life. Phagocytosis requires controlled movements of cytoplasmic contents, matters that are now and perhaps always have been the business of actin, myosin, tubulin, and dynein. (An alternative possibility, preferred by Margulis, is that predatory aerobic bacteria akin to *Bdellovibrio* initiated the symbiosis from which mitochondria derive.) In any case, it has been generally taken for granted that the hypothetical host was itself a bacterium, if an unusual one: ultimately all parts of the eukaryotic cell must derive from prokaryotic precursors.

This presumption, self-evident on first sight, has been seriously shaken by recent discoveries in the molecular biology of eukaryotic cells. The differences between eukaryotes and prokaryotes (with the latter column including both mitochondria and chloroplasts) have come to loom so large that direct descent of eukaryotes from prokaryotes is increasingly doubtful. Eukaryotic nuclear DNA is in pieces, and the products of its transcription require splicing before translation. The proteins of eukaryotic cytoplasmic ribosomes (indeed, cytoplasmic proteins in general) show no homology to prokaryotic ones. The RNAs of eukaryotic and prokaryotic ribosomes belong to distinct molecular families that do not overlap (Darnell, 1978; Fox et al., 1980; Gray and Doolittle, 1982). All this strengthens the claim, first put forth by Woese, that the protoeukaryotic host (as represented by its ribosomal RNA) derives from an independent evolutionary lineage that is as ancient as those of the eubacteria and the archaebacteria. Whatever the host looked like, it was not itself a bacterium but a significantly different beast.

Did archaebacteria make any contribution to the early stages of the eukaryotic consortium? Enthusiastic investigators have pointed out several traits in which archaebacteria resemble eukaryotic cells more closely than prokaryotic cells do: the compartmentalized structure of some methanogens, the prevalence of sodium-linked porters in halobacteria, a sequence homology in a particular ribosomal protein. Searcy (1982) has reported that the DNA of the archaebacterial *Thermoplasma acidophilum* is associated with histonelike proteins and that its cytoplasm contains proteins bearing a functional resemblance to actin. Some of these resemblances may prove to be fortuitous or misleading, but one hopes that they will entice a few researchers who prefer blazing trails to paving roads. And it may not be beyond belief that some anaerobic niche still harbors organisms neither quite prokaryotes nor quite eukaryotes, but a little of both.

When Bigger Is Better.   The origin of eukaryotic cells is a problem in biological prehistory, and the course of events may never be known with certainty, but their rise to dominance can be seen as a matter of economics. In a thoughtful essay, Carlile (1980) has analyzed the costs and benefits of eukaryotic organization in terms of what ecologists call *r* and *K* selection. Prokaryotes are the supreme *r* strategists, multiplying quickly when abundant resources suddenly become available. Food spoilage, for instance, is typically due to bacteria. Prokaryotes also predominate in anaerobic environments, thanks to their diverse modes of anaerobic metabolism, and in the extreme niches favored by archaebacteria. The eukaryotic world, though in itself extremely diverse, is by and large governed by

K selection, which favors organisms able to make efficient use of scarce but dependable resources. In crowded conditions and in places where resources are thinly spread, rapid growth is not necessarily the most useful attribute. Greater selective advantage may lie with such eukaryotic characteristics as large size and speed and precision of movement and with innovative technologies of prey capture, digestion, excretion, and superior cellular homeostasis. These capacities would benefit not only animals but also those that adopted the plant's way: eukaryotes are better equipped to monitor the level of light or nutrients, to respond with directed movement or growth, and to adjust the timing of physiological activity to coincide with the time of day. The invention of mechanisms for intracellular transport was necessary for organisms to grow beyond bacterial size, but also made it possible to collect resources from a wide area at a central location, such as a fruiting body or a flower.

In the remaining chapters of this book we shall be primarily concerned with the workings of eukaryotic cells. Energy generation by mitochondria and chloroplasts, in what is basically a prokaryotic mode, will have its due. But I shall put greater emphasis on the essentially eukaryotic functions associated with the cytoskeleton and the plasma membrane, for these underpin the advances in motility, behavior, growth, and development that sustain the splendors of natural history.

References

BERKNER, L. V., and MARSHALL, L. C. (1965). History of major atmospheric components. *Proceedings of the National Academy of Sciences USA* **53**:1215–1225.

BRODA, E. (1975). Op. cit., Chapter 2.

BRODA, E., and PESCHEK, G. A. (1979). Did respiration or photosynthesis come first? *Journal of Theoretical Biology* **81**:201–212.

BROWN, I. I., GALPERIN, M. YU., GLAGOLEV, A. N., and SKULACHEV, V. P. (1983). Op. cit., Chapter 5.

CARLILE, M. J. (1980). Op. cit., Chapter 4.

CLOUD, P. (1974). Evolution of ecosystems. *American Scientist* **62**:54–66.

CLOUD, P. (1983). The biosphere. *Scientific American* **249**(3):176–189.

CRICK, F. H. C. (1981). *Life Itself: Its Origin and Nature*. MacDonald, London.

DANIELS, L., SPARLING, R., and SPROTT, G. D. (1984). The bioenergetics of methanogenesis. *Biochimica et Biophysica Acta* **768**:113–163.

DARNELL, J. E., JR. (1978). Implications of RNA:RNA splicing in evolution of eukaryotic cells. *Science* **202**:1257–1260.

DICKERSON, R. E. (1980). Cytochrome-*c* and the evolution of energy metabolism. *Scientific American* **242**(March):137–153.

DODDEMA, H. J., VAN DER DRIFT, C., VOGELS, C. D., and VEENHUIS, M. (1979). Chemiosmotic coupling in *Methanobacterium thermoautrophicum*. Hydrogen-dependent adenosine-5-triphosphate synthesis by subcellular particles. *Journal of Bacteriology* **140**:1081–1089.

EIGEN, M., and SCHUSTER, P. (1982). Stages of emerging life—Five principles of early organization. *Journal of Molecular Evolution* **19**:47–61.

EIGEN, M., GARDINER, W., SCHUSTER, P., and WINKLER-OSWATITSCH, R. (1981). The origin of genetic information. *Scientific American* **244**(April):78–94.

EISENBACH, M., and CAPLAN, S. R. (1979). The light-driven proton pump of *Halobacterium halobium*. Mechanism and function. *Current Topics in Membranes and Transport* **12**:165–248.

FOX, G. E., STACKEBRANDT, E., HESPELL, R. B., GIBSON, J., MANILOFF, J., DYER, T. A., WOLFE, R. S., BALCH, W. E., TANNER, R. S., MAGRUM, L. J., ZABLEN, L. B., BLAKEMORE, R., GUPTA, R., BONEN, L., LEWIS, B. J., STAHL, D. A., LUEHRSEN, K. R., CHEN, K. N., and WOESE, C. R. (1980). The phylogeny of prokaryotes. *Science* **209**:457–463.

FOX, S. W., and DOSE, K. (1977). *Molecular Evolution and the Origin of Life,* revised ed. Decker, New York.

FUCHS, G., and STUPPERICH, E. (1982). Autotrophic CO_2 fixation pathway in *Methanobacterium thermoautotrophicum*. *Zentralblatt für Bakteriologie und Hygiene, I. Abteilung Originale* **63**:277–288.

GEST, H. (1980). Evolution of biological energy-transducing systems. *FEMS Letters* **7**:72–77.

GIDDINGS, T. H., JR., WHITHERS, N. W., and STAEHELIN, L. A. (1980). Supramolecular structure of stacked and unstacked regions of the photosynthetic membranes of *Prochloron* sp., a prokaryote. *Proceedings of the National Academy of Sciences USA* **77**:352–356.

GILLHAM, N. W., and BOYNTON, J. E. (1981). Evolution of organelle genomes and protein-synthesizing systems. *Annals of the New York Academy of Sciences* **361**:20–40.

GRAY, M. W., and DOOLITTLE, W. F. (1982). Has the endosymbiont hypothesis been proven? *Microbiological Reviews* **46**:1–42.

HENDERSON, R. (1977). The purple membrane from *Halobacterium halobium*. *Annual Review of Biophysics and Bioengineering* **6**:87–109.

HOLLÄNDER, R. (1978). The cytochromes of *Thermoplasma acidophilum*. *Journal of General Microbiology* **108**:165–167.

JOHN, P., and WHATLEY, F. R. (1975). *Paracoccus denitrificans* and the evolutionary origin of the mitochondrion. *Nature* **254**:495–498.

KELL, D. B., DODDEMA, H. J., MORRIS, J. G., and VOGELS, G. D. (1981). Energy coupling in methanogens, in *Microbial Growth on C-1 Compounds,* H. Dalton, Ed. Heyden, London, pp. 159–170.

LANYI, J. K. (1978). Light energy conversion in *Halobacterium halobium*. *Microbiological Reviews* **42**:682–706.

LANYI, J. K. (1979). Op. cit., Chapter 5.

MAHLER, H. R., and RAFF, R. A. (1976). Evolutionary origin of the mitochondrion: A nonsymbiotic model. *International Review of Cytology* **43**:1–124.

MARGULIS, L. (1970). *Origin of Eucaryotic Cells*. Yale University Press, New Haven, Conn.

MARGULIS, L. (1981). *Symbiosis in Cell Evolution*. W. H. Freeman and Co., New York.

MARGULIS, L., TO, L., and CHASE, D. (1978). Microtubules in prokaryotes. *Science* **200**:1118–1123.

MILLER, S. L., and ORGEL, L. E. (1974). *The Origin of Life on the Earth*. Prentice-Hall, Englewood Cliffs, N.J.

MOROWITZ, H. J. (1967). Biological self-replicating systems. *Progress in Theoretical Biology* 1:35–58.

OPARIN, A. I. (1953). *The Origin of Life*, 2d ed. Dover, New York.

RACKER, E., and STOECKENIUS, W. (1974). Reconstitution of membrane vesicles catalyzing light-driven proton uptake and adenosine triphosphate formation. *Journal of Biological Chemistry* 249:662–663.

RAFF, R. A., and MAHLER, H. R. (1972). The non-symbiotic origin of mitochondria. *Science* 177:575–582.

RAVEN, J. A., and SMITH, F. A. (1976). The evolution of chemiosmotic energy coupling. *Journal of Theoretical Biology* 57:301–312.

RAVEN, J. A., and SMITH, F. A. (1981). H^+ transport in the evolution of photosynthesis. *Biosystems* 14:95–111.

RAVEN, J. A., and SMITH, F. A. (1982). Solute transport at the plasmalemma and the early evolution of cells. *Biosystems* 15:13–26.

RAVEN, P. H. (1970). A multiple origin for plastids and mitochondria. *Science* 169:641–645.

RIEDL, R. (1978). Op. cit., Chapter 1.

SCHOBERT, B., and LANYI, J. K. (1982). Halorhodopsin is a light-driven chloride pump. *Journal of Biological Chemistry* 257:10306–10313.

SCHOPF, J. W., Ed. (1983). *Earth's Earliest Biosphere: Its Origin and Evolution*. Princeton University Press, Princeton, N.J.

SCHWARTZ, R. M., and DAYHOFF, M. O. (1978). Origins of prokaryotes, eukaryotes, mitochondria and chloroplasts. *Science* 199:395–403.

SEARCY, D. G. (1982). Thermoplasma: A primordial cell from a refuse pile. *Trends in Biochemical Sciences* 7:183–185.

STANIER, R. Y. (1970). Some aspects of the biology of cells and their possible evolutionary significance. *Symposia of the Society of General Microbiology* 20:1–38.

STANIER, R. Y., and COHEN-BAZIRE, G. (1977). Phototrophic prokaryotes: The cyanobacteria. *Annual Review of Microbiology* 31:225–274.

STOECKENIUS, W., and BOGOMOLNI, R. A. (1982). Bacteriorhodopsin and related pigments of halobacteria. *Annual Review of Biochemistry* 51:587–616.

STOECKENIUS, W., LOZIER, R. H., and BOGOMOLNI, R. A. (1979). Bacteriorhodopsin and the purple membrane of halobacteria. *Biochimica et Biophysica Acta* 505:215–278.

THAUER, R. K., JUNGERMANN, K., and DECKER, K. (1977). Op. cit., Chapter 4.

UZZELL, T., and SPOLSKY, C. (1974). Mitochondria and plastids as endosymbionts: A revival of special creation? *American Scientist* 62:334–343.

VOGELS, G. D., KELTJENS, J. T., HUTTON, T. J., and VAN DER DRIFT, C. (1982). Coenzymes of methanogenic bacteria. *Zentralblatt für Bakteriologie und Hygiene. I. Abteilung Originale* 63:258–264.

WHATLEY, J. M., JOHN, P., and WHATLEY, F. R. (1979). From extracellular to intracellular: The establishment of mitochondria and chloroplasts. *Proceedings of the Royal Society of London, ser. B*, 204:165–187.

WILSON, T. H., and LIN, E. C. C. (1980). Evolution of membrane bioenergetics. *Journal of Supramolecular Structure* 13:421–446.

WITHERS, N. W., ALBERTE, R. S., LEWIN, R. A., THORNBER, J. P., BRITTON, G., and GOODWIN, T. W. (1978). Photosynthetic unit size, carotenoids and chlorophyll-protein composition of

Prochloron sp., a prokaryotic green alga. *Proceedings of the National Academy of Sciences USA* **75**:2301–2305.

WOESE, C. R. (1979). A proposal concerning the origin of life on the planet earth. *Journal of Molecular Evolution* **13**:95–101.

WOESE, C. R. (1981). Archaebacteria. *Scientific American* **244**(June):94–106.

WOLFE, R. S. (1979). Methanogens: A surprising microbial group. *Antonie van Leeuwenhoek* **45**:353–364.

Mitochondria and Oxidative Phosphorylation

An expert seldom gives an objective view; he gives his own view.

Morarji Desai

• • •

The Mitochondrion at Work
Secondary Functions of the Proton Circulation
Stoichiometry and Thermodynamics of Phosphorylation
Are Proton Currents Localized?
Proton Translocation by the Respiratory Chain
How Cells Make ATP

To biochemists, mitochondrial oxidative phosphorylation has always been the core of bioenergetics. The long quest for a chemical explanation divides naturally into three historical phases, rather like those that G. S. Stent (1968) discerned in the development of molecular biology.

The heroic phase, from approximately 1925 to 1966, belongs to the pioneers who marked the outlines of the subject. Its landmarks include the discovery of the respiratory chain, oxidative phosphorylation, and the ATP synthase; the isolation of mitochondria and the recognition of their biological functions; and the first stages in the biochemical dissection of energy coupling. Research during this phase was generally guided by the expectation that respiration and phosphorylation were chemically linked, and the isolation of energized intermediates was one of its primary goals.

The dogmatic phase opened with Mitchell's proclamation of the chemiosmotic hypothesis in 1961. This was a time of strife, dominated by the controversy over the essential nature of energy coupling whose flavor was at times almost Byzantine. Chemical, chemiosmotic, and conformational models, each in several versions, were vigorously promoted and roundly condemned. The debate concluded with the general (albeit not universal) acceptance of the chemiosmotic central dogma, that a current of protons is the sole link between respiration and phosphorylation. Consensus was fittingly celebrated by the issuance of a joint communiqué (Boyer et al., 1977) in which leading investigators spelled out areas of agreement while staking out positions for the next round of conflict, and by the award of the Nobel prize to Peter Mitchell in 1978.

We are now well into the academic phase, with oxidative phosphorylation still the cockpit of controversy. A casual observer, baffled by an abstruse and often hypertechnical literature, may gather the impression that little progress has been made during the past quarter century. This impression would be mistaken, for the debate has shifted from general principles to mechanistic particulars and from the cellular level to the molecular. We do not at present understand how respiratory chains, ATP synthase, or the equivalent elements of photosynthetic energy transduction couple chemical catalysis to proton translocation, nor are we entirely certain how protons travel between the primary chemiosmotic assemblies. These subtle and exceedingly demanding problems are, from the molecular point of view, the heart of biological energy transduction, and they have bewitched the majority of investigators who consider themselves to be bioenergeticists.

The purpose of this chapter is to provide an overview of mitochondrial energetics, and then to focus more sharply on the molecular mechanics of proton translocation and energy transduction. I have taken for granted the essential unity of mitochondria and bacteria, and have adopted the broadly chemiosmotic viewpoint to which the majority of investigators in this field now subscribe. The limited objectives of this chapter require that many topics of current interest be given short shrift, including mitochondrial genetics, biogenesis, and most of their ancillary functions. More to my regret, by taking the chemiosmotic theory for a point of departure, this presentation distorts the historical unfolding of our understanding, and it unintentionally diminishes the contributions of many scientists who, working with mitochondria or chloroplasts, laid the foundations of bioenergetics in general.

The mass and density of the literature on mitochondrial oxidative phosphorylation numbs the mind. Albert Lehninger's classic, *The Mitochondrion* (1964), and Efraim Racker's *A New Look at Mechanisms in Bioenergetics* (1976) give the reader a clear sense of how the subject evolved. Recent books by Prebble (1981) and Tzagaloff (1982) bring the reader more nearly up to date, while Nicholls (1982) provides a rigorous and contemporary chemiosmotic framework on which to hang the experimental data. In addition to these sources, and to Mitchell's writings (Mitchell 1966, 1976*a*, 1979*a*, 1979*b*, 1981*a*), I have found the following reviews especially helpful: Boyer et al. (1977), De Pierre and Ernster (1977), Hinkle and McCarty (1978), Kell (1979), Wikström and Krab (1979), Fillingame (1980*a*), Cross (1981), Maloney (1982), and Senior and Wise (1983). Ernster and Schatz (1981) should be consulted for a valuable historical review. This is an eclectic list, and Morarji Desai would probably scorn what follows as an expert's view.

The Mitochondrion at Work

Structure and Organization.   Mitochondria are organelles, part of a larger whole rather than independent entities, and this must be kept in mind when one compares mitochondrial architecture with that of bacteria. If mitochondria are in fact remote descendants of prokaryotic endosymbionts, their integration with the genome and cytoplasm of the host called for deep-seated transformations of both structure and function.

Although mitochondria contain circular DNA, ribosomes of a generally bacterial type, and transfer RNAs, most of the mitochondrial proteins are specified by nuclear genes and produced on cytoplasmic ribosomes. Mitochondrial DNA typically codes for fewer than a dozen proteins, including components of cytochrome oxidase and of the F_0 sector of ATP synthase. It is not at all clear why these particular polypeptides are produced locally while all the others are imported.[1]

The mitochondrial envelope consists of two continuous lipid membranes (Figure 7.1). The outer membrane, characteristically eukaryotic in composition and rich in sterols, is pierced by large pores that give free passage to charged and

---

[1] Mitochondria, while not as diverse as bacteria, yet exhibit much more variety than was recognized until recently. A liver cell contains some 5000 small discrete mitochondria, comparable in size to bacteria. Some yeasts have but a single mitochondrion, but it is gigantic and ramifies throughout the cell's volume. Most mitochondrial DNAs are circular, but that of *Paramecium* is linear; most mitochondria encode the DCCD-binding peptide in their own DNA, but in *Neurospora* this sequence is located in the nucleus. I have perforce glossed over such differences except where they are pertinent to the argument.

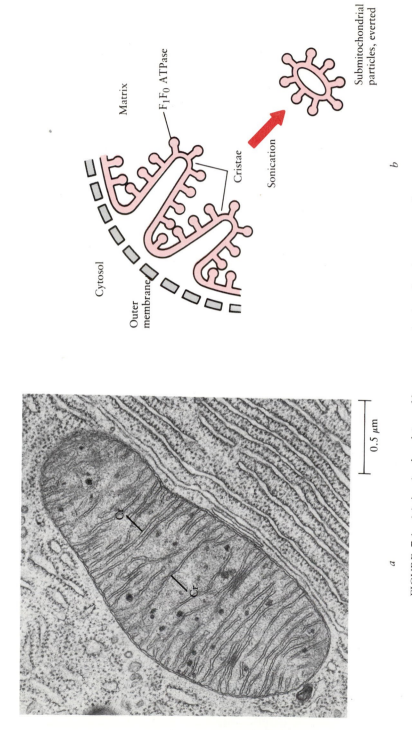

Matrix

F_1F_0 ATPase

Cristae

Sonication

Submitochondrial particles, everted

Cytosol

Outer membrane

0.5 µm

Cr

Cr

a

b

FIGURE 7.1   (*a*) A mitochondrion of bat pancreas in situ, showing outer and inner membranes, the latter folded into cristae (Cr). (Courtesy of K. R. Porter.) (*b*) The basic features of mitochondrial structure. (After Metzler, 1977, with permission of Academic Press.)

uncharged solutes up to 5 kdal. The inner membrane, whose lipids resemble those of bacterial plasma membranes, is concerned with energy transduction; it is essentially impermeable to ions and can sustain large gradients of electrochemical potential. The surface area of the inner mitochondrial membrane is greatly enlarged by deep infoldings called cristae. These bear the respiratory assemblies, which show up in freeze-fracture electron micrographs as abundant intramembranous particles. Negative staining reveals numerous small knobs, 9 to 10 nm in diameter, that project from the inner membrane into the matrix; these are the headpieces of the F_1F_0 ATPase.

Mitochondria are specialized for the rapid oxidation of a limited number of substrates, particularly NADH, fatty acids, and succinate, with conservation of the free energy of oxidation in the form of ATP (Figure 7.2). The substrates are generated within the matrix by the Krebs cycle and by the enzymes of fatty acid oxidation. Cytoplasmic NADH is not oxidized as such by animal cell mitochondria. Instead, an array of transport systems embedded in the inner membrane give passage to certain cytoplasmic metabolites such as pyruvate, glutamate, α-ketoglutarate, and α-glycerophosphate and thus enable the mitochondrion to process reduced substrates generated in the cytoplasm. (Plant mitochondria generally oxidize NADH directly, at the outer aspect of the inner membrane.) ATP produced in the matrix is exported to the cytoplasm, where it is consumed, while ADP and P_i are imported. Mitochondrial oxidative phosphorylation, unlike that of bacteria, therefore depends on transport systems that admit P_i and exchange ATP for ADP. The respiratory chain itself has not escaped evolutionary tinkering. Formally, all mitochondrial respiratory chains are much alike and closely resemble those of *Paracoccus denitrificans* and *Rhodopseudomonas sphaeroides* (Chapter 6). However, mitochondria do not reduce nitrate or oxidize hydrogen. Instead, there appear to be additional redox carriers and ancillary polypeptides within the core of the chain (Figure 7.3). The evolutionary, or functional, significance of these "extra" components is not yet clear.

For all their adaptations to an intracellular existence, mitochondria have unmistakably retained an essentially prokaryotic organization. Some of their ancillary functions may likewise be ancestral, including mitochondrial production of aminolevulinic acid (the first precursor on the pathway of heme biosynthesis) and of cardiolipin (an acidic phospholipid characteristic of the inner membrane of mitochondria and of bacterial plasma membranes). Other ancillary functions have no clear parallel in the prokaryotic world. In certain animal tissues, mitochondria are responsible for the metabolic transformation of steroids. The mitochondria of brown adipose tissue in animals and of certain reproductive structures in plants can switch from ATP production to the generation of heat. And some animal tissue mitochondria accumulate calcium ions and play an important role in the regulation of the cytosolic calcium level. The restricted distribution of the latter capacities clearly suggests that they are relatively late adaptations, superimposed on a central core of functions concerned with respiratory ATP production.

The principal functions of mitochondria revolve around energy transduction

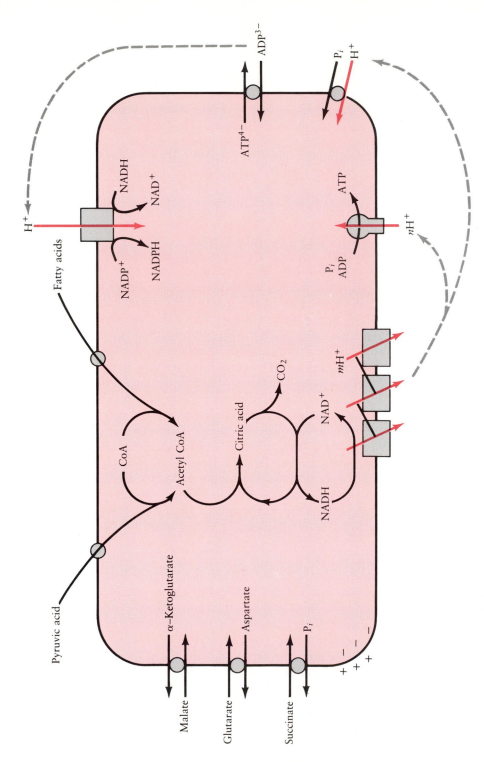

FIGURE 7.2   Diagrammatic view of the mitochondrion as a whole, showing the major metabolic pathways and porters. Red arrows identify the proton-translocating systems.

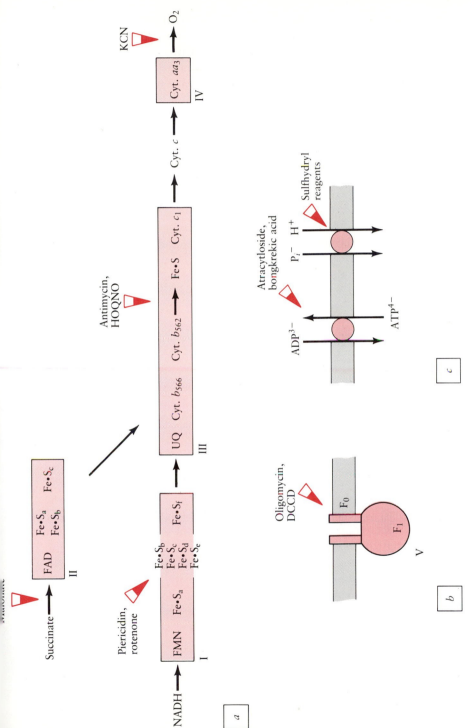

FIGURE 7.3   The machinery of oxidation and phosphorylation. (*a*) The respiratory chain is composed of four relatively independent blocks, identified by Roman numerals. Fe·S designates various iron-sulfur centers detectable by spectroscopic methods. (After Boyer et al., 1977.) (*b*) The F_1F_0 ATP synthase. (*c*) ATP-ADP exchange carrier and P_i porter. The pointers indicate the approximate sites of action of inhibitors mentioned in the text.

by a proton current (Figure 7.2). As in bacteria, the oxidation of NADH and other substrates by the respiratory chain is accompanied by the translocation of protons out of the matrix and by the generation of a proton potential, matrix electronegative and alkaline. This in turn is the immediate driving force for ATP production by the F_1F_0 ATP synthase. Let us now examine the chief components of the working machinery and consider how they collaborate to harness the free energy of respiration to the generation of ATP and other work functions.

The Respiratory Chain. The oxidation chain of mitochondria consists of about 35 distinct polypeptides and cofactors, some integral components of the inner membrane and some peripherally bound to it. The chain can be arranged into a linear sequence, but in fact consists of four independent macromolecular complexes, as shown in Figure 7.3. These correspond approximately to the traditional "coupling sites" of oxidative phosphorylation.

The segmental organization of the chain was originally inferred from the judicious use of artificial electron donors and acceptors, together with specific inhibitors whose sites of action are indicated in the scheme. For example, oxidation of NADH (or of NADH-generating substrates) is said to be accompanied by the phosphorylation of three molecules of ADP (P/O = 3); oxidation of succinate has a P/O ratio of 2, suggesting that one molecule of ATP is formed in the region labeled I in Figure 7.3. When cytochrome oxidase is blocked with cyanide, ATP can still be formed by the oxidation of NADH with added cytochrome *c*; this partial reaction yields two ATP per electron pair (P/O = 2), suggesting that one ATP is normally formed in the region of cytochrome oxidase. The assignment is confirmed by the finding that the oxidation of ascorbate (with the aid of tetramethylphenylenediamine as artificial redox mediator), which donates electrons to cytochrome *c*, generates ATP with a P/O ratio of 1. Such experiments led to the definition of three coupling sites: one in the oxidation of NADH, the second between ubiquinone and cytochrome *c*, and the third associated with cytochrome oxidase. Traditionally, each of these coupling sites was thought to be individually linked to ATP synthase. The chemiosmotic interpretation is somewhat different: each coupling site denotes a sector of the respiratory chain that translocates protons across the mitochondrial inner membrane. We may thus think of the respiratory chain between NADH and oxygen as consisting of three proton pumps in series.

This tripartite conception of the respiratory chain has received strong support from the biochemical dissection of the mitochondrial inner membrane. When treated gently with appropriate solvents and detergents, the membrane dissociates into five complexes of reproducible composition that collectively contain all the proteins required for coupled respiration and phosphorylation. Complexes I through IV, together with ubiquinone and cytochrome *c*, comprise the respiratory chain, while V designates the F_1F_0 ATPase. The constituents of the five blocks are listed in Table 7.1. Except for complex II, the succinate-ubiquinone reductase, each is an individual proton-translocating unit that is but loosely associated with the other blocks. The complexes are thought to be

TABLE 7.1 Enzyme Complexes of Mitochondrial Respiration and Phosphorylation

	Complex	Subunits	Prosthetic groups	Topology*
I	NADH-ubiquinone reductase, 850 kdal	16	1 FMN; 16–24 Fe · S in 5 to 7 centers	Spans the membrane; NADH site on M side; Q site in membrane; translocates H^+
II	Succinate-ubiquinone reductase, 97 kdal	4	1 FAD; 8 Fe · S in 3 centers	Succinate site on M side; Q site in membrane; not proton translocating
III	Ubiquinone–cytochrome-c reductase, 280 kdal	8	2 b-type hemes, 1 c-type heme (c_1); 2 Fe · S	Spans the membrane; cytochrome c_1 and Fe · S protein on C side; cytochromes b in membrane; translocates H^+
	Cytochrome c, 13 kdal	1	1 c-type heme	Peripheral; C side
IV	Cytochrome-c oxidase, 200 kdal	7	2 a-type hemes ($a_1 a_3$); 2 Cu	Spans the membrane; cytochrome c site on C side; O_2 site uncertain; translocates e^- and/or H^+
V	$F_1 F_0$-ATP synthase, ~450 kdal			Translocates H^+
	F_0	3–5		Spans the membrane
	F_1	5	(3 bound nucleotides)	Peripheral; M side
	Inhibitor peptide, 10 kdal	1		On F_1, dissociable
	OSCP (18 kdal) and F_6 (8 kdal)	1 each		Stalk, peripheral to F_0 (?)

After De Pierre and Ernster, 1977; von Jagow and Sebald, 1980; Wikström et al., 1981; Futai and Kanazawa, 1983.
* M and C sides refer to matrix and cytosol, respectively.

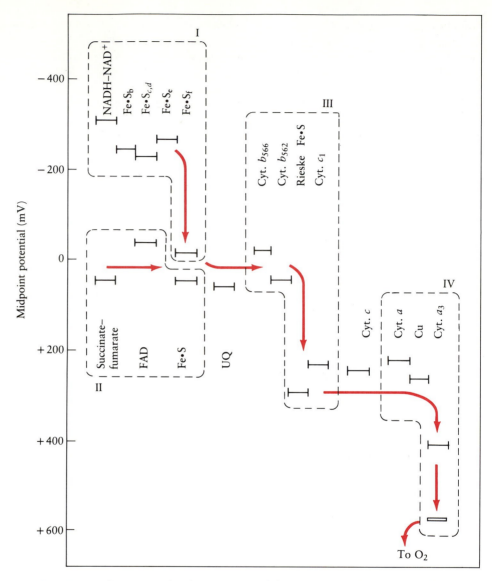

FIGURE 7.4   Redox potentials of components of the mitochondrial respiratory chain. Thin horizontal bars indicate E_m values measured in deenergized beef heart mitochondria in situ. As a rule, the actual redox potentials in respiring mitochondria are not very different from E_m, but the redox potential of cytochrome oxidase is much more positive (thick horizontal bar). The dashed lines indicate respiratory complexes I to IV, the red line traces the path of electrons over the chain. (After Nicholls, 1982, with permission of Academic Press.)

individually mobile in the fluid lipid bilayer, interacting chiefly by random collision. This is not as implausible as it may appear, since integral proteins make up 70 percent of the membrane's mass and are capable of rapid diffusion.

The work of proton translocation must be paid for by the loss of free energy as electrons pass through the chain. Potentiometric analysis has proved to be a powerful method of locating the proton-transport steps and measuring the free-energy changes involved (Dutton and Wilson, 1974). Briefly, during respiration electrons pass from the NADH-NAD$^+$ couple ($E_m = -320$ mV) to oxygen at $E_m = +820$ mV via a series of reversible redox reactions. In order to function efficiently in a cyclic manner, both the reduced and the oxidized form of the carrier must be plentiful; redox catalysts, therefore, normally operate in the vicinity of their E_m, or midpoint potential. It is possible to determine the E_m of the respiratory catalysts by redox titration in situ and also to measure the proportions of the reduced and oxidized species during respiration, from which the actual operating redox potential, E_h, can be computed. A set of values obtained in this manner is shown in Figure 7.4. It will be noted that each of the proton-transporting respiratory complexes displays an abrupt drop in redox potential, on the order of 200 mV, which is accentuated by the line that traces the pathway of electron flow. These transitions reflect the decrease in free energy that is conserved in the form of the proton potential across the membrane. In other words, proton translocation must take place between iron-sulfur centers Fe·S$_e$ and Fe·S$_f$, between cytochromes b and c_1, and in the cytochrome oxidase reaction. The amount of free energy available from each of these reactions will be considered below. The titration procedure also allows one to determine whether a given carrier transfers one electron or two, while the effect of pH on the midpoint potential may reveal carriers that undergo protonation during electron transport.

How many protons does each complex translocate? This important number can in principle be obtained quite directly by monitoring the pH of the external medium during a brief burst of respiration, as illustrated in Figure 7.5. A small, accurately calibrated pulse of oxygen is injected into an anaerobic suspension of mitochondria supplied with the requisite substrates. The protons extruded appear in the external medium but flow back into the mitochondrial matrix on exhaustion of the oxygen, in response to the protonic potential generated during the respiratory pulse. Now, since protons are expelled electrogenically, $\Delta\Psi$ builds up quickly and tends to inhibit further proton extrusion. Net acidification of the external phase will be observed only when conditions are so arranged as to prevent the buildup of $\Delta\Psi$; for instance, if the medium contains both K$^+$ ions and the ionophore valinomycin, to let K$^+$ influx compensate for the protons extruded. Many additional precautions must be taken for accurate results: any pH changes due to scalar enzymatic reactions must be subtracted, and allowance must be made for the backflow of protons into the matrix by symport with metabolites or by leakage. Such experiments, in the hands of Mitchell and Moyle (1967b) and of Hinkle and Horstman (1971), indicated that six protons were translocated out of the matrix for each pair of electrons passing from NADH to

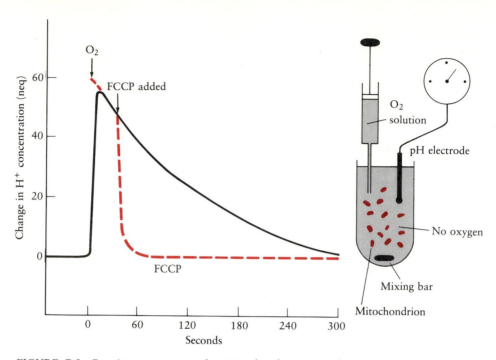

FIGURE 7.5   Respiratory proton pulse. Mitochondria are incubated under anaerobic conditions in a lightly buffered medium containing an oxidizable substrate, valinomycin, and a high concentration of KCl. When a small, accurately calibrated pulse of oxygen is introduced into the suspension, the pH declines sharply due to the extrusion of protons. When the oxygen has been used up, the protons leak slowly back into the matrix. The proton-conducting uncoupler FCCP facilitates their return. The total number of protons ejected during the pulse is obtained by extrapolating the decay curve to correct for protons that reentered during the pulse. (From "How Cells Make ATP," by P. C. Hinkle and R. E. McCarty. Copyright © 1978 by Scientific American, Inc. All rights reserved.)

oxygen: four protons per electron pair derived from succinate, and therefore two protons at each coupling site (Figure 7.14a). These numbers, which coincided with expectations from the hypothesis that the respiratory chain consists of three successive redox loops, have recently come under heavy fire; we shall return to this matter shortly.

ATP Production.   The mechanism of ATP formation will be considered in some detail in the final section of this chapter. Suffice it for the present to emphasize that the mitochondrial ATP synthase is closely allied to the bacterial one (Chapter 4) both in structure and in function, but the numbers and masses of the subunits may be somewhat different (Table 7.1) and are still somewhat uncertain.

Once again, the complex consists of two modules, F_1 and F_0, whose names recall Efraim Racker's seminal contributions to the study of oxidative phosphorylation. F_1 bears the catalytic site or sites; isolated F_1 is a potent ATPase. (The designation stands for coupling factor 1, a protein that could be extracted from mitochondrial membranes by agitation with certain buffers; depleted membranes were uncoupled, but ATP synthesis could be restored by re-addition of F_1.) The F_1 headpiece, whose mass is about 360 kdal, is composed of five subunits with a probable stoichiometry of $\alpha_3\beta_3\gamma\delta\varepsilon$. It is noteworthy that the α and β subunits of beef heart mitochondrial ATPase show extensive sequence homology to the corresponding subunits from *E. coli*, but the smaller subunits have been less stringently conserved (Senior and Wise, 1983). The basepiece F_0, which spans the membrane, contains three to five kinds of subunits. One of these binds the antibiotic oligomycin, a potent and highly specific inhibitor of the ATP synthase; hence the designation of the entire sector. Most bacterial F_1F_0 ATPases are insensitive to oligomycin; significantly *Rhodospirillum rubrum*, putatively related to the progenitor of mitochondria, has an oligomycin-sensitive ATPase. In total, at least 10 polypeptides collaborate in the reversible coupling of proton movements to the hydrolysis or synthesis of ATP.

The stoichiometry of proton translocation is a controversial subject. Mitchell and Moyle (1968) originally reported $2H^+$ per ATP, a ratio confirmed by subsequent experiments in their laboratory and in others. This stoichiometry was based on experiments like the one shown in Figure 7.6, in which the number of protons translocated during ATP hydrolysis is measured directly with submitochondrial particles. (Recall that the polarity of these particles is everted, as shown in Figure 7.1; protons are therefore taken up.) This procedure is likely to underestimate the number of protons, and less direct approaches increasingly

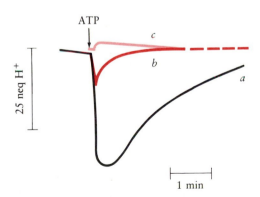

FIGURE 7.6   Proton uptake during ATP hydrolysis by submitochondrial particles. The particles are incubated in lightly buffered medium containing antimycin A (to block respiration), valinomycin, and KCl. Addition of a small, calibrated pulse of ATP at the arrow induces uptake of protons by the particles. When the ATP is exhausted, the protons leak back out of the particles. Trace *a*, ATP only; trace *b*, ATP plus CCCP; trace *c*, ATP plus oligomycin. (After Thayer and Hinkle, 1973, with permission of *Journal of Biological Chemistry*.)

suggest that the true stoichiometry may be $3H^+$ per ATP (see below). The matter is far from settled, but I shall tentatively take it that each cycle of the F_1F_0 ATPase translocates three protons across the mitochondrial membrane. Proton translocation has been documented with highly purified preparations reconstituted into artificial phospholipid vesicles.

The ATP synthase catalyzes a reversible reaction, yet carefully prepared mitochondria hydrolyze ATP only after a substantial delay. The effect is ascribed to a sixth subunit of F_1, a peptide of relatively small size, which binds to F_1 and inhibits ATP hydrolysis. When a protonic potential exists across the membrane (generated, as a rule, by respiration), this peptide dissociates from F_1 and allows the ATP-ADP couple to come into equilibrium with the proton potential. The function of this regulatory subunit need not be sought afar: it prevents the futile hydrolysis of ATP if respiration slows down due to anaerobiosis or for other reasons. (For a full discussion of regulatory subunits see Pedersen et al., 1981.) Whether bacterial F_1F_0 ATPases also contain a special regulatory subunit of this kind is not clear.

The catalytic headpiece faces the matrix, and the ATP produced there must be exported across the membrane. By the same token, ADP and P_i produced in the cytosol must enter the matrix. The carriers that mediate these translocations have been studied extensively, but differences of opinion persist about the molecular species that traverse the membrane. The majority view holds that P_i is transported by electroneutral symport with protons, effectively as H_3PO_4, with the aid of a specific porter that is particularly sensitive to the sulfhydryl reagent N-ethylmaleimide. A separate porter, originally recognized by its susceptibility to the inhibitors atractyloside and bongkrekic acid, mediates electrogenic antiport of ATP^{4-} for ADP^{3-} (see below). The stoichiometry of the ATP synthase and of the ATP-ADP antiporter has important implications for the thermodynamics of oxidative phosphorylation.

The Electrochemical Potential Gradient.   The protonic potential across the inner mitochondrial membrane, matrix alkaline and electronegative, is the central parameter in mitochondrial energy transduction, and much effort has been expended on measuring its magnitude. The methods used for this purpose are indirect, relying on the distribution of permeant cations and weak acids between the mitochondria and the surrounding fluid, and are always open to some question (Chapter 4). However, the results obtained with liver and heart mitochondria in different laboratories are reasonably consistent, most of them falling between -190 and -220 mV (Mitchell and Moyle, 1969; Nicholls, 1974; Berry and Hinkle, 1983; reviewed by Mitchell, 1976a; Fillingame, 1980a; Nicholls, 1982; Ferguson and Sorgato, 1982). Disappointingly, attempts to demonstrate the expected large membrane potential by impaling mitochondria with microelectrodes in situ have not met with success (Tedeschi, 1980). The difficulty is probably technical rather than conceptual, for mitochondria in living cells do stain brightly with rhodamine dyes that accumulate in these organelles in response to the membrane potential (Johnson et al., 1981).

It should be recalled here that the proton-motive force, Δp, is the sum of the driving forces due to the membrane potential and the pH gradient:

$$\Delta p = \frac{\Delta \tilde{\mu}_{H^+}}{F} = \Delta \Psi - 59 \Delta pH \qquad \text{(Eq. 3.4)}$$

In fully coupled respiring mitochondria Δp remains quite constant, but the contributions of ΔpH and $\Delta \Psi$ can be varied by manipulating the conductance of the membrane to ions other than protons (Figure 7.7). For example, on the addition of K^+ ions plus valinomycin, one observes net electrophoretic accumulation of K^+ balanced by the extrusion of additional protons. The internal pH rises, and with it ΔpH; since the respiratory chain can only achieve the same Δp as before, $\Delta \Psi$ falls by an equivalent amount. The result is a redistribution of the contributions from ΔpH and $\Delta \Psi$ in favor of the former. By the same token, nigericin allows K^+ from the matrix to exit by exchange for protons; ΔpH collapses, but $\Delta \Psi$ rises and Δp remains the same (Mitchell and Moyle, 1969).

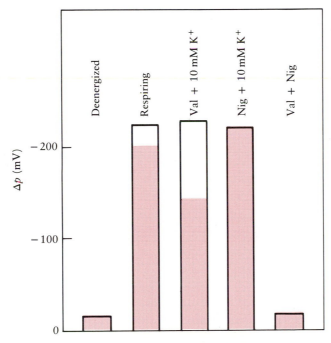

FIGURE 7.7   Ionophores modulate the composition of the proton potential. Respiring mitochondria generate a Δp of -220 mV, consisting chiefly of $\Delta \Psi$. Valinomycin plus 10 mM K^+ enhances ΔpH and reduces $\Delta \Psi$ but does not alter Δp. Nigericin dissipates ΔpH and enhances $\Delta \Psi$, with little effect on Δp. Valinomycin and nigericin in combination allow K^+ ions to move both electrogenically and by antiport for protons; this is equivalent to proton conduction, which elicits collapse of both ΔpH and $\Delta \Psi$. Color indicates $\Delta \Psi$; open parts of bars, ΔpH. (Data from Mitchell and Moyle, 1969.)

These compensatory changes depend on the membrane retaining its native resistance to the diffusion of protons (or H_3O^+) as such. Proton-conducting uncouplers such as FCCP (carbonylcyanide p-trifluoromethoxyphenylhydrazone), which allow protons to equilibrate across the membrane, collapse both ΔpH and $\Delta \Psi$; Δp falls to zero and phosphorylation ceases. The same is true when both valinomycin and nigericin are added, since the two ionophores together effectively transport protons.

The proton potential provides a simple and quite compelling explanation not only for the conservation of the free energy of respiration during ATP synthesis but also for the dependence of the *rate* of respiration on that of phosphorylation. Every runner knows that the rate of respiration rises when ATP consumption increases as a result of exercise; in the blowfly, for instance, oxygen consumption increases sixtyfold when the insect takes flight (cited in Hansford, 1980). Figure 7.8 illustrates this response at the mitochondrial level, where it is described as "respiratory control." A suspension of mitochondria supplied with substrate but lacking ADP consumes oxygen slowly. Addition of a small amount of ADP elicits a burst of rapid respiration that lasts until all the ADP has been phosphorylated. Addition of the uncoupler FCCP induces accelerated ("uncoupled") respiration, which persists indefinitely. In the same vein, rapid respiration by phosphorylating mitochondria is inhibited by agents that block phosphorylation, such as oligomycin or atractyloside; here again, uncouplers overcome the inhibition of oxygen consumption.

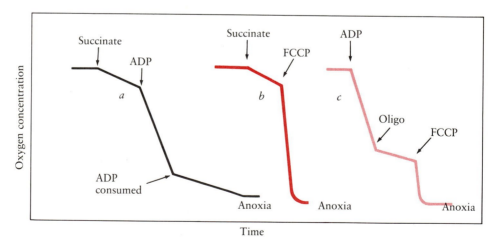

FIGURE 7.8   Respiratory control. The rate of oxygen consumption depends on concurrent phosphorylation. Trace *a*, mitochondria in buffer containing substrate respire slowly; addition of a small amount of ADP elicits a burst of respiration, which terminates when all the ADP has been phosphorylated. Trace *b*, rapid respiration can also be elicited by addition of FCCP, a proton-conducting uncoupler. Trace *c*, oligomycin, an inhibitor of the F_1F_0 ATP synthase, inhibits both phosphorylation and oxygen consumption; FCCP overcomes the block.

The chemiosmotic explanation is that with phosphorylation blocked by an inhibitor or by lack of ADP, $\Delta\tilde{\mu}_{H^+}$ rises and approaches ΔG of oxidation; respiration then slows to a rate determined by the leakage of protons across the membrane into the matrix. With the onset of phosphorylation, protons begin to flow through the F_1F_0 ATP synthase; $\Delta\tilde{\mu}_{H^+}$ falls below ΔG, the "back pressure" is relieved, and respiration accelerates. Uncouplers increase the rate of proton leakage; the rate of respiration is now set by the rate of proton conduction or (at high uncoupler concentrations) by the capacity of the respiratory chain itself, but is no longer controlled by the rate of phosphorylation. It must be added that although this interpretation is widely accepted, there are data that point to control of respiration by either the phosphorylation potential, ΔG_p, or the extramitochondrial ADP/ATP ratio; we shall return to this matter below.

Another phenomenon that reflects the thermodynamic coupling of respiration and phosphorylation via the electrochemical potential of protons is the reversal of the direction of electron flow by ATP hydrolysis. For example, mitochondria supplied with succinate and ATP, while electron transfer to oxygen is blocked by antimycin (Figure 7.3), can reduce NAD^+ to NADH. This process depends on the generation of a protonic potential across the membrane and is abolished by proton conductors and by inhibitors of the F_1F_0 ATPase. Reversed electron transport is probably of no physiological significance in mitochondria, but as outlined in Chapter 5, it assumes importance in many photosynthetic and chemolithotrophic bacteria.

Porters.   Like the cytoplasmic membrane of bacteria, the mitochondrial inner membrane is intrinsically impermeable to ions and polar metabolites. The integration of mitochondrial processes with those of the cytosol depends therefore on a battery of porters, some of which are shown in Figures 7.2 and 7.9 (LaNoue and Schoolwerth, 1978). It is worth noting here that all the known mitochondrial transport systems are secondary. Presumably the primary transport systems that bacteria require (Chapter 5) became dispensable in the sheltered environment provided by the cytosol.

The consequences of metabolite transport can be quite dramatic because mitochondria, which lack a rigid cell wall, swell or shrink in response to the movements of osmotically active metabolites. Figure 7.10 shows an example, together with the fluxes that account for the observations. Mitochondria whose respiration was blocked with antimycin were osmotically stable when incubated in 0.1 M ammonium succinate. Since all biological membranes are permeable to NH_3, succinic acid must be impermeant. Addition of small amounts of phosphate induced swelling, suggesting that succinate anion becomes permeant in the presence of phosphate. The explanation is that phosphate enters the matrix together with a proton left over from the entry of NH_3; the phosphate then exits again by exchange for succinate. The net result is that ammonium succinate can pass across the inner membrane and water follows in its train.

Porters for anionic organic metabolites are key elements in the oxidation of reducing equivalents generated in the cytosol, such as the NADH produced

Porter	Function	Inhibitors
(1) Adenine nucleotides	ATP^{4-} / ADP^{3-}	Attactyloside, bongkrekic acid
(2) Phosphate	P_i^- / H^+	SH reagents
(3) Dicarboxylate	$Succ^{2-}$ / P_i^{2-}	Butylmalonate
(4) Tricarboxylate	Citrate / Mal	
(5) α-Ketoglutarate	α-Kg / Mal	Phenylsuccinate
(6) Glutamate-aspartate	Glu / Asp	
(7) Pyruvate	Pyr / H^+	Cyanohydroxycinnamic acid
(8) Fatty acid	Acylcarn / Carn	
(9) Potassium	K^+	
(10) Potassium-proton	K^+ / H^+	Physiological control by Mg^{2+}
(11) Sodium-proton	Na^+ / H^+	

FIGURE 7.9   Porters for organic metabolites and inorganic ions in the inner mitochondrial membrane.

during glycolysis. As mentioned before, animal cell mitochondria are impermeable to NADH and cannot oxidize it when added to the external phase. In vivo the barrier is circumvented by one or another "shuttle" that involves reduction of some metabolite in the cytosol, followed by its oxidation in the mitochondrion. Figure 7.11 illustrates one of these, the malate-aspartate shuttle of liver and heart cells, in which malate serves as a carrier for reducing equivalents from the cytosol to the respiratory chain. Briefly, cytosolic NADH reduces cytosolic oxaloacetate to malate; the malate enters the mitochondrial matrix by antiport for α-ketoglutarate and is oxidized by the respiratory chain to oxaloacetate. The latter returns to the cytosol in the form of aspartate, after transamination by glutamate, while the α-ketoglutarate exits by exchange for more malate. The net result is the transfer of two reducing equivalents from the cytosol to the mitochondrial matrix with each turn of the cycle. Other porters for organic metabolites primarily mediate export. For example, exchange of glutamate for citrate makes the latter, produced in the mitochondrial matrix, available for biosynthetic reactions in the cytosol.

Porters for K^+ and Na^+ are required for mitochondrial housekeeping, particularly for the control of volume, and their existence was one of Mitchell's initial postulates. As mentioned above, mitochondria swell when they take up K^+ or Na^+ together with an anion, and the pH of the matrix rises. Both tendencies are countered by porters that extrude Na^+ and K^+ by electroneutral antiport for protons (Brierley et al., 1978; Garlid, 1980). The evidence suggests that two distinct porters are involved: one that is highly specific for Na^+, and

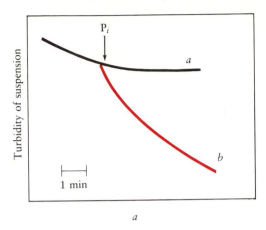

a

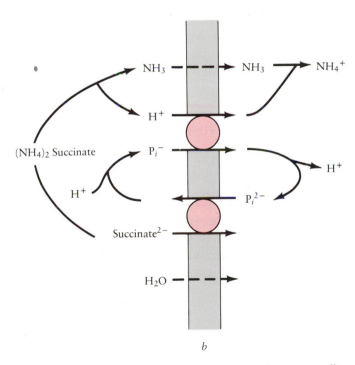

b

FIGURE 7.10 Metabolite uptake by mitochondria elicits swelling. (*a*) Swelling in ammonium succinate requires P_i. Mitochondria incubated in 0.1 M ammonium succinate, with respiration blocked, did not swell (trace *a*). Addition of 2 mM P_i elicited swelling (trace *b*), which was monitored by the decrease in turbidity. (*b*) Swelling results from the interaction of three transport processes: passive diffusion of NH_3, P_i^--H^+ symport, and P_i^{2-}-succinate^{2-} antiport. The net result is the passage of ammonium succinate across the membrane, followed by osmotic influx of water. (After Chappell and Haarhoff, 1966, with permission of Academic Press.)

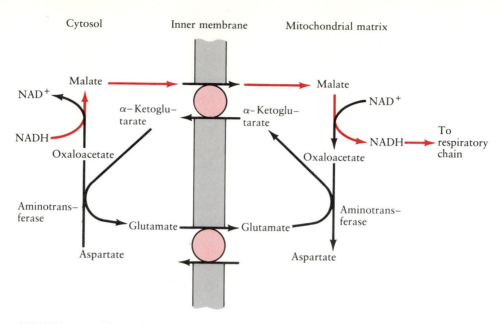

FIGURE 7.11   The malate-aspartate shuttle. This pathway for the oxidation of cytosolic NADH occurs in heart and liver mitochondria. Colored lines emphasize the effective transfer of electrons from the cytosol to the matrix. For explanation see text. (After Metzler, 1977, with permission of Academic Press.)

another that prefers K^+ to Na^+ and is activated by swelling of the mitochondrion. Whether mitochondria exercise homeostatic control over the pH of the matrix, as bacteria do, does not appear to be known; this would be a plausible second function for mitochondrial cation porters.

Much of the present excitement in mitochondrial transport centers not so much on the role of porters in the physiology of the organelle but on the mechanism of transport per se. The ATP-ADP antiporter is a major constituent of the mitochondrial inner membrane; in heart mitochondria it is the single most abundant protein. Martin Klingenberg and his associates have isolated it, reconstituted it into lipid bilayer membranes, and characterized its modus operandi in detail (Klingenberg, 1980). Thanks to this splendid work the ATP-ADP antiporter is at present the best-studied secondary carrier, a paradigm for the entire field of biological transport.

The monomeric protein as isolated has a molecular mass of 32 kdal but its functional form is probably the dimer. The evidence (Klingenberg, 1980; Krämer and Klingenberg, 1982) strongly suggests that the porter has but a single nucleotide-binding site that alternates between two conformations; in one the binding site faces the matrix, in the other the cytosol (Figure 7.12). ATP-ADP exchange is not energy-dependent; it proceeds rapidly in fully uncoupled mitochondria, and under these conditions, the porter does not discriminate between the two nucleotides. When the mitochondria are energized by respiration, the characteristics of transport are sharply altered: ADP is drawn inward while ATP is expelled from the matrix, each with a twentyfold preference over the other nucleotide. The selectivity is clearly due to the establishment of an electric potential, matrix negative: ATP is effectively extruded by electrophoresis. Kinetic data suggest that no changes in affinity are involved, only the preferential reorientation of carrier centers loaded with ATP. Klingenberg's interpretation (Figure 7.12) calls for a symmetrical porter whose carrier center bears three positive charges. When ATP^{4-} binds to the site, the complex as a whole bears one negative charge and will tend to reorient away from the electronegative (matrix) surface. By contrast, binding of ADP^{3-} produces an uncharged complex that does not sense the gradient of electric potential. However, due to the accumulation of carrier centers at the cytosolic aspect of the membrane, ADP uptake will also be enhanced.

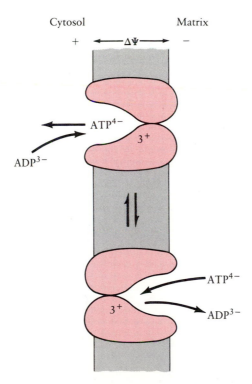

FIGURE 7.12 The ATP-ADP antiporter. Binding of ATP^{4-} to the carrier center at the matrix surface produces a negatively charged complex, which reorients in response to the electric field, expelling ATP. Binding of ADP^{3-} to the center produces an electroneutral complex, which does not sense the field. (After Klingenberg, 1980, with permission of Springer-Verlag.)

Secondary Functions of the Proton Circulation

The object of the collaboration between the respiratory proton pumps, the F_1F_0 ATP synthase, and the porters for nucleotides, P_i, and organic metabolites is the efficient transduction of the free energy of oxidative reactions into the free energy of ATP hydrolysis. However, oxidative phosphorylation is not the only service that mitochondria render to the cells that harbor them. Some of their secondary functions deserve mention here, if only to illustrate once again the adaptability of the proton circuitry.

The Regulation of Cytosolic Calcium Levels.   It has been known for a quarter of a century that mitochondria from liver and heart accumulate Ca^{2+} ions avidly. This activity takes precedence over ATP synthesis, and Ca^{2+} was, for a time, considered an uncoupler of oxidative phosphorylation. Today this puzzling effect fits neatly into our growing comprehension of the central role of calcium ions in regulating a multitude of cellular functions (Chapters 11 to 14). Recent reviews on mitochondrial calcium transport and its physiological significance have been done by Saris and Åkerman (1980) and Nicholls and Åkerman (1982).

Calcium uptake is almost certainly mediated by a specific uniporter, characteristically sensitive to inhibition by the dye ruthenium red, which mediates electrophoretic transport in response to the electric potential across the inner membrane (Figure 7.13). Since Ca^{2+} is divalent, a membrane potential of

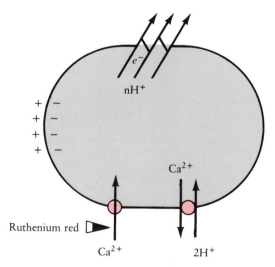

FIGURE 7.13   Calcium cycling in rat liver mitochondria. Ca^{2+} accumulates electrophoretically in response to $\Delta\Psi$ but is discharged by the electroneutral antiporter, which tends to balance the concentration gradient of Ca^{2+} against that of protons. The steady-state distribution of Ca^{2+} between cytosol and matrix is a function of the activities of both porters.

−180 mV can in principle sustain a concentration gradient of 10^6 between matrix and cytosol. An electrophoretic uniporter accounts for the enormous calcium capacity of mitochondria and for the high rate of uptake, but it poses another problem. The concentration of free calcium ions in the matrix is known to be about 1 mM; it cannot be much larger, or Ca^{2+} would form an insoluble precipitate with the P_i that is also present there. If matrix Ca^{2+} were in electrochemical equilibrium with that of the cytosol, the latter should be vanishingly low, around 10^{-9} M. In fact, the cytosolic free Ca^{2+} concentration is about 10^{-6} M, and so the two calcium pools cannot simply be in equilibrium. The capacity of mitochondria to modulate calcium uptake can also be demonstrated in vitro: respiring liver mitochondria in suspension maintain the free calcium concentration in the medium near 10^{-6} M, accumulating Ca^{2+} when the external level is higher and releasing Ca^{2+} when it is lower. The homeostatic mechanism depends on the presence of a second calcium carrier, a Ca^{2+}-$2H^+$ antiporter. Being electroneutral, this carrier tends toward an equilibrium distribution of Ca^{2+} across the membrane quite different from that favored by the uniporter: as Ca^{2+} accumulates in the matrix and the pH rises with the extrusion of protons, release of calcium by the antiporter is increasingly favored. In consequence, a continuous slow cycling of calcium takes place across the membrane. The kinetic parameters of the two transport systems are such as to stabilize the steady-state level of external free Ca^{2+} between 10^{-6} and 10^{-7} M. (A variation on this theme is found in heart and brain mitochondria, in which calcium efflux is mediated by a Ca^{2+}-$2Na^+$ antiporter.)

The physiological importance of mitochondria in controlling the cytosolic Ca^{2+} level is still under debate, and it is likely to vary from one kind of cell to another. In muscle, cytosolic Ca^{2+} is controlled by the sarcoplasmic reticulum with its ATP-driven calcium pump (Chapter 11). Yeast mitochondria typically lack the capacity to accumulate calcium. But for nerve axons Brinley (1978) obtained clear evidence that mitochondria are the chief agents of calcium regulation. Axons were loaded with Ca^{2+} by repetitive electrical stimulation, while the axoplasmic free Ca^{2+} level was continuously monitored with the indicator dye arsenazo-III. The axoplasmic calcium level rose to 3 μM and then held steady, with excess calcium accumulating in the mitochondria.

Thermogenesis: Keeping Warm with Protons.   When we are cold, we shiver. Shivering is due to fine, uncoordinated muscular contractions that consume ATP, stimulate respiration, and dissipate the free energy as heat. It is an effective response but in many ways an inconvenient one, and many animals supplement it by oxidizing stored fat in response to cold stress through a mechanism that bypasses ATP generation. This specialized metabolic pathway is confined to the brown adipose tissue possessed in abundance by hibernators, small mammals, and also newborn lambs and humans. The cells of brown adipose tissue are packed with mitochondria, each dense with cristae, that have the remarkable capacity to switch from coupled ATP-producing respiration to an uncoupled mode that generates heat. The subject has been reviewed by D. G. Nicholls, whose laboratory has contributed most notably to the clarification of the molecular basis of thermogenesis (Nicholls, 1979; Nicholls and Locke, 1984).

Energy dissipation by brown adipose tissue is a complex business, subject to regulation at several levels. Mitochondria from cold-adapted or hibernating animals were found to respire furiously but made no ATP and exhibited no respiratory control. This is due in part to the presence of free fatty acids; removal of fatty acids by the addition of lipid-free serum albumin improved the performance of the mitochondria. A second control point was revealed by the discovery that the addition of 2 mM GTP, ATP, and many other nucleotides restored full coupling and respiratory control. These exert their effects by modulating the proton conductance of the inner membrane. Mitochondria isolated from the tissues of cold-adapted animals exhibit a proton conductance a hundredfold higher than that of liver mitochondria and maintain no detectable proton potential. Removal of fatty acids allows them to maintain a proton potential of about -80 mV, but on addition of GTP the proton conductance falls to the normal range, allowing Δp to go as high as -220 mV.

The effect is a very specific one. Purine nucleotides bind to sites on the outer face of the inner mitochondrial membrane. Photoaffinity labeling with 8-azido-ATP led to the identification of a peptide of molecular mass 32 kdal, which is thought to provide a proton-conducting pathway that short-circuits the membrane much as uncouplers do. The abundance of this protein is subject to physiological regulation: it is absent from most mitochondria but may constitute as much as 10 percent of the inner membrane protein in mitochondria from the brown adipose tissue of cold-adapted animals. In addition, free fatty acids (released by lipolysis, itself under hormonal control) bind to the 32-kdal protein at a site different from that which recognizes nucleotides, and override the latter's effects. It is the interplay between the antagonistic effects of nucleotides and free fatty acids that modulates the proton conductance and permits the controlled generation of metabolic heat.

The teleonomic utility of what appears, from the thermodynamic viewpoint, a wasteful dissipation of energy has also been exploited by certain plants. The floral spikes of the skunk cabbage (and of other members of the Arum family) maintain a temperature as much as 20°C above ambient. This helps vaporize odoriferous substances that attract insects to fertilize the flower. Here again uncoupled respiration is involved (Meeuse, 1975; Moore and Rich, 1980). Respiratory rates rise sharply at a certain stage of development to intensities comparable to those of hovering birds, but without ATP production. Apparently the effect does not involve changes in proton conductance. Rather, a modified respiratory chain is produced that does not translocate protons and releases all the energy as heat.

Transhydrogenation of Nicotinamide Adenine Nucleotides. Mitochondria, like bacteria, maintain their NADP(H) pool in a more reduced state than their NAD(H) pool. The functional significance of this differential is uncertain, but much attention has been paid to the transhydrogenase that reversibly transfers reducing equivalents between the nicotinamide adenine nucleotides to an extent determined by $\Delta\tilde{\mu}_{H^+}$ across the inner mitochondrial membrane (for a recent review see Rydström et al., 1981). As outlined in Chapter 5, the transhydrogenase spans the membrane. Given a protonic potential of the usual polarity, the enzyme transfers two reducing equivalents from NADH to NADP$^+$ and concurrently translocates two protons (or possibly just one) from the cytosol to the matrix. The exergonic flux of protons provides the driving force for the endergonic reduction step. Mitochondrial transhydrogenase has been purified, reconstituted into liposomes, and shown to mediate both processes in a strictly coupled fashion.

The transhydrogenase, like the respiratory complexes, mediates a chemiosmotic reaction that interconverts the free energies of a redox potential and of the protonic potential, but the former has some distinctive mechanistic features. It has been established that the transhydrogenase transfers 2H directly from one nucleotide to the other, without interaction with the protons of water, and that there are separate catalytic sites for NAD(H) and NADP(H), both of which face the matrix. The reaction cannot be written as a redox loop, and hypothetical mechanisms have been based on redox-linked conformational transitions in the catalytic protein. There is evidence that a proton potential, matrix alkaline and negative, enhances the binding of NADH and NADP$^+$ to the enzyme while favoring the release of NAD$^+$ and NADPH, and also that the enzyme is dimeric. Taken together, these observations suggest that the two catalytic sites function alternately; just what transpires at each site remains to be specified.

It would also be helpful to know the physiological function of the transhydrogenase. The most plausible suggestion at present is that it generates reducing power for biosynthetic reactions in the matrix, such as the production of fatty acids and steroids.

Transition.   The object of this section and of the preceding one was to provide an overview of mitochondrial energetics. Not all investigators would subscribe to every point; still, by and large, the positions stated here reflect a broad consensus of qualified opinion. The remainder of this chapter is concerned with issues on which no consensus exists: the stoichiometry and thermodynamics of oxidation and phosphorylation, the nature of the proton current, and the molecular mechanics of energy transduction by respiratory complexes and by the F$_1$F$_0$ ATP synthase. It is obviously not possible within the confines of a single and relatively nontechnical chapter to do justice to all the points of view that have been expressed or to the experimental findings from which they stem. It is also not my place to adjudicate differences that those most intimately involved have failed to resolve. My object is rather to help the reader gain some sense of what the issues are and why they are important. For however arid these disputes may appear, they revolve around questions that are central to an understanding of the molecular nature of energy coupling.

Stoichiometry and Thermodynamics of Phosphorylation

It has long been accepted that the passage of a pair of electrons from NADH to oxygen is accompanied by the synthesis of three molecules of ATP. Only two molecules of ATP are formed during the oxidation of succinate, which funnels electrons into the respiratory chain at the quinone level (Figures 7.3 and 7.4). This stoichiometry was originally inferred from concurrent measurements of the quantities of substrate oxidized, oxygen consumed, and ADP phosphorylated, and the numbers obtained were rarely integers. As long as the coupling of respiration to phosphorylation was assumed to be chemical, and therefore stoichiometric, nonintegral values could be attributed to experimental shortcomings and rounded off to the next higher integer. The conventional statement that the P/O ratios for NADH and succinate are 3 and 2 respectively is thus somewhat laden with prechemiosmotic conceptions. From the chemiosmotic perspective, P/O ratios may well be nonintegral. They do nevertheless reflect the overall stoichiometry of energy transduction and must be accounted for.

Another touchstone for any comprehensive theory of oxidative phosphorylation is the conservation of free energy. From Table 2.2, $\Delta G^{\circ\prime}$ for the oxidation of NADH is -220 kJ/mol (-52 kcal/mol); under physiological conditions, ΔG is closer to -212 kJ/mol (-51 kcal/mol). Measurements of the maximum phosphorylation potential ΔG_p (Equation 3.10) attained by respiring mitochondria yield a value near 65 kJ/mol (16 kcal/mol; Slater et al., 1973). If three molecules of ATP are produced in the reaction, virtually all the free energy of oxidation must be conserved and phosphorylation must be very nearly in thermodynamic equilibrium with oxidation. The terminal reaction in respiration, catalyzed by cytochrome oxidase, is kinetically irreversible, and therefore true equilibrium over the entire redox span cannot be demonstrated. However, the facile reversal of the normal direction of electron flow between NADH and succinate in the presence of ATP ("reversed electron transport") shows that this segment of the respiratory chain does come into equilibrium with phosphorylation. Two important consequences follow from these observations. First, there is very little room for dissipation of the protonic potential by leakage: proton fluxes must be tightly confined to useful channels. Indeed, direct estimation of the basal proton conductance of the inner mitochondrial membrane shows it to be on the order of 5×10^{-7} siemen, little greater than that of a pure phospholipid bilayer. Second, the magnitude of the proton potential and the stoichiometries of proton flux must be consistent with the high efficiency of energy conservation.

The original measurements of Mitchell and Moyle (1967b), confirmed by Hinkle and Horstman (1971) and by other investigators, led to the conclusion that six protons were translocated out of the matrix during the oxidation of NADH and four during the oxidation of succinate, that is, two protons per coupling site. These numbers were compatible with the accepted P/O ratios of 3 and 2, respectively, and with parallel measurements indicating that the F_1F_0 ATP synthase translocates two protons per cycle (Table 7.2). All the other translocations required for phosphorylation appeared to be electroneutral. Measurements

TABLE 7.2 Stoichiometry of Proton Translocations in Electron Transport and Oxidative Phosphorylation*

Segment	After Mitchell and Moyle		After Wikström and Krab		After Hinkle		After Lehninger et al.	
	Out	In	Out	In	Out	In	Out	In
NADH-UQ	$2H^+$	–	$3H^+$	–	$4H^+$	–	$4H^+$	–
Succinate-UQ	0	0	0	0	0	0	0	0
UQ–cyt. c	$4H^+$	0	$4H^+$	0	$4H^+$	0	$4H^+$	0
Cyt. c–O_2	–	$2e^-$	$2H^+$	0	$2H^+$	0	$4H^+$	0
F_1F_0 ATP synthase	–	$2H^+$	–	$2H^+$	–	$3H^+$	–	$3H^+$
ATP-ADP translocator	0	0	–	$1e^-$	–	$1e^-$	–	$1e^-$
P_i symporter	0	H^0	0	H^0	0	H^0	0	H^0
Overall	$6H^+/2e^-$ $2H^+/ATP$ P/O = 3		$9H^+/2e^-$ $3H^+/ATP$ P/O = 3		$10H^+/2e^-$ $4H^+/ATP$ P/O = 2.5		$12H^+/2e^-$ $4H^+/ATP$ P/O = 3	

* Numbers refer to translocations that accompany the passage of a pair of electrons from NADH to oxygen.

of the protonic potential, first by Mitchell and Moyle and later by other laboratories, gave values close to -220 mV. Given that $\Delta G_p = -nF\,\Delta p$ (Equation 3.12) and that $n = 2$, a Δp of -220 mV could sustain a phosphorylation potential of 42 kJ/mol (10 kcal/mol). However, Slater et al. (1973) demonstrated that ATP synthesis can proceed against a far steeper gradient of chemical potential, up to a ΔG_p of 65 kJ/mol (16 kcal/mol). The various thermodynamic parameters can only be reconciled by assuming that at least one is incorrect. It now appears that part of the explanation is that n, the number of protons translocated during each cycle of ATP synthesis, is not 2 but 3 or even 4.

An important contribution to the debate came from the work of Klingenberg and his colleagues (Klingenberg, 1980), who showed that ATP-ADP exchange is electrogenic, exporting one negative charge. Since one proton enters the matrix electroneutrally by symport with phosphate, formation of each ATP molecule entails the effective uptake of one proton in addition to those translocated by the ATP synthase. Another consequence of electrogenic nucleotide antiport is that the phosphorylation potential of ATP in the mitochondrial matrix should be lower than that in the cytosol; this expectation is borne out by experiment.

In order to determine the number of protons carried by the F_1F_0 ATP synthase itself it is best to work with submitochondrial particles, for this evades complications due to the operation of porters for substrates and nucleotides. One approach is to estimate the number of protons taken up during the hydrolysis of a small pulse of ATP (Figure 7.6; Mitchell and Moyle, 1968; Thayer and Hinkle, 1973). The result, after correction for backflow, is near $2H^+$ per ATP but may still be too low. An alternative approach is to measure both ΔG_p and Δp precisely and then calculate n from the ratio of these parameters using Equation 3.12; the calculation assumes an equilibrium relationship between ΔG_p and Δp and would be invalid if protons traveling from the respiratory assemblies to the ATP synthase avoided equilibration with the bulk water phase. Nicholls (1974, 1982) concluded that n is probably 2, but could be larger. A recent paper by Berry and Hinkle (1983) notes the many pitfalls and describes the corrections and painstaking labor required to obtain consistent results. These authors favor a ratio of $3H^+$ per ATP, and recent data from chloroplasts and bacteria (likewise based on the thermodynamic approach) also converge on this figure. If so, the total number of protons translocated by intact mitochondria for each molecule of ATP synthesized would be 4.

So far, so good. But if H^+/ATP is effectively 3 or 4 and six protons are extruded by the respiratory chain, then the P/O ratio for NADH oxidation must be less than 2. Some investigators do argue that the traditional P/O ratio of 3 is too high, but most authorities take another view. Lehninger's laboratory initiated a systematic reexamination of the number of protons extruded during respiration (Brand et al., 1976). They concluded that this number had been significantly underestimated in all earlier studies because insufficient allowance had been made for the backflow of protons during a respiratory pulse (Figure 7.5). During the period of anaerobiosis that precedes the addition of oxygen, P_i leaks out of the mitochondria but is quickly reabsorbed by symport with protons during the first few seconds of respiration. When the phosphate porter was blocked with sulfhydryl reagents, additional protons could be measured in the external medium, a total of eight or nine during the oxidation of NADH.

These observations were rejected by Mitchell and Moyle, who have stoutly defended their original findings. By now, several laboratories using various methods have reported values ranging from six to 12 protons for each electron pair traversing the respiratory chain, and the subject has become a minefield that it would be foolhardy to enter. On the other hand, the stoichiometries of proton translocation are fraught with implications for both thermodynamics and mechanisms. Table 7.2 summarizes representative data from four of the warring laboratories. The differences turn mostly on matters of methodology, which are analyzed in all recent papers (see for example Alexandre et al., 1978; Mitchell, 1979b, 1981a; Wikström and Krab, 1979; Wikström et al., 1981; Hinkle, 1981; Reynafarje et al., 1982; Costa et al., 1984). Rather than attempting to render judgment, let us see what the numbers may mean.

Mitchell and Moyle maintain their original position, whose implications were noted above. According to Wikström, Nicholls, and their collaborators, three protons are extruded by the NADH-ubiquinone reductase, four by the ubiquinone–cytochrome-c reductase, while cytochrome oxidase translocates two more. Taking H^+/ATP to be 3, these numbers are consistent with measured values for the protonic potential, with the free energy available from the three redox spans and with the traditional P/O ratios of 3 and 2 for NADH and succinate oxidation, respectively. Hinkle (1981; Berry and Hinkle, 1983) finds H^+/ATP to be 4 but argues that the number of protons extruded during respiration is closer to 10; this makes the P/O ratio for NADH oxidation 2.5. Lehninger and his colleagues find that 12 protons are ejected between NADH and O_2. Assuming that the correct value for Δp is -180 mV, their data fall just within the limits prescribed by Equation 3.13, namely, $2 \times 1100/180 \simeq 12$. With an H^+/ATP stoichiometry of 4, Δp would suffice (by Equation 3.12) to sustain the maximum observed phosphorylation potential of 16 kcal/mol. The experimental data thus range to the limits set by thermodynamics. I shall not pretend to know the truth of the matter, but must caution with James Thurber that there is no safety in numbers, or anywhere else.

Are Proton Currents Localized?

That energy is transmitted between the respiratory chain and the F_1F_0 ATP synthase by means of a proton current is now all but universally acknowledged. However, students of bioenergetics remain at odds over a more subtle question: Is the proton current through the aqueous phases the sole link between the exergonic and the endergonic limbs of the energy-transducing machinery, or could there be some more intimate mode of coupling that is not reflected in the protonic potential across the mitochondrial membrane?

The roots of this issue reach back more than a quarter of a century. About the time that Mitchell was formulating his chemiosmotic hypothesis, R. J. P. Williams (1962) independently developed the idea that respiratory chains generate protons by charge separation and that they are coupled to phosphorylation via those protons. However, whereas Mitchell's protons were osmotically active in the bulk water phase, Williams's were confined to anhydrous locales within the membrane proper. "Energized" protons would be channeled to a proximal ATP synthase, where they effect the elimination of water from ADP and P_i to form ATP. In his more recent writings, Williams (1978) has shifted emphasis from the location of the protons to the teleonomic importance of restricting the pathways of proton flux. He argues that equilibration with spatially extensive water phases must dissipate much of the free energy of the coupling protons; efficient energy coupling requires tight channeling, subject to kinetic control rather than to thermodynamic constraints. Williams's views were never as explicitly formulated as Mitchell's; they were not as amenable to experimental testing and have had less influence on the evolution of bioenergetics. Nevertheless, they answer to a persistent feeling among many mitochondriologists that the association between respiratory complexes and F_1F_0 ATP synthases is closer than chemiosmotic dogma would allow. In recent years, a growing number of investigators have proposed that protons translocated during respiration may be guided to the synthase without equilibrating with protons in the bulk phase. There is little experimental support for intramembrane protons in Williams's original sense, but much interest in "localized" protons or protonic microcircuits.

The observations that sustain this somewhat vague notion usually take the form of quantitative discrepancies. There are more than a few reports of experimental protocols in which the measured proton potential, $\Delta\tilde{\mu}_{H^+}$, is too low to account for the phosphorylation potential, ΔG_p, if three protons are allowed for each ATP produced. To cite but one of these papers, Holian and Wilson (1980) found that their mitochondrial suspensions generated a Δp of -160 mV, while ΔG_p attained 67 kJ/mol (16 kcal/mol). These could be in thermodynamic equilibrium via Equation 3.12 only if n were 4 to 5 and variable to boot. To be sure, these measurements may be at fault or the H^+/ATP ratio may in fact equal 4, but localized proton microcircuits between respiratory proton pumps and ATP synthases specifically affiliated with them are a legitimate alternative. Let us recall that discrepancies of this kind are not confined to mitochondria: analogous findings have been made with chloroplasts, chromatophores, and halobacteria. The extreme case of the alkalophilic bacteria was noted in Chapter 4.

Further inconsistencies come from experiments with inhibitors. Classical chemiosmotic theory predicts that on addition of proton-conducting uncouplers or of inhibitors of respiration, ΔG_p should decline in parallel with $\Delta\tilde{\mu}_{H^+}$. But this is often not what one observes; instead, certain inhibitor concentrations largely collapse $\Delta\tilde{\mu}_{H^+}$ but have distinctly less effect on ΔG_p. The phosphorylation potential may be more closely correlated with the *rate* of respiration (Wilson and Forman, 1982; Zoratti et al., 1982; additional examples are cited in reviews by

Kell, 1979; Ferguson and Sorgato, 1982; and Westerhoff et al., 1984). Such findings can be best accommodated within the framework of energy transduction by a proton current if each respiratory complex donates protons to an associated ATP synthase, without loss of potential energy by equilibration with the bulk phase.

Localized proton currents are so appealing that one must guard against swallowing them whole. Other investigators, using different protocols, showed that $\Delta\tilde{\mu}_{H^+}$ builds up as rapidly as ΔG_p and is both kinetically and thermodynamically competent to be the driving force for phosphorylation, that one or at most a few molecules of ionophore suffice to uncouple an entire phosphorylating vesicle, and that mitochondria can store energy in the form of a protonic potential under conditions that exclude localized coupling (Thayer and Hinkle, 1975; Lemasters and Hackenbrock, 1980). Besides, the physical basis of the alleged proton microcircuit is not evident: just what would confine the localized protons to a restricted pathway and prevent them from equilibrating with the bulk phase? Finally, in chloroplasts and also in halobacteria, proton pumps are often spatially distant from any F_1F_0 ATP synthase; in these cases there is no plausible alternative to a delocalized proton current in the classical sense.

For my part, I suspect that there is less to this issue than meets the eye. Calculations show that mitochondrial membranes and the adjacent matrix are tightly packed with proteins (Srere, 1982), leaving little open space for diffusion. Perhaps one should envisage the mitochondrial membrane as a loose multienzyme complex in which protons may be quickly whisked from source to adjacent sink. Whether they enter the bulk phase or remain within a more localized domain could depend on such factors as the extent of ordered water, the degree of swelling, and the local distribution of charged lipids and proteins (Kell, 1979; Skulachev, 1982; Haines, 1983). Experimental conditions may well exert subtle effects on the partitioning of protons between localized and delocalized pathways. The persistent discrepancies do not seriously challenge the principle of energy transduction by a current of free protons, but they do remind us that there is still much to be learned about the structural context in which it takes place.

Proton Translocation by the Respiratory Chain

The reductionist road descends inexorably from the respiring cell to the phosphorylating mitochondrion, thence to the coupling membrane with its chemiosmotic proton circuits, and at last confronts us with a skeletonized version of the original enigma: What are the molecular mechanisms by which redox complexes and the F_1F_0 ATPase couple chemical transformations to the movement of protons across a membrane? The central chemiosmotic dogma, that energy is transmitted by a current of protons, does not depend on the mechanisms by which the current is generated or utilized. Historically, however, the chemiosmotic hypothesis grew out of Mitchell's fascination with the concept of chemical reactions organized in space (Chapter 3), and its formal presentation (Mitchell, 1966) included explicit ligand-conduction mechanisms for proton translocation by both redox reactions and the ATP synthase. Contemporary debates over molecular mechanisms turn on the question, how useful is the principle of ligand conduction for the analysis of real osmoenzymes?

Redox Loops and the Q Cycle.   Originally Mitchell proposed that the mitochondrial respiratory chain is bent into three redox loops, as shown in Figure 7.14a. (This particular version, which differs somewhat from Mitchell's own scheme, is from V. P. Skulachev.) Each loop consists of two limbs. In the first limb a hydrogen carrier undergoes a protolytic reaction. Two protons are discharged while two electrons are conducted back across the membrane by an electron-carrying "wire" that constitutes the second limb. The heart of the matter is that the reaction pathway of each loop is supposed to pass twice across the membrane (more precisely, across the barrier phase) and that proton translocation is a necessary consequence of this topological arrangement.

Redox loops make explicit predictions concerning the chemistry of respiratory carriers and their positions in space. The chain must consist of alternating carriers for hydrogen and for electrons, the topology must be such that catalytic sites are accessible alternately from the matrix and from the cytosol, and the stoichiometry cannot exceed two protons per electron pair passing over the loop. Several of these predictions have been confirmed. For example, dehydrogenases are (usually) found on the matrix side of the membrane, while cytochromes c_1 and c are exposed to the cytosol. More spectacular corroboration came from the work of P. C. Hinkle with purified cytochrome oxidase inlaid into liposomes; the cytochrome aa_3 complex was clearly shown to span the membrane and to translocate electric charge across it. Moreover, as outlined in the preceding section, measurements of the number of protons translocated during respiration supported the redox loop model.

However, the 1970s also saw the accumulation of findings inconsistent with Figure 7.14a. Most glaringly, the hypothetical hydrogen carrier in the third loop, designated Z, failed to materialize. Instead, there were two spectroscopically distinguishable b-type cytochromes, b_{566} and b_{562}, that appeared to be involved in energy transduction but had no obvious place in the scheme. In order to resolve these and other difficulties, Mitchell proposed a fiendishly clever arrangement that effectively fuses the second and third loops to produce the proton-motive Q cycle (Mitchell, 1976b). Figure 7.14b shows his modified respiratory chain in summary fashion, while a current version of the Q cycle is displayed in Figure 7.15.

The Q cycle was devised ("while suffering from insomnia . . . at about 3 a.m. on May 20th, 1975") to account for three cardinal points. First, the only potential hydrogen carrier identifiable in the segment that contains cytochromes b and c_1 (complex III of Table 7.1) is ubiquinone. Second, four protons are extruded for each pair of electrons traversing this span (most investigators in the field agree on this point). Finally, cytochromes b_{566} and b_{562} carry electrons across the membrane but exhibit peculiar redox behavior: oxidation of cytochrome c promotes the *reduction* of the b cytochromes, particularly of b_{566}. The effect is enhanced by antimycin A, a potent inhibitor of electron flow in the b-c_1 segment.

Let us trace a turn of the Q cycle during steady-state operation (Figure 7.15). Electron transfer through the b-c_1 segment begins with the partial reduction of a molecule of ubiquinone (UQ) to the semiquinone level,[2] on the matrix side of the membrane (the semiquinone is thought to be in the anionic form, shown as $U\dot{Q}_i^-$). The electron comes from cytochrome b_{562} (see below). The semiquinone is next reduced to UQH_2, accepting a second electron from a dehydrogenase (or from complex I), with the uptake of two protons from the matrix. UQH_2 formed on the matrix side diffuses to the cytosolic surface, where it is reoxidized to the semiquinone level ($U\dot{Q}_o^-$) with the discharge of two protons to the exterior. One electron is accepted by an iron-sulfur center, named the Rieske protein after its discoverer, which is a well-characterized constituent of complex III. This electron subsequently proceeds to cytochrome c and ultimately to oxygen. Oxidation of the semiquinone $U\dot{Q}_o^-$ to UQ is completed by donation of the second electron to cytochrome b_{566}. The two b-type cytochromes span the membrane, returning the second electron to the matrix side. Overall, in each turn of the Q cycle one electron is transferred to cytochrome c and two protons are translocated across the membrane. By contrast, in an orthodox redox loop a single proton is translocated per electron.

This somewhat simplified account omits several important details that we can barely touch on here. The two semiquinone pools, $U\dot{Q}_i^-$ and $U\dot{Q}_o^-$, must be kept apart, each at the appropriate redox potential; these substances are presumably not free but bound to quinone-binding proteins that have recently been isolated. One must account for the partitioning of energy between the two electrons produced by UQH_2 oxidation at the cytosolic surface, since one electron flows downhill to cytochrome c while the other flows uphill to cytochrome b_{562} and the matrix side. The precise sequence of events during the protonation and deprotonation of quinones is debatable. Finally it is necessary to reconcile the Q cycle and its confined, dedicated quinones with earlier work that

---

[2] Ubiquinone has the structure

where R stands for a side chain of isoprene units, 10 in mitochondrial ubiquinone. Reduction of both keto groups to enols requires two electrons and is accompanied by the uptake of two protons. The semiquinone has only one enol group; free semiquinones undergo rapid dismutation but can be stabilized by binding to proteins.

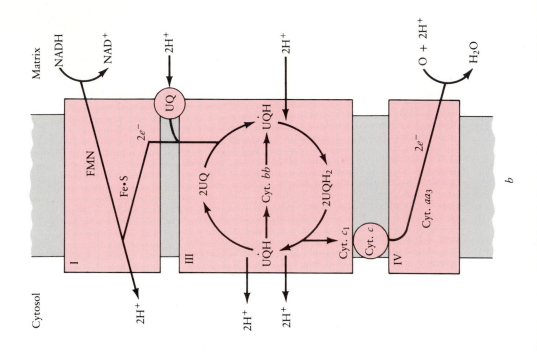

b

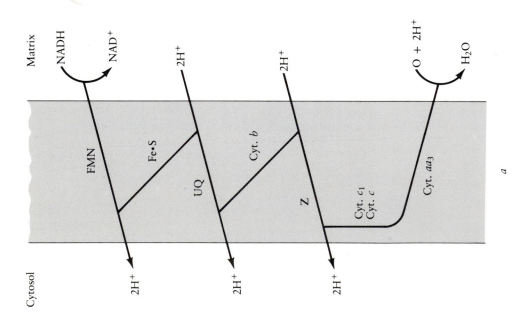

a

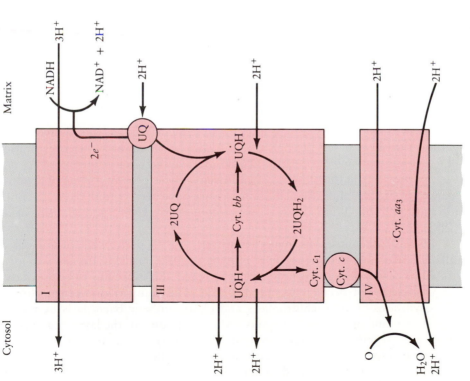

FIGURE 7.14   Changing views of the proton-motive respiratory chain of mitochondria. (*a*) Three redox loops, as envisaged around 1970. Each loop translocates two protons out of the matrix; cytochrome oxidase is the electron-carrying limb of the third loop. (*b*) Three respiratory complexes in series, after Mitchell (1976*a*, 1979*a*). The NADH-ubiquinone reductase is shown as a redox loop ($2H^+$), the ubiquinone–cytochrome-*c* reductase as a Q cycle ($4H^+$); cytochrome oxidase returns electrons to the matrix, where oxygen is reduced to water. (*c*) Revisionist model, after Wikström, Nicholls, and others. Complex I is shown as a proton pump extruding $3H^+$; complex III is shown as a Q cycle, extruding $4H^+$. Cytochrome oxidase is shown as a proton pump, transporting $4H^+$ out of the matrix for each pair of electrons; $2H^+$ are consumed in reducing oxygen to H_2O at the cytosolic surface, and $2H^+$ appear in the medium. It bears repeating that the true sidedness of oxygen reduction and all translocation stoichiometries are uncertain.

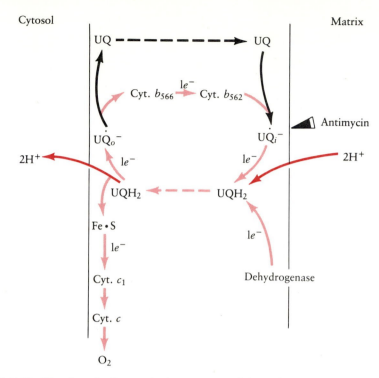

Cytosol   Matrix

FIGURE 7.15   The Q cycle. Events during one turn of the cycle are traced in the text. UQH_2 and UQ represent ubiquinol and ubiquinone; $U\dot{Q}^-$, the semiquinone anions bound at the inner and outer centers. Pink arrows, electron flow; dashed lines, diffusion or translocation; red arrows, proton flow.

identified ubiquinone as a mobile pool of reducing equivalents derived from various dehydrogenases. But one can intuitively work out how the configuration of a Q cycle explains the peculiar redox interplay between cytochromes c_1 and b_{566}, particularly in the presence of antimycin, whose probable locus of action is marked in Figure 7.15. These and other issues have been considered in detail by Mitchell (1976b), Trumpower (1981), Slater (1981, 1983), and Hauska et al. (1983).

Like the chemiosmotic hypothesis itself, the Q cycle was formulated in advance of the evidence, but it is proving to be a resilient framework for a growing body of data and is now widely accepted. It is especially to be noted that the Q cycle is not peculiar to mitochondria. The pathway of electron flow in photosynthesis by both chloroplasts and photosynthetic bacteria is best interpreted as a kind of Q cycle (Chapters 4 and 8); in some of the latter organisms a given b-c_1 segment can function both in photosynthetic and in respiratory electron transport. On the other hand, nitrate reduction and some other electron-transport pathways that include quinones and cytochrome b are at present better explained by linear redox loops. Could it be that the Q cycle evolved in a cluster of lines that gave rise to both respiration and photosynthesis?

Cytochrome Oxidase as a Proton Pump.   In the Q cycle, as in all redox loops, charge separation and proton translocation are brought about by metabolic reactions whose pathway traverses the membrane; this conception should be clearly distinguished from its alternative, a redox-driven proton pump. According to the latter hypothesis, the pathway of electron flow bears no obligatory topological relationship to the membrane in which the redox catalysts are embedded. Instead, the cyclic reduction and oxidation of particular carriers would be accompanied by sequential changes in the orientation of a proton-binding site and in its proton affinity (that is, its dissociation constant pK_a); each cycle would result in the transport of one or more protons from the matrix to the cytosol. Thus whereas redox loops effectively translocate protons thanks to the topological placement of the catalytic centers, pumps pick up protons on one side and release them on the other by virtue of conformational changes in the catalytic proteins. Historically, the redox-pump hypothesis can be traced to the recognition (by B. Chance and others) of a possible analogy between proton transport by mitochondria and the dissociation of protons during the oxygenation of hemoglobin. It owes its present popularity to the mounting evidence that protons are translocated during the reduction of oxygen by cytochrome oxidase.

In the classical view of the proton-motive respiratory chain, the cytochrome oxidase (cytochrome aa_3; complex IV of Table 7.1) serves as the electron-conducting limb of the third loop (Figure 7.14a and b). Observations in several laboratories demonstrated that cytochrome oxidase does translocate electric charge across the membrane with the expected polarity and were entirely consistent with the view that the enzyme conducts electrons from cytochrome *c* (known to be exposed at the cytosolic surface) back into the matrix, where oxygen is reduced to water with the consumption of four protons. By contrast, a series of papers by Mårten Wikström, beginning in 1972, reported that the spectrum of cytochromes aa_3 varies as a function of the protonic potential in a manner suggesting that these hemes may be involved in the transport of protons rather than electrons. Mitchell and others rejected this interpretation, attributing the observations to technical artifacts, but Wikström and his colleagues stuck to their guns and have put forward ever more compelling evidence (Wikström and Krab, 1979; Wikström et al., 1981; Wikström, 1984).

Figure 7.16 shows two of their experiments, to illustrate what is at issue. Mitochondria were incubated with inhibitors to functionally isolate the segment of the respiratory chain between cytochrome *c* and oxygen; ferrocyanide served as electron donor and K^+ plus valinomycin were present to provide charge compensation and allow proton movements to be monitored in the external medium. Oxygen reduction is described by Equation 7.1, and the reaction catalyzed by cytochrome oxidase is shown in Equation 7.2:

$$O_2 + 4H^+ + 4e^- \longrightarrow 2H_2O \qquad \text{(Eq. 7.1)}$$

$$2\text{cyt. } c^{2+} + 2H^+ + \tfrac{1}{2}O_2 \longrightarrow 2\text{cyt. } c^{3+} + H_2O \qquad \text{(Eq. 7.2)}$$

If cytochrome oxidase translocated only electrons, one would expect to see a

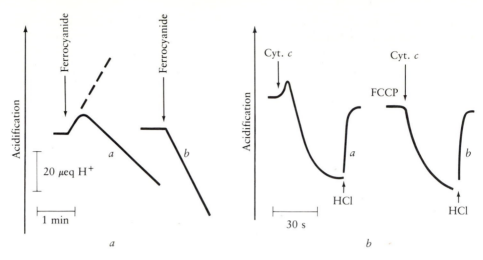

FIGURE 7.16   Proton extrusion by cytochrome oxidase. (*a*) Mitochondria were incubated in air with KCl, antimycin A, rotenone, and valinomycin. Respiration was initiated by the addition of ferrocyanide at the arrow. The trace shows the pH of the suspension. Trace *a*, addition of the electron donor elicits a transient acidification whose initial rate (dashed line) equals that of electron transport. Within less than a minute, proton extrusion is overwhelmed by the alkalinization produced by oxygen reduction. Trace *b*, in the presence of FCCP the transient extrusion of protons is not observed, but alkalinization continues. (*b*) Proton extrusion by purified cytochrome oxidase reconstituted into liposomes. The reaction mixture contained valinomycin and excess K^+; oxygen reduction was initiated by the addition of reduced cytochrome *c* at the arrow. Trace *a*, mixture as above. The blip of external protons corresponds to a pH change of less than 0.01 unit. Trace *b*, with FCCP present as well. (After Wikström and Krab, 1979, with permission of Elsevier Biomedical Press.)

monotonic alkalinization of the suspension as protons, consumed in the matrix, were replaced from the medium by diffusion across the membrane. Instead, the pH trace shows a transient *acidification* of the medium (that is, proton expulsion) whose rate is equal to that of electron flow from substrate to oxygen. The acidification soon bends off, overwhelmed by the overall consumption of protons according to Equation 7.1 (Figure 7.16*a*, trace *a*). The initial acidification was abolished by proton-conducting uncouplers, confirming that it reflects net transport of protons outward (trace *b*). Very recently these findings were supplemented by the demonstration that protons are removed from the matrix. The clincher came when it was shown that purified cytochrome oxidase reconstituted into phospholipid vesicles transports protons in the same manner as intact mitochondria do (Figure 7.16*b*): proton transport appears to be an intrinsic capacity of the cytochrome oxidase (Wikström and Krab, 1979; Casey et al., 1980; Wikström et al., 1981; Wikström, 1984).

   The issue is by no means settled to everyone's satisfaction, but I shall provisionally adopt the majority view that cytochrome oxidase translocates protons outward with a stoichiometry of $2H^+$ per electron consumed. Reduction of one oxygen molecule requires four electrons; it entails the transport of eight protons out of the matrix, of which four are consumed in the formation of water while four appear in the medium. The redox span of the reaction, $\Delta E_h \simeq 500$ mV, is sufficient to support the transport of two protons per electron; the free energy of cytochrome c oxidation would then be efficiently conserved in the form of the protonic potential. One difficulty is that it is not certain whether the site of oxygen reduction is on the matrix surface or on the cytosolic one. Figure 7.14c shows a conception of the respiratory chain that incorporates the proton-translocating cytochrome oxidase; in this instance the site of oxygen reduction is shown on the cytosolic (exterior) aspect of the inner membrane. The proton- and electron-transport stoichiometries listed in column 2 of Table 7.2 correspond to this model.

Now, the outward transport of protons by the cytochrome oxidase cannot be attributed to a redox loop, nor is there any other obvious vectorial reaction. Instead, cytochrome oxidase must be seen as a proton pump: the cycle of reduction and oxidation is linked to concurrent changes in protein conformation that in turn control the orientation and affinity of a proton-binding site (Figure 7.17). In truth, as Wikström and Krab (1979) have cogently pointed out, the distinction between loop and pump is neither as sharp nor as basic as it first appears. In a redox loop, protons are transferred as H^+ plus e^- bound directly to the redox center; in a proton pump, protons are bound to some other site whose properties are closely coupled to the redox reaction and which may be in close proximity to the redox center. A hair, perhaps, divides the false and true. . . .

We have still to specify just how this particular redox-driven proton pump works, and that cannot quite be done, but cytochrome oxidase is already so well characterized that it is likely to be among the first primary transport systems to be understood at the molecular level. Cytochrome oxidase is a large and complex structure (Table 7.1) that spans the membrane. It contains two hemes, cytochromes a and a_3, which serve different functions: cytochrome a_3 catalyzes oxygen reduction proper, while cytochrome a is part of the proton pump. The functional system is probably a dimer, (cytochrome aa_3)$_2$. These and other data led Wikström and his colleagues to propose a reciprocating-site mechanism in which the two monomers alternately occupy one of two states: the "input state" (reduction of cytochrome a, with proton binding on the inside) and the "output state" (transfer of electron to cytochrome a_3, with proton release to the outside). Analogous reciprocating mechanisms have been drawn up for the F_1F_0 ATPase (see below); they may prove to be a common feature of primary transport systems. Curiously, the linkage between electron transport and proton movements is dissociable: the inhibitor DCCD blocks proton translocation by mitochondrial cytochrome oxidase but does not interfere with oxygen reduction (Casey et al., 1980). Does DCCD plug some kind of proton channel in the oxidase, analogous to that found in the F_1F_0 ATPase?

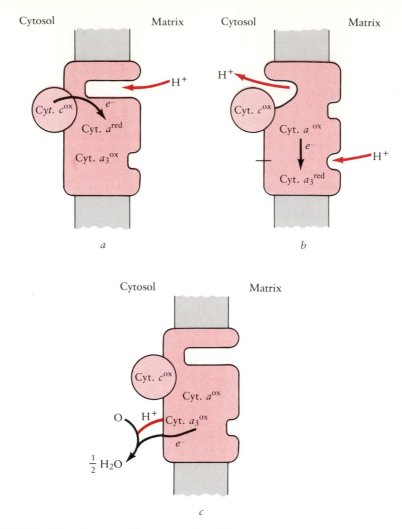

FIGURE 7.17 Cytochrome oxidase as a redox-driven proton pump. (*a*) When cytochrome *c* donates an electron to cytochrome *a*, a proton from the matrix binds to the site marked. (*b*) Intramolecular transfer of the electron to cytochrome a_3 is accompanied by a conformational change such that the proton is released to the exterior. (*c*) Reduction of oxygen by a proton derived from the matrix side. Overall, for each electron transferred by the pump two matrix protons are consumed, one of which is liberated into the cytoplasm.

How far should one generalize the conclusion that mitochondrial cytochrome oxidase is a proton pump? Von Jagow and Sebald (1980) have outlined a proton-pump mechanism for the ubiquinone–cytochrome c_1 segment of the respiratory chain as a possible alternative or complement to the Q cycle. There is at present no compelling evidence that discriminates between ligand conduction and conformational pumping of protons in this region. The NADH dehydrogenase (complex I) has proved to be particularly intricate, structurally as well as mechanistically (Ragan et al., 1981). If, as many investigators now believe, the NADH-ubiquinone segment translocates three protons per cycle, alternatives to the redox loop must be seriously considered, but the stoichiometry in this segment is particularly doubtful. The situation in bacteria was touched on in Chapter 4. It now appears likely that many bacterial cytochrome oxidases transport protons, even though their subunit structure is simpler than that of the mitochondrial oxidase. Other redox reactions, however, clearly traverse the plasma membrane as expected for a redox loop. What seems to be emerging is that both mechanisms have their place; redox loops and conformational pumps may prove to be the ends of a continuum rather than mutually exclusive alternatives.

How Cells Make ATP

With the recognition that the F_1F_0 ATP synthase is a primary transport system that couples the movement of protons across a membrane to the hydrolysis or generation of ATP, oxidative phosphorylation ceased to be a mystery, but it remains a baffling problem in vectorial enzymology. The F_1F_0 ATPases of bacteria, mitochondria, and chloroplasts comprise a distinctive family of enzymes that differ from one another only in detail. They are set apart from other ion-translocating ATPases (Chapters 5 and 9) by their large size and structural complexity, by their modular construction from F_1 and F_0 sectors, and by the absence of a phosphorylated enzyme intermediate. It seems sensible to assume that these features bespeak a common molecular mechanism, whose nature is the ultimate issue. The following status report draws on observations made with various members of the family. Recent reviews particularly concerned with the unsolved mechanistic questions have been contributed by Boyer et al. (1977), McCarty (1978), Fillingame (1980a, 1980b), Cross (1981), Mitchell (1981b), Maloney (1982), Kozlov and Skulachev (1982), and Senior and Wise (1983).

Structure of the F_1F_0 ATPase.   We begin by recalling from Chapter 4 that the functional holoenzyme is composed of two modules: the headpiece, F_1, which bears the catalytic site or sites, and the membrane sector, F_0, which channels protons across the membrane. Each of these modules is quite intricate, and their structure and assembly is at present an area of intense research effort. Much of what we know comes from work with the relatively heat-stable ATPase from the thermophilic bacillus SP3 (Kagawa et al., 1979) and with *E. coli* (Dunn and Heppel, 1981; Futai and Kanazawa, 1983). The ATP synthase of mitochondria and chloroplasts has proved somewhat less tractable (McCarty, 1978; Baird and Hammes, 1979; Nelson, 1981; Amzel and Pedersen, 1983).

In all cases the headpiece contains five subunits (Table 7.1). Microbiologists agree on a stoichiometry of $\alpha_3\beta_3\gamma\delta\varepsilon$; students of mitochondria and chloroplasts have in the past preferred $\alpha_2\beta_2\gamma\delta\varepsilon_2$, but it is increasingly likely that the differences result from technical problems and that $\alpha_3\beta_3\gamma\delta\varepsilon_{1-2}$ is the correct stoichiometry throughout. This interpretation is reinforced by recent kinetic studies indicating that each mitochondrial F_1 bears three catalytic sites. Nucleotide-binding sites are present on both the α and the β subunits, but reconstitution of ATP hydrolysis by recombination of the purified subunits calls for α and β together. The γ subunit is required as well, perhaps because it contributes the core on which the entire headpiece is assembled. Purified F_1 also contains several molecules of tightly bound ATP or ADP. These do not participate in the catalytic activities and are assigned an ill-defined structural role. The δ and ε subunits are not required for ATPase activity but are involved in attaching the headpiece to the basepiece; whether they have a functional role as well is still uncertain.

Removal of the headpiece from the membrane exposes the F_0 sector as a proton-conducting channel; proton conductance is blocked when F_1 reassociates with F_0. Isolated subunits δ and ε bind to F_0, but proton conductance is suppressed only when γ is added as well. These and other findings suggest that γ forms the "gate" that couples the proton channel to the catalytic headpiece and plays a central functional role. Biochemical data combined with electron micrographs suggest the kind of structure shown in Figure 7.18. It is intended to represent the F_1F_0 ATPase of bacteria, mitochondria, and chloroplasts, but the only assurance one can give is that it will require substantial modifications in the light of new data.

The composition of F_0, with its integral membrane proteins, is particularly uncertain. Even the number of subunits is in doubt: there are three kinds in the

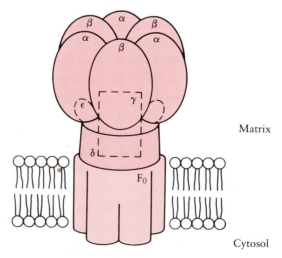

FIGURE 7.18   Structure of the F_1F_0 ATP synthase of mitochondria, chloroplasts, and bacteria. (After several authors, none of whom should be held responsible.)

thermophile SP3 and in *E. coli*; possibly four or five in mitochondria and chloroplasts. Interest centers on the smallest subunit, a proteolipid of about 8 kdal, which is present in at least six copies per F_0 module. Both chemical and genetic studies have identified the proteolipid as a major component of the proton channel. Proton conductance through F_0 is suppressed by *N,N'*-dicyclo-hexylcarbodiimide (DCCD), an inhibitor that reacts covalently with the carboxyl group of a particular residue in the proteolipid (aspartate in *E. coli*, glutamate in other organisms). Mutations that result in the replacement of that same residue by a neutral amino acid also suppress proton conductance. Only a fraction of the proteolipid in each sector needs to be modified in order to block the movement of protons. This bolsters the suggestion, again tentative, that six (possibly as many as 10) proteolipid monomers contribute to an oligomeric structure that forms the proton channel.

The term "proton channel" conjures up the image of a pore, but this simple view is certainly misleading. The evidence indicates that F_0 conducts protons (not hydroxyl ions) with high specificity but at a relatively low rate. One possible model for the F_0 sector is the ionophore antibiotic gramicidin (Figure 3.6), a cylinder whose central hole is filled with water molecules. If the proteolipids in F_0 collaborate to form an analogous structure, specificity for protons may be conferred by one of the other subunits acting as a "filter." An alternative view sees the protons as moving along a "proton wire," formed by the peptide links in helical transmembrane sections of the F_0 proteins (Nagle and Morowitz, 1978); the role of that special aspartyl residue is not obvious.

Energy Coupling and the Proton Well.   So far we have spoken of the F_0 sector as a conductor for protons, and that is certainly its most elementary function. However, the true role of the F_0 sector in energy coupling is more subtle and better expressed by Mitchell's concept of the proton well (Mitchell 1968, 1976a).

According to the chemiosmotic theory, the driving force for ATP synthesis is the protonic potential across the membrane, $\Delta\tilde{\mu}_{H^+}$. This in turn is the sum of two factors, ΔpH and $\Delta\Psi$ (Equation 3.4), whose relative contributions vary greatly. In mitochondria, $\Delta\tilde{\mu}_{H^+}$ consists largely of $\Delta\Psi$ while ΔpH is small; in chloroplasts, ΔpH predominates; and in bacteria the distribution of $\Delta\tilde{\mu}_{H^+}$ between $\Delta\Psi$ and ΔpH depends on circumstances. Numerous experiments have confirmed that either $\Delta\Psi$ or ΔpH alone can sustain ATP synthesis and that the two are additive as required by Equation 3.4 (Maloney, 1982). Yet at the mechanistic level it is far from obvious how a pH gradient and a membrane potential can exert thermo-dynamically equivalent molecular forces. This problem was solved by recogniz-ing that thanks to its structure, F_0 converts the $\Delta\Psi$ component into an equivalent ΔpH, thereby allowing $\Delta\tilde{\mu}_{H^+}$ to exert its full effect as a pH gradient across the bottom of the proton well.

In the intact F_1F_0 ATP synthase the passage of protons through F_0 is blocked by F_1 at the matrix end, but the length of the channel is open to the exterior. We may therefore think of the permeability barrier to protons as looping around F_0, turning it into a "well" (Figure 7.19). Protons residing within the well are in thermodynamic equilibrium with those in the exterior phase; therefore, so long as there is no electric potential across the membrane, the effective pH within the well will be the same as that in the cytosol. Now let us impose an electric potential, matrix negative. This tends to pull protons into the well, causing them to "accumulate" at the bottom of it; the result is to lower the effective pH in that region (Figure 7.19). In this fashion a proton well serves to transduce a difference in electric potential between the bulk phases into a pH gradient across the bottom of the well. A $\Delta\Psi$ of -180 mV, interior negative, corresponds to approximately three units of pH (Equation 3.4, setting $\Delta p = 0$); if the external pH were 7.5, the effective pH at the bottom of the well would be about 4.5. The immediate driving force for the generation of ATP is therefore not $\Delta\tilde{\mu}_{H^+}$ but a highly localized gradient of proton concentration. What remains to be done is to specify just what those protons are up to, and there's the rub.

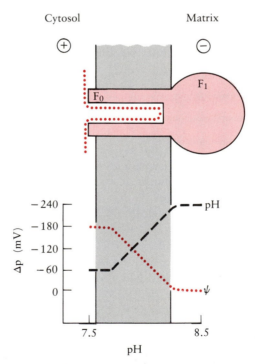

FIGURE 7.19   The proton well. The upper register shows the position of the permeability barrier to protons (dotted line), assuming that F_0 forms a channel whose matrix end is blocked by F_1. The lower part of the drawing shows in a general way the profile of electric potential Ψ and of pH across the F_0 sector, assuming a Δp of -240 mV across the membrane. At the cytosolic surface $\Delta\Psi$ contributes -180 mV and ΔpH -60 mV; at the bottom of the well Δp consists almost entirely of ΔpH. (After Mitchell, 1968, with permission.)

The ATPase Puzzle. We left the catalytic headpiece, F_1, positioned across a sharp but local gradient of pH. Given a sufficient protonic potential across the entire structure plus a supply of ADP and P_i, F_1 mediates a concerted reaction whereby ATP is formed with the elimination of water; concurrently two or three protons make the passage from the positive (cytosolic) to the negative (matrix) side of the membrane. The chemical and osmotic limbs are strictly coupled and can operate in either direction depending on the free-energy gradients. There is no evidence for the formation of any phosphorylated intermediate; the first stable product is ATP itself. Just what transpires at the headpiece remains a topic for research and argument. Two major models have emerged, each displaying traces of its ancestry in the almost philosophical principles of vectorial metabolism and conformational coupling (Chapter 3).

In Mitchell's view, the vectorial movement of protons is integrated with the chemical events at the catalytic site (Mitchell 1974, 1977*b*, 1981*b*). Figure 7.20 depicts one such direct chemiosmotic mechanism, modified to allow for the participation of three protons. The catalytic site is assumed to be situated at the interface between F_0 and F_1. The function of the proteins that constitute these sectors is to specify the particular species admitted to the catalytic site and to impose a particular geometry on them once they are there. ADP and P_i are thought to enter the catalytic site in particular states of salt formation and protonation, symbolized by $ARPP^-$ and P_i^{2-} to indicate that each of these species bears one and two fewer protons respectively than ADP and P_i do in the free state. Thus as ADP and P_i interact with F_1, three protons are liberated into the matrix space. The catalytic site serves to orient $ARPP^-$ and P_i^{2-} in a particular manner, in line with the gradient of protonic potential, as indicated in Figure 7.20*b*. The protons attack one of the oxygen atoms of P_i, forming water and leaving an extremely reactive species that can react directly with ADP to form ATP. The overall result is that during the production of one molecule of ATP, three protons are consumed at the cytosolic surface (two of which end up in water) and three protons are released into the matrix.

The major alternative mechanism, championed by Paul Boyer and his disciples and also by Kozlov and Skulachev, assigns the protons passing through the F_1F_0 complex an altogether different role (Boyer, 1975; Kayalar et al., 1977; Cross, 1981; Kozlov, 1981; Kozlov and Skulachev, 1982). Instead of participating directly in the catalytic mechanism, the protons would effect a cyclic sequence of protonations and deprotonations at unspecified sites on the F_1 sector (or even on F_0). These drive the headpiece through a sequence of conformational states, accompanied by changes in the binding affinity of the catalytic site for nucleotides, that are the salient events in energy transduction. The general principle, which is akin to the redox-pump formulation of cytochrome oxidase, is illustrated in Figure 7.21*a*. ADP and P_i are first loosely bound to a catalytic site on the headpiece, facing the matrix. On protonation of an allosteric site on or near the F_0 sector, the catalytic site is so altered as to greatly increase its affinity for ATP. This renders the spontaneous combination of ADP and P_i exergonic, producing tightly bound ATP. The enzyme then undergoes a conformational

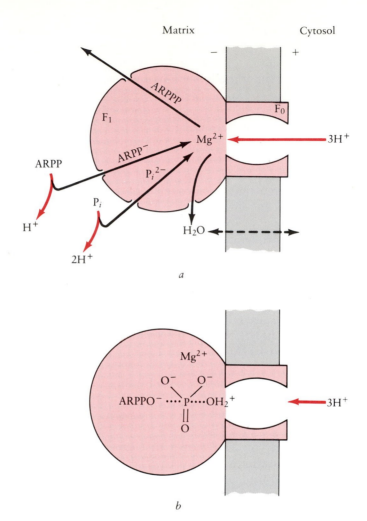

FIGURE 7.20   Ligand-conduction mechanism for ATP generation. (*a*) ADP and P_i enter the catalytic site in states of protonation corresponding to $ARPP^-$ and P_i^{2-} respectively, leaving three protons in the matrix. ATP formation entails the passage of three protons across the membrane. (*b*) Proposed orientation of ADP and P_i at the catalytic site, which is located at the interface between the F_1 and F_0 sectors. (After Mitchell, 1974, 1977*b*; Maloney, 1982.)

change that translocates the protons to the matrix side, where their electrochemical activity is much lower. As the protons dissociate, the catalytic site reverts to its original conformation, releasing ATP. The heart of the matter is that the energy-requiring step is not the formation of ATP per se, but the release of ATP from its tightly bound state.

Current versions of the mechanism are considerably more elaborate. Kinetic studies in several laboratories indicate that there are three catalytic sites per F_1

and that these interact cooperatively while passing through the sequence of energy-dependent changes in affinity (Figure 7.21b). First ADP and P_i bind to one site. On protonation of the enzyme this site is activated, producing bound ATP. Concurrently ATP is released from a second site, which is thereby inactivated, while a third site becomes available to ADP and P_i. The newly formed ATP in the first site will be released on the third turn of the cycle. It is tempting to speculate that the catalytic cycle entails the rotation of the headpiece on its core (the γ subunit) in a manner analogous to the operation of the flagellar motor (Chapter 5; Cox et al., 1984).

Without permitting ourselves to become enmeshed in the argument over the relative merits of such models, we may briefly consider the kinds of data that bear on them. Models that invoke changes in binding affinity are in principle consistent with any H^+/ATP stoichiometry; Mitchell's model, though it fits a stoichiometry of $2H^+/ATP$ most neatly, can be reformulated to accommodate three or even four protons. Certain data suggest that the catalytic site is buried within the body of F_1; it goes without saying that the placement of the catalytic site at the surface of F_1 in Figure 7.21 is intended only for clarity, since both schemes demand sequestered catalytic sites. It has been known for a decade that the headpiece undergoes substantial changes in conformation during its operation. For example, in "energized" chloroplasts the exchange of ^3H between water and certain amino acid residues is accelerated, and the ATPase becomes susceptible to various inhibitors and reagents that modify its structure. Such findings have been cited in support of Boyer's model, but conformational transitions are equally to be expected with Mitchell's model, as part of the process by which nucleotides and P_i enter or leave the catalytic site. Surprisingly, recent studies on ATP synthesis in D_2O suggest that changes in protein configuration are not an essential part of energy coupling (Khan and Berg, 1983). Perhaps the most persuasive data stem from studies on various isotope-exchange reactions catalyzed by F_1F_0 ATPase, which may be regarded as partial reactions that take place at the catalytic site. In particular, exchange of ^{18}O between P_i and water, unlike net ATP synthesis, is not inhibited by proton-conducting uncouplers (Boyer et al., 1973). This discovery suggested that the reaction $ADP + P_i \rightleftharpoons ATP$ proceeds readily in the absence of a protonic potential and prompted the formulation of the model shown in Figure 7.21. Recent researches (see for example Chernyak et al., 1981; Cross et al., 1982; Cross and Nalin, 1982) strongly reinforce the view that the binding and release of nucleotides, rather than the chemical step of ATP formation as such, are the sites of energy input. The balance of the evidence presently favors the view that the passage of protons through the F_1F_0 ATP synthase drives cyclic changes in the nucleotide affinity of the catalytic sites, as suggested in Figure 7.21. Nothing is known about the spatial relationship between the sites that bind nucleotides and those that bind protons.

There is certainly a significant conceptual distinction between "direct" chemiosmotic mechanisms (Figure 7.20) and "indirect" ones (Figure 7.21), but I suspect that a new generation of investigators equipped with different spectacles will come to see it as being more semantic than physical. The chemical and

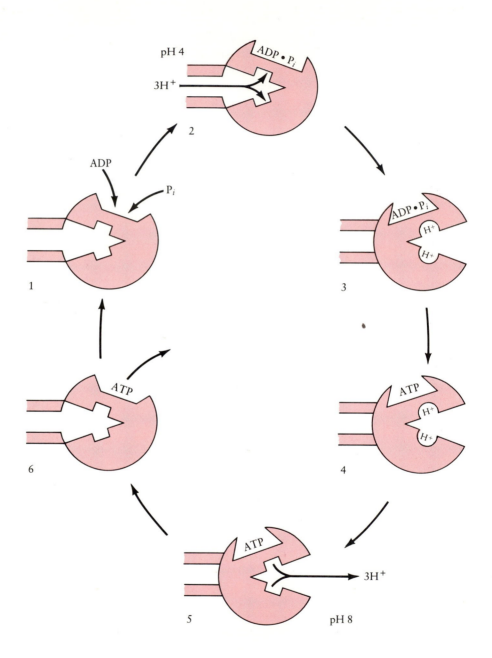

pH 4

ADP · P_i

3H⁺

2

ADP

P_i

1

ADP · P_i

H⁺

H⁺

3

ATP

H⁺

H⁺

4

ATP

6

ATP

5

ATP

3H⁺

pH 8

a

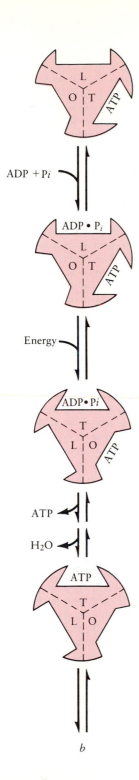

FIGURE 7.21   Generation of ATP by proton-linked changes in nucleotide binding affinity. (*a*) The Boyer mechanism in principle. A turn of the cycle is traced in the text. (After Nicholls, 1982, with permission of Academic Press.) (*b*) Catalytic cooperativity among three interacting sites. The scheme shows one-third of an enzyme cycle. Sites on F_1 are interconverted among three forms: O, open site with very low affinity for ligands and catalytically inactive; L, loose binding of ligands and catalytically inactive; T, tight binding of ligands and catalytically active. (After Cross, 1981; reproduced with permission of the *Annual Review of Biochemistry,* vol. 50, 1981, by Annual Reviews.)

genetic dissection of the ATP synthase that is now proceeding in several laboratories should in time define unambiguously just what groups undergo protonation and what roles they play in the reaction. Only when that has been accomplished can we be fully satisfied that the long quest for a chemical explanation of oxidative phosphorylation has attained its objective. In the meanwhile, however, cell physiologists can take comfort from what has already been achieved: in principle, albeit not in detail, we understand even now how cells make ATP.

References

ALEXANDRE, A., REYNAFARJE, B., and LEHNINGER, A. L. (1978). Stoichiometry of vectorial H$^+$ movements coupled to electron transport and to ATP synthesis in mitochondria. *Proceedings of the National Academy of Sciences USA* **75**:5296–5300.

AMZEL, L. and M., PEDERSEN, P. L. (1983). Proton ATPases: Structure and mechanism. *Annual Review of Biochemistry* **52**:801–824.

BAIRD, B. A., and HAMMES, G. G. (1979). Structure of oxidative and photophosphorylation coupling factor complexes. *Biochimica et Biophysica Acta* **549**:31–53.

BERRY, E. A., and HINKLE, P. C. (1983). Measurement of the electrochemical proton gradient in submitochondrial particles. *Journal of Biological Chemistry* **258**:1474–1486.

BOYER, P. D. (1975). A model for conformational coupling of membrane potential and proton translocation to ATP synthesis and to active transport. *FEBS Letters* **58**:1–6.

BOYER, P. D., CHANCE, B., ERNSTER, L., MITCHELL, P., RACKER, E., and SLATER, E. C. (1977). Op. cit., Chapter 3.

BOYER, P. D., CROSS, R. L., and MOMSEN, W. (1973). A new concept for energy coupling in oxidative phosphorylation based on a molecular explanation of the oxygen exchange reactions. *Proceedings of the National Academy of Sciences USA* **70**:2837–2839.

BRAND, M. D., REYNAFARJE, B., and LEHNINGER, A. L. (1976). Stoichiometric relationship between energy-dependent proton ejection and electron transport in mitochondria. *Proceedings of the National Academy of Sciences USA* **73**:437–441.

BRIERLEY, G. P., JURKOWITZ, M., CHAVEZ, E., and JUNG, D. W. (1978). Energy-dependent contraction of swollen heart mitochondria. *Journal of Biological Chemistry* **252**:7932–7939.

BRINLEY, F. J., JR. (1978). Calcium buffering in squid axons. *Annual Review of Biophysics and Bioengineering* **7**:363–392.

CASEY, R. P., THELEN, M., and AZZI, A. (1980) Dicyclohexylcarbodiimide binds specifically and covalently to cytochrome oxidase while inhibiting its H$^+$-translocating activity. *Journal of Biological Chemistry* **255**:3994–4000.

CHAPPELL, J. B., and HAARHOFF, K. N. (1966). The penetration of the mitochondrial membrane by anions and cations, in *Biochemistry of Mitochondria*, E. C. Slater, Z. Kaniuga, and L. Wojtczak, Eds. Academic, New York, pp. 75–91.

CHERNYAK, B. V., CHERNYAK, V. YA., GLADYSHEVA, T. B., KOZHANOVA, Z. E., and KOZLOV, I. A. (1981). Structural rearrangements in soluble mitochondrial ATPase. *Biochimica et Biophysica Acta* **635**:552–570.

COSTA, L. E., REYNAFARJE, B., and LEHNINGER, A. L. (1984). Stoichiometry of mitochondrial H^+ translocation coupled to succinate oxidation at level flow. *Journal of Biological Chemistry* **259**:4802–4811.

COX, G. B., JANS, D. A., FIMMERL, A. L., GIBSON, F., and HATCH, L. (1984). The mechanism of ATP synthase: Conformational change by rotation of the β subunit. *Biochimica et Biophysica Acta* **768**:201–208.

CROSS, R. L. (1981). The mechanism and regulation of ATP synthesis by F_1-ATPases. *Annual Review of Biochemistry* **50**:681–714.

CROSS, R. L., GRUBMEYER, C., and PENEFSKY, H. L. (1982). Mechanism of ATP hydrolysis by beef heart mitochondrial ATPase. *Journal of Biological Chemistry* **257**:12101–12105.

CROSS, R. L., and NALIN, C. M. (1982). Adenine nucleotide binding sites on beef heart F_1-ATPase. *Journal of Biological Chemistry* **257**:2874–2881.

DE PIERRE, J., and ERNSTER, L. (1977). Enzyme topology of intracellular membranes. *Annual Review of Biochemistry* **46**:201–262.

DUNN, S. D., and HEPPEL, L. A. (1981). Properties and functions of the subunits of the *Escherichia coli* coupling factor ATPase. *Archives of Biochemistry and Biophysics* **210**:421–436.

DUTTON, P. L., and WILSON, D. F. (1974). Redox potentiometry in mitochondrial and photosynthetic bioenergetics. *Biochimica et Biophysica Acta* **346**:165–212.

ERNSTER, L., and SCHATZ, G. (1981). Mitochondria: A historical review. *Journal of Cell Biology* **91**:227S–255S.

FERGUSON, S. J., and SORGATO, M. C. (1982). Proton electrochemical gradients and energy-transduction processes. *Annual Review of Biochemistry* **51**:185–217.

FILLINGAME, R. H. (1980*a*) Op. cit., Chapter 4.

FILLINGAME, R. H. (1980*b*). Op. cit., Chapter 4.

FUTAI, M., and KANAZAWA, H. (1983). Op. cit., Chapter 4.

GARLID, K. D. (1980). On the mechanism of regulation of the mitochondrial K^+/H^+ exchanger. *Journal of Biological Chemistry* **255**:11273–11279.

HAINES, T. F. (1983). Anionic lipid headgroups as a proton-conducting pathway along the surface of membranes. *Proceedings of the National Academy of Sciences USA* **80**:160–164.

HANSFORD, R. G. (1980). Control of mitochondrial substrate oxidation. *Current Topics in Bioenergetics* **10**:217–278.

HAUSKA, G., HURT, E., GABELLINI, N., and LOCKAU, W. (1983). Comparative aspects of quinol–cytochrome *c*/plastocyanin oxidoreductases. *Biochimica et Biophysica Acta* **726**:97–133.

HINKLE, P. C. (1981). Coupling ratios of proton transport by mitochondria, in *Chemiosmotic Proton Circuits in Biological Membranes,* V. P. Skulachev and P. C. Hinkle, Eds. Addison-Wesley, Reading, Mass., pp. 49–58.

HINKLE, P. C., and HORSTMAN, L. L. (1971). Respiration-driven proton transport in submitochondrial particles. *Journal of Biological Chemistry* **246**:6024–6028.

HINKLE, P. C., and MCCARTY, R. E. (1978). Op. cit., Chapter 3.

HOLIAN, A., and WILSON, D. F. (1980). Relationship of transmembrane pH and electrical gradients with respiration and adenosine 5′-triphosphate synthesis in mitochondria. *Biochemistry* **19**:4213–4221.

JOHNSON, L. V., WALSH, M. L., BOCKUS, B. J., and CHEN, L. B. (1981). Monitoring of relative mitochondrial membrane potential in living cells by fluorescence microscopy. *Journal of Cell Biology* **88**:526–535.

KAGAWA, Y., SONE, N., HIRATA, H., and YOSHIDA, M. (1979). Structure and function of H$^+$-ATPase. *Journal of Bioenergetics and Biomembranes* **11**:39–78.

KAYALAR, C., ROSING, J., and BOYER, P. D. (1977). An alternating site sequence for oxidative phosphorylation suggested by measurement of substrate binding patterns and exchange reaction inhibitions. *Journal of Biological Chemistry* **252**:2486–2491.

KELL, D. B. (1979). Op. cit., Chapter 3.

KHAN, S., and BERG, H. C. (1983). Isotope and thermal effects in chemiosmotic coupling to the membrane ATPase of *Streptococcus*. *Journal of Biological Chemistry* **285**:6709–6712.

KLINGENBERG, M. (1980). The ADP-ATP translocation in mitochondria, a membrane potential controlled transport. *Journal of Membrane Biology* **56**:97–105.

KOZLOV, I. A. (1981). How does membrane potential drive ATP synthesis? in *Chemiosmotic Proton Circuits in Biological Membranes*, V. P. Skulachev and P. C. Hinkle, Eds. Addison-Wesley, Reading, Mass., pp. 407–420.

KOZLOV, I. A., and SKULACHEV, V. P. (1982). An H$^+$-ATP synthetase: A substrate translocation concept. *Current Topics in Membranes and Transport* **16**:285–301.

KRÄMER, R., and KLINGENBERG, M. (1982). Electrophoretic control of reconstituted adenine nucleotide translocation. *Biochemistry* **21**:1082–1089.

LANOUE, K. F., and SCHOOLWERTH, A. C. (1979). Metabolite transport in mitochondria. *Annual Review of Biochemistry* **48**:871–922.

LEHNINGER, A. L. (1964). *The Mitochondrion*. Benjamin, New York.

LEMASTERS, J. L., and HACKENBROCK, C. R. (1980). The energized state of rat liver mitochondria. ATP equivalence, uncoupler sensitivity and decay kinetics. *Journal of Biological Chemistry* **255**:5674–5680.

MALONEY, P. C. (1982). Op. cit., Chapter 4.

MCCARTY, R. E. (1978). The ATPase complex of chloroplasts and chromatophores. *Current Topics in Bioenergetics* **7**:245–278.

MEEUSE, B. J. D. (1975). Thermogenic respiration in aroids. *Annual Review of Plant Physiology* **26**:117–126.

METZLER, D. E. (1977). Op. cit., Chapter 2.

MITCHELL, P. (1966, 1968, 1976a, 1979a, 1979b, 1981a). Op. cit., Chapter 3.

MITCHELL, P. (1974). A chemiosmotic molecular mechanism for proton-translocating adenosine triphosphatases. *FEBS Letters* **43**:189–194.

MITCHELL, P. (1976b). Possible molecular mechanisms of the protonmotive function of cytochrome systems. *Journal of Theoretical Biology* **62**:327–367.

MITCHELL, P. (1977b). A commentary on alternative hypotheses of protonic coupling in the membrane systems catalysing oxidative and photosynthetic phosphorylation. *FEBS Letters* **78**:1–20.

MITCHELL, P. (1981b). Biochemical mechanism of protonmotivated phosphorylation in F$_0$F$_1$ adenosine triphosphatase molecules, in *Mitochondria and Microsomes*, C. P. Lee, G. Schatz, and G. Dallner, Eds. Addison-Wesley, Reading, Mass., pp. 427–457.

MITCHELL, P., and MOYLE, J. (1967b). Respiration-driven proton translocation in rat liver mitochondria. *Biochemical Journal* **105**:1147–1162.

MITCHELL, P., and MOYLE, J. (1968). Proton translocation coupled to ATP hydrolysis in rat liver mitochondria. *European Journal of Biochemistry* **7**:530–539.

MITCHELL, P., and MOYLE, J. (1969). Estimation of membrane potential and pH difference across the cristae membrane of rat liver mitochondria. *European Journal of Biochemistry* **7**:471–484.

MOORE, A. L., and RICH, P. R. (1980). Plant mitochondria. *Trends in Biochemical Sciences* 5:284–288.

NAGLE, J. F., and MOROWITZ, H. J. (1978). Molecular mechanisms for proton transport in membranes. *Proceedings of the National Academy of Sciences USA* 75:298–302.

NELSON, N. (1981). Proton-ATPase of chloroplasts. *Current Topics in Bioenergetics* 11:1–33.

NICHOLLS, D. G. (1974). The influence of respiration and ATP hydrolysis on the proton-electrochemical gradient across the inner membrane of rat-liver mitochondria as determined by ion distribution. *European Journal of Biochemistry* 50:305–315.

NICHOLLS, D. G. (1979). Brown adipose tissue mitochondria. *Biochimica et Biophysica Acta* 549:1–29.

NICHOLLS, D. G. (1982). Op. cit., Chapter 1.

NICHOLLS, D. G., and ÅKERMAN, K. (1982). Mitochondrial calcium transport. *Biochimica et Biophysica Acta* 683:57–88.

NICHOLLS, D. G., and LOCKE, R. M. (1984). Thermogenic mechanisms in brown fat. *Physiological Reviews* 64:1–64.

PEDERSEN, P. L., SCHWERZMANN, K., and CINTRON, N. (1981). Regulation of the synthesis and hydrolysis of ATP in biological systems: Role of peptide inhibitors of the H^+-ATPases. *Current Topics in Bioenergetics* 11:149–199.

PREBBLE, J. N. (1981). Op. cit., Chapter 2.

RACKER, E. (1976). Op. cit., Chapter 3.

RAGAN, I. C., SMITH, S., EARLY, F. G. P., and POORE, V. M. (1981). NADH dehydrogenase, in *Chemiosmotic Proton Circuits in Biological Membranes,* V. P. Skulachev and P. C. Hinkle, Eds. Addison-Wesley, Reading, Mass., pp. 59–68.

REYNAFARJE, B., ALEXANDRE, A., DAVIES, P., and LEHNINGER, A. L. (1982). Proton translocation stoichiometry of cytochrome oxidase: use of a fast-responding oxygen electrode. *Proceedings of the National Academy of Sciences USA* 79:7218–7222.

RYDSTRÖM, J., LEE, C. P., and ERNSTER, L. (1981). Energy-linked nicotinamide nucleotide transhydrogenase, in *Chemiosmotic Proton Circuits in Biological Membranes,* V. P. Skulachev and P. C. Hinkle, Eds. Addison-Wesley, Reading, Mass., pp. 483–508.

SARIS, N. E., and ÅKERMAN, K. E. O. (1980). Uptake and release of bivalent cations in mitochondria. *Current Topics in Bioenergetics* 10:103–179.

SENIOR, E., and WISE, J. G. (1983). Op. cit., Chapter 4.

SKULACHEV, V. P. (1982). The localized $\Delta\tilde{\mu}_{H^+}$ problem: The possible role of the local electric field in ATP synthesis. *FEBS Letters* 146:1–4.

SLATER, E. C. (1981). The cytochrome-*b* paradox, the BAL-labile factor and the Q-cycle, in *Chemiosmotic Proton Circuits in Biological Membranes,* V. P. Skulachev and P. C. Hinkle, Eds. Addison-Wesley, Reading, Mass., pp. 69–104.

SLATER, E. C. (1983). The Q cycle, an ubiquitous mechanism of electron transfer. *Trends in Biochemical Sciences* 8:239–242.

SLATER, E. C., ROSING, J., and MOL, A. (1973). The phosphorylation potential generated by respiring mitochondria. *Biochimica et Biophysica Acta* 292:534–553.

SRERE, P. A. (1982). The structure of the mitochondrial inner membrane-matrix compartment. *Trends in Biochemical Sciences* 7:375–378.

STENT, G. S. (1968). That was the molecular biology that was. *Science* 160:390–396.

TEDESCHI, H. (1980). The mitochondrial membrane potential. *Biological Reviews of the Cambridge Philosophical Society* 55:171–180.

THAYER, W. S., and HINKLE, P. C. (1973). Stoichiometry of adenosine triphosphate-driven proton translocation in bovine heart mitochondrial particles. *Journal of Biological Chemistry* **248**:5395–5402.

THAYER, W. S., and HINKLE, P. C. (1975). Kinetics of adenosine triphosphate synthesis in bovine heart submitochondrial particles. *Journal of Biological Chemistry* **250**:5336–5342.

TRUMPOWER, B. L. (1981). New concepts on the role of ubiquinone in the mitochondrial respiratory chain. *Journal of Bioenergetics and Biomembranes* **13**:1–24.

TZAGALOFF, A. (1982). *Mitochondria*. Plenum, New York.

VON JAGOW, G., and SEBALD, W. (1980). *b*-Type cytochromes. *Annual Review of Biochemistry* **49**:281–314.

WESTERHOFF, H. V., MELANDRI, B. A., VENTUROLI, G., AZZONE, G. F., and KELL, D. B. (1984). Mosaic protonic coupling hypothesis for free energy transduction. *FEBS Letters* **165**:1–5.

WIKSTRÖM, M. (1984). Pumping of protons from the mitochondrial matrix by cytochrome oxidase. *Nature* **308**:558–560.

WIKSTRÖM, M., and KRAB, K. (1979). Proton-pumping cytochrome-*c* oxidase. *Biochimica et Biophysica Acta* **549**:177–222.

WIKSTRÖM, M., KRAB, K., and SARASTE, M. (1981). Proton-translocating cytochrome complexes. *Annual Review of Biochemistry* **50**:623–655.

WILLIAMS, R. J. P. (1962). Op. cit., Chapter 3.

WILLIAMS, R. J. P. (1978). Op. cit., Chapter 3.

WILSON, D. F., and FORMAN, N. G. (1982). Mitochondrial transmembrane pH and electrical gradients: Evaluation of their energy relationships with respiratory rate and adenosine 5′-triphosphate synthesis. *Biochemistry* **21**:1438–1444.

ZORATTI, M., PIETROBON, D., and AZZONE, G. F. (1982). On the relationship between rate of ATP synthesis and H^+ electrochemical gradients in rat-liver mitochondria. *European Journal of Biochemistry* **126**:443–451.

Harvesting the Light

It is a century now since Darwin gave us the first glimpse of the origin of species. We know now what was unknown to all the preceding caravan of generations: that men are only fellow-voyagers with other creatures in the odyssey of evolution. This new knowledge should have given us, by this time, a sense of kinship with fellow creatures; a wish to live and let live; a sense of wonder over the magnitude and duration of the biotic enterprise.

Aldo Leopold, *A Sand County Almanac*

• • •

In the course of a year some 5×10^{20} kcal of light energy (2×10^{21} kJ) impinges on the earth's surface, of which 1 to 2 percent is absorbed by photosynthetic organisms and put to work. A small minority of these creatures, all prokaryotes, carry out the anaerobic, cyclic mode of photosynthesis described in Chapter 4, with H_2S or organic compounds as reductants. The great majority, including the cyanobacteria, eukaryotic algae, and higher plants, live by a more elaborate kind of photosynthesis in which water is the ultimate source of reducing power and molecular oxygen is liberated. The invention of oxygenic photosynthesis made available unlimited stores of reducing power and thereby underwrote first the proliferation of aerobic bacteria and later that of the eukaryotes (Chapter 6). It remains the paramount process that supplies energy for the uses of life.

The scale of photosynthetic energy conversion is staggering. Some 7×10^{14} kg of CO_2 is fixed annually, roughly half on land and half in the sea, while 5.1×10^{14} kg of oxygen is liberated. "If a year's yield of photosynthesis were amassed in the form of sugar cane, it would form a heap over two miles high and with a base of 43 square miles" (Fogg, 1972). By the same token, since the composition of the atmosphere is in almost perfect balance, respiration must consume the same amount of oxygen and restore the CO_2. The free oxygen turns over every 2000 years, CO_2 in a few hundred, making animal life immediately dependent on photosynthesis. By one estimate, if photosynthesis were to cease, all higher forms of life would be extinct within 25 years. A milder version of this catastrophic scenario has been invoked to explain the mass extinction of animals at the end of the Cretaceous period. There is evidence that some 65 million years ago the earth collided with a monstrous asteroid or comet; this may have kicked up enough dust to reduce the incidence of sunlight to the point of limiting primary photosynthetic productivity, with disastrous effects on animal life (Alvarez et al., 1980).

Organisms that live by oxygenic photosynthesis range in size from unicellular prokaryotes to the giant redwoods and flourish in almost every habitat to be found on earth. Plants or algae grow on bare rock in the glare of the desert sun and in the dusk of the rain forest as well as on arctic tundra, in hot springs and in the open sea. It is scarcely surprising that they exhibit much diversity with respect to the pigments that absorb the light, the structures that bear these pigments, and even the pathways of carbon assimilation. What is remarkable is that near-perfect uniformity prevails at the mechanistic level, where light energy is transduced into useful forms. The proteins involved have, of course, diverged considerably over the two or three billion years that oxygenic photosynthesis has been in operation, but the mechanism of energy coupling appears to have been rigorously conserved.

The common denominator of oxygenic photosynthesis, whether in prokaryotes or chloroplasts, is the reduction of $NADP^+$ to NADPH with water as the electron donor and ATP and oxygen gas as additional products. This reaction is strongly endergonic. The driving force is supplied by the absorption of two quanta of light by separate photosystems acting in series. The reaction centers of both photosystems contain chlorophyll a, never bacteriochlorophyll. The two light reactions are linked by an electron transport chain that contains quinones, cytochromes, iron-sulfur centers, and ferredoxin. The entire ensemble is built into specialized intracellular membranous lamellae enclosing an internal space and is so arranged that when light absorption drives electrons over the chain, protons are translocated across the membrane. The protonic potential generated thereby supports the production of ATP by an F_1F_0 ATP synthase. Light energy is thus transduced into the chemical potentials of ATP and NADPH, both of which participate in subsequent dark reactions that reduce CO_2 to either carbohydrates or organic acids. The diversity of photosynthesis is chiefly a matter of accessory light-harvesting pigments, structural arrangements, and regulation; it has little to do with the core processes of energy transformation.

The object of this chapter is to survey how far we have come toward understanding how light energy is captured, transformed into the common currencies of cellular economics, and then fed into the metabolic web. Many readers will be less familiar with photosynthesis than with other subjects treated in this book. To ease their way, I shall first recount how the standard conception of photosynthesis developed and later turn to topics of current research interest. Most of our knowledge, particularly at the mechanistic level, stems from research with the chloroplasts of higher plants, and they dominate this chapter. But oxygenic photosynthesis is more widely distributed than chloroplasts are and is far more ancient. Some attention must be paid to its prokaryotic practitioners and to the manner in which photosynthetic bacteria surrendered their liberties and inherited the earth.

Photosynthesis is far too broad a subject for a chapter; it demands a book, and several recent ones are available, including elementary introductions by Fogg (1972) and Gregory (1977), a comprehensive reference work edited by Hatch and Boardman (1981), and two technical treatises on the bioenergetics of photosynthesis edited by Govindjee (1975, 1982). I am particularly indebted to several chapters in Prebble's recent book (1981) and to the lucid review of photosynthesis from the biophysical viewpoint by Clayton (1980). Obviously, only a small fraction of the available information can be covered in this chapter, which should be taken as the appetizer for a gargantuan feast.

Chloroplasts and Thylakoids

If you gently grind up a few leaves in half-molar sucrose and examine the mash under a microscope, you can readily make out the chloroplasts that have escaped from broken cells. Most leaf cells harbor several dozen of them, bright-green oblong globules 5 to 10 μm in length but without much personality; it takes an effort of will to believe that the biosphere revolves around them.

Electron micrographs are considerably more impressive (Figures 8.1 and 8.2). Their most conspicuous features are the elaborate internal membranes called thylakoids (Greek for saclike). Each thylakoid consists of a flattened vesicle, with the lipid bilayers so closely apposed that the internal space is but 5 nm across.

The thylakoids anastomose, collectively making up a continuous internal phase (sometimes called the loculus) sealed off from the chloroplast matrix (Figure 8.3). In the chloroplasts of higher plants most of the thylakoid disks are stacked like coins, one on top of the other, a configuration called grana. Individual grana stacks are connected by single thylakoids (stroma thylakoids). Later on, we shall have something to say about the physiological roles of these two thylakoid classes. The essential point now is that thylakoid membranes bear the chlorophyll, electron-transport catalysts, and ATP synthase; they are the locus of photosynthetic energy transduction.

Prominent grana stacks are typical of higher-plant chloroplasts but are absent from those of certain algae; they are obviously not an obligatory feature. The thylakoids of algal chloroplasts assume a variety of configurations, from

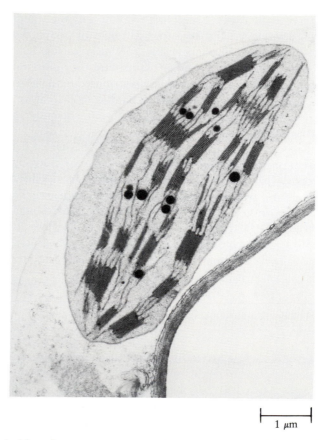

1 μm

FIGURE 8.1 A chloroplast in situ, in a cell of spinach leaf tissue. Note the double plastid envelope and the arrangement of thylakoids in grana stacks connected by single stroma lamellae. The dense round bodies are lipid globules. (Electron micrograph courtesy of L. A. Staehelin.)

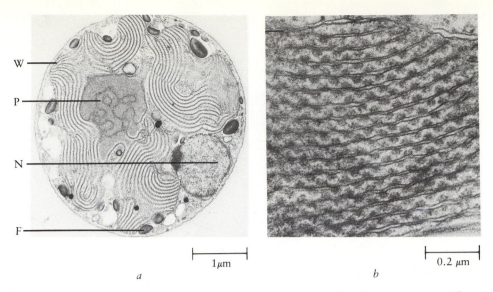

Labels on figure (a): W, P, N, F

1μm

a

0.2 μm

b

FIGURE 8.2 (a) Section through the unicellular red alga *Porphyridium aerugineum*. The single chloroplast occupies much of the cell's volume. The thylakoids are loose, rather than arranged in grana. P, pyrenoid, an inclusion body consisting of precipitated ribulose bisphosphate carboxylase; N, nucleus; W, cell wall; F, starch deposit in cytoplasm. (Reproduced from the *Journal of Phycology* **29**:423, 1966.) (b) Portion of a cell, enlarged. The granules associated with the external surfaces of the thylakoids are phycobilisomes. (Reproduced from the *Journal of Cell Biology* with copyright permission of The Rockefeller University Press. Both electron micrographs courtesy of Elizabeth Gantt.)

tight whorls of concentric lamellae to loose formations of individual thylakoids; Figure 8.2 shows one of the latter. In the unicellular red alga *Porphyridium* (Rhodophyta), the large chloroplast occupies most of the cell's volume. At high magnification the thylakoids are seen to be studded with granules 30 nm in diameter. These are the phycobilisomes, which contain ancillary light-harvesting pigments called phycobilins; they are the hallmark of cyanobacteria and their descendants. Neither granules nor these particular pigments occur in other algae or in higher plants; other pigments take their place. But thylakoids of some kind are almost universal among organisms that live by oxygenic photosynthesis, both prokaryotic and eukaryotic; we shall consider a possible reason for this below.

The chloroplast as a whole is bounded by an envelope consisting of at least two lipid-bilayer membranes in close apposition (Figure 8.3). The inner membrane forms the permeability barrier between the chloroplast matrix and the host cytoplasm; communication across it is effected by an array of porters for particular metabolites. The other membrane is freely permeable to small molecules. How they pass across an apparently continuous barrier is not known, but by analogy with the outer membranes of mitochondria and bacteria one may expect aqueous channels made of specialized proteins, or "porins." The various chloroplast membranes differ in composition from one another and from the other membranes of the plant cell. Both thylakoid and envelope membranes contain characteristic galactosyl lipids and sulfolipids that are not found outside the plastid (these lipids also occur in cyanobacteria). The envelope lacks

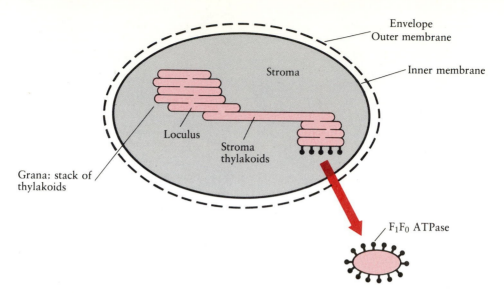

FIGURE 8.3   Diagram of chloroplast structure. Note that the thylakoids form a common internal space, the loculus. The polarity of an individual thylakoid vesicle is shown by the stalked F_1F_0 ATPase knobs facing into the stroma.

chlorophyll, cytochromes, and the other proteins of energy transduction. It has a distinctive set of at least 75 polypeptides, and the outer envelope membrane, at least, contains sterols. It has been argued that the inner envelope membrane is homologous to the bacterial plasma membrane while the outer envelope membrane is an extension of the host cell's endoplasmic reticulum, but recent data implicating the entire envelope in the biosynthesis of the chloroplasts' galactosyl lipids (Douce and Joyard, 1981) suggest that this dichotomy is overly simplistic.

Rupture of the envelope releases the soluble contents of the matrix, or stroma: ions, small metabolites, and a variety of proteins. These include all the enzymes and intermediates required for the reduction of CO_2 to the carbohydrate level and for the conversion of hexoses into starch. The key enzyme of the reductive pentose cycle, ribulose bisphosphate carboxylase, is a particularly abundant constituent; part is loosely bound to the thylakoid surfaces, but the bulk occurs in the stroma. This may be the place to mention that chloroplasts perform a variety of metabolic functions besides energy transduction and CO_2 fixation: biosynthesis of proteins, galactolipids, fatty acids, carotenoids, and chlorophyll. The enzymes are either free in the stroma or else associated with one or another of the membrane systems. In all, chloroplast proteins may contribute as much as 70 percent of the total protein complement of higher plant cells.

The chloroplast matrix also contains the instruments of molecular genetics: DNA, several kinds of RNA, and ribosomes. Chloroplast DNA suffices to code for some 125 polypeptides of molecular weight 50,000. Chloroplasts contain distinctive ribosomes, transfer RNAs, aminoacyl tRNA synthases, and RNA polymerase, but not all these macromolecules are manufactured within the chloroplast. Chloroplast DNA is known to code for the large subunit of ribulose bisphosphate carboxylase (but not for the small subunit), for some of the subunits of the F_1F_0 ATP synthase, and for several proteins associated with chlorophyll. The remaining chloroplast proteins are coded on nuclear DNA, translated by cytoplasmic ribosomes, and imported across the chloroplast envelope by mechanisms that are not well understood. The genetic autonomy of chloroplasts is greater than that of mitochondria, but sharply circumscribed all the same.

Unless special precautions are taken, chloroplast envelopes tend to rupture during their isolation from leaves, and the material with which most of the research on photosynthetic energy transduction has been done consists of more or less native aggregates of thylakoids. These carry out all the stages of the process: light absorption, electron transport from water to $NADP^+$, generation of a protonic potential, and production of ATP. In the following sections we shall examine each stage in turn, followed by a brief summary of CO_2 fixation. The object is to provide an overview of chloroplast energetics, leaving the thornier issues of molecular mechanism for later.

Energy Conversion in Oxygenic Photosynthesis

Light Absorption.   Chlorophyll is the linchpin of photosynthesis. This cardinal fact was recognized long ago, for only organisms that contain chlorophyll can grow autotrophically in the light. Loss of chlorophyll, whether by mutation or as the result of a period of growth in the dark, entails the loss of photosynthetic CO_2 fixation. In plants, eukaryotic algae, and cyanobacteria the chlorophyll responsible for energy transduction is chlorophyll *a*, whose structure is shown in Figure 8.4*a*. In prokaryotes that do not generate oxygen the critical molecule is bacteriochlorophyll. Of all the organisms known to us, only the halobacteria carry out a mode of photosynthesis that does not depend on chlorophyll (Chapter 6).

However, not all of the chloroplast's chlorophyll complement is directly involved in photochemistry, nor is it all chlorophyll *a*. In the 1930s R. Emerson and W. Arnold carried out studies on the efficiency of photosynthesis in the green alga *Chlorella*. Oxygen generation was taken as a measure of the photosynthetic yield, and light was presented in flashes brief and intense enough to excite every chlorophyll molecule in the suspension once, but only once. Their main finding was that the amount of chlorophyll present far exceeded that required for the production of oxygen. At light intensities sufficient to saturate the system, only one molecule of O_2 was produced for 2480 molecules of chlorophyll excited, suggesting that a photosynthetic unit is made up of a large number of chlorophyll molecules. The modern interpretation of these classical observations is as follows. The oxidation of $2H_2O$ to O_2 requires four electron-transfer steps, each

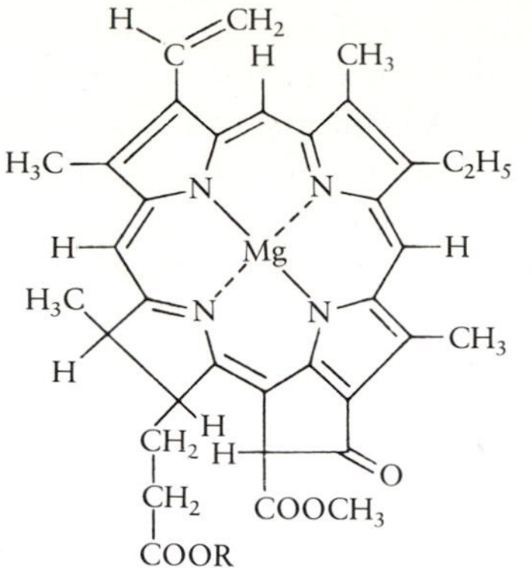

(a) Chlorophyll *a*

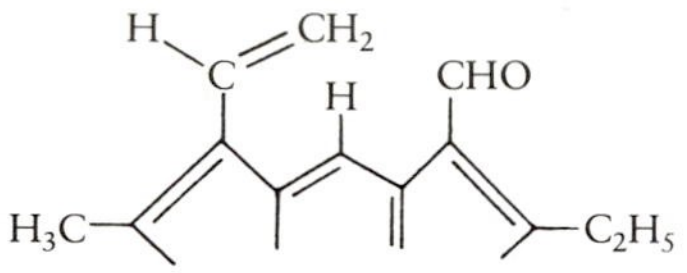

(b) Chlorophyll *b*

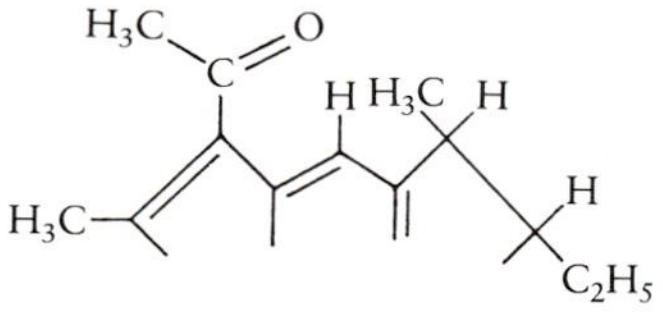

(c) Bacteriochlorophyll *a*

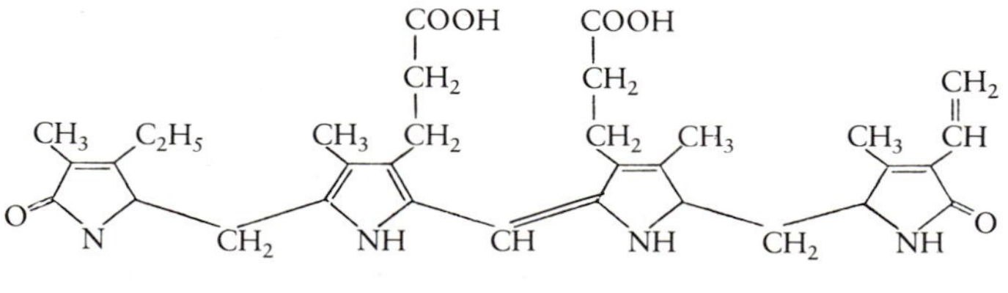

(d) Phycoerythrobilin

FIGURE 8.4   Some of the major pigments involved in photosynthesis. Chlorophyll *a* (*a*) occurs in reaction centers and collecting antennas, chlorophyll *b* (*b*) only in the latter; bacteriochlorophyll *a* (*c*) is characteristic of purple bacteria. For the latter two, only the portions of the structure that differ from chlorophyll *a* are shown. Phycoerythrobilin (*d*) is one of the antenna pigments of cyanobacteria and red algae.

of which involves two light reactions in series; eight quanta of light must be processed for each O_2. This suggests that 2480/8 or about 310 chlorophyll molecules form a photosynthetic unit. Each unit contains two kinds of chlorophyll: a small amount of reaction-center chlorophyll[1] and about 300 molecules of light-collecting or antenna chlorophyll. A quantum of light absorbed anywhere in the unit is funneled to the reaction center with near-perfect conservation of energy. Only the reaction centers are engaged in photochemistry, and they sum eight quanta to yield one molecule of O_2.

As was mentioned already in the bacterial context, antenna chlorophylls perform a vital service. Reaction centers can turn over about 100 times per second, but under low natural light intensities they can expect to be excited only about once per second. The addition of a collecting antenna brings the light-gathering capacity of the unit into line with its energy-transducing power. In general, adaptation of organisms to weak light relies on amplification of the light-collecting pigments, not of reaction centers.

Antenna pigments perform a second function, namely, to widen the spectral range available for use. We know that the chlorophyll found in reaction centers is chlorophyll a, whose absorption spectrum in solution is shown in Figure 8.5a; note the sharp maxima at about 660 nm (red light) and 420 nm (blue light). Chlorophyll b is a widespread antenna constituent; it is closely related to chlorophyll a in structure (Figure 8.4b), and its absorption spectrum is similar in general form, but the peaks are somewhat displaced (Figure 8.5b). These spectra should be compared with those in Figure 8.6, which depicts the absorption spectrum of a suspension of whole cells of $Chlorella$ together with the action spectrum[2] of photosynthesis in this organism ($Chlorella$ is a green alga containing both chlorophyll a and chlorophyll b). The action spectrum is far broader than that of chlorophyll a, in fact nearly identical with the spectrum of the cell as a whole. Evidently light absorbed by pigments besides chlorophyll a can be efficiently harvested; one of these is chlorophyll b.

Accessory light-harvesting pigments contribute strikingly to the diversity of organisms that live by oxygenic photosynthesis, in keeping with their far-flung habitats. In green algae (Chlorophyta, including $Chlorella$) the main pigments are chlorophylls a and b. Each is present in several spectroscopically distinguishable species, apparently as a result of their association with proteins in the antenna, further broadening the range over which each pigment absorbs light. Chlorophylls a and b also predominate in the higher plants and in the prokaryote $Prochloron$, which, partly for that reason, is a favorite candidate for the

---

[1] Reaction center refers to the chlorophyll molecules that participate in energy transduction. Terms such as P_{700} or P_{680} designate particular reaction centers, in this instance those of photosystems I and II (PS I and PS II). Photosystem is a broader conception, including pigments such as quinones and cytochromes that are intimately involved in the photochemical process.

[2] An action spectrum is a plot of the effectiveness of light as a function of its wavelength. In the present context it may be defined as the reciprocal of the number of quanta required to produce a photochemical effect, such as the evolution of oxygen.

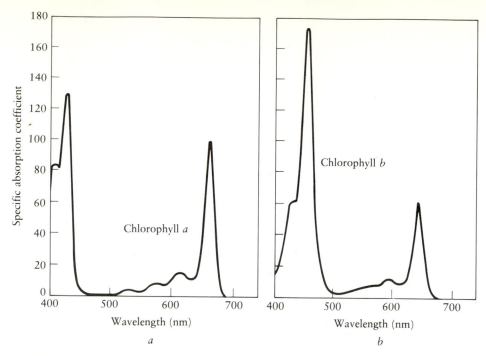

FIGURE 8.5   Absorption spectra of chlorophylls *a* and *b*, dissolved in petroleum ether.

evolutionary precursor of green algae and higher plants (Chapter 6). In the red algae (*Rhodophyta*, including *Porphyridium*, Figure 8.2) the chief light-harvesting pigments are phycobilins: open-chain structures that arise by an oxygen-dependent cleavage of chlorophyll (Figure 8.4*d*). The physiological importance of light absorption by these pigments is illustrated in Figure 8.7, which shows that at wavelengths below 600 nm most of the photosynthetic activity of red algae is supported by phycobilin pigments. (Nevertheless, energy transduction proper is effected by chlorophyll *a*.) As mentioned before, phycobilins also dominate in the cyanobacteria. In some marine diatoms and dinoflagellates light is collected by chlorophyll *c* with absorption peaks at 450 and 630 nm. Finally, all photosynthetic membranes are rich in carotenoids. These also serve as accessory pigments, and in some of the marine brown algae light absorbed by carotenoids is efficiently channeled to the reaction center. In general, however, the primary role of carotenoids is not to gather light but to protect the thylakoid membranes from photo-oxidation by free-radical derivatives of oxygen. Mutants that lack carotenoids are quickly killed when exposed to light in the presence of oxygen.

How is one to imagine a collecting antenna that funnels light from hundreds of pigment molecules to a central biochemical transducer? The physics of energy transfer is subtle and beyond our scope (see Clayton, 1980, and Prebble, 1981, for lucid explanations). Suffice it to say that before light energy can be manipulated in any manner, it must first be absorbed by chlorophyll or by another pigment molecule with an appropriate absorption spectrum. Absorption of a quantum of light raises an electron from the ground state to that of the first excited (singlet) level. In the case of chlorophyll *a*, with an absorption band in the red at 660 nm, this first excited state corresponds to a gain in energy of 1.87 eV per quantum, or 179 kJ/mol. Excited chlorophyll molecules have a short life, on the order of 10^{-9} second and may lose their excess energy in several ways.

1. Energy may be lost as heat, by a series of small deexcitation steps. This is what happens when chlorophyll absorbs a quantum of blue light, which corresponds to an energy gain of 262 kJ/mol for light at 450 nm. The excited electron quickly drops down to the first excited singlet state, dissipating its excess energy. The useful energy available to the organism is therefore no greater for blue light than for red.

2. The energy may be reemitted by fluorescence, in the form of a quantum of light of slightly longer wavelength and therefore of lower energy. As a rule, only a small fraction of the excited molecules decay in this manner.

3. Energy may be transferred to an adjacent molecule with little loss. Two pigment molecules readily exchange energy by a resonance mechanism provided the fluorescence band of the donor closely overlaps the absorption band of the recipient and provided the molecules are closely apposed. Indeed, antenna pigments are tightly packaged together with particular proteins that serve as a

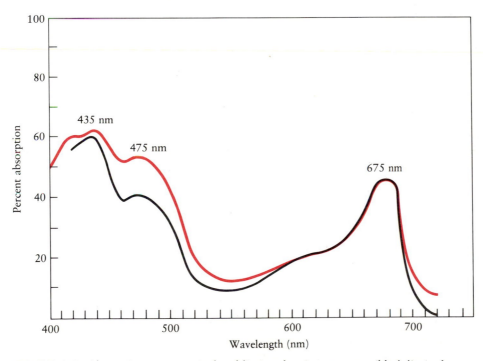

FIGURE 8.6   Absorption spectrum (colored line) and action spectrum (black line) of photosynthesis in the green alga *Chlorella*. The absorption maxima are due chiefly to chlorophylls *a* and *b*; most of the light absorbed by the whole cell is effective in photosynthesis. (After Haxo, 1960, with permission of Academic Press.)

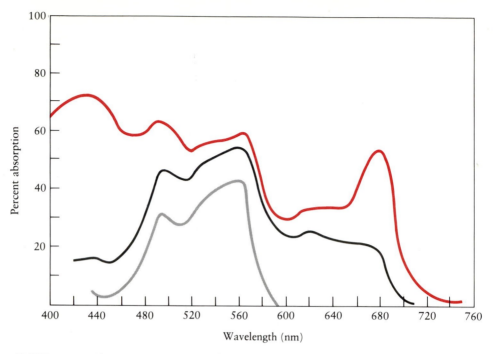

FIGURE 8.7 Absorption spectrum (colored line) and action spectrum (black line) of photosynthesis in whole cells of the red alga *Porphyra*. Over the range from 460 to 580 nm, the action spectrum matches the absorption spectrum of phycobilin pigments (gray line). Light absorbed by chlorophyll *a* (660 to 710 nm) is less efficiently used than that absorbed by phycobilins. (After Haxo and Blinks, 1950, by copyright permission of The Rockefeller University Press.)

scaffold, allowing excitation energy to migrate from one pigment molecule to another until it reaches the reaction center. Transmission of energy to that center is facilitated by the spatial organization of the antenna. As Figure 8.8 suggests, light is first absorbed by pigments with absorption bands at the shorter wavelengths (higher energy level). Excitation energy migrates through successive "shells" of pigments whose absorption bands lie at progressively longer wavelength. It thus travels "downhill" on the free-energy scale, as in a funnel, toward the reaction center.

4.   Finally, chlorophyll may return to the ground state by participating in a photochemical act. Reaction centers are designed to capitalize on this option by providing an acceptor for the excited electron and an electron donor to reduce the oxidized chlorophyll instantly. At this point energy absorption ends and energy transduction begins.

Two Photosystems Cooperate.   The characteristic feature of oxygenic photosynthesis is the use of two light reactions in series to drive electrons from water to

$NADP^+$, up the thermodynamic hill. This conception evolved over a period of two decades, from 1940 to 1960, out of two lines of research; the quantum efficiency of photosynthesis and the effectiveness of light of various colors as energy donor.

The reduction of CO_2 to the carbohydrate level, with concurrent evolution of oxygen (Equation 8.1), is strongly endergonic: ΔG is about 470 kJ/mol (112 kcal/mol).

$$CO_2 + H_2O \xrightarrow{\text{light}} O_2 + (CH_2O) \qquad \text{(Eq. 8.1)}$$

Red light of wavelength 700 nm, the upper limit of effectiveness for plant photosynthesis, supplies 171 kJ per einstein (that is, per "mole" of photons). In principle, then, $470/171 = 2.8$ einsteins should suffice to produce one mole of oxygen, or 2.8 quanta per molecule, assuming that none of the energy is wasted. In the 1940s a major controversy over the quantum efficiency of photosynthesis embroiled much of the research community: Otto Warburg argued vociferously that four quanta sufficed, while his opponents found so high an efficiency implausible on thermodynamic grounds and inconsistent with reasonable mechanisms. In the end, Warburg lost the war. It is now generally agreed that eight quanta, or even a little more, are required to reduce one molecule of CO_2 and generate one molecule of oxygen. Each of the four reductive steps must therefore result from the cooperation of two quanta.

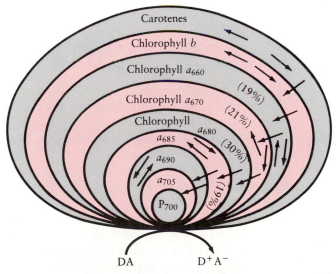

FIGURE 8.8   Antenna pigments funnel energy toward the reaction center. This diagrammatic representation of the chloroplast pigment bed of PS I is intended to suggest how excitation energy absorbed anywhere will tend to migrate toward adjacent pigments absorbing at longer wavelengths, ending up in the reaction center. Numbers in parentheses indicate the percentage of each spectral form. D stands for donor, A for acceptor. (From Prebble, 1981, with permission of Longman Group, Ltd.)

In the 1950s, after the preceding conclusions had been largely accepted, a number of clues began to emerge that hinted at the nature of this collaboration. The most dramatic came again from R. Emerson's laboratory. Far-red light, of wavelength longer than 680 nm, is well absorbed by chlorophyll *a* but proved to be quite ineffective for photosynthesis as judged by oxygen production. However, it could be used by the cells provided some light of shorter wavelength was supplied as well; a mixture containing, for example, light of 600 and 700 nm was more effective in photosynthesis than either sort alone. Emerson therefore proposed that oxygen production requires the cooperation of two light reactions; light below 680 nm can drive both, light of longer wavelength only one. Significantly, the two light reactions need not take place simultaneously: red and green light cooperate even if presented several seconds apart.

Biochemical evidence for two light reactions came more than a decade later, through the use of artificial electron carriers and acceptors. Robert Hill had discovered in 1939 that chloroplast preparations evolved oxygen when illuminated in the presence of certain artificial electron acceptors such as ferricyanide. Another example of the "Hill reaction" is the reduction of dichlorophenol indophenol; this is supported by light of 650 nm or less, with concurrent evolution of oxygen. Far-red light (700 nm) did not drive the Hill reaction; it did, however, support the reduction of $NADP^+$ with reduced dichlorophenol indophenol as electron donor:

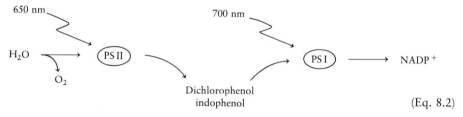

$$(Eq. 8.2)$$

This experiment, reported by Daniel Arnon's laboratory in 1961, is but one of many that gradually fleshed out the concept that two separate photosystems, differing slightly in their absorption spectra, collaborate in photosynthesis.

In 1960, R. Hill and F. Bendall proposed a scheme to account for the cooperation of two quanta and also for the growing evidence that quinones and cytochromes were part of a photosynthetic electron-transport chain (Hill and Bendall, 1960). A contemporary version of their Z scheme is shown in Figure 8.10. It envisages two sequential photosystems. One, now known as photosystem I (PS I or P_{700}), is activated by red and far-red light and produces a reductant powerful enough to reduce $NADP^+$ to NADPH. Activation of this system is accompanied by a characteristic bleaching of the absorption at 700 nm, hence the designation P_{700}. Spectroscopic studies indicate that P_{700} is a form of chlorophyll *a* in a special environment that alters its spectral properties. Photosystem II (PS II or P_{680}) is activated by light of shorter wavelength and produces an oxidant powerful enough to oxidize water to oxygen. This system was elusive for a time, but by 1970 research in H. T. Witt's laboratory in Berlin had identified absorption changes at 680 nm with photosystem II. This, too, appears to be a form of chlorophyll *a*. The two systems supply the strong oxidant and the strong reductant postulated in van Niel's unifying theory (Chapter 2). They are connected by an electron transport chain whose nature and composition will be taken up shortly.

Supporting evidence came almost immediately from L. N. M. Duysens and his colleagues (1961) in Amsterdam, who used differential absorption spectrophotometry to follow the oxidation and reduction of cytochromes in the red alga *Porphyridium* in response to light. A weak beam of light at 420 nm served to monitor the absorption of the cell suspension: oxidation of cytochromes is accompanied by increased transmittance at this wavelength. A beam at right angles to the monitoring beam was used to excite either photosystem I primarily (680 nm), photosystem II (562 nm), or both together. The essential results are shown in Figure 8.9, their interpretation in Figure 8.10. Excitation of PS I caused oxidation of the cytochrome, which serves as a donor of electrons to PS I. Excitation of PS II (in addition to PS I) induced partial re-reduction of the cytochrome due to a flow of electrons from PS II to PS I. The sequence was then repeated in the presence of the herbicide 3-(3,4-dichlorophenyl)-1,1-dimethyl-urea (DCMU), a reagent known to abolish photosynthetic oxygen production. Light of wavelength 680 nm still excited PS I, with oxidation of the cytochrome; however, with PS II effectively blocked, light of 562 nm did not cause cytochromes to be re-reduced. On the contrary, with DCMU present all one sees is enhanced oxidation due to further stimulation of PS I.

This experiment nicely illustrates another point: while the two photosystems can and must collaborate to drive electrons from water to $NADP^+$, each can harvest light separately. The nature of the light reactions and their structural basis will be considered more explicitly below.

Electron Transport and Photophosphorylation.   During the 1960s it became apparent that chloroplast membranes contain a variety of redox carriers similar to those found in mitochondria. Among these are plastoquinones,[3] whose structure differs only in detail from ubiquinone; cytochrome *f* (for frond), a *c*-type cytochrome also referred to as cytochrome c_{552}; several *b*-type cytochromes;

---

[3] Plastoquinones have the general structure

Plastoquinones A, B, and so on differ in the number of isoprenoid groups in the side chain; $n = 9$ in plastoquinone A, the most abundant member in green plants.

several iron-sulfur centers; and a copper protein called plastocyanin. The amounts of cytochromes present are comparable to those of the reaction centers, but plastoquinone is present in excess. We now know that the electron-transport chain comprises some 35 polypeptides.

The general sequence of these redox carriers was established by electron-transfer experiments of the kind illustrated in Figure 8.9, coupled with measurements of their midpoint potentials. The termini of the chain were also clarified. The long-known Hill reaction, in which chloroplasts evolve oxygen when illuminated in the presence of artificial electron acceptors (Equation 8.2), was clearly identified as a manifestation of photosystem II. At the reducing end, $NADP^+$ had been identified some years before as a physiological terminal electron acceptor. However, $NADP^+$ is not the first reduced product generated by photosystem I. Daniel Arnon and his colleagues discovered that the iron-sulfur protein ferredoxin serves as a low-potential electron carrier between photosystem I and $NADP^+$; electron transfer is mediated by a membrane-bound flavoprotein enzyme, ferredoxin-$NADP^+$ reductase. Ferredoxin was known already, albeit under another name; it is by no means unique to chloroplasts. A similar iron-sulfur protein takes part in many redox reactions of anaerobic bacteria.

The current textbook version of the photosynthetic electron-transport chain is shown in Figure 8.10, which illustrates graphically how the two light reactions boost electrons from their highly oxidized state in water ($E_{m,7} = +820$ mV) to

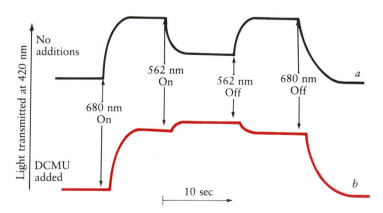

FIGURE 8.9   An experiment illustrating the cooperation of PS I and PS II. A suspension of algal cells was monitored with a beam of light at 410 mn; increased transmittance reports the oxidation of a cytochrome in these cells. Trace *a*, no additions. Illumination with red light (680 nm) activates chiefly PS I, with oxidation of the cytochrome. Illumination with blue light (562 nm) activates PS II and causes partial reduction of the cytochrome. Trace *b*, when DCMU is present, electron flow from PS II to cytochrome is blocked. Red light still activates PS I and oxidizes the cytochrome, but its reduction by blue light is not seen. (Data from Duysens et al., 1961, idealized by Clayton, 1980. Reproduced with permission of Cambridge University Press.)

the reduction level of NADPH ($E_{m,7} = -320$ mV), thus establishing a gradient of chemical potential at the expense of light energy. This sequence is nearly universal among oxygen-producing photosynthetic organisms, except that in certain eukaryotic algae a c-type cytochrome takes the place of plastocyanin. The evidence that supports the scheme is extensive; it includes both biochemical and spectroscopic data as well as the analysis of mutants deficient in one or another link in the chain. Nevertheless, Figure 8.10 is by no means the end of the quest. We shall see below that in order to link electron transport to phosphorylation the plastoquinone–cytochrome-f segment must be reformulated into a Q cycle, and the pathway of electron transport must be related to the structure of the membrane in which it is housed.

Before proceeding, let us recall that photosynthetic electron transport serves a dual purpose: to generate NADPH and also ATP. In their pioneering demonstration of photophosphorylation, Arnon and his colleagues showed that certain chloroplast preparations produced ATP in the light without either oxygen evolution or the net reduction of any electron acceptor. They attributed this to cyclic electron flow, driven (we now know) by photosystem I such that the excited electron eventually returns to its starting place, as in bacteria (Arnon et al., 1958). By contrast, when $NADP^+$ was present, oxygen was evolved as electrons flowed from water to $NADP^+$ with concurrent generation of ATP; Arnon (1959) designated this noncyclic photophosphorylation. The two modes of photophosphorylation are summarized as follows:

$$ADP + P_i \xrightarrow[\text{chloroplasts}]{\text{light}} ATP + H_2O \qquad \text{(Eq. 8.3)}$$

$$2H_2O + 2NADP^+ + nADP + nP_i \xrightarrow[\text{chloroplasts}]{\text{light}} O_2 + 2NADPH + nATP + H^+$$

$$\text{(Eq. 8.4)}$$

It is not immediately apparent how these two modes of phosphorylation mesh with the pathway of electron transport, and this is one of the chief topics of current research in light-energy transduction. Cyclic electron flow is supported by PS I alone. Spectrophotometric data indicate that the excited electron reduces ferredoxin and then cascades back to the reaction center via a pathway that includes cytochrome b_{563}, the plastoquinone pool, cytochrome f, and plasto-cyanin as suggested by the dashed line in Figure 8.10. Noncyclic photo-phosphorylation requires both photosystems I and II but has no obvious role for cytochrome b_{563}. Somehow photosystem I functions both as a member of a linear electron-transport chain and as the center of a cyclic one, both coupled to ATP production. The distribution of light energy between these two modes is under physiological control. We shall return to the resolution of these perplexities shortly, but first we must incorporate the discovery that the electron-transport chain of chloroplasts, like that of mitochondria, translocates protons across a membrane.

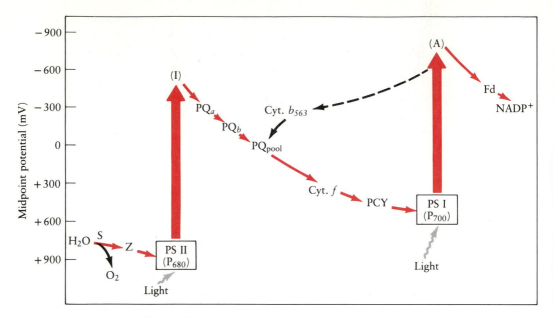

FIGURE 8.10   Pathways of electron transport in plant photosynthesis. The colored line traces electron flow in noncyclic photosynthesis. The ordinate shows the approximate midpoint potential of each redox carrier. S, water-splitting enzyme; Z, electron donor to P_{680}; I, pheophytin a; Q_a and Q_b, special bound quinone molecules; PCY, plastocyanin; Fd, ferredoxin; A, a special $Fe \cdot S$ center that serves as the primary acceptor for PS I. The black line shows cyclic electron transport around PS I alone.

Proton Translocation and ATP Generation.   The Z configuration of the electron-transport chain in Figure 8.10 is a thermodynamic abstraction, but with the application of the chemiosmotic theory to photosynthesis the Z scheme acquired a concrete spatial dimension. From the beginning, Mitchell (1961, 1966) maintained that electron flow is coupled to the translocation of protons across the thylakoid membrane and that the protonic potential thereby generated drives ATP production by an F_1F_0 ATP synthase. Extensive research during the past 20 years has amply confirmed that energy conservation in chloroplasts is fundamentally chemiosmotic. The light reactions mediated by photosystems I and II are the primary electrogenic steps that separate charges across the thylakoid membrane and generate the electric field that drives all subsequent events.

The experiments that first lent credibility to Mitchell's proposal, performed by André Jagendorf and his colleagues in the 1960s, remain landmarks in the evolution of bioenergetics. Their point of departure was the discovery that if chloroplast membranes were illuminated for a while and subsequently incubated in the dark with ADP and P_i, ATP was produced. Apparently illumination induced the accumulation of an "energized intermediate" that was reasonably stable (with a half-life of seconds) and plentiful enough to yield significant amounts of ATP. The yield of ATP was far greater than the amount of redox carriers present, and Hind and Jagendorf (1963) could find no chemical $\sim$P donor. However, they observed that illuminated thylakoid suspensions took up protons, making the pH more alkaline (Figure 8.11), and noted a correlation between the extent of proton absorption and the amount of ATP produced. Might the pH gradient between membranes and medium possibly be the driving force for ATP synthesis, as Mitchell had just proposed? The test devised by Jagendorf and Uribe (1966) became known as the acid bath experiment. Thylakoids were loaded with protons by soaking the membrane preparations in a permeant buffer at pH 4 in the dark. The membranes, still in the dark, were then abruptly mixed with a buffer containing ADP, P_i and sufficient alkali to raise the pH of the suspension to 8; under these conditions there was a burst of ATP synthesis (Figure 8.12). The pH gradient across the membrane generated by the alkalinization corresponded in all testable respects to the "energized intermediate" that had earlier been detected by the studies on "postillumination ATP synthesis." It was the first major success of the chemiosmotic viewpoint, and what had seemed merely an outlandish notion became a force to be reckoned with.

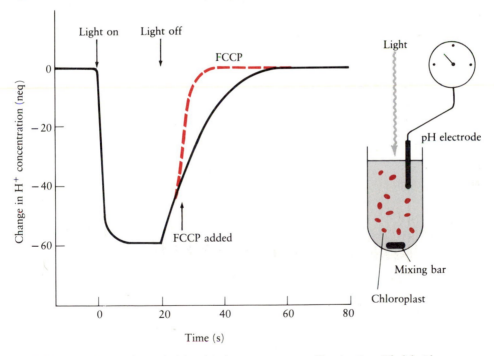

FIGURE 8.11 Chloroplast thylakoids take up protons on illumination. Thylakoids suspended in medium containing $NADP^+$ were kept in the dark. When the light was switched on, the pH of the suspension rose due to the uptake of protons; when the light was turned off, the pH gradient slowly decayed. The proton conductor FCCP accelerated the leakage of protons out of the thylakoids into the medium. (From "How Cells Make ATP," by P. C. Hinkle and R. E. McCarty. Copyright ©1978 by Scientific American, Inc. All rights reserved.)

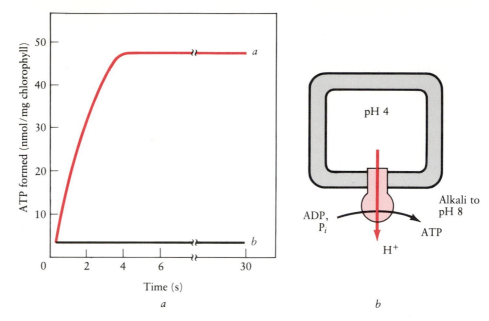

FIGURE 8.12   An artificial pH gradient drives ATP production by chloroplast thylakoids. (*a*) The experiment. The membranes were first loaded with acid by prolonged incubation in the dark with a permeant buffer, pH 4. The following additions were then made at $t = 0$ seconds: Trace *a* (colored line): ADP, P_i, Mg^{2+}, and alkali to pH 8; the ATP content of the suspension was assayed at intervals. Trace *b* (black line): omission of ADP, P_i, or alkali all precluded ATP synthesis. Addition of ADP and P_i 30 seconds after the alkali also did not support ATP synthesis. (After Jagendorf and Uribe, 1966.) (*b*) The principle. Addition of alkali imposed a pH gradient across the thylakoid membrane ($\Delta pH = 4$ units), which drives ATP synthesis by the F_1F_0 ATPase.

A few years later H. T. Witt, W. Junge, and their associates in Berlin demonstrated the electrogenic nature of the primary light reactions (Junge and Witt, 1968). This conclusion was drawn from changes in the absorption spectrum around 520 nm in response to illumination, which they correctly attributed to an electrochromic shift in the spectrum of carotenoids when an electrical potential develops across the membrane. Their extensive and meticulous studies (to which we shall return in the final section of this chapter) led to the following interpretation (Figure 8.13). When photosystems I and II absorb light, electrons are driven from a relatively positive redox potential to a more negative one; at the same time they are driven vectorially across the thylakoid membrane. Reaction-center chlorophylls are buried in the barrier phase; the primary acceptors for excited electrons lie near the stroma surface, while the electron donors that donate electrons to the oxidized chlorophyll reside at the luminal surface. Excitation of the photosystem thus induces a charge separation across the membrane, with the lumen positive. Subsequent proton movements, coupled to the flow of electrons from water to $NADP^+$, transform the electric

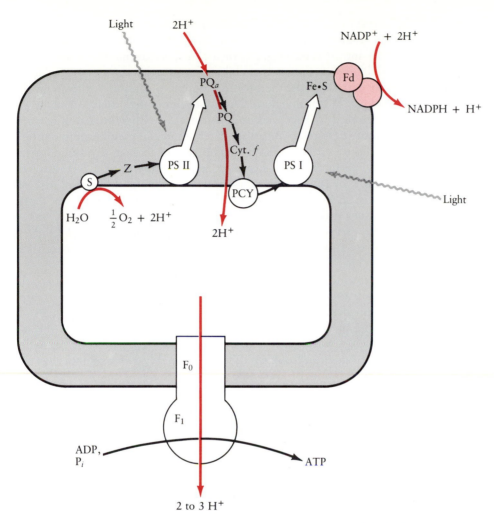

FIGURE 8.13   The chemiosmotic thylakoid vesicle. Absorption of light by PS I and PS II elicits electron flow from water to NADP$^+$. The pathway is vectorially looped across the membrane such that for each pair of electrons donated by water, four protons appear in the loculus. The protonic potential across the membrane, loculus acid and positive, drives the synthesis of ATP by the CF_1F_0 ATP synthase. Colored arrows, proton flow; black arrows, electron flow.

field into an electrochemical potential gradient of protons, loculus positive and acid (Junge, 1977; Witt, 1979; Figure 8.13).

Note that whereas mitochondria and bacteria expel protons, chloroplast thylakoids accumulate protons: ΔpH is interior acid, $\Delta \Psi$ is interior positive, and the topology corresponds to that of everted submitochondrial particles. But the topological difference between thylakoids and cristae is not so profound as it first appears. If the cristae of the inner mitochondrial membrane were to pinch off and seal (Figure 7.2), they would acquire the everted topology of thylakoids. Fundamentally, the intrathylakoid space, or loculus, is external to the matrix. A more significant difference between thylakoids and other energy-transducing membranes is that in thylakoids, the protonic potential is almost entirely in the form of a pH gradient. When thylakoids are first illuminated, a large membrane potential can be detected; within a second, however, the thylakoid lumen turns acid by three to four units relative to the matrix, and the membrane potential, lumen positive, falls to $+10$ to $+40$ mV. The reason is that thylakoid membranes are readily permeable to several ions; movement of Cl^- into the thylakoids and of K^+ or Mg^{2+} out compensates for the uptake of protons and largely converts the $\Delta\Psi$ component of the protonic potential into ΔpH. Thylakoid membranes provide buffering for 99.9 percent of the protons taken up. In the end, the protonic potential developed by chloroplast thylakoids is equivalent to 200 to 250 mV, much like that of other proton-translocating membranes.

Generation of a protonic potential consisting chiefly of a pH gradient would seem to offer one advantage: the energy-storage capacity of ΔpH is far greater than that of an equivalent membrane potential (Chapter 3). But there is also a price to be paid: the photosynthetic apparatus cannot be located in a surface membrane, as is the case with bacteria and mitochondria, but must be placed in vesicles segregated from the cytoplasm. This presumably is what underlies the almost universal association of oxygenic photosynthesis with discrete thylakoids; just what teleonomic use plants make of the energy-storage capacity of thylakoids is not at all clear.

Conversion of the protonic potential into the chemical potential of ATP proceeds along quite orthodox lines. As early as 1965, Efraim Racker and his colleagues isolated an ATPase from thylakoid membranes and showed that depleted membranes lost the capacity to make ATP. Reassociation of the ATPase with the membranes restored ATP synthesis as well as the characteristic stalked knobs that stud the cytoplasmic surface of thylakoids (Figure 8.14). The enzyme was designated coupling-factor ATPase, CF_1, with the C indicating that it derives from chloroplasts. Later research, to which we shall shortly return, has confirmed the basic identity of the chloroplast ATP synthase with F_1F_0 ATPases from other sources (Chapter 7). Like other members of this family, CF_1F_0 ATP synthase reversibly translocates protons across the thylakoid membrane. The stoichiometry is still uncertain, but the evidence presently favors $3H^+$ per ATP. We saw in previous chapters that $3H^+$ per ATP, rather than the $2H^+$ per ATP originally proposed, may hold for the mitochondrial and bacterial enzymes as well.

The overall modus operandi of the chemiosmotic thylakoid is summarized in Figure 8.13. On illumination, photosystems I and II eject electrons, generating an electric field across the thylakoid membrane at the expense of light energy. The light reactions also generate a scalar gradient of redox potential, which causes electrons to flow from water to $NADP^+$. Electron flow is coupled to the

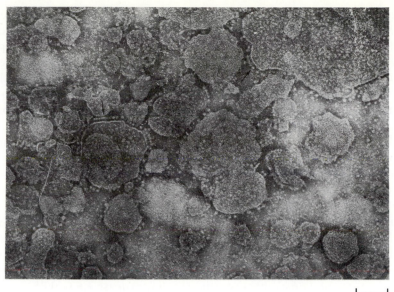

0.1 μm

FIGURE 8.14   Thylakoid vesicles are studded with stalked knobs that correspond to the CF_1 ATPase (arrows). (Electron micrograph courtesy of R. E. McCarty.)

translocation of protons into the vesicle; the consequent protonic potential, lumen acid and positive, supports ATP generation by the CF_1F_0 ATP synthase. Since the F_1 headpiece faces the stroma, there is no need for an ATP-ADP antiporter, and none exists in thylakoids.

Thermodynamics of Photosynthesis

By way of a summary of energy conversion in photosynthesis, let us review the main features from a quantitative viewpoint. Readers should keep in mind that the thermodynamic parameters remain open to revision by new experiments. Noncyclic and cyclic photosynthesis will be treated in turn.

The energy content of a quantum of light at 680 nm is 2.9×10^{-19} joule (174 kJ, or 42 kcal, per einstein) but is most conveniently expressed as 1.82 electron volts (eV). An electron volt is the energy acquired by an electron when it moves over a potential difference of one volt; 1 eV is equivalent to 23 kcal/mol. Light energy is initially captured as a vectorial redox reaction. As shown in Figure 8.10, both photosystems I and II establish a difference in redox potential of about 1300 mV between the primary electron acceptor and the primary electron donor. In addition, they generate an electric potential difference of, say, 100 mV across the thylakoid membrane. With each photoact, then, the electron ejected from the reaction center acquires an energy content of about 1.4 eV.

Eight quanta of light are required to drive four electrons from water with the generation of 2NADPH and O_2 (Equation 8.4). Eight quanta correspond to $8 \times 1.82 = 14.6$ eV. The redox span from the O_2-H_2O couple to $NADP^+$-NADPH is 1150 mV (Table 2.2). Therefore we can describe the effect of four electrons moving over this span as the "conservation" of $4 \times 1.15 = 4.6$ eV in the form of NADPH (equivalent to 442 kJ or 106 kcal). This is not quite accurate because of the use of the standard free-energy change, $\Delta G^{\circ\prime}$, in place of ΔG, but the error is small.

The number of protons transported across the thylakoid membrane per quantum absorbed is not certain, but lies between one and two; let us provisionally take it to be one. We shall also assume that three protons pass through the CF_1F_0 ATPase with each cycle. Eight photons will then translocate eight protons and drive the synthesis of 8/3 molecules of ATP. By actual measurement, the free energy of ATP synthesis during photophosphorylation, ΔG_p, reaches 14 kcal/mol (0.6 eV), but to be conservative we shall use the standard free-energy value $\Delta G^{\circ\prime} = 7.3$ kcal/mol (0.32 eV). The total energy "conserved" in the form of ATP is then $(8/3) \times 0.32 = 0.85$ eV (19.5 kcal, 82 kJ). In sum, of 14.6 eV put into the system, at least $4.6 + 0.85 = 5.5$ eV is recovered in the chemical potential of the NADPH-$NADP^+$ and ATP-ADP couples. This is often expressed by assigning photosynthetic energy transduction an "efficiency" of $(5.5/14.6) \times 100 = 38$ percent.

As mentioned above, the phosphorylation potential, ΔG_p, of chloroplast ATP synthesis can reach 0.6 eV; that is, it is equivalent to the energy acquired by a unit charge in passing over a potential span of 600 mV. In order to sustain this chemical potential with three protons, each must pass over a protonic potential span of $600/3 = 200$ mV (see also Equation 3.11). Experimentally one finds ΔpH to be around 3.5 units ($3.5 \times 59 = 210$ mV) and $\Delta\Psi$ less than 100 mV, for a total of about 300 mV. The data are thus quite compatible with the protonic potential being the driving force for ATP synthesis.

Cyclic photosynthesis requires only photosystem I. Here again the number of protons translocated per cycle is still subject to debate; for reasons to be developed later, let us take it to be $2H^+$. If so, one quantum sustains the translocation of $2H^+$ and the synthesis of $\frac{2}{3}$ATP. If the quantum carries 1.82 eV of energy and we take that of the ATP-ADP couple to be 0.32 eV, the "efficiency" works out to be $(2/3) \times (0.32/1.82) \times 100 = 12$ percent. The absolute efficiencies are not to be taken seriously, but they do emphasize the importance of NADPH generation in maximizing the harvest of light energy. The advantages gained by improved energy capture and by access to limitless supplies of reducing power underlie the biological success of oxygenic photosynthesis.

Chloroplast Metabolism and Work

CO_2 Fixation.   One of the interesting differences between mitochondria and chloroplasts is that the latter export not the raw output of energy transduction

but a finished product of CO_2 fixation. In plants this is a triose phosphate, generated by the reductive pentose cycle shown in Figure 8.15.

The discovery and elucidation of this metabolic pathway in the 1950s by Melvin Calvin, J. A. Bassham, and A. A. Benson is one of the milestones in the development of both biochemistry and cell physiology. It is covered in every textbook of biochemistry and so, despite its central role in photosynthesis, must receive perfunctory treatment here. We can summarize Figure 8.15 by stating that CO_2 is "fixed" by reaction with a phosphorylated pentose, ribulose bisphosphate, to yield two molecules of 3-phosphoglyceric acid. This reaction is catalyzed by the enzyme ribulose bisphosphate carboxylase, to which we shall return shortly. Each of the two molecules of phosphoglyceric acid is then reduced to glyceraldehyde-3-phosphate. The reaction is in effect the reversal of one of the key reactions in glycolysis (Equation 2.12), except that the chloroplast enzyme is specific for NADPH rather than NADH. Reduction of two molecules of phosphoglyceric acid consumes two molecules of ATP and two of NADPH. The glyceraldehyde-3-phosphate then enters into an intricate set of rearrangements involving sugars of three, four, five, six, and seven carbon atoms; in the end six carbon atoms of phosphoglyceric acid have been redistributed such that five wind up in ribulose-5-phosphate and one in fructose-6-phosphate. Ribulose-5-phosphate is phosphorylated to ribulose bisphosphate, with consumption of a third molecule of ATP. Fructose-6-phosphate is converted into glucose-6-phosphate, a precursor for the deposition of starch. Overall, the cycle assimilates one molecule of CO_2 into glucose-6-phosphate at the expense of two molecules of NADPH and three of ATP, supplied by the energy-transducing machinery of the thylakoids. All of CO_2 assimilation takes place within the bounds of the chloroplast.

The Calvin cycle requires ATP and NADPH in the ratio $3:2 = 1.5$; the ratio produced by the thylakoids seems to be lower, $\frac{8}{3}:2 = 1.33$. Additional demands for ATP arise from metabolic reactions within the chloroplast envelope and, perhaps, from transport processes. Since the requirements for ATP and NADPH are variable, it is of great physiological importance that thylakoids can operate in either a cyclic or a noncyclic mode, with the allocation of light energy to each mode controlled by the demand for its products.

The key transaction in CO_2 fixation is the carboxylation of ribulose bisphosphate by the reaction shown in Figure 8.16, which is effectively irreversible. Ribulose bisphosphate carboxylase from spinach is composed of two kinds of subunits, 54 and 3.5 kdal. The functional complex contains eight large subunits, which bear the catalytic sites, and eight small subunits. Note that CO_2, rather than bicarbonate ion, is the source of the new carboxyl group that arises by cleavage between C_2 and C_3 of ribulose bisphosphate. When properly activated, ribulose bisphosphate carboxylase has a K_m for CO_2 of around 10 μM, comparable to the K_m of intact chloroplasts and also to the concentration of CO_2 in water equilibrated with air.

Curiously, ribulose bisphosphate carboxylase is not entirely specific for CO_2; it can accept oxygen as an alternative substrate and catalyze the conversion of ribulose bisphosphate into one molecule of 3-phosphoglyceric acid and one of

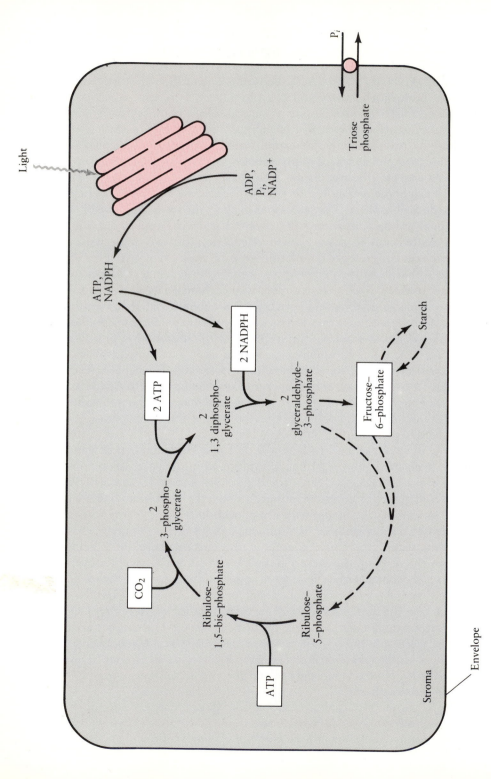

FIGURE 8.15 CO_2 fixation by the reductive pentose cycle. A simplified version, emphasizing ribulose bisphosphate carboxylase and the steps that consume NADPH and ATP but omitting the rearrangement of carbon atoms between C_3 and C_5 sugars. The NADPH and ATP are supplied by the thylakoids; fixed carbon is exported primarily in the form of glyceraldehyde-3-phosphate, which exits by antiport for P_i.

Light

ATP, NADPH

ADP, P_i, NADP+

2 ATP

2 NADPH

CO_2

2 3-phospho-glycerate

2 1,3 diphospho-glycerate

2 glyceraldehyde-3-phosphate

Fructose-6-phosphate

Starch

Ribulose-1,5-bis-phosphate

ATP

Ribulose-5-phosphate

Triose phosphate

P_i

Envelope

Stroma

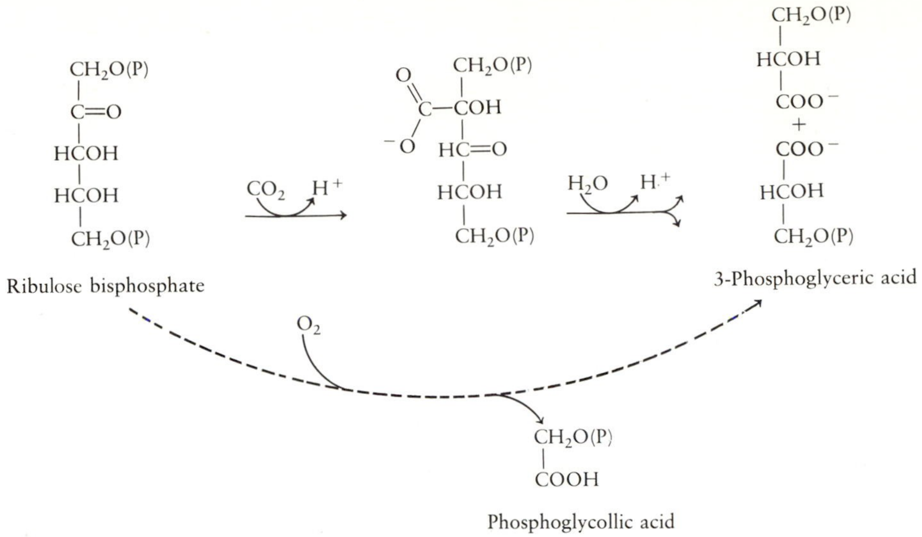

FIGURE 8.16   Reactions catalyzed by ribulose bisphosphate carboxylase. Upper part of drawing, CO_2 fixation. Lower part, oxygenase reaction, producing 3-phosphoglyceric acid and phosphoglycollic acid. (From Prebble, 1981, with permission of Longman Group, Ltd.)

phosphoglycollic acid (Figure 8.16). The latter is metabolized further, mostly outside the chloroplast, with release of CO_2. Together these reactions constitute a light-dependent pathway that consumes oxygen and liberates CO_2; it has long been familiar to plant physiologists under the name photorespiration. Note that unlike true respiration by the mitochondria, photorespiration is a process that dissipates energy.

It is not clear whether the oxygenase reaction of ribulose bisphosphate carboxylase is an irremediable imperfection in its specificity or if it has a function of its own. It has been argued, for example, that photorespiration is one of several processes that help protect chloroplasts from damage by photo-oxidation (Foyer and Hall, 1980). But there is no doubt that photorespiration reduces the yield of fixed carbon per quantum of light. This can be a handicap when CO_2 levels are low, which is often the case in hot, sunny weather, when stomata close to conserve moisture. Plants have responded to the shortcomings of ribulose bisphosphate carboxylase in two ways. One is to raise the concentration of the enzyme in the stroma to extraordinary levels, ensuring an adequate rate of CO_2 fixation. Ribulose bisphosphate carboxylase constitutes half of the soluble protein of chloroplasts, where its concentration is about 0.5 mM; it is almost certainly the most abundant protein on earth. Besides, many tropical plants have evolved ancillary pathways to keep the CO_2 level in the stroma high and thus minimize the flow of carbon and energy into glycollate.

The best known of these routes, and the only one that will be mentioned here, is characteristic of many tropical grasses and of such major crop plants as maize and sugar cane. These first assimilate CO_2 not into phosphoglyceric acid but into four-carbon dicarboxylic acids. The pathway involved, elucidated by the Australian investigators M. D. Hatch and R. C. Slack in 1966, is shown in Figure 8.17. The heart of the matter is the conversion of pyruvate into phosphoenolpyruvate at the expense of two molecules of ATP, followed by carboxylation to yield oxaloacetate. Subsequent reactions convert oxaloacetate into malate or aspartate. Now, the C_4 pathway takes place in one class of leaf cells, the mesophyll. The malate (or aspartate) migrates to another group of cells, those of the bundle sheath, which are adjacent to the phloem tubes that carry the products of photosynthesis out of the leaf. In the bundle sheath cells malate is decarboxylated, releasing CO_2 once again. The point of this apparently futile exercise is that it serves as a "CO_2 pump." Stomata are closed in the daytime to reduce water loss, restricting the entry of CO_2 into the leaf. But there is plenty of sunlight to make ATP. The reactions of the C_4 pathway make available high concentrations of CO_2 to the bundle sheath cells and thereby tilt the balance between the alternative catalytic functions of ribulose bisphosphate carboxylase in favor of CO_2 fixation. The added metabolic cost pays off in biomass.

The subsequent metabolism of phosphoglycollate, which completes photorespiration, cannot be discussed here. Suffice it to state that it returns carbon to the reductive pentose cycle and thus compensates to a degree for the unavoidable diversion of carbon resulting from the competition of O_2 and CO_2 for ribulose bisphosphate carboxylase.

Exports and Imports.   The end product of photosynthesis is usually said to be sucrose. This is indeed the form in which plants transport fixed carbon, via the phloem, from the leaves to shoots and roots. But sucrose is produced in the cytoplasm, not the chloroplast. Alternatively we can take the end product of photosynthesis to be starch. This is generated in chloroplasts and deposited there, but it can only be made available to the remainder of the cell by first being broken down again. From the viewpoint of chloroplast economics the production line ends at the triose level: chloroplasts export chiefly glyceraldehyde-3-

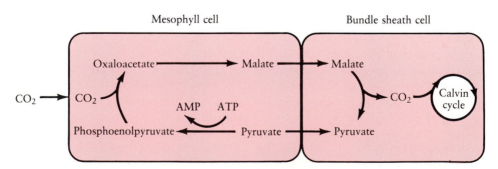

FIGURE 8.17   Essential features of the C_4 pathway of CO_2 fixation. For discussion see text. (From L. Stryer, *Biochemistry*. Copyright ©1981 by W. H. Freeman and Co.)

phosphate and to a lesser degree dihydroxyacetone phosphate or 3-phosphogly-ceric acid (Figure 8.15).

This subject was first explored by U. Heber, H. W. Heldt, and their associates in studies on the permeability of the chloroplast envelope (for a recent comprehensive review see Heber and Heldt, 1981). The original observation was that triose phosphate, phosphoglycerate, and also inorganic phosphate penetrate readily into chloroplasts. It has now been established that a specific porter present in abundance in the inner envelope membrane catalyzes the tightly coupled exchange of the triose phosphates for P_i. Strict antiport is not an accidental feature: export of sugar phosphates must be balanced by P_i import to prevent depletion of the chloroplast's phosphate stores. The porter has been solubilized as a protein of 29 kdal; it will be interesting to see how far it resembles the superficially similar ATP-ADP antiporter of mitochondria.

It is curious that illuminated chloroplasts export chiefly triose phosphate, even though the stroma level of 3-phosphoglycerate is generally much higher. The reason is probably that in response to illumination, the pH of the stroma rises to about pH 8. The carrier prefers the divalent anionic species. Unlike the other substrates, phosphoglycerate has three ionizable groups and at pH 8 the doubly ionized species is scarce and less available for export.

Heber and Heldt (1981) list a number of additional porters. One carries dicarboxylic acids (malate, oxaloacetate, α-ketoglutarate, glutamate, and aspartate), usually exchanging one for another. This porter (or porters) is thought to be involved in the transfer of carbon skeletons and reducing equivalents between cytoplasm and stroma by a shuttle analogous to the malate aspartate shuttle of mitochondria (Chapter 7). There is also a weak but specific ATP-ADP antiporter, which may supply ATP to the chloroplast during the night; a sugar porter, which allows glucose produced by starch degradation to exit; and perhaps others for various ions.

One of the unsolved mysteries of chloroplast function is the mechanism by which chloroplasts excrete protons on illumination and raise the pH of the stroma. An ATPase of unknown function is present in the chloroplast envelope; prejudice assures one that it will turn out to be a proton pump, but this has yet to be demonstrated. Alternatively, one could imagine that protons, pumped into the thylakoid at the onset of photosynthesis, are somehow excreted by an unknown route (Heber and Heldt, 1981). The effect is teleonomically important, for several enzymes of the reductive pentose cycle are sufficiently pH sensitive that little CO_2 fixation can occur without alkalinization of the stroma.

Finally, a word is in order about the movements of those all-important gases CO_2 and O_2. It is generally agreed that CO_2 travels as such, by diffusion, and that chloroplast membranes are largely impermeable to bicarbonate. Oxygen, thought to arise at the luminal surface of the thylakoid membrane, diffuses readily across lipid phases.

Regulation.   One should not leave the subject of chloroplast metabolism without at least passing mention of the elaborate controls that link it to energy transduction. Two general mechanisms can be distinguished, the regulation of stroma pH and the control of enzyme activity by light-dependent, covalent modification.

As noted above, the pH of the chloroplast matrix rises by as much as one unit in response to illumination, thereby activating the Calvin cycle. A second, independent effect of light is the activation of a number of enzymes by reduction of critical disulfide groups; examples are fructose-1,6-bisphosphatase, $NADP^+$-glyceraldehyde-3-phosphate dehydrogenase, phosphoribulokinase, and even the CF_1 headpiece of the ATP synthase. In many cases, reduced ferredoxin (itself produced with the aid of photosystem I) donates electrons to a small redox protein called thioredoxin, which in turn serves as the reductant for the enzyme proteins. We can think of thioredoxin as a regulatory signal: it alerts key enzymes of biosynthesis that the light is on, ATP and NADPH are available, and biosynthesis is to proceed (Buchanan et al., 1979). Thioredoxin, again, is not unique to photosynthetic organisms; it was first characterized by P. Reichard as a component of the pathway by which ribonucleoside diphosphates are reduced to the deoxyribonucleoside level, and is widely distributed among prokaryotes and eukaryotes.

Digression: Oxygenic Photosynthesis in Prokaryotes

Readers who are not bacteriologists may well encounter the cyanobacteria (or blue-green algae, as they used to be called) only as the greenish scum that forms on the surface of stagnant water. However unappetizing its smell or appearance, the slime repays inspection with a microscope, for it teems with minute forms of life: rods, cocci, conspicuous filaments, and long chains of cells. They are large by bacterial standards and often encased in a gelatinous capsule; many glide over the surface in a characteristic stately fashion. The cyanobacteria are the largest and most diverse group of photosynthetic prokaryotes. Thanks to their capacity for nitrogen fixation they are of ecological, even economic, importance, but they have been somewhat neglected until recently (for a comprehensive survey see Carr and Whitton, 1982). Their prokaryotic nature and place in the evolution of photosynthesis were the special province of the late Roger Stanier[4] (Stanier, 1974; Stanier and Cohen-Bazire, 1977).

The cyanobacteria "can be succinctly defined as organisms that harbor, within a prokaryotic cell, a photosynthetic apparatus remarkably similar in functional, structural, and molecular respects to that contained in the eukaryotic chloroplast" (Stanier and Cohen-Bazire, 1977). They are clearly distinguished from the two other major groups of photosynthetic prokaryotes, the purple and

---

[4] I would like to record here a personal tribute to this extraordinary scholar, investigator, and teacher who revealed the bacterial world to generations of graduate students at Berkeley. I am one of many who, regardless of what our diplomas say, were inspired by him to become microbiologists.

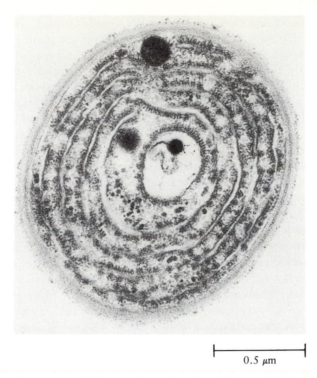

$\vdash\!\!-\!\!-\!\!-\!\!-\!\!-\!\!\dashv$
0.5 µm

FIGURE 8.18   Thin section of the cyanobacterium *Synechococcus lividus*. Note the concentric whorls of thylakoids studded with phycobilisomes. (Electron micrograph courtesy of M. R. Edwards and E. Gantt.)

the green bacteria (Chapter 4), by their pigment complement and by their capacity to perform oxygenic photosynthesis. Their anatomy already hints broadly at homology with chloroplasts. Whereas most bacteria house their energy-transducing machinery in the plasma membrane or its extensions, cyanobacteria contain quite typical thylakoids (Figure 8.18) that are not continuous with the plasma membrane and bear the photosynthetic machinery.[5]

The essential identity of energy transduction in cyanobacteria and chloroplasts is plain to be seen at the molecular level (for recent reviews see Binder, 1982; Ho and Krogmann, 1982). Despite their diversity of form and habitat, cyanobacteria contain the same ensemble of components that were listed for chloroplasts in the preceding sections: photosystems I and II, with reaction centers identifiable as P_{700} and P_{680}; chlorophyll *a* (neither chlorophyll *b* nor bacteriochlorophyll is to be found); a water-splitting complex; plastoquinone;

[5] This generalization must immediately be qualified. One genus of cyanobacteria, *Gloeobacter*, lacks thylakoids and houses its photosynthetic pigments in the plasma membrane (Stanier and Cohen-Bazire, 1977).

several cytochromes including types b_6 and f; plastocyanin; iron-sulfur centers; and ferredoxin (Figure 4.4c). Even the finer details of the photosystems and the electron-transport chain, which will be the subject of the following section, seem to be very nearly the same in cyanobacteria as in chloroplasts (Stewart and Bendall, 1981; Guikema and Sherman, 1982). Electron transport is accompanied by the generation of a protonic potential across the thylakoid membrane, consisting chiefly of a pH gradient (interior acid), and ATP synthesis is mediated by an F_1F_0 ATP synthase of the standard kind. CO_2 assimilation takes place by the reductive pentose cycle; even the oxygenase function of ribulose bisphosphate carboxylase has been documented.

This should not be construed to mean that cyanobacteria are, from the bioenergetic viewpoint, merely free-living chloroplasts: they are far more diverse and versatile, equipped to cope with a range of circumstances and environments. Many are facultative chemotrophs and some grow well by anoxygenic photosynthesis, using photosystem I in the bacterial fashion with H_2S as reducing agent. Cyanobacteria, which normally generate oxygen during photosynthesis, are also capable of respiration. Some can grow chemotrophically under aerobic conditions in the dark, although photoautotrophy is their preferred mode; one teleonomic purpose of respiration may be to supply the cells with ATP during the night. Respiration depends on a conventional chain of redox carriers that terminates in a cytochrome oxidase of the aa_3 type, but there are two unique features: plastoquinone takes the place of the ubiquinone or menaquinone usually found in bacteria; and the chief electron donor is not NADH, but NADPH. (Curiously, it has recently been claimed that even chloroplasts retain a vestigial capacity for respiration, with NADPH as substrate; Bennoun, 1982.)

The spatial and biochemical relationships between the respiratory and photosynthetic machinery are still uncertain. A priori one might expect them to be separate, the former in the plasma membrane and the latter in the thylakoids, but there is growing evidence that respiration and photosynthesis share at least some redox intermediates: plastoquinone, cytochrome f, and cytochrome c_{553}, which takes the place of plastocyanin in a number of cyanobacteria and also in some algal chloroplasts. The F_1F_0 ATP synthase may also function interchangeably in both oxidative and photophosphorylation. If so, one must explain how cyanobacteria, like other bacteria, extrude protons across the plasma membrane, generating a protonic potential, cytoplasm alkaline and negative, that provides energy for the uptake of metabolic substrates during chemotrophic growth in the dark. One possibility is the presence of a second proton-translocating ATPase in the plasma membrane. Circumstantial evidence for the existence of such an enzyme has just come to hand (Scherrer et al., 1984), and it may be pertinent that an ATPase is also present in the chloroplast envelope. Alternatively, cyanobacteria may expel protons by respiratory redox reactions in the plasma membrane (Matthijs et al., 1984), as other bacteria do. A speculative model of the bioenergetic organization of cyanobacterial cells is shown in Figure 8.19.

Cyanobacteria take their bluish-green color and their name from their antenna pigments, the phycobilins. These are linear tetrapyrroles, structurally

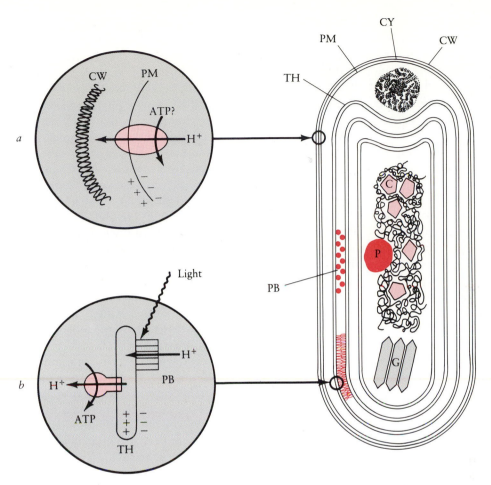

FIGURE 8.19   Schematic diagram of a cyanobacterial cell. PM, plasma membrane; CW, cell wall; TH, thylakoid; PB, phycobilisomes shown in front and side views; P, polyphosphate granule; CY, cyanophycin storage granule; G, gas vacuole; C, carboxysome consisting of ribulose bisphosphate carboxylase. Inset *a*, enlarged view of the cell envelope, showing a proton pump of uncertain nature that expels protons outward across the plasma membrane. Inset *b*, enlarged view of a thylakoid, with phycobilisomes attached to its external surface; light absorption leads to translocation of protons into the thylakoid. (Modified from Stanier and Cohen-Bazire, 1977, with permission of Annual Reviews.)

related to the bile pigments of animals (Figure 8.4), which serve as the prosthetic groups of a special class of light-harvesting proteins called phycobiliproteins. In the intact cell, phycobiliproteins are assembled into remarkably intricate structures called phycobilisomes. These are attached in rows to the outer surface of each thylakoid and funnel light energy primarily to photosystem II. Most of the energy required for oxygen production is captured by the phycobilisomes; carotenoids and even chlorophyll *a* make little contribution. The phycobilisomes are cleverly arranged so that quanta are captured by peripheral pigment molecules with absorption bands at the shorter wavelengths and then migrate to pigments that absorb at longer wavelengths, ending up in the reaction center. The structure of phycobilisomes and their role in energy capture have been reviewed by Gantt (1980), Cohen-Bazire and Bryant (1982), and Glazer (1982).

Phycobilisomes occur in the chloroplasts of but a single phylum of eukaryotes, the Rhodophyta (red algae; these are common along rocky shores, particularly in warm waters). The pigments of rhodophytan chloroplasts are identical with those of the cyanobacteria, and their polypeptides are structurally related. Recent work on the primary sequences of ribosomal RNAs (Chapter 6) confirmed what the majority of students had believed for some years: that rhodophytan chloroplasts must be the lineal descendants of a cyanobacterial line that took up residence in the cytoplasm of an early eukaryote and evolved from endosymbionts into organelles. The cyanelles of *Cyanophora paradoxa* (Chapter 6) capture this transformation in the act: they employ phycobilisomes for light harvesting and retain the envelope structure of free-living cyanobacteria (Giddings et al., 1983), but they are permanently integrated into the cytoplasm of their eukaryotic host.

What about the majority of chloroplasts, those of green algae and higher plants, which lack phycobiliproteins and whose antennas contain chlorophylls *a* and *b* instead? The first light was shed on this question in 1976, when Ralph Lewin reported the discovery of *Prochloron,* an endosymbiont of certain marine ascidians found along tropical shores, which represents a new phylum of photosynthetic prokaryotes. The cells are spherical, up to 30 μm across, bright green, and contain thylakoids with prominent grana stacks (Withers et al., 1978; Giddings et al., 1980). There are no phycobilisomes or phycobilin pigments, but both chlorophylls *a* and *b*. As yet, *Prochloron* has not been grown in culture and information about its physiology and bioenergetics is sparse. But it is clear that *Prochloron* exemplifies both the kind of prokaryote that may have been ancestral to plant chloroplasts and the manner in which the transformation into chloroplasts came about; it stands midway between a free-living organism that was and an organelle that may be.

Current Issues in Energy Coupling

The energetics of photosynthesis is presently an area of intense research activity, with the focus on molecular organization and mechanisms. How do photosystems I and II convert electronic excitation into vectorial electron flow? How is water oxidized to protons and oxygen? Just how is electron transport coupled to proton translocation, and the latter to ATP production? How is light collected and allocated to the two photosystems? The object of this section is to examine selected topics in a little detail.

The flavor of the photosynthesis literature is quite unlike that dealing with mitochondria. Chemiosmotic principles were assimilated early and without much anguish; separation of charges across a membrane and the generation of an electric field have been recognized as the primary act of photosynthesis for nearly two decades. Instead, the photosynthesis community has been preoccupied with the elaboration of its biophysical armamentarium, particularly the refinement of flash spectrophotometry. The reason is that the most interesting components, the reaction centers, constitute less than 1 percent of the total pigment bed, and progress toward their isolation and functional reconstitution has been slow. New insights have come primarily from advanced spectrophotometric methods, whose time resolution and signal-to-noise ratio are equal to the task of dissecting single turnovers of the photosynthetic machinery. This has fostered a predominantly biophysical conception of photosynthesis, couched in esoteric language that tends to isolate this subject from the remainder of bioenergetics.

This tradition has lately been given a more biochemical slant through the development of methods for dealing with the catalytic entities as supramolecular units. The photosynthetic machinery of chloroplasts has been dissected into five such complexes, at least four of which span the thylakoid membrane.

1. The PS II complex, including the enzymatic machinery for splitting water;

2. The PS I complex;

3. A complex containing cytochromes b_{563} and f, which links the two photosystems;

4. The CF_1F_0 ATP synthase;

5. A supramolecular complex (probably multiple rather than singular) that has no catalytic functions but contains antenna chlorophyll arranged on a protein framework.

The light-harvesting complex is thought to be coupled chiefly to PS II (hence the designation LHC II), but there is probably a second light harvesting antenna dedicated to PS I (LHC I). These supramolecular complexes provide a framework for the discussion that follows (Figure 8.20).

Among the numerous review articles that survey recent developments, I have found the following most helpful. The contributions of flash spectrophotometry to the analysis of kinetics and energetics have been reviewed by Junge (1977), Witt (1979), and Junge and Jackson (1982). Electron transport, including the splitting of water and the perplexities of a possible Q cycle, are covered by Crofts and Wood (1978), Velthuys (1980), Cramer and Crofts (1982), Hauska et al. (1983), Haehnel (1984), and with admirable clarity by Bendall (1982). Proton translocation and chemiosmotic aspects are discussed in almost all of the articles cited above and also by Avron (1977) and Hinkle and McCarty (1978). The chloroplast CF_1F_0 ATP synthase has been the subject of several reviews, for instance those of McCarty (1978), Nelson (1981), McCarty and Carmeli (1982), and Strotmann and Bickel-Sandkötter (1984).

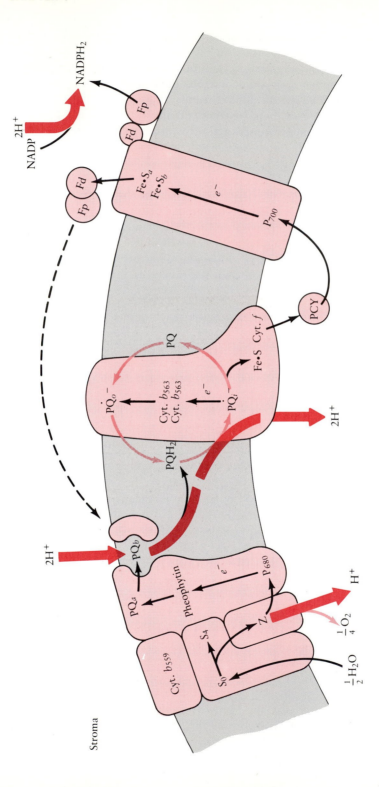

FIGURE 8.20 The chloroplast electron-transport chain. The scheme shows the chain composed of three proton-translocating complexes, each of which carries out an electrogenic reaction. For each electron that passes from H_2O to NADP, three protons are translocated. Red arrows, proton movements; black arrows, electron transfer; pink arrows, chemical transitions. (After Cramer and Crofts, 1982, with permission of Academic Press.)

Photosystem II and Oxygen Evolution.   Photosystem II absorbs visible light extending into the red, but not much beyond 690 nm. Excitation of the reaction center brings about the transfer of one electron to a primary electron acceptor, leaving behind a positively charged "electron hole." The oxidized reaction center is a powerful oxidant, strong enough to draw electrons from so weak a reductant as water with the generation of protons and oxygen gas; the E_m of this oxidant must be $+900$ mV or more. The electron acceptor in turn constitutes a reductant, E_m around -500 mV, which ultimately passes the electron forward to the plastoquinone pool (Figure 8.10). The PS II complex is juxtaposed to a set of antenna chlorophylls and spans the thylakoid membrane; excitation generates an electric field across that membrane, lumen positive. This is a consequence of the location of the primary electron acceptor toward the matrix surface, while the electron donor resides near the luminal surface; protons released by the oxidation of water appear in the lumen (Figure 8.13). One would now like to clothe this biophysical skeleton with chemical detail, but in fact many issues remain to be resolved.

It is generally accepted that the photosynthetic reaction center consists of one or two special chlorophyll *a* molecules, designated P_{680}. The primary photochemical event, transfer of an electron from P_{680} to an acceptor, is completed in less than 20 nanoseconds (ns) even in liquid nitrogen; it is clearly a solid-state reaction, not a diffusion process. The electron acceptor is designated X320 in the literature. Like P_{680} itself, X320 was first identified by T. H. Witt and his colleagues in Berlin; it is now thought to be a special plastoquinone molecule that undergoes a one-electron cycle, oscillating between the oxidized form PQ and the semiquinone anion $P\dot{Q}^-$; in Figures 8.10 and 8.20 this molecule is designated PQ_a. Electrons pass right on to another special plastoquinone, PQ_b, which is thought to undergo full reduction to PQH_2, thereby mediating between the one-electron chemistry of the reaction center and the two-electron chemistry of the bulk plastoquinone pool. The photochemical events associated with the excitation of P_{680} remain controversial. There is evidence that the initial acceptor of the excited electron is not PQ_a but bound pheophytin (a chlorophyll in which protons take the place of the central magnesium), which may be the (I) of Figure 8.10. The spatial relationship between these various "primary" electron acceptors is not well worked out, but photosystem II appears to be organized along the lines of the bacterial reaction centers described in Chapter 4.

Reaction-center chlorophylls make up but a small fraction of the pigment complement of photosystem II. The bulk of the chlorophyll collects light and funnels it into the reaction center. The light-harvesting complex LHC II, is built around two kinds of protein subunits (25 kdal each) and contains half the total chlorophyll *a* and most of the chlorophyll *b*. At least four additional polypeptides contribute to photosystem II, including two that bind the reaction-center chlorophylls and a *b*-type cytochrome whose function is quite uncertain. This somewhat inchoate description should leave no doubt that the structural organization of the PS II region remains a challenging topic for biochemical research.

This is true a fortiori for the system that splits water, the ultimate electron donor for photosystem II (Figure 8.10). There is evidence that the immediate electron donor to P_{680} is an unidentified chromophore called Z, intimately associated with the reaction center; Z itself accepts electrons from the enzyme that splits water and is designated S. The two systems can be separated to some extent: thylakoids washed with Tris buffer, for example, lose the capacity to split water but still generate a powerful oxidant on illumination. It has also long been known that manganese is intimately involved in the oxidation of water, since algal cells grown in manganese-deficient medium specifically lack this capacity. However, biochemical data remain sparse and to date, at least, most of what is known or surmised has been inferred from studies on oxygen evolution and proton release in response to flash illumination.

Fifteen years ago, P. Joliot and A. Joliot discovered the remarkable fact that the release of oxygen by algal cells or chloroplasts subjected to repetitive flashes of light occurred with a distinct periodicity of 4. The explanation developed by the late Bessel Kok and the Joliots proposed that the photosystem II complex $S \cdot Z \cdot P_{680} \cdot Q_a$ can exist in four successive photoactive states S_1 to S_4 (and in a basal state S_0), as suggested schematically in Figure 8.21. Each flash induces a single transfer of electrons, placing an increasing number of charges on enzyme S. Four charges must accumulate before a molecule of oxygen can be released. Concurrently, protons are formed and released into the lumen of the thylakoids. The latter was neatly demonstrated by W. Junge and his colleagues, who took advantage of the ability of the indicator dye neutral red to penetrate the thylakoids. Protons are produced in bursts of one or two, not all at once like oxygen. But the details of the sequence and mechanism have remained elusive and controversial; readers anxious to delve into this abstruse mystery should consult the reviews by Crofts and Wood (1978), Velthuys (1980), and Cramer and Crofts (1982). No single aspect of photosynthesis seems more urgently in need of biochemical clarification than the enzyme that splits water.

Photosystem I.   Red and far-red light ($\lambda > 700$ nm) excite the more reducing of the two photosystems, designated P_{700} for the prominent change in its absorption spectrum that occurs on excitation. Unexcited P_{700} has an E_m of about $+450$ mV. Absorption of a quantum of light causes the extremely rapid transfer of an electron, on a picosecond time scale, to a primary acceptor group or groups; the electron ultimately reduces ferredoxin, a small iron-sulfur protein that is loosely attached to the stroma surface of the thylakoid membrane. Oxidized P_{700} is quickly re-reduced, accepting an electron from plastocyanin; the latter is a blue copper protein loosely associated with the luminal surface of the thylakoid (Figures 8.10, 8.13, and 8.20). Photosystem I as a whole spans the membrane and generates an electric field across it: each quantum transfers one negative charge out of the thylakoid, producing an electric potential, lumen positive.

The reaction center P_{700} probably consists of a special pair of chlorophyll a molecules. The electron acceptor, however, is quite unlike the pheophytin and

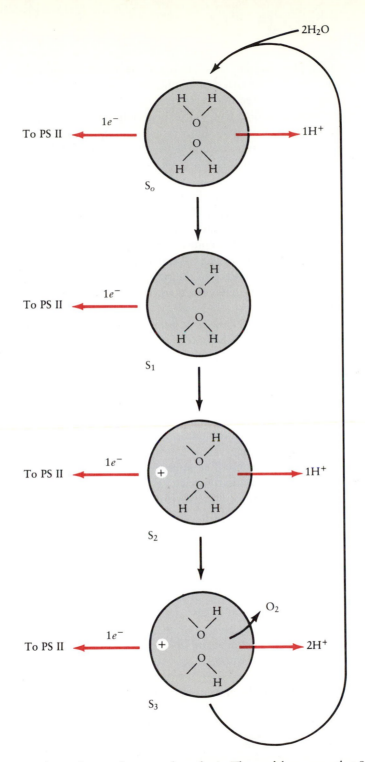

FIGURE 8.21   The oxidation of water, a hypothesis. The model proposes that S, the enzyme that splits water, catalyzes successive oxidative steps. One electron is released at each step while protons are released in groups. Oxygen is produced only when four charges have accumulated on S. (After a sketch by H. T. Witt, with permission.)

quinones found in photosystem II and in bacterial reaction centers. In 1971, A. J. Bearden and R. Malkin reported that photosystem I contains "bound ferredoxin," an iron-sulfur center that was reduced on illumination even in liquid nitrogen. Since then a second iron-sulfur center has been identified. There are about 10 to 12 atoms of iron and sulfur per P_{700} dimer. The iron-sulfur centers appear to form a molecular electron-transport chain that whisks the electron instantaneously to ferredoxin. The first of these centers has an E_m as low as -730 mV.

The molecular architecture of photosystem I is complex, uncertain, and in dispute (Malkin, 1982). At least six polypeptides are present, of which the largest (70 kdal) binds P_{700} itself. In addition there is a peripheral light-harvesting complex. LHC I, which contains some 40 molecules of chlorophyll *a* per P_{700} center. One of the most interesting developments in this area is the discovery that photosystem I tends to be localized in the stroma lamellae, while photosystem II prefers the grana stacks. We shall return to this point in the final section.

How Is Proton Translocation Coupled to Electron Flow?   The classic Z scheme of photosynthetic electron transport (Figure 8.10) can easily be written in the form of two proton-translocating redox reactions that traverse the thylakoid membrane, as was done in Figure 8.13. In the first reaction, absorption of two photons by photosystem II extracts two electrons from water, leaving two protons in the thylakoid lumen. The electrons reduce plastoquinone at the external surface with the consumption of two protons from the matrix. The plastoquinone diffuses across the membrane in an electroneutral manner. The second loop is driven by photosystem I, which on absorption of two photons again ejects a pair of electrons. The source of the electrons is PQH_2, which undergoes oxidation at the luminal surface; the electrons are donated to PS I via cytochrome *f* and plastocyanin, and two protons remain in the lumen. At the external surface the electrons reduce $NADP^+$ to NADPH, with the consumption of two more matrix protons[6]. This hypothesis predicts that overall four photons drive $2e^-$ from H_2O to $NADP^+$; four protons should appear in the loculus and four be consumed in the matrix. In effect, the passage of a pair of electrons over the chain translocates four protons into the thylakoid; in crisp professional notation, $H^+/e^- = 2$. An analogous argument leads one to expect that $H^+/e^- = 1$ for cyclic photosynthesis, in which photosystem I performs solo.

Experimental determination of the H^+/e^- ratio is fraught with uncertainties, but on the whole most of the data are consistent with the preceding analysis. For example, Izawa and Hind (1967) monitored simultaneously the rates of proton uptake by thylakoids and of ferricyanide reduction, a reaction that

---

[6] It will be appreciated that this is a formal stoichiometry, corresponding to the equation $NADP + 2e^- + 2H^+ \rightleftharpoons NADPH_2$. The number of protons consumed in the reaction $NADP^+ + 2e^- + 2H^+ \rightleftharpoons NADPH + H^+$ depends on the pH.

depends on both PS I and PS II; after applying suitable corrections they concluded that $H^+/e^- \simeq 2$. When photosystem II was blocked with DCMU and an artificial electron donor for photosystem I was provided, H^+/e^- fell to about 1 (Junge, 1977; Witt, 1979). However, under certain conditions larger numbers were obtained. As a result, most investigators have concluded that the Z scheme, while basically sound, must be amended by the addition of a third electrogenic process.

This belief stems, inter alia, from the time course of the development of the electric potential across the thylakoid membrane. As mentioned previously, some 15 years ago T. H. Witt, W. Junge, and their colleagues discovered that illumination induces a significant shift in the absorption spectrum of chloroplast membranes in the vicinity of 515 nm. They attributed this to an electrochromic effect on the spectrum of carotenoids. Given an electric potential of 50 mV and a membrane thickness of 5 nm, the electric field across the membrane would be about 100,000 V/cm, a field quite strong enough to induce electrochromic shifts in the spectra of carotenoid molecules oriented across the dielectric. Despite initial reservations, this interpretation is now generally accepted and the "515-nm shift" has come to serve as a built-in molecular voltmeter of formidable sensitivity and response time. (For a fuller exposition of the evidence and of the nature of electrochromism see Witt, 1979, and Junge and Jackson, 1982.) By use of this technique, several research groups have monitored the rise in $\Delta\Psi$ following flash illumination. Most of the electric potential has a rise time of 20 ns or less, reflecting the rapid ejection of electrons by photosystems I and II. However, under certain conditions (for example, when matters are so arranged that the plastoquinone pool is largely reduced to PQH_2) one can discern a slower phase as well, with a rise time of several milliseconds (Figure 8.22). This is the candidate for the third electrogenic step in the transport of both electrons and protons (Velthuys, 1978; Slovacek et al., 1979).

What could this process be? Mitchell (1976b) was quick to suggest that the Q cycle, originally put forward as an explanation for findings with the mitochondrial respiratory chain (Chapter 7), might be applicable to light-driven electron transport as well. One possible version of a photosynthetic Q cycle is illustrated in Figure 8.20. By tracing the path of an electron through the cycle, as was done in Figure 7.15, the reader will be convinced that for each electron passing from plastoquinone to plastocyanin, two protons are translocated from the matrix into the thylakoid. If this scheme or one of a number of possible variants is correct, H^+/e^- for noncyclic electron transport equals 3, while for the cyclic mode $H^+/e^- = 2$.

It is not practicable here to pursue the controversies that currently swirl around this proposal (for reviews see Velthuys, 1980; Cramer and Crofts, 1982; Bendall, 1982; Slater, 1983; Hauska et al., 1983), but let me mention two points that illustrate the difficulties. One is that the mitochondrial $b-c_1$ segment has two distinguishable b-type cytochromes to mediate the electrogenic transfer of electrons from one side to the other; there ought to be two in the photosynthetic $b-f$ segment as well, but only one kind of cytochrome b_{563} is known at present. A

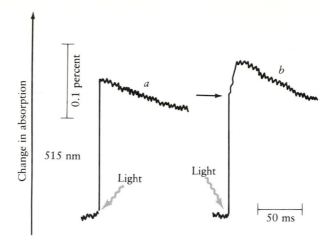

FIGURE 8.22   Development of the membrane potential as monitored by the 515-nm shift. Trace *a*, chloroplasts were dark adapted and then exposed to a saturating flash of light; the absorption at 515 nm rises abruptly, indicating the rapid generation of an electric field by PS I and PS II. Trace *b*, chloroplasts were preilluminated for five seconds in the absence of an electron acceptor, in order to reduce the pool plastoquinone. Under these conditions, the rapid initial rise in $\Delta\Psi$ is followed by a slower phase (arrow); the latter corresponds to the third electrogenic process, probably a Q cycle. (After Velthuys, 1978.)

deeper conundrum is that in noncyclic electron transport, the Q cycle seems to operate only at the beginning of a period of illumination and may then be "switched off." Clearly there are differences between the mitochondrial and the chloroplast pathways that are probably of functional significance.

On the other hand, the obvious parallels between the two patterns have been reinforced by the physical isolation and reconstitution of the chloroplast *b–f* complex. Hurt and Hauska (1981) found it to contain one molecule of cytochrome *f*, two of cytochrome b_{563}, a Rieske iron-sulfur center similar to that of the mitochondrial $b–c_1$ segment, and five polypeptides; neither plastoquinone nor chlorophyll was present. The isolated *b–f* complex catalyzed reduction of plastocyanin by plastoquinol and when reconstituted in liposomes, translocated one proton for each electron. Thus while the details of the chloroplast Q cycle should be taken with the proverbial grain of salt, the existence of some such pathway seems all but certain.

Figure 8.20 shows our present conception of photosynthetic electron transport and its coupling to proton movements. As outlined above the chain results from the interaction of three supramolecular complexes, each of which spans the thylakoid membrane: photosystem II, photosystem I, and the cytochrome *b–f* complex, which functions as a plastoquinol-plastocyanin reductase. The three complexes are linked by three mobile carriers that shuttle electrons (plastocyanin, ferredoxin) or H^+ plus e^- (the pool plastoquinone). In a recent paper Lam and Malkin (1982) described the reconstitution of electron transport from water to $NADP^+$ in an artificial lipid membrane. In agreement with Figure 8.20, the reaction requires the three complexes plus plastoquinone, plastocyanin, ferredoxin, and the ferredoxin-NADP$^+$ reductase.

The plastoquinol-plastocyanin reductase is shown operating as a Q cycle, translocating two protons for each electron that traverses this segment. The overall stoichiometry, then, is one electron transported and three protons pumped for a single turnover of both reaction centers (H$^+$/e^- = 3). Alternatively, the chain can function in a cyclic manner: as indicated by the dotted line, ferredoxin then donates its electron to plastoquinone rather than to NADP$^+$. The factors that determine whether the complexes function in a linear or a cyclic manner will be considered in the final section.

Photophosphorylation.   The chain of causality is clear enough. Light elicits electron flow, which drives proton translocation and the generation of a protonic potential; this in turn supports ATP production by the CF$_1$F$_0$ ATP synthase. But here, as for mitochondria, the devil lurks in the details. Is the gross proton-motive force across the thylakoid membrane thermodynamically and kinetically adequate to account for energy coupling? Is there a more intimate, localized pathway for current to flow from the photosynthetic current generators to the ATP synthase? What is the H$^+$/ATP stoichiometry of that enzyme, how is it constructed, and how does it work? These matters were all discussed in goodly detail in Chapter 7, and there is no need to grind that grist twice. Suffice it here to mention a handful of findings that bear specifically on the workings of chloroplasts or on the contributions that their study has made to bioenergetics in general.

Thanks to the extreme sensitivity of modern spectrophotometric techniques, students of photosynthesis have been able to exploit approaches denied to other bioenergeticists: for example, it is possible to monitor the electrical events that attend a single turnover of the photosynthetic machinery (Witt, 1979). Figure 8.23 illustrates a set of such experiments. In response to a single flash, $\Delta\Psi$ rises on a time scale of nanoseconds, as reported by the shift in absorption at 515 nm; the rise reflects the turnover of photosystems I and II. Subsequent decay of the 515-nm signal in the dark is much slower; decay is due to the redistribution of various ions (H$^+$, K$^+$, Mg^{2+}, Cl$^-$) across the membrane, moving in response to the primary electric field. Ion movements and dissipation of the field are accelerated by ionophores. As little as one molecule of gramicidin per 10^5 molecules of chlorophyll, or about one gramicidin dimer per thylakoid, doubles the rate of signal decay; evidently the entire vesicle is a single electrical unit. Decay of the 515-nm signal can also be accelerated by the addition of ADP plus P$_i$. In this case the extra ion conductance is due to proton flow through the CF$_1$F$_0$ ATP synthase, and once again the unit of phosphorylation is the entire vesicle. These findings (Junge et al., 1970; Junge and Witt, 1968), which demonstrated that energy coupling is delocalized and depends on the topological integrity of the vesicles, provided strong support for the chemiosmotic view of photophosphorylation.

The kinetics and thermodynamics of ATP production have also been

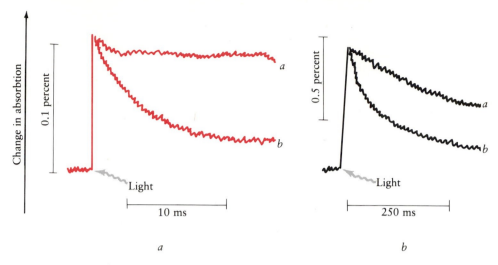

FIGURE 8.23   Electrical events during a single turnover of the photosynthetic machinery. (*a*) Gramicidin accelerates the decay of the electric field. The chloroplast suspension was illuminated with a single flash of light at the arrow; the absorption shift at 515 nm monitors the electric potential across the thylakoid membrane. A flash of light causes the potential to rise quickly, followed by slow decay (trace *a*); decay was accelerated in the presence of gramicidin (trace *b*). (*b*) ATP synthesis accelerates decay of the electric potential. The procedure was the same as in *a*. Trace *b* shows the dissipation of $\Delta\Psi$ in the presence of ADP and P_i. (After Witt, 1979.)

explored at the level of single turnovers. From estimates of $\Delta\Psi$ and ΔpH, the number of protons translocated in a single cycle of the ATP synthase was calculated to be 2.5 Note that thylakoids have no need for an ATP-ADP translocator since the headpiece faces the matrix; the proton stoichiometry pertains to the CF_1F_0 ATP synthase itself. For a decade now, quantitative studies on this issue have consistently indicated that H^+/ATP is close to 3 (for thorough recent studies see Davenport and McCarty, 1981, 1984; Hangarter and Good, 1982), and recent work with mitochondria and bacteria favors the same stoichiometry (Chapter 7). A universal stoichiometry of $3H^+/ATP$ seems increasingly probable.

Some of the plainest evidence for conformational changes in the F_1F_0 ATPase comes from research with chloroplasts. The ATPase activity of thylakoids isolated in the dark is latent, becoming manifest on illumination. Activation of the ATPase is a reductive process, mediated by thioredoxin; it is accompanied by the release of bound nucleotides, probably as a result of changes in the structure of the enzyme (Schlodder and Witt, 1981). In addition, there are numerous indications that each cycle of ATPase activity is accompanied by conformational changes that expose groups on the protein to exchange with protons of water or to attack by inhibitors. Unfortunately these studies have not helped to discriminate between the possible models of ATPase operation outlined in Chapter 7.

How do protons get from the photosystems to the ATP synthase? The finding that the whole thylakoid vesicle serves as a unit of phosphorylation argues against obligatory association between particular electron-transport complexes and CF_1F_0 synthases. Indirect coupling is also demanded by the discovery that an electric field imposed artificially across the thylakoid membrane in the dark supports ATP synthesis, and by the finding that the F_1F_0 ATPase of chloroplast lamellae may be spatially separated from the particles that bear the reaction centers (Miller and Staehelin, 1976). But these data do not necessarily imply that energy coupling is effected by the protonic potential between the bulk aqueous phases in loculus and stroma, and several authors have put forward arguments for more localized proton currents. The evidence has been reviewed by McCarty (1981), whose verdict was "not proven." I shall forgo the subject here, but would take note of some very recent papers. According to Davenport and McCarty (1981, 1984), stoichiometries of $3H^+/ATP$ and $2H^+/e^-$ account satisfactorily for photophosphorylation by isolated thylakoids. By contrast, Hangarter and Good (1984) report that the energized state produced by illumination of thylakoid lamellae is kinetically very different from that which results from the imposition of a pH gradient, and they conclude that whatever drives ATP synthesis, it is not simply the electrochemical potential gradient of free protons. Hong and Junge (1983) claim that the characteristics of the space into which protons are released on illumination depend on the history of the thylakoid preparation. In fresh membranes, gently prepared, the protons seem to remain in a special phase close to the membrane surface. Thylakoids subjected to a cycle of swelling and shrinking are still capable of photophosphorylation, but protons in the loculus now appear to be freely diffusible in the bulk water phase. If this interpretation of the data proves correct, the issue of localized versus bulk-phase protons turns on surface organization rather than on bioenergetic principle.

Lateral Organization of the Photosynthetic Machinery

The prevailing view of photosynthetic energy transduction rests squarely on the Z scheme, with two photosystems in tandem. It is therefore more than a little disconcerting to discover at this late stage that the Z scheme may be a somewhat misleading representation. The iconoclastic view emerging from recent work is that the supramolecular complexes that transduce light energy and translocate protons are relatively independent units that may be spatially segregated and do not necessarily link up into the continuous chain depicted in figures 8.10 and 8.20. Moreover, regulation of energy transduction may depend in part on the mobility of these complexes in the membrane. Emerging views of the functional architecture of photosynthetic membranes are discussed by Kaplan and Arntzen (1982), Anderson and Andersson (1982), Staehelin and Arntzen (1983), and Haehnel (1984).

The reexamination of received doctrine began with the question, Why are the thylakoids of higher plants arrayed in grana stacks? It is quite clear that stacks are not obligatory for oxygenic photosynthesis: neither cyanobacteria nor the chloroplasts of red algae display grana. Even chloroplasts of higher plants lose their grana stacks when thylakoids are allowed to swell, regaining them under proper ionic conditions; yet they remain photosynthetically active throughout. On the other hand, the differences in chloroplast organization between sun-loving and shade plants underscore its adaptive significance. Sun plants, which have high rates of photosynthesis, have a high proportion of stroma-exposed membrane. Shade plants, which must optimize the efficiency of light harvesting, have giant grana stacks but limited areas of stroma thylakoids. Maximum rates of photosynthesis are lower than those of sun plants, and the rates saturate at lower light intensities. Such observations suggest that grana stacks have to do with the efficiency of energy capture and its allocation to the two photosystems.

The important finding is that the two photosystems are not uniformly distributed in thylakoid membranes. Numerical analysis of the particles revealed by freeze fracture indicated that PS I is located primarily in the stroma thylakoids, while PS II is concentrated in the grana stacks. Miller and Staehelin (1976) applied this technique to the CF_1F_0 ATP synthase and found it to be localized in the stroma thylakoids and on the exposed surfaces of grana stacks, but absent from the appressed membranes of the grana stacks. The $NADP^+$-ferredoxin reductase, a peripheral membrane protein, shows the same distribution.

More extensive data have recently come to hand thanks to the development of procedures to separate not only grana from stroma thylakoids but also the appressed regions of grana from their exposed margins and surfaces. Anderson and Andersson (1982) have lucidly summarized the results, many of which are the fruit of a long-term research program at Australia's CSIRO laboratories. The latest findings (Anderson and Melis, 1983) suggest that there are two forms of PS II, distinguishable by their absorption spectra, which differ in their complement of antenna pigments. PS II_α, the major component, appears to be localized exclusively in grana thylakoids. These also contain the bulk of the light-collecting antenna LHC II and of plastoquinone but possess little or no PS I. Stroma thylakoids hold most or even all of the PS I, together with small amounts of PS II_β. The third electron-transfer complex, cytochrome b-f, is uniformly distributed among the various membrane fractions (Figure 8.24). These observations accord with earlier ones suggesting that the stoichiometric ratio of photosystems I and II in chloroplasts from different plants is not unity (as the Z scheme implies) but varies with the organisms and the growth conditions over a range of nearly tenfold (Melis and Brown, 1980; Thielen and van Gorkom, 1981).

One can hardly avoid the conclusion that the textbook picture of photosynthetic electron transport stands in need of revision. In place of continuous redox chains carrying electrons from water to $NADP^+$, one must envisage independent energy-transducing complexes, more or less widely separated, perhaps linked by such mobile carriers as protons and free plastoquinone. Continuous chains

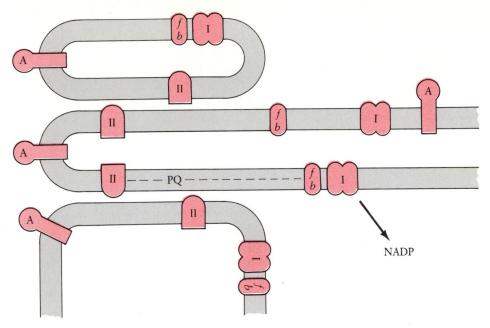

FIGURE 8.24   Lateral organization of the photosynthetic machinery. PS II clusters preferentially in appressed regions of a grana stack, PS I and the ATP synthase in exposed regions; the distribution of the cytochrome *b-f* complex is uncertain. The scheme shows long-range electron transfer from PS II to PS I by diffusion of plastoquinone. (After Anderson and Andersson, 1982; Haehnel, 1984.)

containing both photosystems I and II$_\beta$ may exist in stroma thylakoids and produce NADPH and ATP as outlined before. The function of grana stacks, rich in antenna LHC II and photosystem II, would be to produce ATP (with the aid of CF_1F_0 ATP synthase plugged across the end margins of grana stacks) and perhaps reduced plastoquinone for export to stroma thylakoids. Incidentally, the segregation of photosystems should be kept in mind when evaluating the evidence for "localized" proton currents.

These changing views of thylakoid organization bear directly on the question, How is light energy allocated to the two photosystems? Sunlight is chiefly absorbed by the abundant antenna complex LHC II and fed into photosystem II. Yet plants are known to redistribute the energy, diverting part to photosystem I. Organisms exposed to light absorbed preferentially by either photosystem adapt within minutes so as to balance the energy allocation between both PS I and PS II. One possible mechanism of energy redistribution is termed spillover: excitons would migrate through a pigment bed that is in physical contact with both reaction centers. The discovery that PS I and PS II are physically segregated renders this hypothesis implausible but favors an alternative one: that redistribution of light energy depends on the redistribution of antenna chlorophylls physically associated with the two photosystems.

Recent findings suggest that part of the LHC II complement is mobile and moves bodily between grana and stroma lamellae (Staehelin and Arntzen, 1983; Bennett, 1983). Migration is elicited by phosphorylation or dephosphorylation of the structural proteins, effected by a protein kinase. Phosphorylation induces LHC II to migrate from grana thylakoids to the stroma thylakoids, probably in response to the altered charge density; increased association of LHC II with PS I should favor light allocation to the latter. Conversely, dephosphorylation of LHC II lets it return to the grana lamellae, tilting energy allocation in favor of PS II. The extent of LHC II phosphorylation is determined by the activity of the kinase, which is regulated by the redox status of the plastoquinone pool.

Why, then, do plant chloroplasts have grana? Staehelin and Arntzen (1983) propose the following answer. If PS I, PS II, and their antenna pigments were uniformly distributed, light energy would tend to end up in PS I, which absorbs at the longest wavelengths and is therefore the deepest energy sink. Segregation of PS II and PS I into different membranes prevents excessive transfer of excitation energy from PS II to PS I and improves energy capture. But if the pigment bed were simply divided between grana and stroma lamellae, its light-absorbing capacities would be fixed: a disadvantage in the real world, where both light intensity and the demand for ATP and NADPH are subject to change. The mobile antenna system mitigates this disadvantage and optimizes both energy capture and its distribution.

References

ALVAREZ, L. W., ALVAREZ, W., ASARO, F., and MICHEL, H. V. (1980). Extraterrestrial cause for the Cretaceous-Tertiary extinction. *Science* 208:1095–1108.

ANDERSON, J. M., and ANDERSSON, B. (1982). The architecture of photosynthetic membranes: Lateral and transverse organization. *Trends in Biochemical Sciences* 7:288–292.

ANDERSON, J. M., and MELIS, A. (1983). Localization of different photosystems in separate regions of chloroplast membranes. *Proceedings of the National Academy of Sciences USA* 80:745–749.

ARNON, D. I. (1959). Conversion of light into chemical energy in photosynthesis. *Nature* 184:10–21.

ARNON, D. I., WHATLEY, F. R., and ALLEN, M. B. (1958). Assimilatory power in photosynthesis. *Science* 127:1026–1034.

AVRON, M. (1977). Energy transduction in chloroplasts. *Annual Review of Biochemistry* 46:143–155.

BENDALL, D. S. (1982). Photosynthetic cytochromes of oxygenic organisms. *Biochimica et Biophysica Acta* 683:119–151.

BENNETT, J. (1983). Regulation of photosynthesis by reversible phosphorylation of the light-harvesting chlorophyll *a/b* protein. *Biochemical Journal* 212:1–13.

BENNOUN, P. (1982). Evidence for a respiratory chain in the chloroplast. *Proceedings of the National Academy of Sciences USA* 79:4352–4356.

BINDER, A. (1982). Respiration and photosynthesis in energy-transducing membranes of cyanobacteria. *Journal of Bioenergetics and Biomembranes* **14**:271–286.

BUCHANAN, B., WOLOSIUK, R. A., and SCHÜRMANN, P. (1979). Thioredoxin and enzyme regulation. *Trends in Biochemical Sciences* **4**:93–96.

CARR, N. G., and WHITTON, B. A., Eds. (1982). *The Biology of Cyanobacteria*. University of California Press, Berkeley.

CLAYTON, R. K. (1980). *Photosynthesis: Physical Mechanisms and Chemical Patterns*. Cambridge University Press, London.

COHEN-BAZIRE, G., and BRYANT, D. A. (1982). Phycobilisomes: Composition and structure in *The Biology of Cyanobacteria*, N. G. Carr and B. A. Whitton, Eds. University of California Press, Berkeley, pp. 143–190.

CRAMER, W. A., and CROFTS, A. R. (1982). Electron and proton transport, in *Photosynthetic Energy Conversion in Plants and Bacteria*, Govindjee, Ed., vol. 1, Academic, New York, pp. 387–467.

CROFTS, A. R., and WOOD, P. M. (1978). Photosynthetic electron-transport chains of plants and bacteria and their role as proton pumps. *Current Topics in Bioenergetics* **7**:175–244.

DAVENPORT, J. W., and MCCARTY, R. E. (1981). Quantitative aspects of adenosine triphosphate–driven proton translocation in spinach chloroplast thylakoids. *Journal of Biological Chemistry* **256**:8947–8954.

DAVENPORT, J. W., and MCCARTY, R. E. (1984). An analysis of proton fluxes coupled to electron transport and ATP synthesis in chloroplast thylakoids. *Biochimica et Biophysica Acta* **766**:363–374.

DOUCE, R., and JOYARD, J. (1981). Does the plastid envelope derive from the endoplasmic reticulum? *Trends in Biochemical Sciences* **6**:237–239.

DUYSENS, L. N. M., AMESZ, J., and KEMP, B. M. (1961). Two photochemical systems in photosynthesis. *Nature* **190**:510–511.

FOGG, G. E. (1972). *Photosynthesis*. 2d ed. Elsevier, New York.

FOYER, C. H., and HALL, D. O. (1980). Oxygen metabolism in the active chloroplast. *Trends in Biochemical Sciences* **5**:188–191.

GANTT, E. (1980). Structure and function of phycobilisomes: Light-harvesting pigment complexes in red and blue-green algae. *International Review of Cytology* **66**:45–80.

GIDDINGS, T. H., JR., WASMAN, C., and STAEHELIN, A. E. (1983). Structure of the thylakoids and envelope membranes of the cyanelles of *Cyanophora paradoxa*. *Plant Physiology* **71**:409–419.

GIDDINGS, T. H., WITHERS, N. W., and STAEHELIN, L. A. (1980). Supramolecular structure of stacked and unstacked regions of the photosynthetic membranes of *Prochloron sp.*, a prokaryote. *Proceedings of the National Academy of Sciences USA* **77**:352–356.

GLAZER, A. N. (1982). Phycobilisomes: Structure and dynamics. *Annual Review of Microbiology* **36**:173–198.

GOVINDJEE, Ed. (1975). *Bioenergetics of Photosynthesis*. Academic, New York.

GOVINDJEE, Ed. (1982). *Photosynthesis*, vol 1: *Energy Conversion by Plants and Bacteria*. Academic, New York.

GREGORY, R. P. F. (1977). *Biochemistry of Photosynthesis*. 2d ed. Wiley, New York.

GUIKEMA, J., and SHERMAN, L. (1982). Protein composition and architecture of the photosynthetic membranes from the cyanobacterium *Anacystis nidulans R2*. *Biochimica et Biophysica Acta* **681**:440–450.

HAEHNEL, W. (1984). Photosynthetic electron transport in higher plants. *Annual Review of Plant Physiology* **35**:659–693.

HANGARTER, R. P., and GOOD, N. E. (1982). Energy thresholds for ATP synthesis in chloroplasts. *Biochimica et Biophysica Acta* **681**:347–404.

HANGARTER, R. P., and GOOD, N. E. (1984). Energized state responsible for adenosine 5′-triphosphate synthesis in preilluminated chloroplast lamellae. *Biochemistry* **23**:122–130.

HATCH, M. D., and BOARDMAN, N. K., Eds. (1981). *Photosynthesis*, vol. 8, in *The Biochemistry of Plants*, P. K. Stumpf and M. D. Hatch, Series Ed. Academic, New York.

HAUSKA, G., HURT, E., GABELLINI, N., and LOCKAU, W. (1983). Op. cit., Chapter 7.

HAXO, F. T. (1960). In *Comparative Biochemistry of Photoreactive Pigments*, M. B. Allen, Ed. Academic, New York, pp. 339–360.

HAXO, F. T., and BLINKS, L. R. (1950). Photosynthetic action spectra of marine algae. *Journal of General Physiology* **33**:389–422.

HEBER, U., and HELDT, H. W. (1981). The chloroplast envelope: Structure, function, and role in leaf metabolism. *Annual Review of Plant Physiology* **32**:139–168.

HILL, R., and BENDALL, F. (1960). Function of two cytochrome components in chloroplasts: A working hypothesis. *Nature* **186**:136–137.

HIND, G., and JAGENDORF, A. T. (1963). Separation of light and dark stages in photophosphorylation. *Proceedings of the National Academy of Sciences USA* **49**:715–722.

HINKLE, P. C., and MCCARTY, R. E. (1978). Op. cit., Chapter 3.

HO, K. K., and KROGMANN, D. W. (1982). Photosynthesis, in *The Biology of the Cyanobacteria*, N. G. Carr and B. A. Whitton, Eds. University of California Press, Berkeley, pp. 191–214.

HONG, Y.-Q., and JUNGE, W. (1983). Localized or delocalized protons in phosphorylation? *Biochimica et Biophysica Acta* **722**:197–208.

HURT, E., and HAUSKA, G. (1981). A cytochrome f/b_6 complex of five polypeptides with plastoquinol-plastocyanin oxidoreductase activity from spinach chloroplasts. *European Journal of Biochemistry* **177**:591–599.

IZAWA, S., and HIND, G. (1967). The kinetics of the pH rise in illuminated chloroplast suspensions. *Biochimica et Biophysica Acta* **143**:377–390.

JAGENDORF, A. T., and URIBE, E. (1966). ATP formation caused by acid-base transition of spinach chloroplasts. *Proceedings of the National Academy of Sciences USA* **55**:170–177.

JUNGE, W. (1977). Membrane potentials in photosynthesis. *Annual Review of Plant Physiology* **28**:503–516.

JUNGE, W., and JACKSON, J. B. (1982). Op. cit., Chapter 4.

JUNGE, W., RUMBERG, B., and SCHRODER, H. (1970). The necessity of an electric potential difference and its use for photophosphorylation in short flash group. *European Journal of Biochemistry* **14**:575–581.

JUNGE, W., and WITT, H. T. (1968). On the ion transport system of photosynthesis: Investigations on a molecular level. *Zeitschrift für Naturforschung* **B23**:244–254.

KAPLAN, S., and ARNTZEN, C. J. (1982). Photosynthetic membrane structure and function in *Photosynthesis*, vol I: *Energy Conversion by Plants and Bacteria*, Govindjee, Ed. Academic, New York, pp. 65–151.

LAM, E., and MALKIN, R. (1982). Reconstruction of the chloroplast noncyclic electron transport pathway from water to NADP with three integral protein complexes. *Proceedings of the National Academy of Sciences USA* **79**:5494–5498.

MALKIN, R. (1982). Photosystem I. *Annual Review of Plant Physiology* **33**:455–479.

MATTHIJS, H. C. P., LUDÉRUS, E. M. E., SCHOLTS, M. J. C., and KRAAYENHOF, R. (1984). Energy metabolism in the cyanobacterium *Plectonema boryanum*: oxidative phosphorylation and respiratory pathways. *Biochimica et Biophysica Acta* 766:38–44.

MCCARTY, R. E. (1978). Op. cit., Chapter 7.

MCCARTY, R. E. (1981). Intramembrane versus transmembrane pH gradients in photophosphorylation, in *Chemiosmotic Proton Circuits in Biological Membranes*, V. P. Skulachev and P. C. Hinkle, Eds. Addison-Wesley, Reading, Mass., pp. 271–281.

MCCARTY, R. E., and CARMELI, C. (1982). Proton translocating ATPases of photosynthetic membranes, in *Photosynthesis*, vol. I: *Energy Conversion by Plants and Bacteria*, Govindjee, Ed. Academic, New York, pp. 647–695.

MELIS, A., and BROWN, J. S. (1980). Stoichiometry of system I and system II reaction centers and of plastoquinone in different photosynthetic membranes. *Proceedings of the National Academy of Sciences USA* 77:4712–4716.

MILLER, K. R., and STAEHELIN, L. A. (1976). Analysis of the thylakoid outer surface. Coupling factor is limited to unstacked membrane regions. *Journal of Cell Biology* 68:30–47.

MITCHELL, P. (1961). Op. cit., Chapter 3.

MITCHELL, P. (1966). Op. cit., Chapter 3.

MITCHELL, P. (1976b). Op. cit., Chapter 7.

NELSON, N. (1981). Op. cit., Chapter 7.

PREBBLE, J. N. (1981). Op. cit., Chapter 2.

SHERRER, S., STÜRZEL, E., and BÖGER, P. (1984). Oxygen-dependent proton efflux in cyanobacteria (blue-green algae). *Journal of Bacteriology* 158:609–614.

SCHLODDER, E., and WITT, H. T. (1981). Relation between the initial kinetics of ATP synthesis and of conformational changes in the chloroplast ATPase studied by external field pulses. *Biochimica et Biophysica Acta* 635:571–584.

SLATER, E. C. (1983). Op. cit., Chapter 7.

SLOVACEK, R. E., CROWTHER, D., and HIND, G. (1979). Cytochrome functions in the cyclic electron transport pathway of chloroplasts. *Biochimica et Biophysica Acta* 547:138–148.

STAEHELIN, L. A., and ARNTZEN, C. J. (1983). Regulation of chloroplast membrane function. Protein phosphorylation changes spatial organization of membrane components. *Journal of Cell Biology* 97:1327–1337.

STANIER, R. Y. (1974). The origins of photosynthesis in eukaryotes. *Symposia of the Society for General Microbiology* 24:219–240.

STANIER, R. Y., and COHEN-BAZIRE, G. (1977). Op. cit., Chapter 6.

STEWART, A. C., and BENDALL, D. (1981). Properties of oxygen-evolving photosystem-II particles from *Phormidium laminosum*, a thermophilic blue-green alga. *Biochemical Journal* 194:877–887.

STROTMANN, H., and BICKEL-SANDKÖTTER, S. (1984). Structure, function and regulation of chloroplast ATPase. *Annual Review of Plant Physiology* 35:97–120.

STRYER, L. (1981). Op. cit., Chapter 2.

THIELEN, A. P. G. M., and VAN GORKOM, M. J. (1981). Quantum efficiency and antenna size of photosystems II_α, II_β and I in tobacco chloroplasts. *Biochimica et Biophysica Acta* 635:111–120.

VELTHUYS, B. R. (1978). A third site of proton translocation in green plant photosynthetic electron transport. *Proceedings of the National Academy of Sciences USA* 75:6031–6034.

VELTHUYS, B. R. (1980). Mechanisms of electron flow in photosystem II and towards photosystem I. *Annual Review of Plant Physiology* **31**:545–567.

WITHERS, N. W., ALBERTE, R. S., LEWIN, R. A., THORNBER, J. P., BRITTON, G., and GOODWIN, T. W. (1978). Photosynthetic unit size, carotenoids and chlorophyll-protein composition of *Prochloron sp.*, a prokaryotic green alga. *Proceedings of the National Academy of Sciences USA* **75**:2301–2305.

WITT, H. T. (1979). Energy conversion in the functional membrane of photosynthesis. Analysis by light pulse and electric pulse methods: The central role of the electric field. *Biochimica et Biophysica Acta* **505**:355–527.

9

Carriers, Channels, and Pumps

Understanding a thing is to arrive at a metaphor for that thing by substituting something more familiar to us. And the feeling of familiarity is the feeling of understanding.

Julian Jaynes, *The Origin of Consciousness in the Breakdown of the Bicameral Mind*

• • •

Speaking of Transport
Electrical and Osmotic Corollaries
The Sodium Circulation of Animal Cells
The Proton Circulation of Fungi, Algae, and Plants
The Calcium Circulation
A Diversity of Patterns

Students sometimes regard membrane transport as a peripheral aspect of cellular physiology, concerned with the machinery that supplies cytoplasmic enzymes with substrates and removes the waste. There is some truth in this, of course, but the preceding chapters should have demonstrated that the narrow perspective misses the point. Most membranes define not merely a compartment but an integrated unit of energy transduction and work, and transport plays a major role in orchestrating the commotion of chemical reactions into unified ensembles. This is true a fortiori of the plasma membrane, for this membrane makes a cell and thereby defines the elementary unit of life.

Much of the transport literature consists of increasingly refined descriptions of particular transport systems, each of which performs some elementary function: the extrusion of Na^+ ions, for example, or the uptake of glucose. But the biological meaning of these entities emerges only when they are seen as members of a team collaborating in the execution of a larger purpose. The integration of separate transport catalysts into coherent assemblages is achieved by linking them to a small number of ion circulations, chiefly those of H^+, Na^+, and Ca^{2+}. Each consists of a primary pump, usually an ion-translocating ATPase, which transduces metabolic energy into a gradient of electrochemical potential, and a bevy of secondary porters that allow the coupling ion to flow down the gradient while harnessing its free energy for use. Such ionic circulations perform two kinds of services. First, ion gradients support the accumulation of nutrients and metabolites and carry out related tasks, such as the control of cellular pH and volume. Second, particularly in eukaryotic cells, ion gradients underlie one of the chief modes of information processing: a brief flux of Na^+ or Ca^{2+} ions, in response to a specific stimulus, often serves as a signal that triggers contraction, secretion, even the initiation of growth or development. The pumps and channels that mediate ionic signals are transport systems and properly belong in the present chapter. But their functions are distinctive, and it seemed best to reserve the latter aspect for Chapter 13.

Eukaryotic cells contain a variety of organelles besides mitochondria and chloroplasts, each bounded by a membrane that selectively passes some substances and excludes others. The nuclear membrane is now thought to be freely permeable to ions and small molecules; its prominent pores are concerned with the selective transport of ribonucleoproteins. But the membranes of other organelles engage in energy transduction. Vacuoles, lysosomes, chromaffin granules, and neurotransmitter storage granules all contain proton-translocating ATPases and proton-linked porters; the tonoplast (vacuolar membrane) of algae and higher plant cells is a major barrier whose functions depend on its complement of pumps and carriers. A cursory survey of so large a territory would merely engender confusion. This chapter therefore concerns itself primarily with two subjects: how eukaryotic cells organize the work of transport across the plasma membrane, and what use they make of it. Molecular aspects of translocation and energy coupling will be treated separately in Chapter 10.

Transport in eukaryotic cells and their organelles is not fundamentally different from transport in prokaryotes, but it is organized in a distinctive fashion. It may be useful to spell out these differences as we presently perceive them.

1. In eukaryotic cells, energy for plasma membrane functions is supplied by a class of ion-translocating ATPases unrelated to the prokaryotic F_1F_0 ATPase. Enzymes that transfer reducing equivalents across the plasma membrane have been described, but these apparently do not transduce metabolic energy into an ionic potential gradient.

2. All the known primary transport systems of eukaryotes carry cations: H^+, Na^+, K^+, and Ca^{2+}. Group translocation across the plasma membrane is rare, if it occurs at all, and the numerous "ATP-linked" transport systems seen in bacteria (Chapter 5) have no known eukaryotic equivalent.

3. Both prokaryotes and eukaryotes expel Ca^{2+} ions from the cytoplasm, but prokaryotes do not appear to make teleonomic use of the resulting gradient. In eukaryotic cells the circulation of Ca^{2+} ions across membranes is a major mechanism for processing signals from the environment (Chapter 13).

There is room for a systematic general text of membrane transport. H. N. Christensen's *Biological Transport* (1975) comes closest, but is more than a little dated. A good introduction to the field from the kinetic and thermodynamic viewpoint is *Transport across Biological Membranes* by M. Höfer (1981). Symposia abound, but most are too specialized for the general reader. I have drawn on three recent ones: *Electrogenic Ion Pumps* (Slayman, 1982), *Plant Membrane Transport: Current Conceptual Issues* (Spanswick et al., 1980), and *Membranes and Transport,* a collection of minireviews edited by Martonosi (1982). Electrophysiological considerations recur in the transport field, but again elementary treatments are hard to find. The classic one is still *Nerve, Muscle and Synapse,* by Bernard Katz (1966); *The Physiology of Excitable Cells,* by D. J. Aidley (1978), is an excellent advanced treatise.

Speaking of Transport

Membrane transport is inherently an abstract and demanding subject whose language relies more obviously on metaphor than most. Molecules pass across membranes in more ways than one, and these have been divided into categories that reflect the historical development of the field from disparate roots in biophysics and in enzymology. Before we embark on a survey of the variety of eukaryotic transport processes and their physiological roles, let us recapitulate the terms most commonly employed and place them in their conceptual and historical context.

Simple Diffusion.   The rate at which molecules in solution diffuse from one region to another is proportional to their concentration. Their net flux across a boundary or membrane from a region of higher to one of lower concentration is described by

$$J = PA([S]_H - [S]_L) \qquad \text{(Eq. 9.1)}$$

in which the flux, J, is the amount of the substance crossing a membrane per unit time, A is the area of the membrane in square centimeters, $[S]_H$ and $[S]_L$ are the respective concentrations of the substance on the high and low sides of the membrane, and P is the permeability constant of the membrane to S, with the dimensions of velocity (centimeters per second). The permeability constant is a measure of the ease with which molecules pass the barrier. For example, since the barrier properties of biological membranes are due to lipids, nonpolar molecules diffuse across them more rapidly than polar ones: permeability constants often reflect the solubility of the substrate in the lipid phase. Furthermore, similar molecules tend to have comparable solubility properties and therefore their permeability constants will be alike; diffusion exhibits little specificity. Net diffusion can occur only from a region of higher concentration to one of lower or, more precisely, from a higher to a lower electrochemical potential. Oxygen, CO_2, NH_3, and probably water are among the small number of molecules of biological importance that traverse membranes by simple diffusion.

Facilitated Diffusion.   This term was originally coined in the 1930s to describe the entry of certain ions and polar molecules into cells far more rapidly than would be expected from their rate of diffusion across a lipid bilayer. In contemporary parlance, facilitated diffusion implies association of the substrate with a catalytic membrane component, a "carrier," which transports it across the membrane. Facilitated diffusion generally exhibits saturation kinetics and a high degree of molecular specificity, just as enzymatic catalysis does. In fact, except for the ionophorous antibiotics, all biological transport systems are made of protein. Facilitated diffusion, like simple diffusion, is thermodynamically passive by definition: net flux can only occur from a region of higher to one of lower electrochemical potential.

Facilitated diffusion is traditionally depicted by the scheme shown in Figure 9.1*a*, which assumes that the substrate traverses the membrane while reversibly associated with the carrier. The rate of transport is not necessarily proportional to the concentration difference $[S]_o - [S]_i$, but can often be described by simple enzyme kinetics:

$$J = \frac{J_{max}[S]}{K_m + [S]} \qquad \text{(Eq. 9.2)}$$

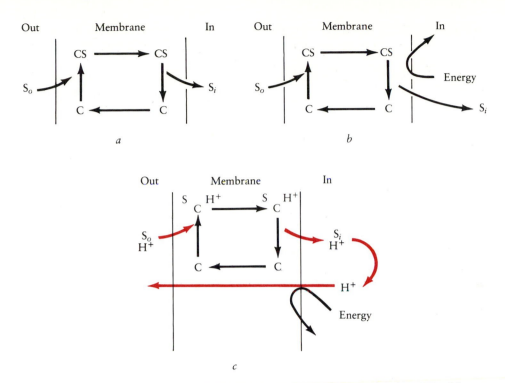

FIGURE 9.1   The carrier concept. (*a*) A mobile carrier, C, that can associate reversibly with the substrate, S. (*b*) Energy coupling to a mobile carrier, favoring dissociation of the substrate on the inside. (*c*) Basic model for a symporter that translocates H^+ and S simultaneously. The asymmetric distribution of H^+, maintained by the proton pump, favors accumulation of S on the inside. The translocation steps mediated by the carrier are reversible, but for clarity the arrows point in one direction.

Here J is again the flux, that is, the velocity of transport; J_{max} is the maximum velocity the system can attain; and [S] is the concentration of the substrate on the side from which transport occurs. The constant K_m can be operationally defined as that concentration of the substrate at which the rate of transport attains half its maximum value. For present purposes we may take it as a measure of the affinity of the carrier for its substrate and a characteristic of the carrier under the conditions of measurement.

The carrier shown in Figure 9.1*a* mediates transport in both directions, and at equilibrium the rate of influx must equal that of efflux. Moreover, for an uncharged substrate at equilibrium, $[S]_i = [S]_o$; more precisely, the activities of S will be the same on both sides of the membrane. (For a charged substrate, the electrochemical potential μ must be the same on both sides.) The carrier need not have equal affinities for its substrate on the two membrane surfaces, nor the same maximum velocity in both directions; indeed, these kinetic parameters are often found to be asymmetric. However, these parameters must be consistent with the constraint that the carrier can only catalyze the movement of S until its electrochemical potential is the same on the two sides of the barrier[1].

Active Transport.   In the 1940s it became apparent that some transport processes in addition to being rapid and specific also generate substantial concentration gradients and even electric potentials. These discoveries led H. H. Ussing and T. Rosenberg to define the category of active transport as the net movement of a metabolite counter to its electrochemical potential gradient (Rosenberg, 1954). By definition, active transport is an endergonic process that cannot take place spontaneously; it implies coupling to some exergonic metabolic reaction, often (but not necessarily) the hydrolysis of ATP. In fact, since intracellular concentrations and membrane potentials are often uncertain, obligatory dependence of a transport process on cellular energy metabolism may serve as prima facie evidence for active transport.

In terms of the elementary carrier scheme, active transport can be ascribed to an alteration in the kinetic parameters, either affinity or velocity, such that the rates of influx and efflux become equal only when $[S]_i$ is much higher than $[S]_o$. Figure 9.1*b* is not intended to specify the manner of energy coupling; what is important is that energy coupling is effected by modifying the carrier, not the substrate. The scheme intentionally implies that energy coupling lowers the affinity of the carrier for efflux; this is often true, but not always.

How much work does an active-transport process entail? This can be calculated with Equations 1.16 to 1.18. At equilibrium, the electrochemical potential of the substrate is balanced against the driving force. As a rule of thumb, 1.37 kcal/mol (5.7 kJ/mol) is required to move an uncharged substance against a tenfold gradient of concentration or a univalent ion against an electric-potential gradient of 59 mV. If we take the free energy of ATP hydrolysis in cytoplasm as -10 kcal/mol, this would support the transport of one mol of uncharged substrate against a gradient of 10^7. The cost of transport depends only on the electrochemical gradient and on the stoichiometry, not on the molecular mechanism of energy coupling.

Primary and Secondary Transport.   The preceding categories continue to be widely used but suffer from serious disadvantages. First, by definition facilitated diffusion is thermodynamically passive; carriers may only mediate diffusion down the gradient of electrochemical potential, and net movement must cease

---

[1] One of the hallmarks of carrier-mediated diffusion is the phenomenon called counterflow. Suppose the carrier shown in Figure 9.1*a* recognizes two substrates, A and B. Cells previously loaded with A and placed in a medium containing B will then transiently accumulate B while A flows out into the medium (see Figure 10.2). At first glance, counterflow appears to violate the constraint that a simple carrier only catalyzes the approach to equilibrium, but in fact there is no conflict. It is the gradient of A that sustains the transient gradient of B, thanks to the coupling of the two fluxes via the carrier, and B soon begins to flow back out, approaching the same equilibrium distribution as that of A.

when the electrochemical potential of the substrate is the same on both sides of the membrane. (Exchange of one substrate molecule for another can, however, continue.) This unexceptional restriction renders the concept of facilitated diffusion ambiguous when two substrates share a single carrier: the driving force on one substrate moving down its electrochemical potential gradient may force the other to move away from its own equilibrium distribution. This of course is precisely the basis of much of biological energy coupling (Chapter 3). Further ambiguities arise from the existence of reactions in which a substrate is simultaneously translocated and chemically altered, so that its chemical nature is not the same on the two sides of the membrane. Uptake of sugars by vectorial phosphorylation (Chapter 5) is a well-known example that does not fit smoothly into the category of active transport, but obviously meets the same needs.

The terminology devised by Mitchell (1967) in the context of the chemiosmotic theory evades these difficulties by emphasizing not the thermodynamic status of the transport substrate but the molecular nature of the process. As was outlined in Chapter 3, there are again three categories: simple diffusion, defined as explained above; primary transport processes, in which translocation across a membrane is directly and obligatorily linked to a concurrent chemical reaction; and secondary transport processes, which involve the exchange of ionic and other secondary links between a carrier and its substrates but no covalent bond exchanges. Each of these categories can be subdivided further (Table 3.1); for example, secondary transport systems, or porters for short, may catalyze uniport, symport, or antiport.

A uniporter merely catalyzes its substrate's movement to electrochemical equilibrium and can do no work on it. But symporters and antiporters couple the movement of one substrate down the electrochemical-potential gradient to the movement of a second substrate uphill (the former substrate is usually a coupling ion, such as H^+ or Na^+). Figure 9.1c shows the kinetic scheme for a symport process, including a primary transport process that maintains the asymmetric distribution of the coupling ion. At equilibrium, the driving force on the substrate S equals that on the coupling ion, provided "slip" can be neglected. The steady-state level of S will then depend on the electrochemical potential of the coupling ion and on the stoichiometry of the carrier (Equation 3.6). In practice, matters are more complex and the steady-state level falls short of that which is thermodynamically possible. We shall return to the factors that determine the steady-state level of transported metabolites in the following chapter.

Mitchell's nomenclature is more comprehensive and has therefore been used throughout this book except for occasional historical references. At times it is convenient to employ a hybrid terminology that distinguishes "primary active" from "secondary active" transport. The Na^+-K^+ ATPase exemplifies the former, Na^+-glucose symport the latter. The colloquial term "pump" is appropriate only for primary active transport and is used here with that intent. But readers should keep in mind that these categories do not exhaust the varieties of transport, and nature is not bound by them: endocytosis, for instance, fits none of these pigeonholes, yet it is assuredly a major mode of membrane transport.

The Carrier Concept.   Two evocative metaphors, "carrier" and "channel," have molded our understanding of membrane transport over the past 40 years; their edges become dulled with use and require periodic sharpening (Wilbrandt and Rosenberg, 1961; LeFevre, 1975; Crane, 1977; Läuger, 1980).

Designation of a transport system as a carrier implies a mobile membrane component that sequentially exposes a substrate-binding site first to one side of the membrane and then to the other, but not to both simultaneously. The ionophorous antibiotic valinomycin is a case in point (Figure 3.6). Valinomycin is a lipid-soluble peptide that forms a specific chelate with K^+; this diffuses across the membrane exactly as shown in Figure 9.1a, allowing K^+ ions to pass from one side to the other. The logical elements of a carrier are complex formation and movement, not necessarily of the entire structure but of some portion, so that the binding site faces alternately the one surface or the other. Transport systems that operate in this manner can be recognized by certain kinetic characteristics, such as the ability to mediate counterflow.

A channel, by contrast, designates a fixed structure containing one or more binding sites arranged in sequence across the membrane and accessible from both sides at the same time. The antibiotic gramicidin is a simple example (Figure 3.6): it does not ferry ions across the membrane but forms a stationary pore through which they pass (see also Figure 10.1). Many channels are more complex, endowed with the capacity to open or close in response to some specific stimulus with the aid of an ancillary device that serves as a "gate." Channels are distinguished from carriers by kinetic criteria, including the high rate of transport and the absence of counterflow; we shall return to this point in Chapter 10.

The various terms used to describe biological transport processes intersect in unexpected ways. Some porters are carriers, some channels; all pumps are carriers; there are barriers and leaks, wells and traps, sources and sinks, and each term means precisely what its author wants it to mean, neither more nor less.

Electrical and Osmotic Corollaries

The transport of most biological metabolites entails the translocation of electric charge, either directly or indirectly. In consequence, a difference in electric potential across the plasma membrane affects almost all traffic across it and impinges on a variety of cellular functions. Conversely, $\Delta\Psi$ itself is modulated by ionic fluxes, both primary and secondary, and may fluctuate dramatically, particularly in cells described as excitable. Electrophysiological concepts loom so large in the workings of eukaryotic cells that the treatment of the underlying physical principles must first be extended.

The Membrane Potential.   There is usually a difference in electric potential between the two surfaces of the plasma membrane, due to an imbalance in the distribution of ions. As a rule, the cytoplasmic surface is electronegative. The electric field hinders the influx of anions but assists that of cations; Equation 1.18 gives the quantitative relationship between the membrane potential and the work of charge translocation. The electric potential difference across the membrane can arise in two ways: by diffusion of ions down their concentration gradients with production of a diffusion potential or by primary electrogenic ion transport (Chapter 1). The electric potential across real membranes usually reflects contributions from both processes.

In the simplest case, where a membrane is passively permeable to only a single ion, the membrane potential is given by the Nernst equation (Equation 1.20). Biological membranes often approximate this condition because most are more permeable to K^+ than to any other ion. In animal cells, which accumulate K^+ with the aid of an ATPase, the distribution of K^+ ions across the plasma membrane approaches its equilibrium potential; in other words, the membrane potential results chiefly from the outward diffusion of K^+.[2] In general, however, the fluxes of several ions contribute to the membrane potential. For example, when Na^+ ions are present in the medium, influx of Na^+ will tend to compensate electrically for the efflux of K^+, thereby diminishing the difference in electric potential. The problem of calculating the diffusion potential resulting from multiple concurrent ion fluxes was solved in the 1940s by A. L. Hodgkin, B. Katz, and D. E. Goldman with the aid of the assumption that the electric field across the membrane is uniform. Equation 9.3 is usually designated the Hodgkin-Katz-Goldman equation:

$$\Delta\Psi = \frac{2.3RT}{F} \log \frac{P_K[K^+]_o + P_{Na}[Na^+]_o + P_{Cl}[Cl^-]_i}{P_K[K^+]_i + P_{Na}[Na^+]_i + P_{Cl}[Cl^-]_o} \qquad \text{(Eq. 9.3)}$$

P_K, P_{Na}, and P_{Cl} are the permeability constants of the major ions that contribute to the membrane potential of animal cells; in other organisms, the movements of other ions must be taken into account. The meaning of the equation may be appreciated by reference to Figure 9.2. Frog muscle is about tenfold more permeable to K^+ than to Na^+. A plot of the membrane potential as a function of the logarithm of the external K^+ concentration yields a straight line with a slope of 58 mV, as predicted by the Nernst equation. However, at low K^+ concentrations $\Delta\Psi$ deviates from the predicted value. The reason is that at low K^+ concentrations $P_{Na}[Na^+]_o$ begins to approach $P_K[K^+]_o$ despite the low

---

[2] Phospholipid bilayers are intrinsically all but impermeable to K^+ ions, whose passage must be mediated by transport systems of one kind or another. The high K^+ permeability of animal cell plasma membranes is due to K^+ channels whose conductance is proportional to the K^+ concentration.

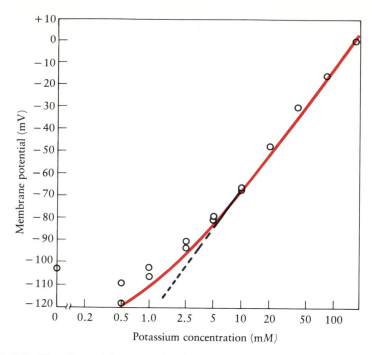

FIGURE 9.2   The effect of the external K^+ concentration on the membrane potential of frog muscle fibers. The circles mark experimental data; the colored line, $\Delta\Psi$ calculated from the Nernst equation, taking the K^+ activity in the cytoplasm as 140 mM; and the black line, $\Delta\Psi$ calculated from Equation 9.3, assuming P_{Na} to be one tenth of P_K. (From Hodgkin and Horowicz, 1959, with permission of the Physiological Society.)

absolute value of P_{Na}, and this diminishes $\Delta\Psi$. What determines $\Delta\Psi$ are the *relative* rates with which the various ions cross the membrane. If the permeability of any one ion is vastly greater than that of all the others, Equation 9.3 reduces to the Nernst equation (Equation 1.20).

　　We now know that many cells contain electrogenic ion pumps in addition to the diffusion pathways designated P_K, P_{Na}, and so on. The membrane potential may be regarded as the sum of contributions from electrogenic pumps and from diffusion, and the magnitudes of these terms vary from one organism to another. In animal cells, as we saw above, the potassium permeability is high and even the Na^+ permeability is significant. Ion diffusion dominates the membrane potential, and the sodium pump, albeit electrogenic, seldom contributes more than a few millivolts. The resting potential of muscle and nerve agrees closely with that calculated from Equation 9.3, and special conditions had to be imposed to document electrogenic sodium transport. By contrast, in bacteria and fungi the ion permeability is low and $\Delta\Psi$ arises largely from the action of electrogenic proton pumps.

　　In practice, the experimenter measures a membrane potential and must then

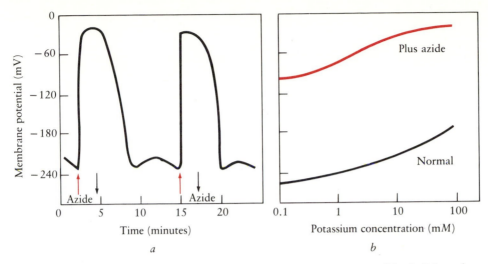

FIGURE 9.3   The membrane potential of *Neurospora crassa.* (*a*) Reversible abolition of the membrane potential by sodium azide. The inhibitor was added at the upward-pointing arrows and removed at the downward-pointing ones. (*b*) The membrane potential is the sum of two components. Control hyphae (black line) maintained a large membrane potential that was little affected by the external K^+ concentration. Addition of sodium azide abolished the contribution of the electrogenic proton pump (colored line), leaving a residual potential due largely to K^+ diffusion. (After Slayman, 1965a, 1965b, with permission of The Rockefeller University Press.)

work out how it arises. The procedure is illustrated in Figure 9.3, based on the classic work of C. L. Slayman with the fungus *Neurospora crassa,* which provided some of the earliest evidence for the existence of electrogenic ion pumps (Slayman, 1965a, 1965b). Slayman observed a large membrane potential, about -200 mV, which could be dissected into two components. The major component of $\Delta\Psi$ was dramatically dependent on oxidative metabolism: inhibition of respiration with azide or cyanide largely abolished the potential gradient, whereas removal of the inhibitor restored it (Figure 9.3a). This metabolic component was later shown to be due to an ATP-driven electrogenic proton pump. The small potential that remained after inhibition of the proton pump responded to changes in the external potassium concentration in a way suggesting that it is chiefly due to K^+ diffusion (Figure 9.3b).

   The transport of ions across a membrane implies that an electric current flows across it as well: a flux of 10 picoequivalents (peq) per second per square centimeter of surface area is approximately equivalent to 1 $\mu A/cm^2$. The characterization of this current provides information that may be unobtainable by other methods, particularly when the ions in question are protons. The electrical properties of the membrane can be described by a current-voltage plot, which summarizes the relationship between the flow of current and the electric potential imposed across a membrane. An example for *Neurospora* is shown in

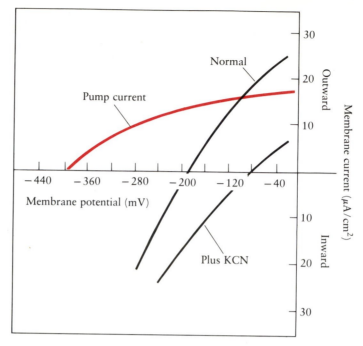

FIGURE 9.4   A current-voltage plot for *Neurospora crassa*. The membrane potential of the living cell was set by the experimenter between 0 and −280 mV with the aid of a voltage clamp, and current across the plasma membrane was determined. The current generated by the proton pump was calculated from the difference between the curves for normal and KCN-inhibited cells. (After Gradmann et al., 1978, with permission of Springer-Verlag.)

Figure 9.4, taken from the work of Gradmann et al. (1978). This is obviously not the place to enlarge on the technical procedures, but a summary of the conclusions that can be drawn from this approach may be useful.

The relationship between the membrane potential (set by the experimenter with the aid of a voltage clamp) and the flow of current across the membrane was determined under two conditions: in normal cells and after inhibition of respiration by KCN. The difference corresponds to the current-voltage plot of the proton pump. Extrapolation indicates that the proton pump reverses at −380 mV; this is the potential at which the pump stalls and therefore the maximum potential that it can generate. The measured maximum current was 17 $\mu A/cm^2$, but corrections suggest that 25 $\mu A/cm^2$ of membrane surface is a better estimate of the rate of proton extrusion by the cells. Finally, the ratio of current to voltage is a measure of membrane conductance, the inverse of its resistance, and gives information about ion fluxes other than those due to the pump. The manner in which data of this kind are utilized will be illustrated below.

Excitability.   Perhaps the most dramatic episode in electrophysiology is the action potential: a sudden short-lived collapse or reversal of the membrane potential, usually triggered by a specific stimulus. Strictly speaking, the term refers to the propagated impulses of nerve axons, whose ionic basis was worked out first. But it is now clear that analogous phenomena occur in many other kinds of animal cells and even in plants and microorganisms; "action potential" will be employed here in a generic sense to encompass all transient, regenerative changes in membrane potential. In most cases these result from temporary changes in the ionic permeability, due to the opening and closing of ion channels.

The principles of the nerve action potential and of cellular excitability in general were worked out by K. S. Cole, A. L. Hodgkin, A. F. Huxley, and B. Katz immediately following World War II. Figure 9.5*a* illustrates the basic mechanism by reference to the squid axon, with which all the original work was done. Thanks to its Na^+-K^+ ATPase, the axon maintains an ionic composition very different from that of the blood, with $[K^+]_i$ higher and $[Na^+]_i$ lower; the membrane potential of the quiescent axon (the "resting potential") is about -60 mV. Now, from the concentration of Na^+ inside the axon and out and from Equation 1.20, one can calculate that $[Na^+]_i$ would be in electrochemical equilibrium with a membrane potential of $+50$ mV. Since the actual membrane potential is negative, there is a large electrochemical gradient that drives sodium ions inward; the magnitude of this electromotive force is $-60 - 50 = -110$ mV. By the same token, $[K^+]_i$ would be in electrochemical equilibrium with a membrane potential of -75 mV. Since $\Delta\Psi$ is actually somewhat more positive, there is an electromotive force that drives K^+ outward, given by $-60 - (-75) = +15$ mV. The concentration gradients can be maintained only because the axon's permeability to Na^+ and K^+ is low.

Figure 9.5*b* illustrates the familiar time course of a propagated action potential in nerve. The explosive or "regenerative" depolarization of the plasma membrane, during which $\Delta\Psi$ turns briefly positive, is due to the opening of a set of sodium channels that are closed in the resting axon. Sodium ions rush in, down their electrochemical gradient, and $\Delta\Psi$ shifts toward the sodium equilibrium potential. This in turn increases the outward driving force on K^+, whose internal concentration is far higher than is compatible with a positive membrane potential. Within milliseconds the sodium channels close spontaneously, while a separate set of potassium channels opens; efflux of K^+ ions restores the resting potential, completing the cycle.

It should be emphasized that the movements of Na^+ and K^+ are entirely passive and involve the transfer of minuscule amounts of matter. From the capacitance of the axonal membrane one can calculate that a change in the membrane potential of 100 mV calls for a flux of about 1 peq of Na^+ or K^+ per square centimeter of membrane area. This is no more than one out of a million ions separated across the axonal membrane, which can therefore fire thousands of times before its ionic gradients are depleted. In the long run, however, ion gradients must be maintained by active transport coupled to cellular energy metabolism.

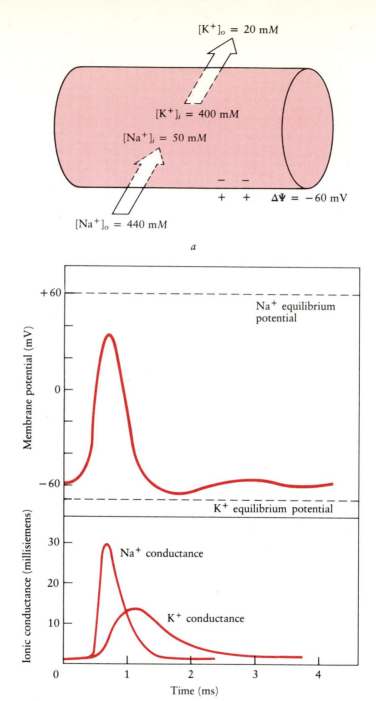

$[K^+]_o = 20 \; mM$

$[K^+]_i = 400 \; mM$

$[Na^+]_i = 50 \; mM$

$[Na^+]_o = 440 \; mM$

$\Delta\Psi = -60$ mV

a

Membrane potential (mV)

+60

Na$^+$ equilibrium potential

0

−60

K$^+$ equilibrium potential

Ionic conductance (millisiemens)

30

Na$^+$ conductance

20

10

K$^+$ conductance

0   1   2   3   4

Time (ms)

b

FIGURE 9.5 Excitability. (*a*) The ionic basis of the nerve action potential; for discussion see text. (*b*) The nerve action potential. The upper illustration shows the time course of the action potential and indicates the sodium and potassium equilibrium potentials. The lower illustration displays the sequence of changes in the sodium and potassium conductances that together account for the shape and time course of the action potential. (From Aidley, 1978, after Hodgkin and Huxley, with permission of Cambridge University Press.)

Axons operate on voltage-controlled channels for Na^+ and K^+, and somewhat similar channels regulated by acetylcholine serve in the transmission of the nerve impulse across neuromuscular junctions. However, most cases of cellular excitability involve calcium fluxes, which control diverse cellular responses such as contraction and exocytosis. Figure 9.6 illustrates one of these, the calcium action potential in the unicellular water mold *Blastocladiella emersonii*. Like other eukaryotes, this organism maintains a cytosolic calcium level in the micromolar range and a negative membrane potential near -100 mV. In the usual media, containing up to 1 mM Ca^{2+}, there is thus a steady electrochemical gradient driving Ca^{2+} ions into the cell. Depolarization of the plasma membrane by current injection beyond a threshold of about -45 mV triggers an abrupt opening of calcium channels; as Ca^{2+} ions flow inward, $\Delta\Psi$ goes positive. The entry of calcium ions triggers the secondary opening of anion channels that allow chloride ions to flow out. Both kinds of channels close spontaneously over the next 300 to 500 milliseconds (ms), and the resting potential is restored by K^+ efflux. Action potentials dependent on Ca^{2+}, chloride, or both are also known from protozoa and plants, not to mention a variety of animal cells from barnacle neurons to lymphocytes. They play a large role in transmembrane signaling, which will be explored in Chapter 13.

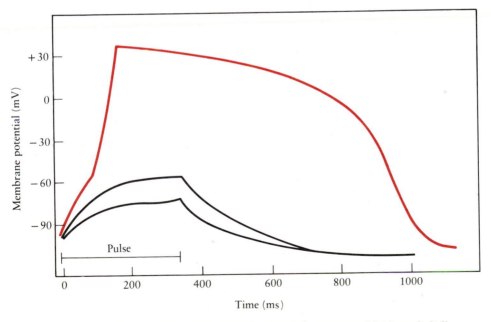

FIGURE 9.6   A calcium action potential in the unicellular water mold *Blastocladiella emersonii*. Injection of small amounts of positive charge into the cells partly depolarizes the membrane for as long as the current flow lasts. Depolarization of the membrane beyond a threshold of -45 mV elicits a prolonged action potential, which results from an initial influx of calcium ions followed by an efflux of chloride. The bottom trace shows the duration of the injected current pulse.

The Flow of Water.   Solutes diffuse from a region of higher concentration to one of lower, and so does water. Whenever there is a difference in solute concentration across a membrane, water tends to flow from the dilute to the concentrated compartment, making the latter swell. The pressure that must be applied in order to stop the net flow of water from a compartment containing pure water into one containing a solution is known as the osmotic pressure.

The quantitative relationship between osmotic pressure and the concentration of dissolved solute was worked out by Jacobus Van't Hoff in the nineteenth century. For dilute solutions it is given by

$$\pi = CRT \qquad \text{(Eq. 9.4)}$$

where π is the osmotic pressure in bars, C is the concentration of a nonionic solute in moles per liter (mol/L), R is the gas constant expressed as 0.083 L·bar/K·mol, and T is the absolute temperature in kelvins (K). (In all but the most dilute solutions, activities must be used in place of concentrations.) For biological purposes, what is usually of interest is the difference in osmotic pressure across a membrane that separates two compartments of different composition, such as cytoplasm and the external medium.

Thanks to the osmotic pressure of their cytoplasmic constituents cells tend to swell, and this must by countered if biological structures are to persist. Two general solutions to this problem have evolved. Bacteria, protists, and plant cells are generally encased in a rigid cell wall, which resists osmotic swelling. There normally is a hydrostatic pressure differential between cytoplasm and medium, called turgor; this is required for the rigidity of the whole organism (plants wilt and sag when they lose water) and provides the driving force for enlargement. Animal cells and other cells lacking walls maintain stability by extruding sodium ions, making Na^+ effectively impermeant; the osmotic pressure of the extracellular sodium salts balances that of cytoplasmic solutes and prevents swelling. Osmotic stability is one of the primary functions of the sodium pump and will be considered more fully below.

There is no known mechanism for the active transport of water per se, and it is very likely that no such mechanism exists. Phospholipid bilayer membranes are permeable to water, probably because water molecules diffuse in between the fatty acyl chains. Net water flow is always secondary to changes in solute concentrations. A rise in cytoplasmic solute levels elicits osmotic water influx, followed either by swelling or by increased turgor; shrinkage or reduction of turgor are achieved by lowering solute levels. Organisms equipped with contractive vacuoles or kidneys obviously expel water actively, but even in these the flow of water across cellular membranes is secondary to solute fluxes.

The Sodium Circulation of Animal Cells

Historically, animal cells and tissues have dominated the field of biological transport. Except for the chemiosmotic theory, most of its basic principles grew out of research with erythrocytes and tumor cells, frog skin and intestinal sacs. The themes that recur throughout this chapter—carriers and specificity, the varieties of active transport, electric potentials and excitability (not to mention those prototypes of biological transport, the ion-translocating ATPases)—are rooted in the labors of animal physiologists and bear directly on the workings of the human body. Surprisingly, there seems to be no contemporary synthetic treatment of the place of membrane transport in animal physiology, apart from chapters in general texts (for example, DeVoe and Maloney, 1980; Eckert and Randall, 1983). My purpose here is not to remedy this lack but to highlight some of the general features that distinguish transport across animal cell membranes. In keeping with the scope of this book, I have chosen examples from the cellular level to the neglect of multicellular tissues and organs.

The heart of the matter is that in animal cells, a circulation of sodium ions across the plasma membrane knits together the multiplicity of transport pathways and makes functional sense of membrane energetics (Figure 9.7). The prime mover is the Na^+-K^+ ATPase, which expels Na^+ by exchange for K^+, generating an electrochemical potential gradient with the cytoplasm low in sodium and electronegative. This gradient gives rise to a "sodium-motive force,"

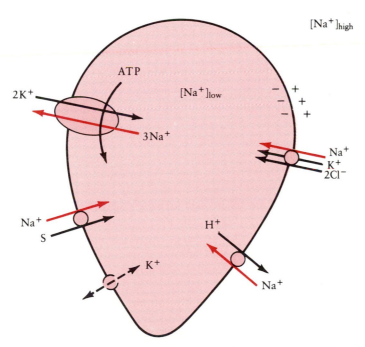

FIGURE 9.7 The sodium circulation of animal cells. The Na^+-K^+ ATPase expels Na^+ by exchange for K^+, generating an electrochemical Na^+ gradient (cytoplasm low in Na^+). Sodium ions return via various porters: Na^+-metabolite symport (S), Na^+-H^+ antiport, and Na^+-K^+-$2Cl^-$ symport are shown.

TABLE 9.1   Representative Electrolyte Concentrations in Animal Cells

	K^+ (mM)	Na^+ (mM)	Cl^- (mM)	Resting potential (mV)
Intracellular				
Squid axon	400	78	100	−70
Erythrocyte	135	15	80	−10
Skeletal muscle	140	10	4	−90
Ehrlich ascites cell	160	40		−50
Extracellular				
Sea water	10	485	566	
Blood (squid)	22	440	560	
Blood (human)	5	143	103	

which drives Na^+ ions back into the cells; it is the immediate energy source for a set of sodium-linked porters and is involved in several homeostatic functions, including the regulation of cellular volume and pH. Animals evolved in salt water, and when they colonized the land they carried a little of the sea ashore. But the origins of the Na^+-K^+ ATPase and the sodium circulation are unknown. The enzyme occurs in mollusks and arthropods, but its distribution among the lowlier metazoan animals has not been worked out.

The Sodium Pump.   In animal cells, as in living things in general, the distribution of Na^+ and K^+ ions across the plasma membrane is markedly asymmetric: Na^+ is the chief cation of sea water and blood, K^+ that of the cytoplasm (Table 9.1). During the first third of this century, this distribution was regarded as an instance of the Donnan equilibrium. Donnan recognized that the presence of impermeant anions in the cytoplasm, such as proteins, nucleic acids, and phosphorylated sugars, impinges on the equilibrium distribution of permeant ions. If the membrane were permeable to K^+ but not to Na^+, K^+ would accumulate in the cytoplasm to an extent determined by the electric potential, while chloride would be largely excluded. This explanation had to be abandoned when early experiments with radioisotopes showed the plasma membrane to be permeable to sodium. Yet the K^+ distribution does accord quite well with expectations from the Nernst equation, while that of Na^+ does not. By 1955 several lines of investigation had corroborated the hypothesis that the discrepancy is due to the active extrusion of Na^+ ions from the cells and implicated ATP as the immediate energy donor. For instance, both erythrocytes and squid axons maintain a sodium gradient when the cytoplasm is supplied with ATP but not in its absence; extrusion of sodium is coupled to the concurrent uptake of K^+.

The discovery of the Na^+-K^+ ATPase by J. C. Skou (1957) must rank as one of the most influential contributions to bioenergetics and cell physiology.

Skou showed that membrane fragments from crab nerve contain an ATPase that requires Mg^{2+} ions and whose activity is synergistically stimulated by Na^+ and K^+ ions (Figure 9.8), and he correctly identified this novel enzyme as the sodium pump of the plasma membrane. This interpretation was brilliantly confirmed a few years later by R. Whittam and I. M. Glynn (Whittam and Ager, 1964; Glynn and Karlish, 1975). By use of a procedure for lysing erythrocytes that permitted them to load the resulting membrane ghosts with either Na^+ or K^+ salts, they showed that stimulation of the ATPase required Na^+ ions on the cytosolic side and K^+ ions on the exterior, and that Na^+ ions were expelled from the ghosts by exchange for K^+. Evidently the reaction pathway traverses the membrane, as chemiosmotic reactions do, although this parallel was not drawn until much later. Both ion transport and ATP hydrolysis were inhibited by ouabain, a steroid alkaloid found in digitalis that had earlier been shown to inhibit active sodium extrusion from erythrocytes. Ouabain, which is active at micromolar concentrations, is a highly specific inhibitor of the Na^+-K^+ ATPase; inhibition by ouabain is diagnostic of the enzyme and the binding of labeled ouabain to membranes can serve as a measure of the amount of ATPase protein.

Erythrocyte ghosts lent themselves to several other exploratory studies that defined the biochemical nature of the sodium pump. Meticulous experiments by R. L. Post, I. M. Glynn, and their colleagues showed that the fluxes of Na^+ and K^+ are unequal, the most probable ratio being $3Na^+/2K^+/ATP$. The same stoichiometry has since been obtained with preparations from a variety of

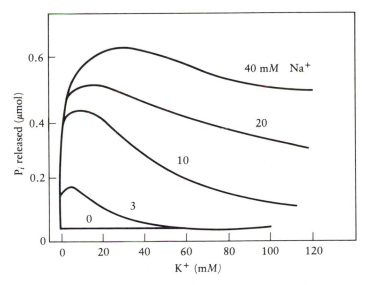

FIGURE 9.8   ATP hydrolysis by the Na^+-K^+ ATPase is stimulated synergistically by Na^+ and K^+ ions. Membrane fragments from crab nerve were incubated in buffer containing 3 mM ATP, 6 mM Mg^{2+}, and Na^+ and K^+ as shown. (After Skou, 1957, with permission of Elsevier Biomedical Press.)

vertebrate and invertebrate tissues; it is likely (albeit not certain) to be a universal feature of the ouabain-sensitive ATPase. Another major advance was the discovery that the hydrolysis of ATP can be described as the sum of two successive reactions. In the first a phosphoryl group from ATP is transferred to an amino acid residue (now known to be aspartate) on the enzyme protein; this step requires Na^+ ions. Subsequently the phosphoryl group is hydrolyzed to P_i; K^+ ions are required and the reaction is inhibited by ouabain. These pioneering studies were followed by a veritable explosion of research, whose progress may be traced in a string of review articles (Skou, 1965; Dahl and Hokin, 1974; Glynn and Karlish, 1975; Robinson and Flashner, 1979; Cantley, 1981; Jørgensen, 1982).

The sequence of reactions given in Figure 9.9 was devised by R. L. Post and his colleagues more than a decade ago; it stands substantially unaltered today, albeit amplified and modified by later work. The description given below follows that by Cantley (1981). In the E_1 configuration the binding sites for monovalent cations face the cytoplasm and have a high affinity for Na^+; in E_2 they face outward and prefer K^+. Exchange of Na_i^+ for K_o^+ is tightly coupled to the hydrolysis of ATP to ADP and P_i. When the cation-binding site of E_1 is occupied by Na^+, ATP phosphorylates an aspartate residue at the active site (step a). The phosphorylated protein is strained in the E_1 configuration and relaxes to E_2,

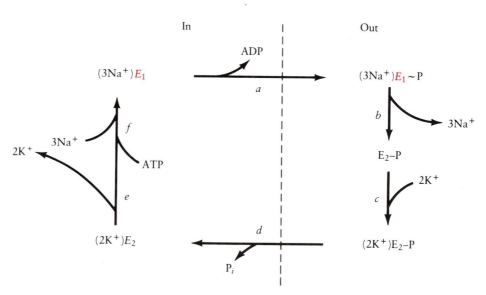

FIGURE 9.9   The Na^+-K^+ ATPase: a diagram of intermediate steps. E_1 and E_2 designate configurations of the transport sites that preferentially bind cytoplasmic Na^+ and external K^+, respectively. The phosphoryl group of $E_1 \sim P$ has a high transfer potential, that of E_2-P a low one. For clarity, the arrows point in the direction of ATP hydrolysis, but in fact all the steps are reversible.

transporting the bound Na^+ ions to the outside (step b). The binding sites now take up K^+; binding of K^+ ions catalyzes the dephosphorylation of the protein (steps c and d). A crucial feature is that the potassium ions are "trapped" as dephosphorylation occurs and can be released to the cytoplasmic side only after the enzyme has changed from the E_2 conformation to E_1. This transition (steps e and f) occurs spontaneously at a low rate but is markedly accelerated by ATP: the nucleotide binds to $(K)E_2$ and forces the enzyme into the E_1 conformation. Each step is individually reversible, but the cycle as a whole drives the vectorial exchange of $3Na^+$ for $2K^+$ because the scalar force of ATP hydrolysis is channeled by spatial constraints: part of the reaction is mediated by the E_1 conformation and part by the E_2, and the conversion of one configuration into the other is obligatorily linked to ion movement.

Several postulates embodied in Figure 9.9 have been confirmed by experiment. There is good evidence for conformational differences between E_1 and E_2, as shown most dramatically by the fact that tryptic digestion of the enzyme in the two states yields different fragments. Sodium ions support the phosphorylation of E_1 with a K_m of about 8 mM, acting from the cytoplasmic side; the affinity of E_2 for sodium is far less. Conversely, K^+ supports the dephosphorylation of E_2 with a K_m near 0.1 mM. During the transition back to E_1, potassium ions pass through a nonexchangeable state, which is thought to represent occlusion of the ion while in transit. Most authorities agree that there is but a single set of cation-binding sites, which change both their orientation and their affinity in the course of a transport cycle. Again, as the model predicts, the protein in the E_2 state can be phosphorylated by P_i; this reaction is inhibited by Na^+, which pulls the enzyme back to E_1. All this is consistent with the implication that $E_1 \sim P$ is a state of higher free energy than is E_2-P.

Since the individual steps are reversible, it should be possible to reverse the transport cycle to generate ATP at the expense of the potential energy stored in an ion gradient. This was accomplished by transferring erythrocyte ghosts containing internal ADP, P_i, and K^+ into buffer containing only Na^+; a burst of ATP synthesis was observed. Moreover, while the ATPase serves physiologically to expel Na^+ by exchange for K^+, it can be made to catalyze alternative reactions under appropriate conditions; for example, exchange of Na^+ for Na^+ or of K^+ for K^+, both of which require the presence of ATP and ADP but no net hydrolysis. These and other alternative modes are again consistent with Figure 9.9.

The Na^+-K^+ ATPase has been solubilized, purified, and inlaid into liposomes. The reconstituted enzyme, which retains all the properties described above, consists of two subunits: α (120 kdal) and β (a glycoprotein, 55 kdal). The ATP binding site, the phosphorylated aspartate residue, and the binding sites for Na^+ and K^+ are all on the α subunit; it is not at all clear what the β subunit does. Until recently the functional transport system was generally thought to be the dimer, $(\alpha\beta)_2$, but new evidence indicates that the monomeric enzyme $\alpha\beta$ can mediate transport (Kyte, 1981; Craig, 1982). The point is important, for it is widely believed that most transport systems are oligomers, with the substrate passing through transient channels between the subunits (Klingenberg, 1981). We shall return to this matter in Chapter 10.

In a cartoon representing ion transport by the Na^+-K^+ ATPase (Figure 3.12), Oleg Jardetzky (1966) depicted a cavity that contained ion-binding sites; the orientation of this cavity and the affinity of the sites for K^+ and Na^+ changed in concert with the cycle of phosphorylation and dephosphorylation. This remains, in all essentials, the view to which most students of the ATPase still subscribe. The only drastic alternative is provided by Mitchell's concept of ligand conduction (Chapter 3), which proposes that the movements of Na^+ and K^+ across the membrane are spatially confluent with the path of the phosphoryl group from ATP to water (P_i) across the catalytic center. This matter will also be deferred to the next chapter.

Energetics. At equilibrium the electrochemical potentials of sodium and potassium ions must be balanced against the free energy of ATP hydrolysis. By analogy with Equation 3.11 we can then write

$$n(\Delta\tilde{\mu}_{Na^+}) + m(-\Delta\tilde{\mu}_{K^+}) = \Delta G_{ATP} \qquad \text{(Eq. 9.5)}$$

where ΔG_{ATP} is the free energy of ATP hydrolysis in the cytoplasm, normally about -10 kcal/mol, and n and m are the number of ions moved per cycle, in this instance 3 and 2; the sign of $\Delta\tilde{\mu}_{K^+}$ is reversed in order to allow for the fact that K^+ is moved inward while Na^+ is expelled.

The electrochemical potential gradients of the two ions, in millivolts, are given by

$$\frac{\Delta\tilde{\mu}_{Na^+}}{F} = \Delta\Psi + \frac{2.3RT}{F} \log \frac{[Na^+]_i}{[Na^+]_o} \qquad \text{(Eq. 9.6)}$$

$$\frac{\Delta\tilde{\mu}_{K^+}}{F} = \Delta\Psi + \frac{2.3RT}{F} \log \frac{[K^+]_i}{[K^+]_o} \qquad \text{(Eq. 9.7)}$$

Now, animal cell membranes are often quite permeable to K^+ ions, so that the K^+ distribution between cytoplasm and medium approaches equilibrium and $\Delta\tilde{\mu}_{K^+}$ is not far from zero. By contrast, the permeability to sodium ions is low; the free energy of the ATPase is therefore conserved chiefly by moving sodium ions out of equilibrium. Given $\Delta G_{ATP} = -10$ kcal/mol and taking $\Delta\tilde{\mu}_{K^+}$ to be zero, the maximum value of $\Delta\tilde{\mu}_{Na^+}$ would be -3.3 kcal/mol ($\Delta\tilde{\mu}_{Na^+}/F = -145$ mV).

In practice, as can be calculated from the data in Table 9.1, $\Delta\tilde{\mu}_{Na^+}/F$ is usually not greater than -120 mV. The table also shows that the concentration term makes up a large part of the "sodium-motive force"; the membrane potential contributes only half, often much less. This is quite unlike the situation in fungi or plants, in which the membrane potential is the major component of the electrochemical ion gradient across the plasma membrane.

Since the sodium and potassium fluxes driven by the ATPase are unequal, the reaction is electrogenic and should make a contribution to the membrane potential. This proved hard to document experimentally, because the plasma membrane of animal cells is sufficiently K^+ permeable that the electric potential is dominated by K^+ diffusion. Eventually R. C. Thomas showed that snail neurons (which are large and readily impaled with multiple electrodes) hyperpolarized when injected with sodium salts; the hyperpolarization, but not the resting potential, was abolished by ouabain (Figure 9.10; Thomas, 1972). Similar findings have since been made with smaller animal cells. Mouse ascites tumor cells, for example, hyperpolarize from -50 to -80 mV in response to Na^+ entry, as judged by fluorescence quenching of voltage-sensitive dyes (Philo and Eddy, 1978). To summarize, the sodium pump generates the concentration gradients for Na^+ and K^+ but normally contributes no more than a few millivolts to the membrane potential. Sodium-loaded cells, however, develop a transient increase in the electric potential due to stimulation of the pump.

Judging by the inhibitory effects of ouabain on the overall rate of metabolism, the sodium pump consumes 15 to 40 percent of the total ATP generated by animal cells. This massive expenditure reflects the central, albeit indirect, role of the Na^+-K^+ ATPase in diverse kinds of work; let us now ask what use the cells make of the $\Delta\tilde{\mu}_{Na^+}$ generated by the pump.

Porters.   Of all the functions that we can assign to the sodium circulation, the best attested is its role in the accumulation of sugars and amino acids. Over 30

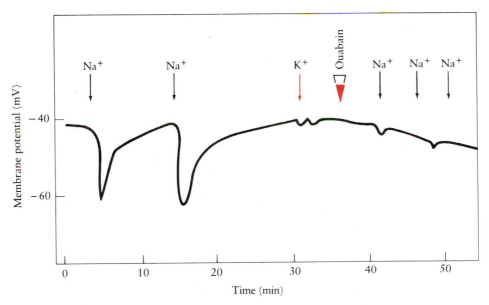

FIGURE 9.10   Injection of sodium ions into snail neurons increases the membrane potential difference. The trace shows the membrane potential. Two successive injections of sodium acetate elicited transient hyperpolarization, but potassium acetate did not. Following injection of ouabain, the effects of Na^+ were no longer observed. (After Thomas, 1982, with permission of Academic Press.)

years ago H. N. Christensen recognized that the uptake of amino acids by ascites tumor cells was a kind of active transport and somehow dependent on Na^+ and K^+ (Christensen and Riggs, 1952). A few years later R. K. Crane explicitly proposed that carriers for organic metabolites mediate cotransport with Na^+ and that the driving force for the accumulation of sugars and amino acids is the electrochemical gradient of sodium ions (Crane et al., 1961; Crane, 1965, 1977). Support for this proposal accumulated rapidly, and the review by Schultz and Curran (1970) made a persuasive case.

Figure 9.11 shows some illustrative data taken from the work of A. A. Eddy and his associates. Mouse ascites tumor cells accumulate glycine from the medium, achieving a concentration gradient of about tenfold, provided sodium ions are present. The rate of glycine uptake, negligible in the absence of Na^+, becomes maximal at about 150 mM Na^+. Cells depleted of metabolic energy by starvation can still accumulate glycine if there is a gradient of sodium ion concentration between the medium and the cytoplasm, and the extent of accumulation is directly proportional to the sodium gradient, as expected if the sodium concentration gradient were the sole driving force for amino acid uptake (Eddy, 1968). However, observations that contradicted this hypothesis also

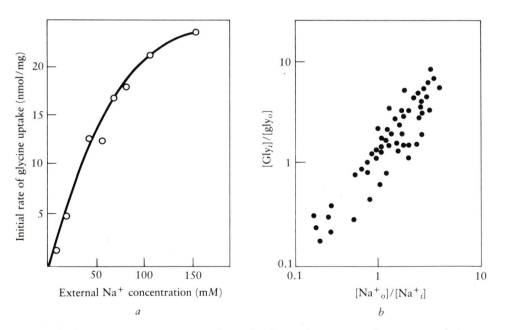

FIGURE 9.11   The concentration gradient of sodium ions supports glycine accumulation by Ehrlich ascites cells. (*a*) The initial rate of glycine uptake (12 mM glycine) rises with the external Na^+ concentration. (*b*) In cells depleted of metabolic energy, the glycine concentration gradient in the steady state is proportional to the sodium concentration gradient. (After Eddy, 1968, by permission of the Biochemical Society, London.)

appeared. It was disturbing, for example, that ascites cells accumulated amino acids even after having been loaded with sodium such that the cytoplasmic concentration exceeded that of the medium. This discrepancy was eventually resolved by the demonstration that these conditions stimulate the sodium pump, making the membrane potential more negative. Since cotransport of glycine with Na^+ is electrogenic, the increase in membrane potential more than makes up for the reversal of the Na^+ ion concentration gradient (Philo and Eddy, 1978; Heinz et al., 1980; Hacking and Eddy, 1981). The consensus now is that $\Delta\tilde{\mu}_{Na^+}$ is both necessary and sufficient to account for sodium-dependent uptake of amino acids and other metabolites. The various transport systems that Christensen and others have identified in these cells (system A, L, ASC, and so on) are evidently secondary porters, analogous at least in principle to the bacterial ones described in Chapter 5.

As in the case of bacteria, some of the most direct evidence for symport with sodium ions was obtained with plasma membrane vesicles. These lack metabolic enzymes and the ionic composition on both membrane surfaces can be varied at will. Figure 9.12 illustrates the effect of a sodium gradient ($[Na^+]_o > [Na^+]_i$) on the uptake of inorganic phosphate by membrane vesicles from the renal brush border (Cheng and Sacktor, 1981). Both the rate and the extent of phosphate uptake were greatly enhanced by the imposition of a Na^+ ion concentration gradient, but manipulation of the electric potential had no effect. Apparently, as has also been reported in several other cell types, the porter mediates electro-neutral symport, either $H_2PO_4^- - Na^+$ or $HPO_4^{2-} - 2Na^+$.

Thanks to their ready availability and metabolic simplicity, erythrocytes have provided many of the best-studied porters. Human red cells generate ATP by glycolysis. The glucose carrier is a uniporter whose kinetics and specificity have been studied for three decades (Widdas, 1980). Recently the porter has been purified and reconstituted into liposomes; it consists of a single polypeptide chain, 45 kdal, whose functional form is thought to be the dimer (Wheeler and Hinkle, 1981). Glycine enters by a symporter that has been extensively studied by Vidaver and his associates; the stoichiometry is probably $Gly/2Na^+$, but Cl^- may be carried as well. Finally, one of the chief functions of erythrocytes is to exchange HCO_3^- for Cl^-, a strictly coupled antiport mediated by one of the most abundant proteins in the erythrocyte membrane. The porter has been isolated (95 kdal) and reconstituted; it is likely that this system also functions as a dimer (Knauf, 1979).

The porters described above are orthodox enough, but a recent addition to the roster is definitely peculiar. It has been recognized for some time that many animal cells contain, in addition to the ouabain-sensitive $Na^+ - K^+$ ATPase, a pathway for the coupled transport of Na^+ and K^+ that is resistant to ouabain but inhibited by furosemide and other diuretics. It now appears that furosemide sensitivity is diagnostic for a widely distributed class of symporters that carry K^+, Na^+, and Cl^-. In Ehrlich ascites cells the process is electrically silent and the probable stoichiometry is $1K^+/1Na^+/2Cl^-$; there is no evidence for ATP

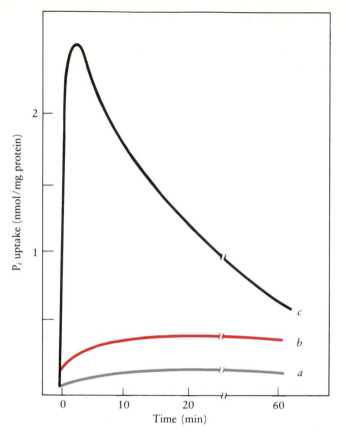

FIGURE 9.12   Uptake of inorganic phosphate by membrane vesicles in response to a sodium gradient. Vesicles of known internal ion content were added to buffer containing $^{32}P_i$ at zero time. Trace *a*, no Na^+ inside or outside; trace *b*, 0.1 *M* Na^+ on both sides; trace *c*, 0.1 *M* Na^+ outside, none inside. (After Cheng and Sacktor, 1981, with permission of the *Journal of Biological Chemistry.*)

involvement. Similar systems have been described in erythrocytes, fish intestine, cultured kidney cells, and squid axons; the latter two require ATP, but ATP may be a regulator rather than an energy donor (Geck et al., 1980; Kregenow, 1981; McRoberts et al., 1982). The function of this unexpectedly elaborate system is not certain, but there is evidence that it is a major element in the ionic regulation of cell volume (see below).

Regulating the Cytoplasmic pH.   Given that animal cells maintain an electric potential difference of about -60 mV and assuming that protons are distributed passively, the cytoplasmic pH ought to be more acid than that of the medium by up to one unit. That the cytoplasmic pH is in fact alkaline by as much as 0.5 unit points to the intervention of transport systems that expel protons. The extensive research on this problem (recently reviewed by Roos and Boron, 1981, and Boron, 1983) has revealed the existence of several kinds (Figure 9.13).

Many animal cells contain a sodium-proton antiporter that is functionally similar to those widespread in bacteria. Examples come from muscles of frog and mouse, from erythrocytes (where exchange of H^+ for Na^+ compensates for the acidification induced by the exchange of respiratory HCO_3^- for Cl^-), and from a variety of tissue culture cells. The system has been well documented in membrane vesicles from intestinal and kidney brush border. It seems to catalyze reversible, electroneutral exchange, with the sodium gradient as the sole driving force for the expulsion of H^+. Antiport is inhibited by the diuretic amiloride, classically regarded as an inhibitor for ill-defined "sodium channels" present in a variety of membranes; many, if not all of these, may prove to be Na^+-H^+ antiporters.

A more elaborate antiport system was discovered by R. C. Thomas (1977) in snail neurons with a pioneering application of intracellular pH-sensitive microelectrodes. When a small pulse of HCl was injected into a snail neuron, the intracellular pH fell from 7.4 to 7.0 and then returned to its former value. Recovery of the pH was shown to depend on the simultaneous presence of HCO_3^-, external Na^+, and internal Cl^-; the membrane potential remained constant during the recovery, indicating an electroneutral process. Fluxes of H^+, Cl^-, Na^+, and (presumably) HCO_3^- are coupled, since all were induced by lowering the cytoplasmic pH and all were blocked by a stilbene derivative called SITS, which also inhibits anion transport by erythrocytes. Thomas attributes the recovery of internal pH to an electroneutral coupled exchange of H^+ and Cl^- for Na^+ and HCO_3^- (Figure 9.13); however, the data do not exclude alternative formulations such as exchange of Na^+ and CO_3^{2-} for Cl^-, which would be thermodynamically equivalent. The energy requirements of the exchange can be met by the sodium gradient, and in snail neurons at least, ATP is not required.

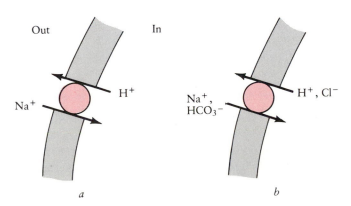

FIGURE 9.13   Regulation of the cytoplasmic pH in animal cells involves two kinds of antiporters: (a) Na^+-H^+ antiport; (b) $NaHCO_3$-HCl antiport.

However, a very similar exchange in squid axons proceeds only if the axoplasm contains ATP; the function of the ATP is unknown.

Both of the preceding mechanisms involve secondary porters, but it must be mentioned in passing that certain epithelial tissues carry out primary transport of protons. Gastric mucosa contains an ATPase that drives an electroneutral exchange of K^+ for H^+. This enzyme, extensively studied by Sachs and his colleagues, plays a major role in the secretion of acid into gastric fluid. Acidification of the urine involves a novel ATPase, possibly related to the F_1F_0 ATPase of mitochondria (Sachs et al., 1982; Steinmetz and Andersen, 1982). We shall return to the diversity of animal ATPases in the following chapter.

The Control of Cell Volume.   One of the most important functions of the sodium pump is to combat the tendency of animal cells to swell due to the osmotic influx of water. By continually expelling whatever Na^+ ions leak in, animal cells establish a countergradient that maintains osmotic balance. Cells poisoned with ouabain ultimately swell since they can no longer eject Na^+.

But volume control is more complex than this. When placed in a medium that is not isotonic with the cytoplasm, many kinds of animal cells first shrink or swell osmotically and then recover their original volume. These corrective changes in volume result from shifts in the water balance that follow ion movements but are not *immediately* dependent on the sodium pump (MacKnight and Leaf, 1977). The responses of nucleated erythrocytes, such as those of ducks, have been examined in detail and the findings reviewed by Kregenow (1981); the principles are summarized in Figure 9.14.

When put into hypertonic medium, duck erythrocytes shrink. This activates the previously quiescent K^+-Na^+-Cl^- symporter ten- to twentyfold; as osmotically active ions enter the cell, water follows and swelling ensues. Interestingly, swelling is observed only if the external K^+ concentration exceeds 2.5 mM. The reason is that the driving force on the porter is the sum of the forces acting on the three ions; inward for Na^+, outward for K^+, and zero for Cl^-, which is in electrochemical equilibrium. Only when $[K^+]_o$ is sufficiently high does net ion influx result. At lower K^+ concentrations K^+, Na^+, and Cl^- exchange rapidly across the membrane, but there is no swelling. Both volume compensation and the symporter are inhibited by the diuretic drug furosemide. How cell shrinkage activates the transport system is not known; interestingly, catecholamines activate it even without prior shrinkage, inducing the cells to swell beyond their original volume.

Duck erythrocytes also respond positively to hypotonic medium. Enlargement of the cell by influx of water is followed by activation of a coupled efflux of K^+ and Cl^-; Na^+ ions do not seem to be involved. Water follows the ions out, and the original volume is restored. Just how the cells sense their original volume remains a mystery.[3]

[3] It should not be taken for granted that the mechanisms outlined above are necessary and sufficient for volume regulation by all animal cells. For example, McRoberts et al. (1983) found that a line of mutant kidney cells lacking the K^+-Na^+-Cl^- symporter is still capable of volume regulation.

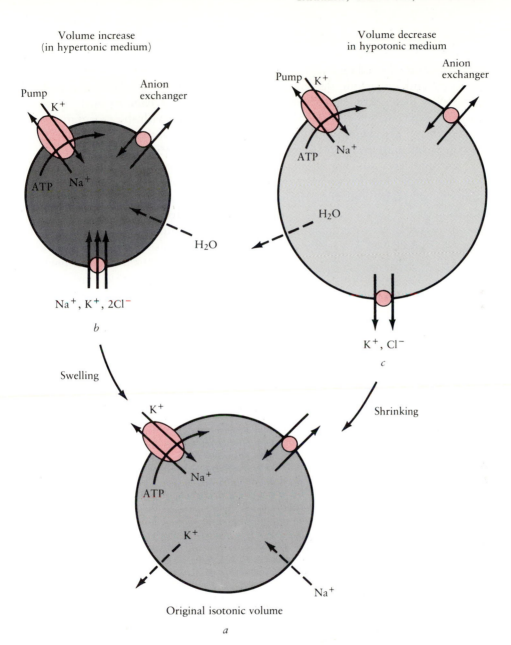

Volume increase
(in hypertonic medium)

Volume decrease
in hypotonic medium

Pump

Anion
exchanger

K^+

ATP Na^+

$Na^+, K^+, 2Cl^-$

b

Anion
exchanger

Pump K^+

ATP Na^+

H_2O

K^+, Cl^-

c

H_2O

Swelling

Shrinking

K^+

ATP Na^+

K^+

Na^+

Original isotonic volume

a

FIGURE 9.14   Ionic control of erythrocyte volume. (*a*) Normally the erythrocyte is isotonic with its medium. (*b*) When transferred to hypertonic medium, the cell shrinks. This activates the Na-K-Cl symporter, water follows the ions in, and the cell swells until its original volume has been restored. (*c*) When placed in hypotonic medium, the erythrocyte swells. This activates a KCl efflux system, water follows the ions out, and the cell shrinks until its original volume has been restored. (After Kregenow, 1981; reproduced with permission from the *Annual Review of Physiology*, vol. 43, 1981, by Annual Reviews.)

Cation Fluxes and the Control of Growth.   The manner in which the sodium current energizes solute transport and the regulation of pH and of volume might have been anticipated by a thermodynamically minded physiologist, but, despite portents and omens in the early literature, the discovery that sodium and potassium fluxes can elicit developmental events has caught most of us by surprise. The pace of discovery has quickened of late with the recognition that sodium fluxes are intimately involved in the transition of tissue culture cells from the quiescent state to one of active growth (Kaplan, 1978; Rozengurt, 1980).

Lymphocytes can be induced to divide by the addition of phytohemagglutinins, the terminal differentiation of mouse erythroleukemia cells is induced by dimethylsulfoxide, and fibroblasts can be made to resume division by the addition of fresh serum or of a variety of growth factors. What all these have in common is that they elicit an influx of sodium ions, followed by activation of the Na^+-K^+ ATPase. A variety of reagents that elicit sodium fluxes serve as mitogens, including the ionophores gramicidin and monensin and the peptides vasopressin and melittin; the mitogenic effects can often be blocked by ouabain. In several instances there is evidence that sodium enters by amiloride-sensitive, electroneutral exchange for protons. This is accompanied by a small increase in the cytoplasmic pH (0.2 to 0.3 unit) (Kaplan, 1978; Rozengurt, 1980; Moolenaar et al., 1982; Schuldiner and Rozengurt, 1982; Frelin and Lazdunski, 1983; Burns and Rozengurt, 1984).

It is far from obvious how sodium influx and activation of the sodium pump, which occur within minutes after addition of the mitogen, are connected to the initiation of DNA synthesis 10 hours later. Quite likely the sequence of events is not always the same. In at least one instance, Ca^{2+} ions may play a critical intermediary role. Mouse erythroleukemia cells differentiate in response to various treatments that raise the cytosolic sodium level; the Na^+ exits by exchange for Ca^{2+}, and Ca^{2+} influx may be the heart of the matter since calcium ionophores are sufficient to induce differentiation (Smith et al., 1982). And how does calcium control differentiation? We do not know, but this bald admission must be elaborated on in Chapter 13.

The Proton Circulation of Fungi, Algae, and Plants

In the 1940s the Irish biochemist E. J. Conway formulated a provocative hypothesis to explain the marked secretion of acid during glucose fermentation by yeast; the suspension may go as acid as pH 1 to 2. Conway proposed that cytochrome chains at the cell surface separate protons from electrons; the protons would be secreted, while the excess of internal OH^- would draw cations into the cell and account for the uptake of K^+ during fermentation (Conway, 1953). Conway's views have long been superseded, if only because we know today that the respiratory chain is not housed in the plasma membrane. But it is evident that he was groping toward what we would now describe as a chemiosmotic interpretation of acid secretion. Indeed, while transport research with animal cells followed its own course, our understanding of the way fungi, algae, and plants conduct their affairs has been profoundly influenced by Mitchell's theory.

The crux of the matter is that fungi, algae, and even higher plants employ protons as the chief coupling ion for membrane bioenergetics. Animal cells lack walls, and the indispensable function of the sodium pump is to maintain osmotic stability by excluding Na^+ ions from the cytoplasm. This mechanism demands that there be a large difference between the concentrations of sodium in cytoplasm and medium; in animal cells $\Delta\tilde{\mu}_{Na^+}$ is dominated by the concentration term and $\Delta\Psi$ is relatively small (Table 9.1, Equation 9.6). By contrast, the walled eukaryotic cells solve their osmotic problems mechanically and can grow in dilute media containing little Na^+ or none. As in the bacteria, protons are expelled from the cytoplasm by a primary proton pump that generates a protonic potential across the plasma membrane, interior alkaline and negative. As a rule, the dominant term in $\Delta\tilde{\mu}_{H^+}$ is the membrane potential which often attains -200 mV and more.

Figure 9.15 illustrates the multiple independent proton circulations that support the workings of a "typical" walled eukaryote, such as a fungal hypha. Note that proton translocation at the expense of redox reactions or of light is confined to mitochondria and chloroplasts. Ejection of protons across the plasma membrane is the business of a novel class of specialized H^+ ATPases; these enzymes are quite unlike the F_1F_0 ATPase of prokaryotes and organelles and seem rather to belong to the same molecular family as the Na^+-K^+ ATPase.

The literature, albeit modest by the standards of animal cell physiology, is still overwhelming. I have found the reviews by Poole (1978), Raven (1980), Spanswick (1981), Goffeau and Slayman (1981), and Eddy (1982) most useful in dealing with recent developments.

The Proton Pump.   In 1965, Clifford Slayman published his landmark studies on the electric potential across the plasma membrane of *Neurospora* hyphae. As was illustrated in Figure 9.3, the large potential could not be accounted for by diffusion of K^+ but was sharply dependent on respiratory metabolism, suggesting the presence of an electrogenic ion pump in the plasma membrane. From the observation that depolarization lagged a little behind the inhibition of respiration by KCN, Slayman inferred that the immediate energy donor for the hypothetical pump was probably ATP. His insight was corroborated by a series of painstaking experiments in which changes in both $\Delta\Psi$ and cellular nucleotide pools were followed after the addition of metabolic inhibitors. The results, summarized in Figure 9.16, showed that $\Delta\Psi$ closely tracks the declining ATP pool and may be regarded as the "product" of an electrogenic ATPase with a K_m near 2 mM (Slayman et al., 1973). A few years later Gradmann et al. (1978) calculated the stoichiometry of the putative proton pump from the current-voltage relationships of the plasma membrane (Figure 9.4). The reversal potential, that is, the potential difference that just stalls the electrogenic pump, was found to be -380 mV, while the free energy available from ATP hydrolysis was about

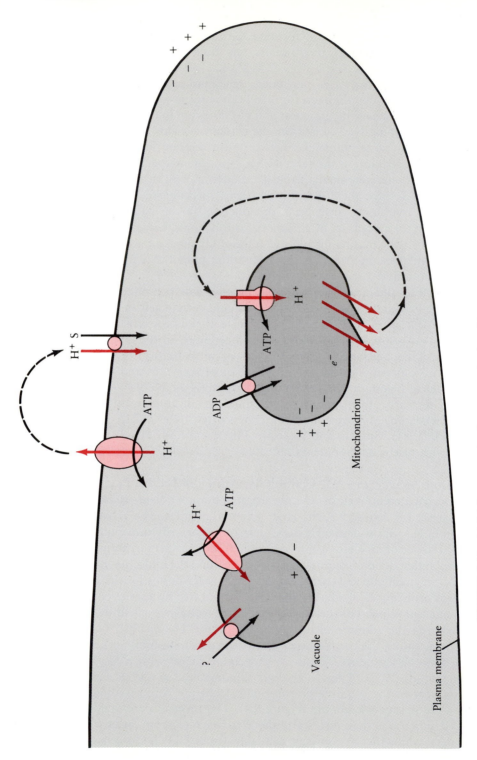

FIGURE 9.15   Multiple proton circulations in a hypha of *Neurospora crassa*. The diagram shows a mitochondrion, a vacuole, and the plasma membrane.

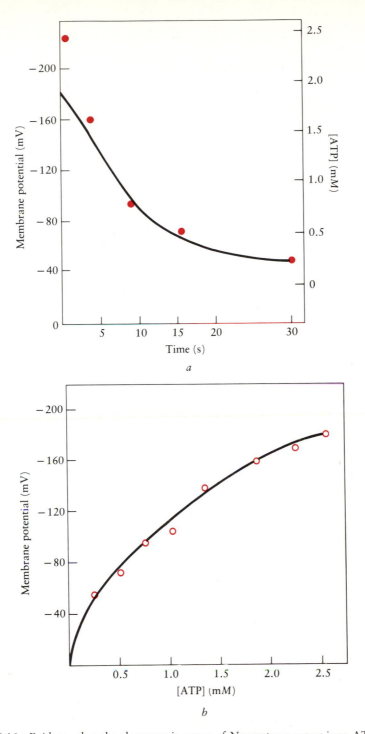

FIGURE 9.16 Evidence that the electrogenic pump of *Neurospora crassa* is an ATPase. (*a*) After addition of KCN to the medium at 0 seconds, decay of the membrane potential (black line) closely tracks the decline of the ATP concentration (circles). (*b*) Plot of the membrane potential as a function of the cell's ATP content. The relationship is hyperbolic, with a K_m of 2 mM. (Data from Slayman et al., 1973; after Goffeau and Slayman, 1981, with permission of Elsevier Biomedical Press.)

12 kcal/mol (50 kJ/mol), equivalent to -500 mV. It follows that the pump ejects one proton for each molecule of ATP hydrolyzed. (It must be added that this stoichiometry holds only when the energy supply is ample; there are indications that under conditions of energy limitation the proton pump shifts to an alternative mode, which ejects two protons per ATP.)

Direct evidence that the pump translocates protons became available when G. A. Scarborough devised an ingenious method for isolating the plasma membrane of *Neurospora* in the form of everted, closed vesicles. Scarborough (1976, 1980) showed that the membrane contains a magnesium-dependent electrogenic ATPase that acidifies the lumen of these vesicles; in the intact hyphae, it must extrude protons from the cytoplasm. Subsequent research in the laboratories of Scarborough and of Carolyn Slayman characterized the plasma membrane ATPase as a single polypeptide, 104 kdal, functionally distinguishable from its mitochondrial counterpart by a relatively acid pH optimum and by a particular spectrum of responses to inhibitors, especially its sensitivity to vandate anion. The H^+ ATPase, unlike the F_1F_0 ATPase, undergoes phosphorylation of an aspartyl residue in the course of a reaction cycle. In these respects, as in molecular size, it clearly resembles the Na^+-K^+ ATPase and the Ca^{2+} ATPase of animal cells (Goffeau and Slayman, 1981; Dame and Scarborough, 1981).

Studies with yeast led to much the same conclusion. Acid secretion is effected by a proton-translocating ATPase that spans the plasma membrane. A. Goffeau and his colleagues have isolated the H^+ ATPase from several species of yeast, characterized it in detail, and reconstituted proton transport in liposomes. The yeast ATPase, like that of *Neurospora*, is a single polypeptide, 103 kdal, containing an aspartyl residue that undergoes phosphorylation in the course of the reaction. The enzyme transports protons electrogenically, probably with a stoichiometry of $2H^+$/ATP. Net acid secretion requires the concurrent movement of anions, of potassium, or of both. So far as we know, movements of protons and of potassium ions are effected by distinct transport systems (Dufour et al., 1982; Amory and Goffeau, 1982).

Algae have proved less tractable than fungi, but it is increasingly clear that they too extrude protons from the cytoplasm with the aid of an electrogenic H^+ ATPase. The giant-celled green algae *Chara* and *Nitella* have long been favorite objects of study by plant electrophysiologists, thanks to their internodal cells, which may be several centimeters long and a millimeter or more in diameter (see Figure 12.12 for an illustration). Internodal cells generate membrane potentials of -170 to -200 mV (Hope and Walker, 1975). Recently it became possible to perfuse the cells so as to remove the vacuole and its contents, as well as the cytoplasm, while leaving the plasma membrane intact. Such preparations still extrude protons when perfused with ATP and Mg^{2+}; the ATPase is vanadate-sensitive and probably has a stoichiometry of $2H^+$/ATP, but it has not yet been characterized biochemically (Shimmen and Tazawa, 1977; Smith and Walker, 1981). The relationship of the plasma membrane H^+ ATPase to other pumps and porters is shown schematically in Figure 9.17, to which we shall return below.

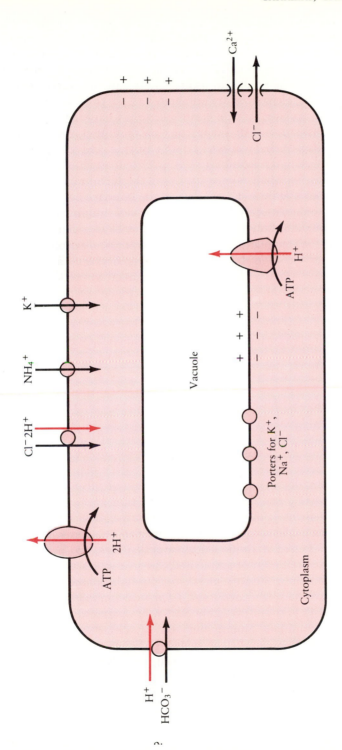

FIGURE 9.17 Pumps and porters in an internodal cell of *Chara corallina*. The diagram omits both chloroplasts and mitochondria in order to emphasize the transport systems of the plasma membrane and the tonoplast mentioned in the text. Not drawn to scale: the central vacuole fills more than 90 percent of the volume.

Higher plant cells, for instance those of oat seedlings and beet roots, are also suitable material for electrophysiology and readily generate electric potentials on the order of -180 mV. A decade ago T. K. Hodges discovered Mg-dependent ATPases in plasma membrane preparations of plant origin and recognized their role in ion transport. These enzymes are stimulated to varying degrees by K^+, Na^+, and even by Cl^-, and it is still not entirely certain what ion is transported. The balance of the evidence, however, favors the view that the essential process is proton transport (Poole, 1978; Spanswick, 1981; Briskin and Leonard, 1982; Vara and Serrano, 1982; Churchill and Sze, 1983; O'Neill and Spanswick 1984). With some reservations, we shall take it here that the membrane potential of higher plant cells is generated by a H^+ ATPase of the standard eukaryotic kind.

Before leaving this topic, it should be pointed out that both fungal and plant cells contain at least one additional ion-translocating ATPase: that of vacuolar membranes, including the tonoplast. These ATPases translocate protons into the vesicular compartment, acidifying the lumen and generating a membrane potential, lumen positive (Figures 9.15 and 9.17). Fungal vesicles accumulate basic amino acids such as arginine, which serve as a nitrogen reserve; most plant vacuoles are filled with salts that are osmotically important. Vacuolar ATPases bear some resemblance to the mitochondrial one (Kakinuma et al., 1981; Bowman and Bowman, 1982), but will probably comprise a separate class of molecules.

Porters.   During the past two decades many individual transport systems have been recognized in fungi, algae, and plants. The data are still fragmentary and should not be overinterpreted, but it is noteworthy that so far at least the only primary transport system found in the plasma membrane is the H^+ ATPase. All the remainder appear to be secondary; in most cases protons serve as the coupling ion but in marine algae Na^+ ions often take their place.

To justify the designation of a proton-linked porter, one must document that a proton potential provides the energy for accumulation of the metabolite in question and that movement of the metabolite is directly and stoichiometrically coupled to that of protons. Examples for which this can be asserted with some confidence include the uptake of amino acids, several sugars, and perhaps P_i by yeast, *Neurospora,* and *Chlorella;* of amino acids by liverworts, duckweed, and some other plants; and the transport of sucrose into the phloem for further distribution. There is good evidence that *Chara* accumulates chloride by electrogenic symport with protons, probably $2H^+\text{-}Cl^-$, and also that *Chara* and *Neurospora* avidly accumulate ammonium ion by electrogenic uniport in response to the membrane potential (Raven, 1980; Slayman, 1980; Komor and Tanner, 1980; Walker, 1980; Sanders, 1980; Felle and Bentrup, 1980; Eddy, 1982; Felle, 1983; Sanders et al., 1983).

Figure 9.18 illustrates the sort of data that underlie the hypothesis of symport with protons. *Neurospora* is known to produce a powerful system for glucose transport when grown under conditions of carbon limitation. Addition of glucose to such cells elicits immediate depolarization, followed by partial

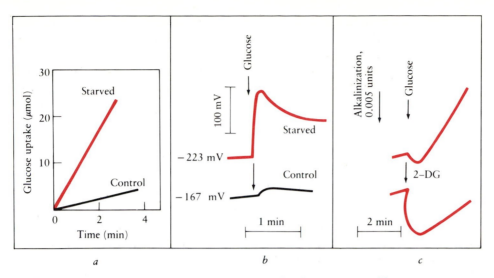

FIGURE 9.18 Sugar uptake by *Neurospora crassa* involves symport with protons. (*a*) Glucose uptake by control and by carbon-starved organisms. (*b*) Addition of 1 m*M* glucose to carbon-starved organisms depolarized the membrane; there was little effect on control cells. (*c*) Protons enter together with sugar. Alkalinization was particularly pronounced when 2-deoxyglucose (2-DG) was added. In the case of glucose, alkalinization was soon overwhelmed by the production of metabolic acid. (After Slayman and Slayman, 1974.)

recovery; no such effect is seen with unstarved organisms, which lack the glucose porter. Concurrently the external pH shifts transiently toward more alkaline values. The latter effect is most pronounced with substrates such as 2-deoxyglucose and 3-O-methylglucose since these, unlike glucose itself, are not catabolized to acidic products. The data point to a simultaneous influx of sugar and protons, with a likely stoichiometry of $1:1$. It may seem puzzling that although sugar uptake continues at a constant rate for several minutes, both depolarization and alkalinization peak within seconds. The reason is that entry of sugar together with a proton stimulates the proton pump; ejection of protons acidifies the medium and repolarizes the membrane. These secondary complications can be avoided by the use of cells whose metabolism has been altogether shut down. For example, in yeast cells given inhibitors of both respiration and glycolysis, uptake of sugars and amino acids is accompanied by the entry of an equivalent amount of protons; exit of K^+ ions from the cells by a separate pathway provides both charge compensation and an electrochemical driving force.

The reader may recall that some bacteria take up sugars by vectorial phosphorylation (Chapter 5), and analogous pathways have been proposed for yeast. Galactose, for example, appears in yeast cytoplasm largely in the form of galactose phosphates; however, the balance of the evidence indicates that this is due not to vectorial phosphorylation but to galactose uniport followed by phosphorylation of the cytoplasmic sugar. A proposal that amino acid uptake is mediated by a group translocation pathway called the γ-glutamyl cycle has been tested and found to be inapplicable to yeast. Future research may reveal actual instances of transport by group translocation, but at present we see only a parade of porters.

With the accretion of further data, the pattern of transport by protists and plants will likely become more complex. For instance, little is presently known about the movements of K^+ and Na^+. Several authors have put forward simple models based on electrogenic K^+ uniport and electroneutral Na^+-H^+ antiport, but there has been no systematic effort to test these models critically. Experience with bacteria suggests that the capacity for K^+ accumulation may often exceed the driving force provided by the membrane potential and that mechanisms more sophisticated than K^+ uniport will have to be invoked. The same may prove true for the accumulation of P_i, a critical metabolite that is often the limiting nutrient in natural waters; it seems to me unlikely that secondary symport with protons is the only means of phosphate uptake. We know next to nothing about the extrusion of calcium ions from the cytoplasm, a process as vital to protists as to animals. And it may be well not to discount recent evidence suggesting electron transport across the plasma membrane (Goldenberg, 1982; Crane et al., 1982), by unknown routes and to an uncertain purpose.

Proton Circulation and Cell Physiology: Neurospora.   It is considerably easier to describe individual pumps and porters than to understand how they collaborate in the execution of physiological functions. This section is intended to illustrate the sophistication and complexity of real chemiosmotic systems with the aid of examples drawn from the extensive researches with *Neurospora* and the giant-celled algae.

The H^+ ATPase of the *Neurospora* plasma membrane consumes between a quarter and a third of the cell's ATP yield, expelling protons from the cytoplasm at a rate of some 250 peq/s·cm² surface area (25 $\mu A/cm^2$). Net proton extrusion requires other ions to flow concurrently in order to compensate for the displacement of electric charge. The measured rate of acid production by the cells is often much less than that of proton extrusion; under such conditions, most of the proton current must return to the cytoplasm across the plasma membrane. Leakage of protons as such is minimal; the only transport processes known to operate at the requisite rate are the expulsion of Na^+ and the uptake of K^+, and these may return as much as 80 percent of the proton current by symport and antiport carriers (Slayman, 1980). Unfortunately the mechanisms of Na^+ and K^+ transport by *Neurospora* are not well understood. Uptake of cations (generally K^+, but Na^+ can take its place in K^+-starved cells) will be coupled to the accumulation of an equivalent amount of anions, either in the form of metabolic acids or by uptake from the medium ($2H^+$-Cl^- symport, for instance). The net result is that the fungus tends to accumulate salts, a key aspect of the maintenance of turgor.

Fungi are saprophytes, and the accumulation of catabolic substrates as well as sources of nitrogen, phosphorus, and sulfur from a dilute solution is essential to their way of life. About a fifth of the proton current returns across the plasma membrane by way of these porters. For example, hyphae are depolarized on addition of glucose, various amino acids, and especially NH_4^+ ions because of the sudden influx of positive charge (Figure 9.18). Note, however, that the initial depolarization is soon followed by partial recovery of the electric potential, suggesting that the organism has the means to stabilize $\Delta\Psi$ (or, more likely, $\Delta\tilde{\mu}_{H^+}$). One mechanism is probably the trans inhibition of metabolite porters: as metabolite concentrations build up in the cytoplasm, they tend to inhibit their porters, reducing the drain on the proton current and allowing the membrane to repolarize. Kinetic regulation of porters is also evident when the organism grows in the presence of high nutrient concentrations. Under these conditions, the cytoplasmic levels of sugars and amino acids are far lower than expected for simple porters that merely allow their substrates to equilibriate with $\Delta\tilde{\mu}_{H^+}$.

The network of regulatory interactions that ensure the stability of the protonic potential can also be seen in *Neurospora*'s response to energy limitation. If porters were to continue at full tilt when the ATP supply diminishes, a catastrophic sequence of events would ensue: the membrane would depolarize, the ATP pool would be squandered in an effort to repolarize, and leakage of ions would be followed by loss of turgor with no hope of recovery. In reality, none of this happens. In response to a metabolic downshift, the H^+ ATPase shifts to an alternative mode in which the H^+/ATP ratio is 2 rather than 1, and all transport systems are throttled down. Within minutes the membrane potential returns to normal and so does the proportion of adenine nucleotides in the pool. The object of this regulatory exercise is to ensure that despite diminution of the *quantity* of energy available for use, its *quality* (that is, its potential) is conserved by reducing consumption. What is not clear is how the fungus puts this admirable prescription into effect. The cell seems to monitor its energy status and to signal a shortfall simultaneously to both pumps and porters. There is evidence that cyclic AMP is part of the mechanism, but perhaps not all of it (Slayman, 1980; Pall et al., 1981).

Fungi, algae, and even higher plants must be prepared to cope with changes in the external pH to a greater extent than do animal cells in their sheltered habitat. Figure 9.19 documents this capacity for a representative selection of walled eukaryotes. Constancy of the cytoplasmic pH, while not unexpected, is remarkable precisely because proton fluxes play so large a role in the operations of these organisms. Regrettably, it is not at all clear how the cytoplasmic pH is stabilized near neutrality. Intracellular buffering is quite inadequate to cope with the acid generated during metabolism. *Neurospora* can produce acid at a rate greater than 100 peq/sc·cm² surface area; given the known dimensions of the hyphae and their buffering capacity, one can calculate that the cytoplasmic pH ought to drop at a rate of 0.6 unit per minute. The observed stability of the internal pH must be attributed to the net extrusion of protons from the cell, to the regulation of metabolic pathways that generate and consume protons, or to a combination of these processes (Smith and Raven, 1979).

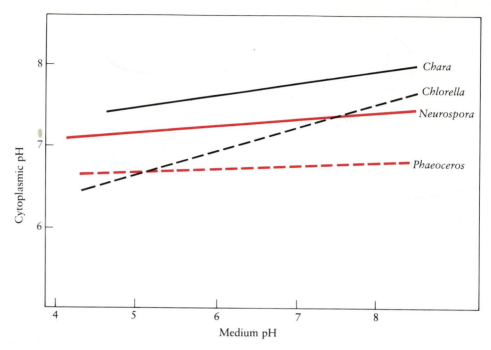

FIGURE 9.19   The constancy of cytoplasmic pH in a selection of walled eukaryotic cells. *Phaeoceros* is the hornwort, a higher plant. (After Smith and Raven, 1979, and Sanders and Slayman, 1982, with permission of Annual Reviews.)

Expulsion of protons by the electrogenic H^+ ATPase will not in itself raise the cytoplasmic pH because the protons will promptly flow back down the electrochemical gradient unless some other ion flux provides charge compensation. This apparently is just what happens. Sanders et al. (1981) reported that acidification of the *Neurospora* cytoplasm (by addition of butyric acid, a permeant weak acid, to the medium) was followed by accelerated proton extrusion and then by an increase in the conductance of the plasma membrane. Neither K^+ nor Na^+ fluxes compensate for proton extrusion; the authors suggest that metabolic anions (succinate, for example, which *Neurospora* produces copiously) may accompany the protons and allow the cytoplasmic pH to rise. In a subsequent study Sanders and Slayman (1982) found that even the H^+ ATPase may be dispensable; vanadate, a powerful inhibitor of the proton pump, did not prevent reextrusion of the protons that had been carried into the cytoplasm by butyric acid. This is an unexpected finding, which points back to the regulation of catabolic pathways as an important aspect of pH stabilization. Some sort of pH sensor would also seem to be required, but at this time we have no information about it.

Algae.   We turn now to the giant-celled green algae, which, for present purposes, may serve as representatives of the plant world. *Chara corallina,* illustrated in Figure 9.17, is a multicellular organism composed of a small number of enormous multinucleate cells. The great bulk of each internodal cell is occupied by the central vacuole, which contains a solution of inorganic salts, normally 0.1 M KCl (NaCl when the medium contains insufficient K^+). The vacuolar sap serves an important function: its high osmotic pressure ensures the influx of water, maintains the turgor of each cell and the rigidity of the plant as a whole, and supplies the driving force for surface expansion during growth. The cytoplasm is no more than a thin film some 10 μm deep, much of it occupied by chloroplasts. The cytoplasm streams incessantly, up one side of the cell and down the other, a matter that will be taken up in Chapter 12. Our concern now is with the integration of transport processes across cellular membranes (see Walker, 1980; Spanswick, 1981).

In *Chara*, as in higher plants, there are two major membrane systems: the plasma membrane and the tonoplast, which bounds the vacuole. Most of the information at hand pertains to the plasma membrane. Under normal conditions, with Ca^{2+} ions present, the cells maintain a membrane potential near -175 mV with the help of an H^+ ATPase. *Chara* grows photosynthetically in dilute salt medium and has no need for porters that catalyze the uptake of organic metabolites. Instead, the central role of the protonic potential is the accumulation of inorganic nutrients: NH_4^+ (carried by a particularly powerful porter), P_i, and the chloride and potassium ions that sustain turgor. There is evidence for a $2H^+$-Cl^- symporter and for uptake of K^+ by uniport; the cytoplasmic K^+ content, about 0.1 M, is not far from electrochemical equilibrium. Both K^+ and Cl^- end up largely in the central vacuole, via ill-defined porters presumably energized by a vacuolar H^+ ATPase (Figure 9.17).

Conventional enough, so far. But it has long been known that *Chara*, like other giant-celled algae, is at least mildly excitable: injection of current elicits a slow action potential that may be propagated from the excited cell to its neighbors. The physical basis of the action potential is complex. Most of the electric current entering the cell is carried by the efflux of chloride anions, but part can be assigned to an early influx of Ca^{2+}; the data suggest that Ca^{2+} influx triggers the opening of chloride channels. Each potential spike reflects the loss of a substantial amount of chloride, accompanied by K^+ or Na^+ ions, so that the turgor pressure falls by about 0.3 percent. This holds the clue to the function of the action potential: it serves to relieve excess turgor by a brief "puff" of KCl. Indeed, action potentials can be elicited experimentally by increasing the turgor pressure (and also by reducing it, an observation not readily rationalized). There is now considerable evidence that both freshwater and marine algae respond to osmotic stress by calcium-controlled emission of KCl pulses. These are signaled electrically by depolarization events, here called action potentials. Figure 9.20 illustrates the relationship for one such alga, *Acetabularia* (Hope and Walker, 1975; Nuccitelli and Jaffe, 1976; Lunevsky et al., 1983; Wendler et al., 1983).

Just how the algae sense their turgor pressure remains unclear. An appealing notion is that ion channels, or an associated receptor protein, are sensitive to

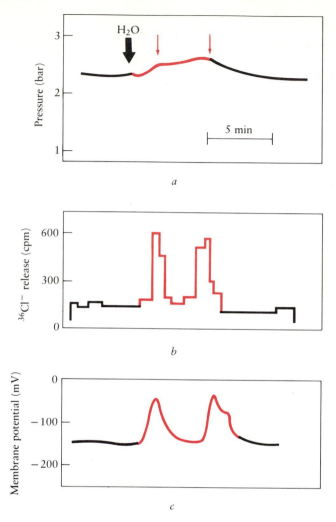

FIGURE 9.20   Turgor regulation and action potentials in *Acetabularia*. The alga is placed in a chamber continuously perfused with fresh sea water. Trace *a*, turgor. Turgor is measured with an intracellular pressure probe. Dilution of the medium with water causes the turgor pressure to increase; the arrows indicate two regulatory episodes that restore the initial turgor. Trace *b*, release of $^{36}Cl^-$ from the alga coincides with the regulatory episodes and also with the action potentials shown in trace *c*. (After Wendler et al., 1983, with permission of Springer-Verlag.)

mechanical deformation: excessive turgor, pressing the membrane into the wall, might elicit opening of the channel (Bisson and Gutknecht, 1980; see also Zimmermann and Steudle, 1980, for an alternative view). Genetic evidence for the existence of a specialized turgor sensor would be welcome, but hard to come by.

The interplay of the proton pump with a set of specialized porters underlies another curious feature of algal cell physiology. Botanists noticed long ago that the cells are often encrusted with annular deposits of calcium carbonate, generating alternating stripes of white and green. The white bands indicate regions where the extracellular pH is alkaline, pH 10 or more, encouraging the deposition of lime; green bands correspond to acidic regions. Figure 9.21*a* shows the bands as displayed when a fine-tipped pH microelectrode is passed along the surface of an internodal cell: adjacent bands differ by as much as three pH units. How does this spatial differentiation arise, and what purpose does it serve?

The pH bands are seen only in the light and when bicarbonate is present, and they have to do with the acquisition of carbon by photosynthesizing cells. *Chara*, like many other plants and algae, grows at pH 8 or above, even though the amount of free CO_2 available at this pH is too low to support growth; the organism makes use of its proton pump to circumvent the CO_2 deficiency. For some years now there has been vigorous (if often obscure) debate about the mechanism by which this is accomplished. According to Walker and his

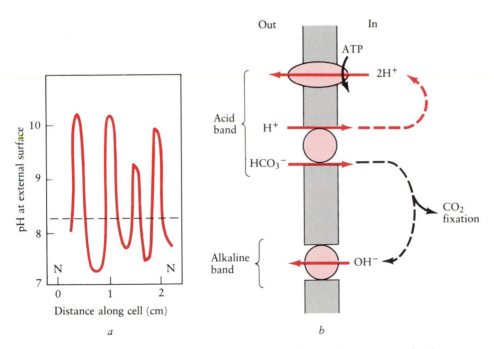

FIGURE 9.21   Bands of alkaline pH at the surface of *Chara* cells. (*a*) Longitudinal variation of the pH along an internodal cell; N designates the nodes, and the broken line shows the pH of the bulk medium. The external pH was determined by passing a miniaturized pH microelectrode along the cell's surface. (After Lucas et al., 1983.) (*b*) Chemiosmotic interpretation of the pH bands. (After Lucas, 1983, and Lucas et al., 1983; reproduced with permission of Springer-Verlag.)

colleagues (Walker, 1980), extrusion of protons by the H^+ ATPase lowers the pH in the unstirred layer adjacent to the cell surface; HCO_3^- is locally converted into CO_2, whose concentration becomes high enough to support photosynthesis by passive diffusion across the plasma membrane. By contrast, Lucas (1983) envisages uptake of bicarbonate ion as such, probably by secondary symport with protons. Both mechanisms require a compensatory flux of H^+ into the cell or of OH^- out (the reason for Lucas's mechanism is shown in Figure 9.21b). The unique aspect of $Chara$ is that the two ion fluxes are segregated into distinct zones. In the acid bands, protons are extruded and carbon is absorbed; compensatory OH^- efflux (or H^+ influx) takes place in the alkaline band. Localization of the fluxes is required so that proton concentrations in the acidic zone can be high enough to do their job.

There may be a second reason for $Chara$ to produce acid bands; an acidic pH in the cell wall itself appears to be a prerequisite for cell extension, perhaps by helping to "soften" the wall. Stimulants of plant growth, such as auxins and the mycotoxin fusicoccin, are thought to exert their effects in part at least by stimulating the proton pump (Marré, 1979). But this is another complex topic that cannot be explored here.

The Calcium Circulation

All animal cells, and probably all eukaryotic cells, expend metabolic energy in order to maintain the cytosolic calcium level as much as four orders of magnitude below that of the extracellular fluid. External calcium concentrations are in the millimolar range (pond water, 0.5 to 1 mM; blood plasma, 3 mM; sea water, 10 mM); by contrast, intracellular levels of free Ca^{2+} ion fall near or below 1 μM. Thus there is a sharp gradient of electrochemical potential driving calcium ions inward across the plasma membrane, sustained partly by the intrinsic impermeability of the membrane to calcium and partly by active calcium extrusion.

The expense of expelling Ca^{2+} ions is one that neither eukaryotes nor prokaryotes can avoid. Given that a membrane potential, cytoplasm negative, is a universal feature of cells, Ca^{2+} will inevitably leak into the cytoplasm and interfere with orderly metabolism by precipitating phosphorylated compounds. However, eukaryotes make a virtue of necessity, putting the calcium gradient to work in processing signals from the external environment. Appropriate stimuli—chemical, electrical, even mechanical—have been found to elicit a sudden rise in the cytoplasmic calcium level, either by opening calcium channels in the plasma membrane or by triggering the release of Ca^{2+} ions from some intracellular reservoir. The rise in cytoplasmic calcium concentration, in turn, is the signal for responses such as contraction, secretion, or changes in behavior (Figure 9.22; Carafoli and Crompton, 1978). The general topic of information processing, including the role of calcium therein, will be taken up in Chapter 13. Our concern here is with the nature of the transport systems that pump Ca^{2+} out and let it flow back in a controlled manner.

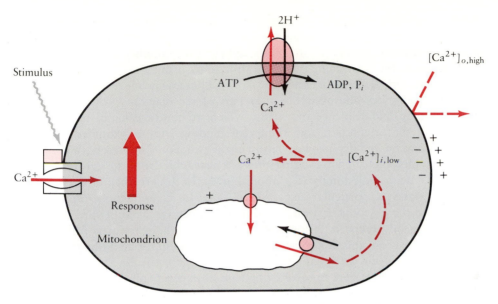

FIGURE 9.22   Role of the calcium circulation in transmembrane signaling. A calcium pump expels Ca^{2+} from the cytoplasm, generating an electrochemical potential gradient. The plasma membrane is intrinsically almost impermeable to Ca^{2+}. However, in response to a particular stimulus a gated Ca^{2+} channel opens transiently; the rise in cytosolic Ca^{2+} triggers the response, and the status quo ante is restored by the pump.

The cytosolic calcium level is set by the interplay of several concurrent processes, one of which is the sequestration and controlled release of Ca^{2+} ions by mitochondria (Chapter 7) and by other internal organelles. Buffering by absorption is probably the key to calcium homeostasis in the short term, but must be supplemented in the longer run by expulsion of Ca^{2+} ions across the plasma membrane. Eukaryotic cells, like bacteria (Chapter 5), employ both primary and secondary transport systems for this purpose. In squid axons, for example, a major pathway for Ca^{2+} extrusion is electrogenic antiport for sodium, with $\Delta\tilde{\mu}_{Na^+}$ as the driving force. There is an analogous Ca^{2+}-H^+ antiporter in the *Neurospora* plasma membrane, although its function is uncertain. Interest has come to center on an array of calcium-translocating ATPases that occur widely, perhaps universally, in the plasma membrane of animal cells.

The Calcium Pump.   In erythrocytes, as in other animal cells, the cytoplasmic calcium level (1 μM or less) is far below that of the blood plasma. The calcium gradient can be ascribed partly to the extraordinary impermeability of the erythrocyte membrane, but ultimately it depends on an ATP-driven calcium pump. The enzyme responsible is an ATPase activated by Ca^{2+} plus Mg^{2+} ions, first reported in 1961 by E. T. Dunham and I. M. Glynn. The enzyme is ouabain-resistant and distinct from the Na^+-K^+ ATPase. Cytosolic Ca^{2+} ions stimulate

ATP hydrolysis but external ones do not; evidently, ATP hydrolysis and calcium transport are closely and necessarily linked. The calcium ATPase of erythrocytes has been studied intensively and is the subject of a number of recent reviews (Schatzmann, 1975; Sarkadi, 1980; Carafoli and Zurini, 1982).

The calcium ATPase, like the Na^+-K^+ ATPase, which it resembles in many respects, undergoes phosphorylation as part of the reaction cycle. Figure 9.23 shows a plausible sequence of events, which attributes calcium transport to changes in the orientation and affinity of binding sites coordinated with the phosphorylation and dephosphorylation of the protein. E_1 represents a configuration in which the calcium-binding sites are accessible from the cytoplasm and bind Ca^{2+} with high affinity (K_m about 2 μM). In the presence of Ca^{2+}, ATP phosphorylates a particular aspartyl residue. In consequence, the enzyme switches to the E_2 configuration; the binding sites now face outward and their affinity is greatly reduced. Release of Ca^{2+} ions to the exterior is followed by dephosphorylation and restoration of the E_1 configuration. Note the virtual identity of this cycle with that shown in Figure 9.9 for the Na^+-K^+ ATPase.

It is not possible here to review the mass of experimental data that buttress this scheme or to discuss the uncertainties that remain, but two of the latter must be pointed out. First, it is not certain whether the erythrocyte ATPase transports

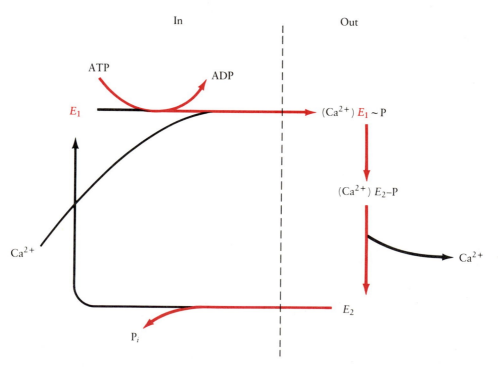

FIGURE 9.23 Major steps in calcium transport by the ATP-driven calcium pump. The conventions are the same as in Figure 9.9.

one calcium ion per cycle or two. Second, only very recently has consensus been achieved that calcium is extruded by electroneutral exchange for protons (Niggli et al., 1982; Smallwood et al., 1983). In any event, ATP hydrolysis provides enough energy to sustain the calcium gradients (up to 10^4) that have been observed experimentally.

The entire catalytic cycle is effected by a single protein, about 145 kdal, which has been purified and reconstituted into liposomes. The rate of cycling and the calcium affinity are greatly enhanced by calmodulin, a ubiquitous calcium-binding regulatory protein that plays an important role in many calcium-regulated responses; its role here is clearly to stimulate the pump in response to calcium influx. This is important to the maintenance of erythrocyte shape and therefore to their usefulness; erythrocytes contain a Ca^{2+}-activated K^+ channel that opens in response to a rise in cytosolic Ca^{2+}, and the ensuing K^+ efflux causes them to shrink and become malformed.

The erythrocyte Ca^{2+} ATPase may be taken as representative of a growing class of plasma membrane enzymes: ascites tumor cells, neutrophils, macrophages, and hepatocytes all possess high-affinity ATPases whose manifest function is to extrude Ca^{2+} from the cytoplasm. However, it is the Ca^{2+} ATPase of the sarcoplasmic reticulum that has drawn the most attention. The sarcoplasmic reticulum is an intracellular network of membranous sacs that controls the activity of skeletal muscle. In relaxed muscle, most of the Ca^{2+} is sequestered thanks to the Ca^{2+} ATPase; discharge of Ca^{2+} in response to a nerve impulse triggers contraction (Chapter 11). The calcium ATPase makes up 80 to 90 percent of the protein complement of sarcoplasmic vesicles and is the object of choice for research into the molecular basis of calcium transport (for reviews see Berman, 1982; Ikemoto, 1982). For present purposes, suffice it to emphasize that the reaction catalyzed by the sarcoplasmic ATPase is essentially identical with that described above, though there may be differences in detail. Most investigators agree on a stoichiometry of $2Ca^{2+}/ATP$; the process is electroneutral, but the counterion is still uncertain.

What about plants, algae, fungi, or protozoa? Scattered reports suggest the presence of calcium ATPases in several of these, but as matters stand we do not know how the great majority of living organisms extrude calcium. Some of the intense effort presently focused on animal systems might be profitably redirected.

Calcium Channels.   There are dozens, possibly hundreds of physiological responses that are known or believed to be triggered by a sudden rise in the cytosolic calcium level (Carafoli and Crompton, 1978; Cheung, 1980). The many ways in which Ca^{2+} ions enter into the physiology of eukaryotes will be taken up elsewhere, but it should be noted that nothing comparable is known from the prokaryotic world: the use of calcium fluxes to transmit signals appears to be a hallmark of eukaryotes.

In general, we know little about the transport systems involved other than that calcium flows down the electrochemical gradient into the cell. There is no a priori reason why calcium must always pass through a channel, and in some cases antiport of Ca^{2+} for Na^+ may be the mechanism of calcium entry. However, there are a number of cases in which calcium does clearly flow through receptor-controlled channels, which are recognizable by their high rate of calcium conduction, and these serve as paradigms for the subject as a whole.

Extensive data come from work with muscles of the giant barnacle, whose fibers may be four centimeters in length and two millimeters in diameter. Action potentials, elicited by depolarization under rather special conditions, do not involve Na^+ fluxes and are not inhibited by tetrodotoxin (a characteristic inhibitor of sodium channels). Instead, they require extracellular Ca^{2+} and are inhibited by La^{3+}, Co^{2+}, and Mn^{2+}. The action potential results from a transient, massive influx of Ca^{2+} ions through a voltage-sensitive selective channel that admits only Ca^{2+}, Sr^{2+}, and Ba^{2+}; both Na^+ and K^+ are excluded. Analogous calcium action potentials and calcium channels have been described in various animal cells including invertebrate as well as vertebrate neurons and muscle, various eggs, and even lymphocytes. All are voltage-sensitive and some are known to be modulated by regulatory metabolites such as adrenalin, acetylcholine, and cyclic AMP. Selectivity varies somewhat, suggesting the existence of an array of such channels, each with its own function and control elements. The recent introduction of the patch-clamping technique made it possible to record from single channels; the finding that some 10^6 Ca^{2+} ions are transported per second leaves no doubt that these are channels, not carriers. Reviews by Hagiwara and Byerly (1981), Kostyuk (1981), and Reuter (1983) cover the field.

More often than not the physiological role of the voltage-sensitive calcium channels is unknown, but some have been clearly identified. The most extensive research has been done on the role of Ca^{2+} ions in synaptic transmission. Nerve impulses are conducted across a synaptic cleft by chemical neurotransmitters, acetylcholine for example; this is stored in granules at the presynaptic terminal and released by exocytosis on arrival of the nerve impulse. The hypothesis that a localized influx of Ca^{2+} ions is the trigger for neurotransmitter release was formulated by B. Katz, A. R. Martin, and their colleagues 30 years ago and has been amply confirmed. Supporting evidence includes the following: removal of Ca^{2+} prevents transmitter release; injection of Ca^{2+} ions into the terminal elicits release, even in the absence of electric impulses; activation of the terminal is accompanied by influx of charge and of Ca^{2+} ions; a rise in the cytosolic Ca^{2+} level occurs as well; and Ca^{2+} ionophores, such as A23187, elicit transmitter release. Apparently, then, arrival of the depolarizing nerve impulse at the presynaptic terminal causes the Ca^{2+} channels to open briefly, allowing Ca^{2+} ions to flow in; this in turn triggers exocytosis (Aidley, 1978; Åkerman and Nicholls, 1983). Other examples of exocytosis in which calcium influx is more or less firmly implicated will be mentioned in Chapters 12 and 13.

Outside the animal world, the best-characterized calcium channel is certainly that which mediates behavioral responses in the ciliated protozoan *Paramecium*. We shall discuss this system in context in Chapter 11. Suffice it here to mention that the ciliary membrane contains voltage-sensitive calcium channels that are transiently activated when the organism collides with an obstacle; the influx of Ca^{2+} causes the direction of ciliary beating to reverse, and the cell to back up. The calcium channel generally resembles that of animal cells (it selects for Ca^{2+}, Sr^{2+}, and Ba^{2+}, for example) but is unaffected by inhibitors such as verapamil, which characteristically block the Ca^{2+} channels of muscle and nerve (Eckert, 1972; Eckert and Brehm, 1979).

A Diversity of Patterns

The preceding sections surveyed in some detail the major ionic networks of eukaryotic plasma membranes: the sodium circulation of animal cells and the proton circulation of plants, algae, and fungi. These two represent different ways of life and must reflect a major phylogenetic cleavage. The existence of others may be inferred from the fact that the sodium and proton circulations are not the only bioenergetic patterns found among the lower eukaryotes. Let me then conclude this chapter with a plea for the protoctists.

Margulis and Schwartz (1982), in their charming atlas of living things, distinguish no fewer than 27 phyla in the kingdom *Protoctista*. Amoebae, ciliates, slime molds, and flagellated algae are reasonably familiar, but few readers will recognize the water molds, and at least a dozen phyla have no common representatives at all. Out of this multitude of creatures barely a handful have been examined from the viewpoint of membrane transport. The data are fragmentary where they exist at all, but it is plain that this kingdom displays a far greater diversity of patterns than do the higher ones and would repay a little attention.

Among the green algae (phylum Chlorophyta) we find *Chlorella* and also the giant algae *Chara* and *Nitella*; these and several other green algae are known to generate a proton circulation by means of an H^+ ATPase. *Acetabularia*, a large unicellular marine alga that has been extensively used for research on morphogenesis and on nucleocytoplasmic relationships, is different. There appears to be no proton pump; instead, the membrane potential induced by light has been attributed to an electrogenic chloride pump whose energy donor remains unidentified. These algae also exhibit spontaneous action potentials due to sudden, transient increases in chloride permeability. These two antagonistic transport pathways probably have to do with the maintenance of turgor. The organism pumps chloride anion into the cytoplasm, accompanied by a passive influx of K^+, thereby increasing the turgor. Excess pressure is relieved by the efflux of chloride with each action potential, again accompanied by K^+ (Gradmann, 1976; Mummert and Gradmann, 1976; Wendler et al., 1983).

The water molds look like fungi and have traditionally been claimed by the mycologists, but there are good reasons for classifying them among the protoctists instead (Margulis and Schwartz, 1982). Only *Blastocladiella emersonii* has been examined so far. There is no evidence for an electrogenic proton pump; the membrane potential can be entirely accounted for by passive diffusion of K^+

out of the cells through a carrier or channel that admits only K^+ and Rb^+. Accumulation of K^+ is mediated by a primary K^+ transport system that apparently exchanges K^+ for H^+; the immediate energy donor is unknown (Van Brunt et al., 1982). The object of the exercise is probably to generate a protonic potential ($\Delta\Psi$, ΔpH, or both) which supports the uptake of phosphate and of amino acids by the rootlike rhizoidal filaments (Harold and Harold, 1980; Kropf and Harold, 1982). Another unexpected feature of these organisms is an action potential (Figure 9.6), mediated by a highly selective, voltage-sensitive calcium channel (Caldwell et al., 1985); like the action potential of algae, it may be involved in osmotic control.

Protozoa come in a bewildering variety of forms and lifestyles, but their habits are those of animals: instead of absorbing nutrients from solution, they engulf their prey and digest it inside the cell. The dominant kind of membrane transport is phagocytosis, which is better regarded as an aspect of cell motility. But ions do cross the membrane. *Paramecium,* for example, maintains a resting potential of $-40\,mV$ with the aid of a calcium-sensitive K^+ channel. The calcium action potential, through which *Paramecium* controls its swimming behavior, has been brilliantly explored by Roger Eckert and his students (Chapter 11). Surprisingly, nothing whatever is known about the primary transport systems for K^+ and Ca^{2+}, which underlie this sophisticated example of excitation and response.

The lower eukaryotes are a neglected field and doubtless will remain so. Yet many pathogens of plants and animals are protoctists, and the dissection of their ionic relations may prove rewarding in more senses than one. The evolutionary precursors of both plants and animals must reside in this remote kingdom; some fortunate explorer may stumble on the Na^+-K^+ ATPase and thus shed new light on the origin of animals, and of humans.

References

AIDLEY, D. J. (1978). *The Physiology of Excitable Cells*, 2d ed., Cambridge University Press, Cambridge.

ÅKERMAN, K. E. O., and NICHOLLS, D. G. (1983). Ca^{2+} transport and the regulation of transmitter release in isolated nerve endings. *Trends in Biochemical Sciences* 8:63–64.

AMORY, A., and GOFFEAU, A. (1982). Characterization of the β-aspartyl phosphate intermediate formed by the H^+-translocating ATPase from the yeast *Schizosaccharomyces pombe. Journal of Biological Chemistry* 257:4723–4730.

BERMAN, M. C. (1982). Energy coupling and uncoupling of active calcium transport by sarcoplasmic reticulum membranes. *Biochimica et Biophysica Acta* 694:95–121.

BISSON, M., and GUTKNECHT, J. (1980). Osmotic regulation in algae, in *Plant Membrane Transport: Current Conceptual Issues,* R. M. Spanswick, W. J. Lucas, and J. Dainty, Eds. Elsevier–North Holland, Amsterdam, pp. 131–142.

BORON, W. F. (1983). Transport of H$^+$ and of ionic weak acids and bases. *Journal of Membrane Biology* 72:1–16.

BOWMAN, E. J., and BOWMAN, B. J. (1982). Identification and properties of an ATPase in vacuolar membranes of *Neurospora crassa*. *Journal of Bacteriology* 151: 1326–1337.

BRISKIN, D. P., and LEONARD, R. T. (1982). Partial characterization of a phosphorylated intermediate associated with the plasma membrane ATPase of corn roots. *Proceedings of the National Academy of Sciences USA* 79:6922–6926.

BURNS, C. P., and ROZENGURT, E. (1984). Extracellular Na$^+$ and initiation of DNA synthesis: Role of intracellular pH and K$^+$. *Journal of Cell Biology* 98:1082–1089.

CALDWELL, J. H., VAN BRUNT, J., and HAROLD, F. M. (1985). Calcium-dependent anion channel in the water mold *Blastocladiella emersonii*. *Journal of Membrane Biology*, in press.

CANTLEY, L. C. (1981). Structure and mechanism of the (Na, K)-ATPase. *Current Topics in Bioenergetics* 11:201–237.

CARAFOLI, E., and CROMPTON, M. (1978). The regulation of intracellular calcium. *Current Topics in Membranes and Transport* 10:151–216.

CARAFOLI, E., and ZURINI, M. (1982). The Ca^{2+}-pumping ATPase of plasma membranes: purification, reconstitution and properties. *Biochimica et Biophysica Acta* 683:279–301.

CHENG, L., and SACKTOR, B. (1981). Sodium gradient-dependent phosphate transport in renal brush border membrane vesicles. *Journal of Biological Chemistry* 256:1556–1564.

CHEUNG, W. Y. (1980). Calmodulin plays a pivotal role in cellular regulation. *Science* 207:19–27.

CHRISTENSEN, H. N. (1975). *Biological Transport*, 2d ed. Benjamin, London.

CHRISTENSEN, H. N., and RIGGS, T. R. (1952). Concentrative uptake of amino acids by the Ehrlich mouse ascites carcinoma cell. *Journal of Biological Chemistry* 194:57–68.

CHURCHILL, K. A., and SZE, H. (1983). Anion-sensitive, H$^+$-pumping ATPase in membrane vesicles from oat roots. *Plant Physiology* 71:610–617.

CONWAY, E. J. (1953). A redox pump for the biological performance of osmotic work and its relation to the kinetics of free ion diffusion across membranes. *International Review of Cytology* 2:419–445.

CRAIG, W. S. (1982). Monomer of sodium and potassium ion activated adenosine-triphosphatase displays complete enzymatic function. *Biochemistry* 21:5707–5716.

CRANE, F. L., ROBERTS, H., LINNANE, A. W., and LÖW, H. (1982). Transmembrane ferricyanide reduction by cells of the yeast *Saccharomyces cerevisiae*. *Journal of Bioenergetics and Biomembranes* 14:191–204.

CRANE, R. K. (1965). Op. cit., Chapter 3.

CRANE, R. K. (1977). Op. cit., Chapter 3.

CRANE, R. K., MILLER, D., and BIHLER, I. (1961). Op. cit., Chapter 3.

DAHL, J. L., and HOKIN, L. E. (1974). The sodium, potassium adenosine triphosphatase. *Annual Review of Biochemistry* 43: 327–356.

DAME, J. B., and SCARBOROUGH, G. A. (1981). Identification of the phosphorylated intermediate of the *Neurospora* plasma membrane H$^+$-ATPase as β-aspartyl phosphate. *Journal of Biological Chemistry* 256:10724–10730.

DEVOE, R. D., and MALONEY, P. C. (1980). Principles of cell homeostasis, in *Medical Physiology*, 14th ed. V. B. Mountcastle, Ed., vol. 1. Mosby, St. Louis, pp. 3–45.

DUFOUR, J. P., GOFFEAU, A., and TSONG, T. Y. (1982). Active proton uptake in lipid vesicles reconstituted with the purified yeast plasma membrane ATPase. *Journal of Biological Chemistry* 257:9365–9371.

ECKERT, R. (1972). Bioelectric control of ciliary activity. *Science* **176**:473–481.

ECKERT, R., and BREHM, P. (1979). Ionic mechanisms of excitation in *Paramecium. Annual Review of Biophysics and Bioengineering* **8**:353–383.

ECKERT, R., and RANDALL, D. (1983). *Animal Physiology: Mechanisms and Adaptations.* W. H. Freeman and Co., New York.

EDDY, A. A. (1968). The effects of varying the cellular and extracellular concentrations of sodium and potassium ions on the uptake of glycine by mouse ascites-tumor cells in the presence and absence of sodium cyanide. *Biochemical Journal* **108**:489–498.

EDDY, A. A. (1982). Mechanisms of solute transport in selected eukaryotic microorganisms. *Advances in Microbial Physiology* **23**:1–78.

FELLE, H. (1983). Driving forces and current-voltage characteristics of amino acid transport in *Riccia fluitans. Biochimica et Biophysica Acta* **730**:342–350.

FELLE, H., and BENTRUP, F. W. (1980). Hexose transport and membrane depolarization in *Riccia fluitans. Planta* **147**:471–476.

FRELIN, C., and LAZDUNSKI, M. (1983). The amiloride-sensitive antiport in 3T3 fibroblasts: Characterization and stimulation by serum. *Journal of Biological Chemistry* **258**:6272–6276.

GECK, P., PIETRZYK, C., BURCKHARDT, B. C., PFEIFFER, B., and HEINZ, E. (1980). Electrically silent cotransport of Na^+, K^+ and Cl^- in Ehrlich cells. *Biochimica et Biophysica Acta* **600**:432–447.

GLYNN, I. M., and KARLISH, S. J. D. (1975). The sodium pump. *Annual Review of Physiology* **37**:13–55.

GOFFEAU, A., and SLAYMAN, C. W. (1981). The proton-translocating ATPase of the fungal plasma membrane. *Biochimica et Biophysica Acta* **639**:197–223.

GOLDENBERG, H. (1982). Plasma membrane redox activities. *Biochimica et Biophysica Acta* **694**:203–226.

GRADMANN, D. (1976). "Metabolic" action potentials in *Acetabularia. Journal of Membrane Biology* **29**:23–45.

GRADMANN, D., HANSEN, U. P., LONG, W. S., SLAYMAN, C. L., and WARNCKE, J. (1978). Current-voltage relationships for the plasma membrane and its principal electrogenic pump in *Neurospora crassa:* I. Steady-state conditions. *Journal of Membrane Biology* **39**:333–367.

HACKING, C., and EDDY, A. A. (1981). The accumulation of amino acids by mouse ascites-tumor cells. *Biochemical Journal* **194**:415–426.

HAGIWARA, S., and BYERLY, L. (1981). Calcium channels. *Annual Review of Neurosciences* **4**:69–125.

HAROLD, R. L., and HAROLD, F. M. (1980). Oriented growth of *Blastocladiella emersonii* in gradients of ionophores and inhibitors. *Journal of Bacteriology* **144**:1159–1167.

HEINZ, E., GECK, P., and PFEIFFER, B. (1980). Energetic problems of the transport of amino acids in Ehrlich cells. *Journal of Membrane Biology* **57**:91–94.

HODGKIN, A. L., and HOROWICZ, P. (1959). The influence of potassium and chloride ions on the membrane potential of single muscle fibers. *Journal of Physiology* **148**:127–160.

HÖFER, M. (1981). *Transport across Biological Membranes.* Pitman, London.

HOPE, A. B., and WALKER, N. A. (1975). *The Physiology of Giant Algal Cells.* Cambridge University Press, London.

IKEMOTO, N. (1982). Structure and function of the calcium pump protein of sarcoplasmic reticulum. *Annual Review of Physiology* **44**:297–317.

JARDETZKY, O. (1966). Op. cit., Chapter 3.

JØRGENSEN, P. L. (1982). Mechanism of the Na, K pump: Protein structure and conformations of the pure (Na$^+$ + K$^+$)-ATPase. *Biochimica et Biophysica Acta* **694**:27–68.

KAKINUMA, Y., OHSUMI, Y., and ANRAKU, Y. (1981). Properties of H$^+$-translocating adenosine triphosphatase in vacuolar membranes of *Saccharomyces cerevisae*. *Journal of Biological Chemistry* **256**:10859–10863.

KAPLAN, J. G. (1978). Membrane cation transport and the control of proliferation of mammalian cells. *Annual Review of Physiology* **40**:19–41.

KATZ, B. (1966). *Nerve, Muscle and Synapse*. McGraw-Hill, New York.

KLINGENBERG, M. (1981). Membrane protein oligomeric structure and transport function. *Nature* **290**:449–454.

KNAUF, P. A. (1979). Erythrocyte anion exchange and the Band 3 protein: Transport kinetics and molecular structure. *Current Topics in Membranes and Transport* **12**:251–363.

KOMOR, E., and TANNER, W. (1980). Proton-cotransport of sugar in plants, in *Plant Membrane Transport: Current Conceptual Issues*, R. M. Spanswick, W. J. Lucas, and J. Dainty, Eds. Elsevier–North Holland, Amsterdam, pp. 247–257.

KOSTYUK, P. G. (1981). Calcium channels in the neuronal membrane. *Biochimica et Biophysica Acta* **650**:128–150.

KREGENOW, F. M. (1981). Osmoregulatory salt transporting mechanisms: Control of cell volume in anisotonic media. *Annual Review of Physiology* **43**:493–505.

KROPF, D. L., and HAROLD, F. M. (1982). Selective transport of nutrients via the rhizoids of the water mold *Blastocladiella emersonii*. *Journal of Bacteriology* **151**:429–437.

KYTE, J. (1981). Molecular considerations relevant to the mechanism of active transport. *Nature* **292**:201–204.

LÄUGER, P. (1980). Kinetic properties of ion carriers and channels. *Journal of Membrane Biology* **57**:163–178.

LEFEVRE, P. G. (1975). The present state of the carrier hypothesis. *Current Topics in Membranes and Transport* **7**:109–215.

LUCAS, W. J. (1983). Photosynthetic assimilation of exogenous HCO$_3^-$ by aquatic plants. *Annual Review of Plant Physiology* **34**:71–104.

LUCAS, W. J., KEIFER, D. W., and SANDERS, D. (1983). Bicarbonate transport in *Chara corallina*: Evidence for cotransport of HCO$_3^-$ with H$^+$. *Journal of Membrane Biology* **73**:263–274.

LUNEVSKY, V. Z., ZHERELOVA, O. M., VOSTRIKOV, I. Y., and BERESTOVSKY, G. N. (1983). Excitation of *Characeae* cell membranes as a result of activation of calcium and chloride channels. *Journal of Membrane Biology* **72**:43–58.

MACKNIGHT, A. D. C., and LEAF, A. (1977). Regulation of cellular volume. *Physiological Reviews* **57**:510–573.

MARGULIS, L., and SCHWARTZ, K. V. (1982). *Five Kingdoms: An Illustrated Guide to the Phyla of Life on Earth*. W. H. Freeman and Co., New York.

MARRÉ, E. (1979). Fusicoccin: A tool in plant physiology. *Annual Review of Plant Physiology* **30**:273–288.

MARTONOSI, A. N., Ed. (1982). *Membranes and Transport*, vols. 1 and 2. Plenum, New York.

MCROBERTS, J. A., ERLINGER, S., RINDLER, M. J., and SAIER, M. H. (1982). Furosemide-sensitive salt transport in the Mardin-Darby canine kidney cell line. *Journal of Biological Chemistry* **257**:2260–2266.

MCROBERTS, J. A., TRAN, C. T., and SAIER, M. H., JR. (1983). Characterization of low potassium-resistant mutants of the Mardin-Darby canine kidney cell line with defects in NaCl/KCl symport. *Journal of Biological Chemistry* 258:12320–12326.

MITCHELL, P. (1967). Translocations through natural membranes. *Advances in Enzymology* 29:33–87.

MOOLENAAR, W. H., YARDEN, Y., DE LAAT, S. W., and SCHLESSINGER, J. (1982). Epidermal growth factor induces electrically silent Na$^+$ influx in human fibroblasts. *Journal of Biological Chemistry* 257:8502–8506.

MUMMERT, H., and GRADMANN, D. (1976). Voltage-dependent potassium fluxes and the significance of action potentials in *Acetabularia. Biochimica et Biophysica Acta* 443:443–450.

NIGGLI, V., SIGEL, E., and CARAFOLI, E. (1982). The purified Ca^{2+} pump of human erythrocyte membranes catalyzes an electroneutral Ca^{2+}/H$^+$ exchange in reconstituted liposomal systems. *Journal of Biological Chemistry* 257:2350–2356.

NUCCITELLI, R., and JAFFE, L. (1976). Current pulses involving chloride relieve excess pressure in *Pelvetia* embryos. *Planta* 131:315–320.

O'NEILL, S. D., and SPANSWICK, R. M. (1984). Characterization of native and reconstituted plasma membrane H$^+$-ATPase from the plasma membrane of *Beta vulgaris. Journal of Membrane Biology* 79:245–256.

PALL, M. L., TREVILLYAN, J. M., and HINMAN, N. (1981). Deficient cyclic adenosine 3′, 5′-monophosphate control in mutants of two genes of *Neurospora crassa. Molecular and Cellular Biology* 1:1–8.

PHILO, R. D., and EDDY, A. A. (1978). The membrane potential of mouse ascites tumor cells studied with the fluorescent probe 3, 3′-dipropyloxadicarbocyanine. *Biochemical Journal* 174:801–810.

POOLE, R. J. (1978). Energy coupling for membrane transport. *Annual Review of Plant Physiology* 29:437–460.

RAVEN, J. A. (1980). Nutrient transport in microalgae. *Advances in Microbial Physiology* 21:47–226.

REUTER, H. (1983). Calcium channel modulation by neurotransmitters, enzymes and drugs. *Nature* 301:569–574.

ROBINSON, J. D., and FLASHNER, M. S. (1979). The (Na$^+$ + K$^+$)-activated ATPase. *Biochimica et Biophysica Acta* 549:145–176.

ROOS, A., and BORON, W. F. (1981). Intracellular pH. *Physiological Reviews* 61:296–434.

ROSENBERG, T. (1954). Op. cit., Chapter 3.

ROZENGURT, E. (1980). Stimulation of DNA synthesis in quiescent cultured cells: Exogenous agents, internal signals and early events. *Current Topics in Cellular Regulation* 17:59–88.

SACHS, G., FALLER, L. D., and RABON, E. (1982). Proton/hydroxyl transport in gastric and intestinal epithelia. *Journal of Membrane Biology* 64:123–135.

SANDERS, D. (1980). The mechanism of Cl$^-$ transport at the plasma membrane of *Chara corallina.* I. Cotransport with H$^+$. *Journal of Membrane Biology* 53:129–141.

SANDERS, D. L, HANSEN, U.-P., and SLAYMAN, C. L. (1981). Role of the plasma membrane proton pump in pH regulation in non-animal cells. *Proceedings of the National Academy of Sciences USA* 78:5903–5907.

SANDERS, D. L., and SLAYMAN, C. L. (1982). Control of intracellular pH. Predominant role of oxidative metabolism, not proton transport, in the eukaryotic microorganism *Neurospora. Journal of General Physiology* 80:377–402.

SANDERS, D. L, SLAYMAN, C. L., and PALL, M. L. (1983). Stoichiometry of H$^+$/amino acid cotransport in *Neurospora crassa* revealed by current-voltage analysis. *Biochimica et Biophysica Acta* **735**:67–76.

SARKADI, B. (1980). Active calcium transport in human red cells. *Biochimica et Biophysica Acta* **604**:159–190.

SCARBOROUGH, G. A. (1976). The *Neurospora* plasma membrane ATPase is an electrogenic pump. *Proceedings of the National Academy of Sciences USA* **73**:1485–1488.

SCARBOROUGH, G. A. (1980). Proton translocation catalyzed by the electrogenic ATPase in the plasma membrane of *Neurospora*. *Biochemistry* **19**:2925–2931.

SCHATZMANN, H. J. (1975). Active calcium transport and Ca^{2+}-activated ATPase in human red cells. *Current Topics in Membranes and Transport* **6**:125–168.

SCHULDINER, S., and ROZENGURT, E. (1982). Na$^+$/H$^+$ antiport in Swiss 3T3 cells: Mitogenic stimulation leads to alkalinization. *Proceedings of the National Academy of Sciences USA* **79**:7778–7782.

SCHULTZ, S. G., and CURRAN, P. F. (1970). Coupled transport of sodium and organic solutes. *Physiological Reviews* **50**:637–718.

SHIMMEN, T., and TAZAWA, M. (1977). Control of membrane potential and excitability of *Chara* cells with ATP and Mg^{2+}. *Journal of Membrane Biology* **33**:167–192.

SKOU, J. C. (1957). The influence of some cations on an adenosine triphosphatase from peripheral nerves. *Biochimica et Biophysica Acta* **23**:394–401.

SKOU, J. C. (1965). Enzymatic basis for active transport of Na$^+$ and K$^+$ across cell membranes. *Physiological Reviews* **45**:596–617.

SLAYMAN, C. L. (1965a). Electrical properties of *Neurospora crassa*. Effects of external cations on the intracellular potential. *Journal of General Physiology* **49**:69–92.

SLAYMAN, C. L. (1965b). Electrical properties of *Neurospora crassa*. Respiration and the intracellular potential. *Journal of General Physiology* **49**:93–116.

SLAYMAN, C. L. (1980). Transport control phenomena in *Neurospora*, in *Plant Membrane Transport: Current Conceptual Issues*, R. M. Spanswick, W. J. Lucas, and J. Dainty, Eds. Elsevier–North Holland, Amsterdam, pp. 179–190.

SLAYMAN, C. L., Ed. (1982). *Electrogenic Ion Pumps, Current Topics in Membranes and Transport*, vol. 16. Academic, New York.

SLAYMAN, C. L., LONG, W. S., and LU, C. Y.-H. (1973). The relationship between ATP and an electrogenic pump in the plasma membrane of *Neurospora crassa*. *Journal of Membrane Biology* **14**:305–338.

SLAYMAN, C. L., and SLAYMAN, C. W. (1974). Depolarization of the plasma membrane of *Neurospora crassa* during active transport of glucose. *Proceedings of the National Academy of Sciences USA* **71**:1935–1939.

SMALLWOOD, J. I., WAISMAN, D. M., LAFRENIERE, D., and RASMUSSEN, H. (1983). Evidence that the erythrocyte calcium pump catalyzes Ca^{2+}:nH$^+$ exchange. *Journal of Biological Chemistry* **258**:11092–11097.

SMITH, F. A., and RAVEN, J. A. (1979). Intracellular pH and its regulation. *Annual Review of Plant Physiology* **30**:289–311.

SMITH, P. T., and WALKER, N. A. (1981). Studies on the perfused plasmalemma of *Chara corallina*. I. Current-voltage curves: ATP and potassium dependence. *Journal of Membrane Biology* **60**:223–236.

SMITH, R. L., MACARA, I. G., LEVENSON, R., HOUSMAN, D., and CANTLEY, L. (1982). Evidence that a Na$^+$/Ca^{2+} antiport system regulates murine erythroleukemia cell differentiation. *Journal of Biological Chemistry* **257**:773–780.

SPANSWICK, R. M. (1981). Electrogenic ion pumps. *Annual Review of Plant Physiology* **32**:267–289.

SPANSWICK, R. M., LUCAS, W. J., and DAINTY, J., Eds. (1980). *Plant Membrane Transport: Current Conceptual Issues.* Elsevier–North Holland Biomedical Press, Amsterdam.

STEINMETZ, P. R., and ANDERSEN, O. S. (1982). Electrogenic proton transport in epithelial membranes. *Journal of Membrane Biology* **65**:155–174.

THOMAS, R. C. (1972). Electrogenic sodium pump in nerve and muscle cells. *Physiological Reviews* **52**:563–594.

THOMAS, R. C. (1977). The role of bicarbonate, chloride and sodium ions in the regulation of intracellular pH in snail neurones. *Journal of Physiology* **273**:317–338.

THOMAS, R. C. (1982). Electrophysiology of the sodium pump in a snail neuron. *Current Topics in Membranes and Transport* **16**:3–16.

VAN BRUNT, J., CALDWELL, J. H., and HAROLD, F. M. (1982). Circulation of potassium across the plasma membrane of *Blastocladiella emersonii*: K^+ channel. *Journal of Bacteriology* **150**:1449–1460.

VARA, F., and SERRANO, R. (1982). Partial purification and properties of the proton-translocating ATPase of plant plasma membranes. *Journal of Biological Chemistry* **257**:12826–12830.

WALKER, N. A. (1980). The transport systems of charophyte and chlorophyte giant algae and their integration into modes of behavior in *Plant Membrane Transport: Current Conceptual Issues*, R. M. Spanswick, W. J. Lucas, and J. Dainty, Eds. Elsevier–North Holland, Amsterdam, pp. 287–300.

WENDLER, S., ZIMMERMANN, U., and BENTRUP, F.-W. (1983). Relationship between cell turgor pressure, electrical membrane potential and chloride efflux in *Acetabularia mediterranea*. *Journal of Membrane Biology* **72**: 75–84.

WHEELER, T. J., and HINKLE, P. C. (1981). Kinetic properties of the reconstituted glucose transporter from human erythrocytes. *Journal of Biological Chemistry* **256**:8907–8914.

WHITTAM, R., and AGER, M. E. (1964). Vectorial aspects of adenosine–triphosphatase activity in erythrocyte membranes. *Biochemical Journal* **93**:337–348.

WIDDAS, W. F. (1980). The asymmetry of the hexose transfer system in the human red cell membrane. *Current Topics in Membranes and Transport* **14**:165–223.

WILBRANDT, W., and ROSENBERG, T. (1961). The concept of carrier transport and its corollaries in pharmacology. *Pharmacological Reviews* **13**:109–183.

ZIMMERMANN, U., and STEUDLE, E. (1980). Fundamental water relations parameters, in *Plant Membrane Transport: Current Conceptual Issues*, R. M. Spanswick, W. J. Lucas, and J. Dainty, Eds. Elsevier–North Holland, Amsterdam, pp. 113–127.

10

Transport Mediators and Mechanisms

But the search for truth is only possible if we speak clearly and simply and avoid unnecessary technicalities and complications. In my view, aiming at simplicity and lucidity is a moral duty of all intellectuals; lack of clarity is a sin, and pretentiousness is a crime.

Karl Popper,
Objective Knowledge

· · ·

Ionophores
Macromolecules as Carriers and Channels
Primary Energy Coupling
Secondary Energy Coupling

A mechanistic understanding of biological transport turns on answers to straightforward questions: What are the pathways by which transport substrates cross the membrane? What determines the selectivity of transport? How do transport systems overcome the energy barrier imposed by the lipid phase? How is energy coupling effected in cases of active transport?

Transport systems are akin to enzymes in chemical constitution and catalytic behavior, but there are clear differences that set them apart. Enzymes catalyze chemical reactions; the interacting species exchange matter and the mechanism of catalysis can be expressed in chemical terms (Chapter 2). The essence of transport is movement of the substrate from one side of a barrier to the other, generally without chemical modification; active transport involves exchanges of free energy between interacting species, but no visible exchange of matter. No transport system is as fully understood in molecular terms as many enzymes are, but it is no longer quite fair to dismiss transport systems as so many black boxes. The general principles of transport across membranes are increasingly scrutable, even though explicit details await future research. The molecular basis of transport deserves a short chapter of its own, for this field is destined to burgeon as insights from novel physical and genetic techniques complement those from the traditional kinetic approach.

Remarkably few contemporary articles grapple effectively and broadly with the molecular basis of translocation and its linkage to sources of free energy. I have drawn particularly on the writings of Crane (1977), Singer (1977), Mitchell (1979b, 1981a), Jencks (1980), West (1980), and Eddy (1982); special acknowledgment is due to a lucid and forceful article by Tanford (1983a).

Ionophores

The ion-conducting antibiotics provide the simplest models for biological transport; the growth of our knowledge can be traced through reviews by Harold (1970), McLaughlin and Eisenberg (1975), Pressman (1976), and Bakker (1979). Ionophores fall into two unambiguous classes: carriers and channels.

Valinomycin is the epitome of a mobile carrier, highly selective for potassium ions. The molecule is a cyclic peptide, only 1.1 kdal, consisting of a trimer of the sequence D-valine, L–lactic acid, L-valine, and D-α-hydroxyisovaleric acid; its form may be likened to a bracelet, about 0.8 nm in diameter and 0.4 nm wide. In the K^+ complex the polar ion is enclosed by a cage consisting of six carbonyl groups; the nonpolar residues form a hydrophobic shell that renders the clathrate lipid-soluble (Figure 10.1a). Valinomycin and its K^+ complex diffuse freely within the membrane's lipid phase, allowing K^+ ions to equilibrate with the aqueous phases on either side (Figure 10.1b). Most of the ionophores commonly employed in cell biology are carriers of this kind: monactin, nigericin, monensin, also the proton-conducting uncouplers.

By contrast, gramicidin forms channels that give passage to all the small univalent cations: H^+, K^+, Rb^+, Na^+, but not Ca^{2+} (Figure 10.1c). Gramici-

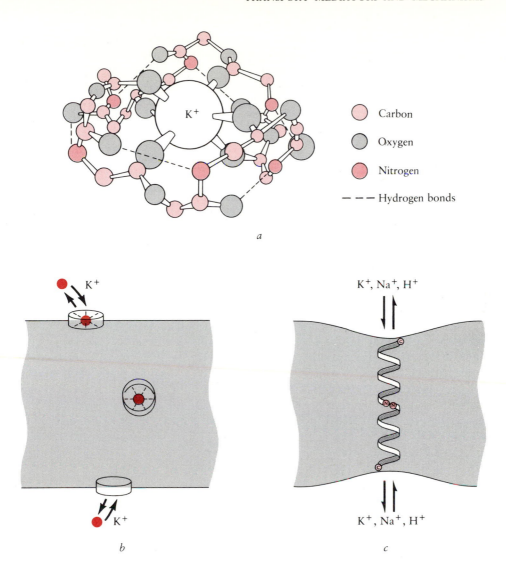

FIGURE 10.1 Ionophores as carriers and channels. (*a*) Valinomycin forms a complex with K^+ such that the K^+ ion is caged by six carbonyl groups. (After Duax et al., 1972; copyright by the American Association for the Advancement of Science.) (*b*) The valinomycin-K^+ complex functions as a mobile membrane carrier for K^+. Bound K^+ ions equilibrate with free K^+ at the two aqueous surfaces. (*c*) Two gramicidin molecules associate head to head, generating a channel that passes all the small cations. Drawn approximately to scale.

dins are linear polypeptides (gramicidin A consists of 15 alternating L and D amino acids, all with hydrophobic side chains and with both ends blocked). In a nonpolar environment the chain coils into a helix, 1.2 to 1.5 nm in length, with a central pore 0.4 nm in diameter. Two gramicidin molecules joined end to end make a cylinder just long enough to span a phospholipid bilayer. Other antibiotics form larger channels made up of multiple antibiotic molecules lined up like the staves of a barrel; monazomycin and the polyenes are of this kind.

Conduction through channels can be operationally distinguished from carrier-mediated transport by three criteria. First, channels are faster: a single gramicidin channel can pass 10^7 to 10^8 ions per second, while the maximum turnover rate of valinomycin is about 10^4 per second. Second, channels are relatively fixed structures that remain open even if the surrounding lipid freezes solid; carriers, which diffuse bodily, are mobile only so long as the lipid phase remains fluid (Krasne et al., 1971; Haydon and Hladky, 1972). Finally, carriers can mediate counterflow but channels cannot.

In thermodynamic terms, ionophores facilitate the passage of a polar ion across a lipid barrier in much the same way that catalysts enhance the rate of a chemical reaction: both reduce the activation energy. In the absence of a carrier or channel, small inorganic ions such as K^+ or Na^+ are virtually insoluble in lipid because of their strong penchant for ion dipole interactions with water. The work required to take K^+ ions out of water and dissolve them in lipid comes to about 40 kcal/mol (170 kJ/mol), a large number compared to the thermal energy of the ion (kT at $25°C = 0.6$ kcal/mol, 2.5 kJ/mol). The partition coefficient between lipid and water is extremely low and diffusion is very slow; this is why lipid membranes have low ionic conductances, about 10^{-9} to 10^{-7} siemen (a resistance of 10^7 to 10^9 ohm/cm^2). Formation of a complex with valinomycin reduces the energy barrier to K^+ diffusion in two ways. First, the size of the complex is larger than that of the ion, delocalizing the charge and reducing the work required to dissolve it in lipid. Second, the outer surface of the clathrate is hydrophobic, and the free energy of moving CH and CH$_3$ groups out of water and into lipid is negative. As a result, the complex is favorably partitioned into the membrane, enhancing the rate of K^+ diffusion by a factor of 10^{30} to 10^{40} (Parsegian, 1969; McLaughlin and Eisenberg, 1975).

Macromolecules as Carriers and Channels

Allosteric Transport Proteins.   The pumps and porters described in the preceding chapters are proteins whose dimensions are comparable to those of the membrane itself. A globular protein 300 amino acid residues in length has a diameter of about 4.3 nm, while the phospholipid bilayer is 5 nm thick; most transport proteins are much larger than that. Macromolecular transport systems must share with the ionophores the capacity to lower the energy barrier to diffusion by forming specific, mobile complexes with selected substrates, but they can hardly emulate the nimble motions of small molecules. Nevertheless, "carriers" and "channels" are significantly different classes of transport. A carrier is mobile to the extent that at any moment, the substrate-binding site is accessible from one side of the membrane or the other, but not from both; it need not move bodily like a ferryboat. By contrast, a channel is a fixed structure whose binding sites, if any, are accessible from both sides at once.

The criteria by which macromolecular transport systems can be assigned to one class or the other are chiefly kinetic. The specificity of transport, its inhibition by structural analogues, and saturation of the rate with rising substrate concentration are often interpreted in favor of a carrier mechanism, but they are in fact compatible with either mode of translocation. There are, however, two criteria on which a distinction can be based. One is the absolute rate of transport: a carrier mechanism is not compatible with a turnover number much above 10^4 per second, whereas channels can conduct ions to the limits allowed by diffusion, 10^7 to 10^8 per second. By this measure it is correct to speak of Na^+, K^+, Ca^{2+}, and Cl^- channels in animal cells, as the conductances of single channels have been shown to be of this order. For most systems, however, absolute transport rates are unavailable. A second generally applicable indicator is the phenomenon called countertransport. When two substrates share a common transport system, the exodus of one substrate downhill can drive the influx of the other uphill without any input of metabolic energy (Figure 10.2).

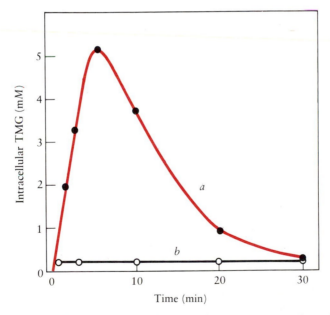

FIGURE 10.2   Counterflow. Cells of *E. coli* poisoned with inhibitors of metabolism were loaded with thiomethyl-β-galactoside (TMG, 30 mM), centrifuged, and resuspended in medium containing 0.5 mM ^{14}C-TMG. Trace *a* shows the transient accumulation of ^{14}C-TMG by these cells. Trace *b* shows that cells not loaded with TMG do not accumulate ^{14}C-TMG. (After Winkler and Wilson, 1966, with permission of the *Journal of Biological Chemistry.*)

Countertransport is strong evidence for a carrier mechanism, for it implies that the two substrates bind alternately to a common, mobile site. The glucose porter of erythrocytes has long been recognized as a carrier by this criterion (LeFevre, 1975; Widdas, 1980); other well-established instances are the glucose porter of *Chlorella,* the Cl^--HCO_3^- antiporter of erythrocytes, the lactose porter of *E. coli*, the ATP-ADP antiporter of mitochondria, and many other porters from animals, plants, and microorganisms.

It is no accident that the great majority of transport systems behave as carriers in the kinetic sense. A channel, even one equipped with a gate that can open and shut, can only mediate downhill fluxes; equilibration of binding sites with both membrane surfaces at once will result in the dissipation of any substrate gradient. Systems that carry out active transport, be it primary or secondary, must generally operate with binding sites that are accessible alternately from the one side or the other, just as a mechanical pump does (Tanford, 1982, 1983a, 1983b).

What kinds of motions can we attribute to ponderous macromolecular carriers? Pumps and porters are integral membrane proteins that span the barrier, with portions projecting into the aqueous phases like the heads of a rivet (Singer, 1974, 1977). Soluble proteins fold in such a way as to expose polar residues on the external surface while nonpolar residues are buried in the interior, shielded from contact with water. Proteins that span a membrane must have a more complex structure, with a hydrophobic midriff consisting of nonpolar amino acids that can associate with lipids. Such "dumbell" molecules can diffuse laterally in the plane of the membrane but can neither flip over nor rotate: the energetic cost of submerging polar amino acid residues in lipid and nonpolar ones in water is so great that such a configuration of the protein will hardly ever occur. It follows that macromolecular transport systems must contain incipient channels lined with polar groups that can communicate with both sides, though not necessarily at the same time.

The simplest general solution to the puzzle of membrane transport is based on the recognition that proteins are dynamic structures, subject to allosteric modification of form and properties (Figure 10.3). This model attributes the selectivity of transport to association of the substrate with binding sites located within those incipient channels that must traverse the transport proteins. These sites remain essentially stationary, but minor conformational changes bring about their exposure to the aqueous phase on one side or the other and may modulate their affinity as well. Ironically, from this standpoint the distinction between channels and carriers ceases to be fundamental, becoming rather a variation on the common theme of a "gated" pore. Where the pore, permanent or transitory, is open to both sides at once, channel kinetics result. In most cases the transport protein oscillates between two configurations that give alternating access to the binding site, and the physiologist marks a carrier. The transport system may be oligomeric, but this is not a necessary feature: a channel may traverse a single subunit and still be subject to allosteric reorientation. And one

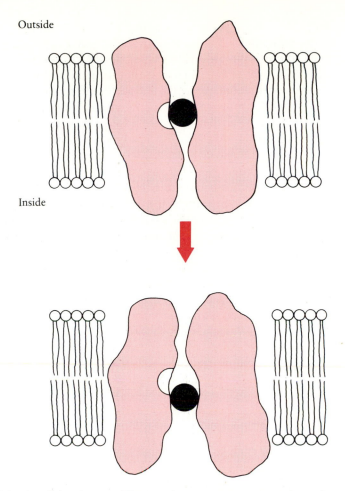

Outside

Inside

FIGURE 10.3 An allosteric pore. The two shapes may represent either subunits of an oligomeric transport protein with a polar channel between the subunits or else domains of a single integral membrane protein. (After Singer, 1974, with permission of Annual Reviews.)

can readily imagine cases in which the affinity of the binding site is not the same in the two configurations, so that active transport ensues (Figure 9.1b).

Translocation Pathways. The hypothesis that transport systems are allosteric proteins with internal conductive channels was drawn from general principles rather than from specific observations (Jardetzky, 1966; Singer, 1974, 1977). Experimental support is now beginning to accumulate from many directions.

If transport required rotation of the catalytic protein or gross reorganization of its structure, one would expect it to be perturbed by combination with bulky ligands. In fact, the Ca^{2+} ATPase and the Na^+-K^+ ATPase were quite unaffected by the attachment of specific antibodies (Kyte, 1974; Dutton et al., 1976). However, there is ample evidence for lesser conformational changes during the transport cycle. The ATPases that translocate H^+, Ca^{2+}, and Na^+-K^+ each exist in two configurations that yield different products on tryptic digestion (Addison and Scarborough, 1982; Jørgensen, 1982). In several systems, fluorescent ligands bound near the active site have given evidence for conformational changes (Karlish, 1980; Moczydlowski and Fortes, 1981; Jørgensen, 1982). Inhibitors may act specifically on the one conformation or the other. A case in point is the ATP-ADP antiporter of mitochondria, which binds atractyloside in one configuration and bongkrekic acid in the other (Klingenberg, 1980). Additional evidence comes from the finding that K^+ and Ca^{2+}, while being transported by their respective ATPases, pass through a state in which the ion is inaccessible to chelation or to exchange; this is thought to result from the transient occlusion of the ion in an enclosed cavity.

How might a transport protein be designed to provide a controllable polar channel across the membrane? There appear to be multiple solutions to this problem. One is exemplified by the F_1F_0 ATPase, which contains a permanent proton-conducting sector that spans the membrane. It has been claimed that other ion-translocating ATPases likewise contain ionophorous sectors, but the status of those channels remains controversial. Bacteriorhodopsin, whose primary and secondary structures are known, has been dubbed an "inside-out" protein (Engelman and Zaccai, 1980), because its nonpolar outer surface associates with membrane lipids while charged and polar amino acids lie in the center. This distribution, the converse of that seen in soluble proteins, may indicate the existence of an internal hydrophilic channel. Proton transport by bacteriorhodopsin is also quite insensitive to low temperatures, suggesting that the mechanism involves some sort of channel rather than a mobile carrier (Racker and Hinkle, 1974). The relationship of the putative proton channel to the seven helical transmembrane segments revealed by X-ray crystallography has yet to be worked out.

An elegant way for proteins to provide both a hydrophobic exterior and a polar pore across the bilayer hinges on their dimeric or oligomeric construction; the polar channel could pass down the center while the binding site (or sites) lies between the monomers. A number of transport systems do in fact appear to be oligomeric in situ, including the Na^+-K^+ ATPase and the antiporters that exchange ATP for ADP and Cl^- for HCO_3^-; cytochrome oxidase is another case in point. Genetic arguments suggest that the lactose porter of *E. coli* functions as an oligomer and that Na^+-linked porters in bacteria may be made up of separate polypeptides that bind Na^+ and the substrate respectively. The "shock-sensitive" transport systems of bacteria consist of four kinds of subunits, one periplasmic and three buried in the membrane; these probably compose some kind of transmembrane channel (Higgins et al., 1982; Hengge and Boos, 1983). The thesis that oligomeric construction is the norm among transport systems (Singer, 1974, 1977; Klingenberg, 1981) is widely accepted but has recently been called into question: bacteriorhodopsin and also the Na^+-K^+ ATPase can apparently carry out the normal transport cycle in the monomeric state (Kyte, 1981; Stoeckenius and Bogomolni, 1982; Craig, 1982). For the present it is prudent to conclude that oligomeric construction is not an obligatory feature of transport proteins, but it seems nevertheless to be a common one.

A word is in order concerning the substrate-binding sites, which, like those of enzymes, play a dual role in transport. They lend the process specificity; furthermore, the free energy of substrate binding may be utilized to drive the conversion of one protein configuration into the other. In several cases, structural analogues of the transport substrate have been used to map the dimensions of the binding site and to learn what chemical groups project into it. The Cl^--HCO_3^- antiporter of erythrocytes and the amino acid porters of tumor cells illustrate the procedure (Knauf, 1979; Christensen, 1979). Other binding sites remain ill-defined. The K^+ selectivity of valinomycin is due to the ring of carbonyl groups that form a tight cage around the ion. Does something analogous hold for the K^+ site of the Na^+-K^+ ATPase? And if so, what accounts for the change in selectivity so as to favor Na^+ in the inward orientation?

Protons are a special case. Protons bind, of course, to groups such as $-COO^-$ and $-NH_2$, as a function of the pK and the pH in the immediate vicinity of the proton-accepting group. However, protons can also hop with alacrity from one acceptor group to an adjacent one; the high electrical conductivity of ice is due to the rapid transfer of protons between neighboring water molecules in an ordered lattice. In principle, the helical segments of an integral membrane protein may provide such a pathway from one membrane surface to the other: a proton wire, so to speak (Nagle and Morowitz, 1978; Nagle and Tristram-Nagle, 1983). The proposal has been developed in detail, but there is no experimental evidence that proton wires do play a role in proton translocation. Until such data are forthcoming, we shall assume that protons, like other ions, are carried by association with mobile binding sites.

Excitable Ion Channels.   Ion channels, controlled either by the electric potential difference or by chemical effectors, underlie much of biological information processing. Albeit specialized, this well-studied class of transport systems no longer seems as esoteric as it did before the allosteric nature of other macromolecular transport systems was recognized: the molecular architecture of ion channels may well embody features that will also show up in pumps and porters. Our conception of what ion channels are and how they work derives largely from the extensive research on the sodium and potassium channels of excitable animal cells (Aidley, 1978; Armstrong, 1981; Latorre and Miller, 1983; Hille, 1984).

Ion channels are protein ensembles that span the membrane and undergo rapid alternations between closed and open configurations. The sodium channel of nerve axons, for example, is closed at the resting potential; it opens in response to partial depolarization, shutting spontaneously within a few milliseconds (Figures 9.5 and 10.4). The open channel is quite selective for Na^+; the sodium permeability coefficient is tenfold greater than that for K^+. Some 10^6 Na^+ ions pass through each channel per second, a number so large as to exclude a carrier mechanism and warrant the designation of a channel. On the other hand, the sodium channel is not simply an aqueous pore. Na^+ fluxes are specifically blocked by tetrodotoxin, a large organic molecule that is thought to plug the mouth of the channel, yet the channel does not pass K^+, an ion whose hydrated radius is smaller than that of Na^+. These and other observations suggest a funnel-shaped cross section, dilated at its mouth but constricted in the interior by a "selectivity filter" that leaves a narrow passage on the order of 0.3×0.5 mn. The nature of the filter is unknown; Bertil Hille has suggested a ring of eight oxygen atoms that form a transitory complex with an unhydrated Na^+ ion but reject K^+, whose unhydrated radius is larger.

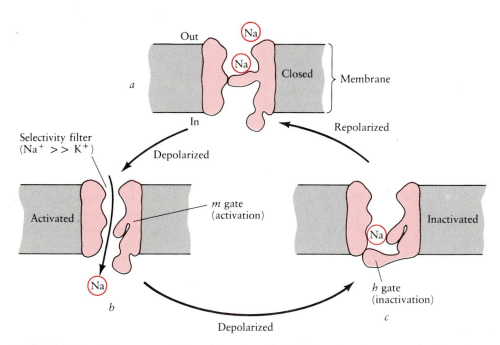

FIGURE 10.4   Major states of the sodium channel. (*a*) Before depolarization, the channel is nonconducting because the *m* gate is closed. (*b*) Depolarization causes the sodium channel to become conducting, owing to the opening of the *m* gate. The conductance of the channel is now largely determined by its selectivity filter, which blocks all anions and has a much greater selectivity for Na^+ than for K^+ or Ca^{2+}. (*c*) While still under depolarization, the *h* gate on the inner end of the channel closes, causing inactivation of the channel. Repolarization to the resting potential opens the *h* gate and closes the *m* gate. The channel is now again ready for activation by a new depolarization. (From *Animal Physiology*, by R. Eckert and D. Randall, 2d ed. Copyright © 1983 by W. H. Freeman and Co.)

How can a small change in the potential difference induce the channel to open briefly and then close? From the time of Hodgkin and Huxley it has been believed that in response to a change in the electric field, particular charged groups move and open the passage; and then relax, letting it close once more. The movement of these charges can be detected experimentally as minuscule "gating currents" that do not involve a flux of Na^+ or any other mobile ion. These are believed to report the successive displacement of two amino acid clusters, comprising the m and h gates marked in Figure 10.4. What these residues are and how they relate to the selectivity filter remain to be discovered.

Channels for potassium ions are distinct from those for sodium, but they embody the same principles. Potassium channels pass 10^7 to 10^8 ions per second; pharmacological blocking agents again suggest a wide entrance, narrowing to a passage only 0.3 nm wide at the selectivity filter. Potassium channels come in considerable variety: some are highly selective for K^+ and others much less so, some are regulated by calcium ions or by phosphorylation, others by voltage alone. Some of these differences can be attributed to the construction of the channel proper, others must reflect the binding of allosteric modifiers to specific sites exterior to the channel.

From the biochemical viewpoint, by far the best known channel is the acetylcholine receptor on the distal side of a synaptic junction (Conti-Tronconi and Raftery, 1982). The receptor protein binds acetylcholine avidly (dissociation constant about 10^{-8} M); one acetylcholine molecule suffices to open a channel that passes indiscriminately all the small monovalent and divalent cations, but no anions, at some 3×10^6 ions per second. The channel is conceived as an aqueous pore, some 0.6 nm across at the narrowest point, and about 10 nm in length. Electron micrographs reveal the acetylcholine receptor as a rosette, 9 nm wide, with a central pit 1.5 to 2 nm in diameter, which presumably represents the mouth of the channel. The complex has been isolated and reconstituted; it consists of four nonidentical subunits that suffice to account for all the properties of the native channel. Thanks to this splendid work one can begin to inquire into the molecular nature of the gate and the manner in which it opens in response to acetylcholine binding.

Primary Energy Coupling

Alternating Access. The majority of transport processes discussed in this chapter and the preceding ones must be classified as active: they involve the movement of an ion or a molecule in such a way that its electrochemical potential is higher on the discharge side than on the uptake side. The requisite work can be performed by linking the uphill flow of the substrate either to a chemical reaction (primary transport) or to the downhill flux of a coupling ion (secondary transport). These two mechanisms differ substantially from the molecular and physiological viewpoints, but are thermodynamically equivalent.

How can a protein drive a substance from a low electrochemical potential to a high one? Investigators who have grappled with this puzzle have almost always had recourse to one or another variant of what Tanford (1983a) aptly called the alternating access model. Its central provisions are the existence of selective binding sites that alternately face inward or outward, and that allosteric reorientation of the binding site or sites is accompanied by changes in binding affinity. Jardetzky, in his classic 1966 paper, spelled out the consequences (Figure 3.12). Suppose the dissociation constant for substrate S were 1 mM in one orientation and 1 M in the other; the two fluxes would then come to equilibrium when the concentration of S on the low-affinity side is a thousandfold higher than that on the high-affinity side. In order to visualize how such a system pumps substrate from a concentration of, say, 1 mM to 1 M, it is helpful to imagine the binding site being located in a tiny cavity, 1 nm^3 or less in volume. A simple calculation shows that up to these limiting concentrations free substrate will tend to diffuse into the cavity on the uptake side and out of it on the discharge side. (The cavity is not an obligatory feature, only a useful heuristic device.) When the system performs work, reorientation must be coupled to some exergonic process, such as the hydrolysis of ATP.

Now, thermodynamics does not insist that affinity changes be the only possible basis for active transport: One can write models in which energy input drives only the reorientation of the binding site in a particular direction (say, from outward to inward) and thereby account for the net accumulation of the substrate without invoking changes in affinity. The question is whether models of the latter kind are realistic; according to Tanford (1983a, 1983b), they entail unacceptable constraints on permissible rate constants, but his view is not universally shared. Be this as it may, experimental data from both pumps and porters do show that the binding affinity at the uptake side is often higher than that at the discharge side by several orders of magnitude. For example, the dissociation constants for the binding of Na$^+$ and K$^+$ ions to the Na$^+$-K$^+$ ATPase are about 1 mM on the uptake side but 100 mM on the discharge side. The same conclusion has been reached for several porters, for instance, the K_m for proton-coupled influx of β-galactosides into *E. coli* or of glucose into *Chlorella* is about a hundredfold lower than that for efflux (see below).

How do such alterations in binding affinity come about? Binding constants for the association of ligands with proteins are determined by the reactive groups that enter into the coordination sphere of the bound ligand. Ca^{2+} ion is one ligand for which detailed information is available, both from model compounds and from calcium-binding proteins of known structure. In this case six coordinating groups, including several carboxyl groups, make a high-affinity site (dissociation constant 10^{-6} M), while three suffice to make a low-affinity site (dissociation constant 10^{-3} M). The binding affinity for protons depends in the first instance on the pK of the acceptor group, but this pK can be modified by as much as two units by charged vicinal groups. It seems likely, then, that changes in binding affinity result from the movement of amino acid residues into or out of the substrate-binding site in the course of the conformational transition between its two orientations.

Just what these transitions consist of is not known for any transport system, and will surely vary from one system to another. Tanford (1982) drew the proposal shown in Figure 10.5 from the structure of bacteriorhodopsin, which consists of seven helical segments that span the membrane. If binding sites are located in cavities *between* the helices, one can imagine a concerted twisting and rocking motion that will alter both their orientation and chemical configuration. A multihelical model also has the merit that it can in principle suggest explanations for the coordinated transport of several ions, such as the exchange of $3Na^+$ for $2K^+$. Moving liganding groups away from a tightly bound ion is work; by itself, the equilibrium envisaged in Figure 10.5 would lie far to the left. In a reversible reaction cycle, this is the place where energy input would be necessary to make the two configurations approximately equal in abundance, a prerequisite for any realistic rate of cycling.

The preceding discussion has tacitly assumed that active transport systems establish concentration differences. But this is not always true: in bacteria, fungi,

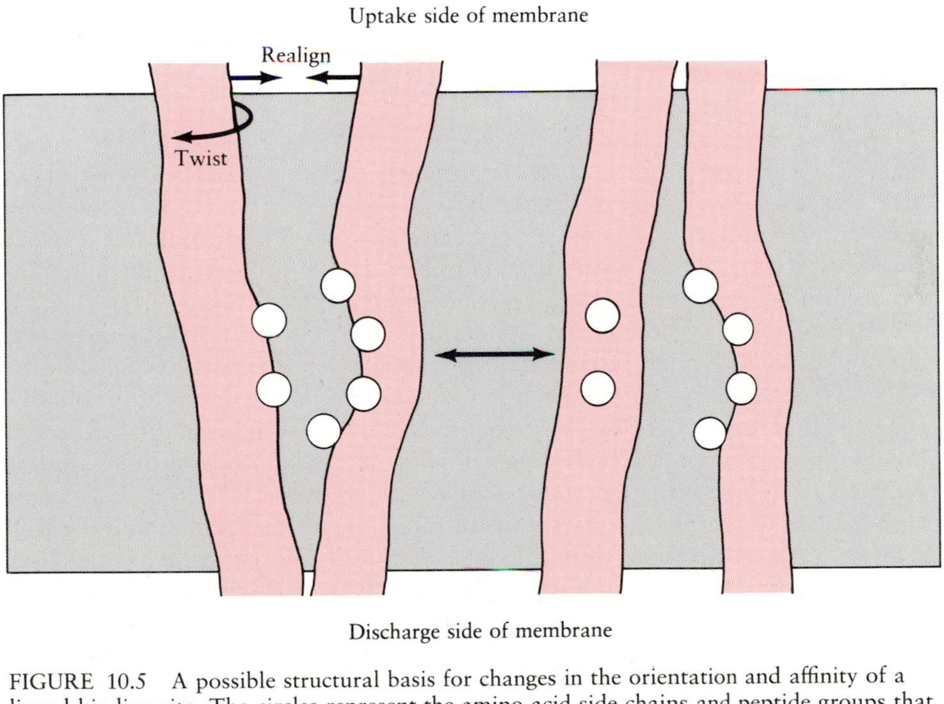

FIGURE 10.5   A possible structural basis for changes in the orientation and affinity of a ligand-binding site. The circles represent the amino acid side chains and peptide groups that collectively make up the binding site. These are part of helical transmembrane segments of the protein molecule. A concerted twisting and rocking motion alters both the orientation of the site and its molecular configuration. (After Tanford, 1982, with permission.)

and plants, $\Delta\Psi$ makes up a large part of the protonic potential. The H^+ ATPases move protons against this potential difference, which, in turn, enters prominently into the driving forces on proton-linked porters. How is the electric field to be brought into calculations that are based on simple association and dissociation of ligands? The most appealing model stems from the F_1F_0 ATPase, for which there is good reason to believe the F_0 sector serves as a proton well that converts $\Delta\Psi$ into ΔpH (Chapter 7); at the interface between F_1 and F_0, the protonic potential consists entirely of a pH difference. It may well be that the polar channels, originally introduced to provide access from the aqueous phase to the binding site (Figure 10.3), also interconvert electric and osmotic potentials. The idea has been invoked on more than one occasion (West, 1980; Komor and Tanner, 1980), but its validity remains to be established.

There is no doubt that the alternating access model, based on energy-linked changes in protein conformation and substrate affinity, provides a satisfying basis for both primary and secondary active transport. But is it the only such basis? In my opinion (*pace* Tanford) Mitchell's concept of ligand conduction remains a legitimate alternative, especially appealing for primary transport. Were the merits of scientific issues decided by counting votes, alternating access would probably win the day. But spatial organization of chemical reactions has too much the ring of truth to be sacrificed to a hasty sweep of Occam's razor; I suspect that these seemingly conflicting views merely emphasize different aspects of a subtle and fluid reality.

Ion-Motive ATPases.   The ion-translocating ATPases exemplify most plainly what is meant by primary active transport. ATPases mediate two distinct processes, the hydrolysis of ATP and the translocation of one or more ions, and the two activities are so coupled that normally the one does not take place without the other. The central question to be resolved is, What is the molecular connection between the chemical reaction and the osmotic one?

During the past decade, the ion-motive ATPases have multiplied prodigiously (Table 10.1); they fall into at least three classes. The first is typified by the F_1F_0 ATPases of bacteria, mitochondria, and chloroplasts: structurally complex enzymes that translocate protons exclusively, by a mechanism that does not involve phosphorylation of the enzyme protein. The second class includes diverse ATPases of animal cells that transport Na^+, K^+, and Ca^{2+}; the electrogenic H^+ ATPase of the plasma membrane of fungi, algae, and plants; and the Kdp ATPase of *E. coli* (Chapter 5). These ATPases consist of a single large polypeptide that undergoes phosphorylation as part of the transport cycle. The third class, as yet somewhat ill-defined, includes the proton-translocating ATPases of various intracellular organelles: fungal and plant vacuoles, tonoplast membranes, endoplasmic reticulum, chromaffin granules and other storage vesicles, as well as membranes of bladder and kidney epithelium (Njus et al., 1981; Kakinuma et al., 1981; Apps, 1982; Bowman and Bowman, 1982; Steinmetz and Andersen, 1982; Bennett and Spanswick, 1983; Cidon et al., 1983; Rees-Jones and Al-Awqati, 1984; Forgac and Cantley, 1984). Many of these

TABLE 10.1 Ion-translocating ATPases

Class	System	Source	Ions translocated	Approximate mass (kdal)	Subunits	Comments
I	F_1F_0 ATPase	Bacteria	$2\text{–}3H^+$	470	8	No ~P intermediate
	F_1F_0 ATPase	Mitochondria	$2\text{–}3H^+$	480	8–10	
	F_1F_0 ATPase	Chloroplasts	$2\text{–}3H^+$	480	8–10	
II	Kdp ATPase	E. coli	K^+; other?	160	3	
	H^+ ATPase	Fungi	$1\text{–}2H^+$ only	105	1	~P intermediate formed; vanadate-sensitive
	H^+ ATPase	Algae	$1\text{–}2H^+$ only	U	U	
	H^+ ATPase	Plants	$1\text{–}2H^+$ only	U	U	
	$Na^+\text{-}K^+$ ATPase	Animals	$3Na^+, 2K^+$	150	2	
	Ca^{2+} ATPase	Animals	$1\text{–}2\ Ca^{2+}, H^+?$	140	1–2	
	$K^+\text{-}H^+$ ATPase	Gastric mucosa	$2K^+, 2H^+$	100	1	
III	H^+ ATPase	Fungal vacuoles	H^+	U	U	Not inhibited by vanadate; similar enzyme in tonoplast
	H^+ ATPase	Chromaffin granules	H^+	U	U	Similar enzymes in other storage granules, lysosomes
	H^+ ATPase	Turtle bladder, renal tubules	$2\text{–}3H^+$	U	U	
Unassigned	Na^+ ATPase	S. faecalis	$Na^+, K^+?$	U	U	Not vanadate-sensitive
	Ca^{2+} ATPase	Cyanobacteria	Ca^{2+}	U	U	
	Ca^{2+} ATPase	Plants	Ca^{2+}	U	U	

U = unknown or uncertain.

ATPases are stimulated by chloride; they show points of resemblance to the F_1F_0 class, but will probably constitute one or more distinct groups. Additional ATPases remain to be classified, including the Na^+-transporting ATPase of *Streptococcus faecalis* (Heefner and Harold, 1982) and calcium ATPases from plant membranes and cyanobacteria (Gross and Marmé, 1978; Lockau and Pfeffer, 1983). It is striking that all the known ATPases transport cations,[1] but the differences in molecular structure and catalytic activities suggest that several distinct mechanisms will eventually be recognized.

ATPases of the F_1F_0 type were discussed at length in Chapter 7, and so I shall concentrate here on those that undergo phosphorylation. Several models have been formulated over the years, but the underlying concepts have remained the same: alternating access to ion-binding sites, ATP processing that is spatially separated from ion movements, and conformational transitions of the catalytic protein as the essential link between the chemical reaction and the osmotic one. These constraints have been incorporated into thermodynamic calculations (Jencks, 1980; Tanford, 1982, 1983a, 1983b), whose upshot can be expressed in terms of "coupling rules": general statements that should hold for all ATPases in which the pathway of ATP processing is separate from that of ion translocation. As I cannot improve on Tanford's formulation of these rules, I quote them almost verbatim.

1.   The reaction pathway must not catalyze the uncoupled hydrolysis of ATP. This is most simply accomplished by requiring the protein to alternate between two conformational states, such that the overall process (which includes substrate binding and product release) is carried part way in one conformation and completed in the other.

2.   At the points of the reaction cycle where ATP, ADP, and P_i are free to exchange between bound and free states (generally distinct points for each substance), the free-energy difference between bound and aqueous species should be small. This means that the overall reaction among *bound* species, ATP (when first bound) $\rightleftharpoons$ ADP (about to be released) + P_i (about to be released), must be accompanied by a large negative free-energy change, -10 to -14 kcal/mol for the examples given in Table 10.1.

3.   A general principle recently suggested by Knowles (1980) for most enzyme-catalyzed phosphoryl transfer reactions is presumably applicable: the difference in free energy between ATP and its hydrolysis products *at the site of bond rupture* should be small.

---

[1] The chloride-stimulated ATPase of *Limonium* (Hill and Hanke, 1980) has been cited as an ATPase that transports anions, but the evidence that this enzyme transports chloride is not compelling. Several proton-translocating ATPases of plant origin are known to be stimulated by chloride.

4.   It follows that there must be a step (or steps) in the reaction cycle where one (or more) of the bound substrate species undergoes a dramatic change in chemical potential, that is, a transition between a "loosely" and a "tightly" bound state (high and low free energy). Only by this device can one reconcile a large ΔG for the overall reaction of the bound species (rule 2) with a small ΔG at the point of bond rupture (rule 3). Reaction steps of this kind would by themselves be essentially irreversible; these are therefore the steps where thermodynamic coupling to the ion translocation pathway must take place (Tanford, 1983a).

The conclusion is important enough to bear restatement. In Tanford's view, an ion-motive ATPase is a thermodynamic engine that cycles between two states. In one, *bound* ATP (or a bound species derived from ATP) has a high group transfer potential; in the other, a low one. This conversion is the crucial exergonic step in the pathway of ATP hydrolysis; it is accompanied by changes in the orientation and affinity of ion-binding sites elsewhere on the protein and can only proceed in the presence of these particular ions.

Supportive evidence comes from work with the Na^+-K^+ ATPase and Ca^{2+} ATPase, some of which was already noted in Chapter 9. All the ATPases of this tribe process ATP via an acyl phosphate intermediate, with the phosphoryl group covalently linked to an aspartyl residue of the protein. This phosphorylated enzyme exists in two very different states. In one, designated $E_1 \sim P$ in Figure 9.9, the phosphoryl group behaves like a compound of high group transfer potential: like the phosphoryl group of soluble aspartylphosphate, the protein-bound one is readily transferred to ADP with production of ATP. In the alternative state of the phosphoenzyme, E_2-P, the phosphoryl group has a far lower group transfer potential, comparable to that of an ester or of P_i in aqueous solution. The difference between $E_1 \sim P$ and E_2-P is on the order of 10 to 14 kcal/mol (42 to 60 kJ/mol). Bond rupture appears to take place at the energy level of ATP, phosphate release at that of P_i, and the transition between the high- and the low-potential forms of the enzyme occurs in the phosphorylated state (rule 4). The features of protein structure that give rise to this remarkable behavior are not known and not easily predictable, but it is reasonable to believe that conformational changes sufficient to alter the thermodynamic behavior of the phosphoryl groups may also alter the orientation and affinities of ion-binding sites. A well-known precedent is the Bohr effect: oxygenation alters the pK of hemoglobin even though the heme groups that bind oxygen are remote from the surface residues that undergo a change in proton affinity.

The preceding discussion was presented in terms of the eukaryotic ATPases whose reaction cycle includes a phosphorylated intermediate. It must be recalled, however, that P. D. Boyer has long championed an analogous mechanism for the F_1F_0 ATPase (Chapter 7). The chief point of difference is that the F_1F_0 ATPase does not undergo phosphorylation; instead, the enzyme would alternate between a state that binds ATP avidly, favoring its formation, and another state that binds ATP loosely, allowing it to be released (Figure. 7.21). Whether ATPases put into class III of Table 10.1 can be accommodated by the coupling rules devised for other ATPases is not clear; at this time we do not even know whether or not their mechanism involves a phosphorylated intermediate.

Ligand conduction does not lend itself so neatly to thermodynamic calculations, but it provides a more graphic sense of the manner in which the free energy of ATP hydrolysis may be transduced into that of an ion gradient. The heart of the matter is that the pathways of ATP processing and ion flow are thought to be spatially confluent rather than separate, and that the movement of charged anionic groups (ATP and the products of its transformation) directly carries cations across the permeability barrier. The concept is illustrated for the Na^+-K^+ ATPase in Figure 10.6. In principle it requires neither alternating access to ion-binding sites nor conformationally induced affinity changes. What is essential is that the phosphoryl group trace a looped path across the permeability barrier while passing from ATP to $E_1 \sim P$, E_2-P, and ultimately to water as P_i, and that chemically specific channels allow K^+ and Na^+ ions access to these ligand-conducting anionic sites at the appropriate stage of the cycle

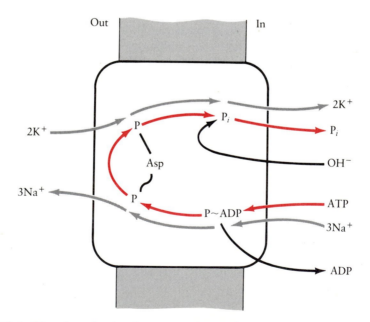

FIGURE 10.6   Ligand-conduction diagram for the Na^+-K^+ ATPase. The colored lines represent the pathway of phosphoryl transfer from ATP to the protein and ultimately to water. The gray lines represent the concurrent flows of Na^+ and K^+ ions associated with the reaction pathway. (After Mitchell, 1981a, with permission of John Wiley and Sons.)

(Mitchell, 1979*b*, 1981*a*; Dupont, 1983). To complete the analogy with the alternating access model, one can think of the cation affinity of the phosphoryl group as changing progressively from sodium-selective to potassium-selective. There is to my knowledge no direct evidence to support this view of the course of events, nor any that refutes it, but it may be pertinent that electron-paramagnetic-resonance measurements locate the phosphorylation site of the Na^+-K^+ ATPase within a few angstroms of the cation-binding site (O'Connor and Grisham, 1979). Most usefully, ligand conduction offers an instant explanation of the fact that all ATPases transport cations.

At first encounter, ligand conduction sounds quite different from the more familiar themes of reorienting binding sites and changes in affinity. But the contrast is deceptive. As Mitchell (1981*a*) points out,

> The notion of cation-phosphoryl symport used here should not be taken to mean that the phosphoryl group would necessarily migrate bodily through the ... ligand conducting regions ... of the enzyme catalytic center during transfer of the phosphoryl group ..., or that the cations must be directly bounded to the oxygens of the catalytic center. The notion of cation-phosphoryl symport ... means that the transfer of the phosphoryl group between ADP and aspartate is tightly linked to the conduction of $3\,Na^+$ through the ... catalytic center, and that the transfer of the phosphoryl group between aspartate and OH^- is tightly linked to the conduction of $2\,K^+$ ....

The chemical steps will inevitably be accompanied by changes in protein conformation. It is therefore not easy to distinguish events that are necessary to unlock the gated pore from those that are contingent on the unlocking. The distinction will blur further when the inquiry is pursued to the atomic level, where bioenergetics melds into protein chemistry.

Secondary Energy Coupling

The hallmark of symporters and antiporters is that they utilize the electrochemical gradient of a coupling ion to generate a gradient of some other substance; the pertinent version of the generalized carrier scheme is shown in Figure 9.1*c*. The thermodynamics of porters follows from the principle that binding sites must be accessible alternately from one side of the membrane or the other, but not from both simultaneously, and from the coupling rules that specify the stoichiometric and vectorial relationships between the linked fluxes, for these determine the maximum gradient the porter can achieve.

As was discussed in Chapter 3, at equilibrium the porter balances the driving force on the coupling ion against the driving force on the substrate S. This is a consequence of the constraints that make the porter mobile only when it is either unoccupied or carries both the substrate and the coupling ion. For the symport of an uncharged molecule with n protons, one should find that at equilibrium

$$\log \frac{[S]_i}{[S]_o} = \frac{n(-\Delta p)}{Z} \tag{Eq. 10.1}$$

where $[S]_i$ and $[S]_o$ are the internal and external substrate concentrations, Δp is the proton potential, and Z stands for $2.3RT/F$, or 59 mV. Table 3.2 lists the maximum gradients attainable by a series of porters that carry S with one or more protons. Since the substrate itself may bear a positive charge, a negative one, or none, a given protonic potential may sustain a range of gradients. Biochemists must now discover what coupling roles govern a particular porter and how they are implemented at the macromolecular level. What we know of these matters comes chiefly from kinetic analysis.

It is by no means easy to determine n, the number of coupling ions translocated with each molecule of S. In principle, given $[S]_i$, $[S]_o$, and Δp, n can be calculated from Equation 10.1. In practice, the observed concentration gradient $[S]_i/[S]_o$ represents a steady state that falls short of the maximum gradient because of substrate efflux, slip, or trans inhibition (see below); consequently n is likely to be underestimated. One can also estimate n directly from simultaneous measurements of the rates of substrate and ion flux under conditions chosen to minimize reextrusion of the coupling ion (Figure 9.18); this procedure is also likely to underestimate n. Finally, in organisms amenable to electrophysiological procedures, n can be deduced from the analysis of current-voltage curves (Sanders et al., 1983). It is interesting that recent experiments with both prokaryotic and eukaryotic cells often favor a stoichiometry of two coupling ions per substrate molecule. Examples include the porters for lactose, proline, and glutamate in *E. coli* (Ramos and Kaback, 1977; Fujimura et al., 1983; Mogi and Anraku, 1984) and the uptake of amino acids with $2H^+$ by *Neurospora* (Sanders et al., 1983) and that of chloride with $2H^+$ by *Chara* (Sanders, 1980). Even the classic sodium-glucose symporter of animal intestine may carry $2Na^+$ per glucose (Kimmich and Randles, 1980; Lever, 1984). Such a stoichiometry makes physiological sense. It ensures that the driving force for substrate uptake will remain ample despite fluctuations in $\Delta\tilde{\mu}_{H^+}$ or $\Delta\tilde{\mu}_{Na^+}$ and lets the substrate gradient be determined by kinetic controls rather than by thermodynamic constraints.

How does a porter couple two fluxes such that the driving force on the coupling ion or ions is transmitted to the desired metabolite? Transport processes, like enzymatic reactions, can be characterized by their kinetic parameters, affinity (K_m), and maximum velocity or flux J_{max}. Analysis of the relationship between the driving forces and the kinetic parameters for influx and efflux is a formidable problem, which must be given short shrift here. Suffice it to say that as in enzyme kinetics, the problem has traditionally been simplified by assuming that the substrates on both sides of the membrane equilibrate rapidly with binding sites on the porter and that the translocation step is rate limiting. For the symporter shown in Figure 9.1c, the accumulation ratio $[S]_i/[S]_o$ should then be related to the kinetic parameters by the relationship

$$\frac{[S]_i}{[S]_o} = \frac{J^o_{max} K^i_m}{J^i_{max} K^o_m} \qquad \text{(Eq. 10.2)}$$

In this equation K^o_m and J^o_{max} pertain to the outer surface or influx, K^i_m and J^i_{max} to the inner surface or efflux. The relationship holds only for low substrate concentrations and states the maximum gradient attainable. (For further discussion see Cuppoletti and Segel, 1975, and Eddy, 1982.) In other words, accumulation of the substrate by the cells results from modulation, by the protonic potential, of the K_m, the J_{max} or both for the influx of S or for its efflux. The most popular version of this hypothesis states that coupling of the fluxes of H^+ and S lowers the K_m for influx of S or raises the K_m for efflux.

For a specific example, consider the glucose transport system of the green alga *Chlorella* (Figure 10.7a), whose status as an electrogenic glucose-H^+ symporter is well documented by the researches of Komor and Tanner (1974, 1980; Eddy, 1982). At micromolar levels of external sugar, the porter can establish a concentration gradient of 1000 or a little over; this accords with estimates of Δp near -180 mV and an approximate H^+/sugar stoichiometry of unity. According to one possible hypothesis, the carrier (more precisely, the binding site) binds both the sugar and a proton when it faces the external medium. The carrier then reorients in response to the change in its charge distribution, effectively translocating CSH^+ to the cytoplasmic surface, where the electrochemical potential of protons is lower than in the medium. As protons dissociate from their binding site on the porter, the latter's affinity for glucose declines and glucose is released as well. The linkage between the two dissociation reactions is best explained by postulating that dissociation of the proton alters the conformation and affinity of the substrate-binding site.

Net accumulation of sugar could result from the cyclic protonation and deprotonation of the carrier center as the latter equilibrates alternately with the pH prevailing in the medium and in the cytoplasm. The experimental data, while generally compatible with these expectations, also indicate that the system is more subtle. The K_m for sugar influx was found to be 0.2 mM, that for sugar efflux 30 mM. Moreover, the affinity for influx was strongly dependent on the pH: the K_m was 0.2 mM at pH 6, but 50 mM at pH 9. In other words, raising the pH titrates the carrier from a high-affinity to a low-affinity form. By Equation 10.2 this effect could account for an accumulation ratio of about 100, provided ΔpH is large enough. In fact, the pH gradient between cytoplasm and medium is only about one unit, cytoplasm alkaline. It follows that in order to account for the observed accumulation ratio of 10^3, the protonic potential must modulate both K_m and J_{max}, and that much of the driving force for sugar accumulation must come from the membrane potential. Surprisingly, changes in $\Delta\Psi$ had no effect on the kinetic parameters of influx and efflux. But they did affect the

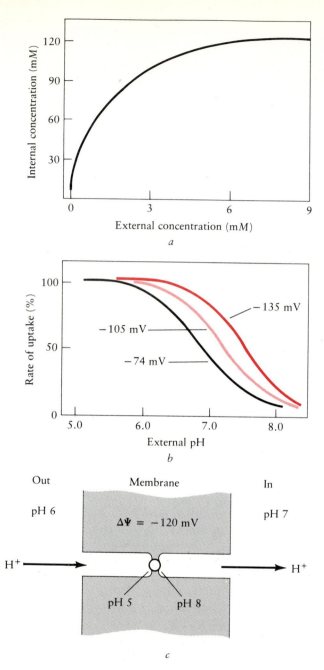

FIGURE 10.7   Aspects of sugar accumulation by *Chlorella*. (*a*) Accumulation of 3-O-methylglucose as a function of the extracellular concentration. Note that the accumulation ratio $[S]_i/[S]_o$ falls markedly as $[S]_o$ increases. (*b*) The pH dependence of sugar uptake is a function of the membrane potential. In this instance the substrate is 6-deoxyglucose. Partial depolarization of the cells was achieved by the addition of K^+ ions. (*c*) A gated pore model to explain the roles of $\Delta\Psi$ and ΔpH in proton-sugar symport. (After Komor and Tanner, 1980, and earlier references; reproduced with permission of Elsevier Biomedical Press and Springer-Verlag.)

protonation of the carrier: depolarization shifted the pH spectrum for sugar influx toward the acid side (Figure 10.7b), indicating that $\Delta\Psi$ facilitates the protonation of the carrier.

In order to account for these findings, and for others that I have omitted for the sake of brevity, Komor and Tanner proposed a gated pore mechanism whose chief features are sketched in Figure 10.7c. Briefly, they postulate that both the proton-binding site and the sugar-binding site are situated deep within a proton-conducting cleft. The external half of this channel functions as a proton well, converting part of the membrane potential into an acidic pH at the bottom of the well, facing the two binding sites. By the same token, the interior half of the channel may generate a locally alkaline pH at the inward face of the carrier center, assisting the discharge of both H^+ and the sugar. Recent experiments (Komor et al., 1983) indicate that H^+ and the sugar do bind simultaneously and close together. Actual movements of the binding sites or of adjacent regions must be minor; this follows both from the high rate of transport and from the minimal effects of changes in $\Delta\Psi$ on the rate of transport.

The foregoing account goes well beyond the popular depiction of a symporter as a circle with two arrows; it must nevertheless be held overly simple, for it is inconsistent with the innocuous observations displayed in Figure 10.7a. According to Equation 10.1, the accumulation ratio $[S]_i/[S]_o$ should be a function of $\Delta\tilde{\mu}_{H^+}$ alone and independent of the absolute concentrations of S in cells and medium. But this is true in practice only when the external substrate concentration is well below the K_m for influx. Figure 10.7a illustrates for *Chlorella* that the steady-state level of internal sugar varies hyperbolically as a function of the external concentration such that when $[S]_o$ is high, the concentration gradient $[S]_i/[S]_o$ is lower than when $[S]_o$ is small. This makes physiological sense: if the concentration gradient were a function of $\Delta\tilde{\mu}_{H^+}$ alone, cells at high substrate concentrations would be liable to swell and burst. But how do porters take account of the absolute substrate concentration?

In their classic work on the lactose porter of *E. coli*, Rickenberg et al. (1956) explained a curve much like that in Figure 10.7a by the balance of two distinct fluxes: influx by a saturable carrier and exodus via a leak whose rate is proportional to $[S]_i$. We know now that this is not correct: leaks play but a minor role in lactose transport and exodus is chiefly mediated by the porter. It follows that the kinetics must be more sophisticated, allowing for what Eddy (1982) calls "slip." Figure 5.4 shows a diagram of the association of the lactose porter with the sugar and a proton. In principle the protein may form a complex with H^+ alone, with S alone, or with both. Equation 10.1 was derived on the assumption that the porter is mobile only when it is either unoccupied or else has bound both H^+ and S. Slip ensues if some of the dashed lines, designated as forbidden in Figure 5.4, become permissible—for example, at high $[S]_i$, efflux via the porter might occur without a proton. Under such circumstances $[S]_i/[S]_o$ will not represent an equilibrium distribution but a dynamic steady state accompanied by a net flow of protons through the system. A second mechanism that often enters into the dissociation of transport from the driving force is called trans inhibition, the allosteric inhibition of a porter by its accumulated substrate $[S]_i$. This mechanism is well established in the accumulation of amino acids by *Neurospora*.

The foregoing argument, which is representative of many others in the transport literature, is rooted in the assumption that reorientation of the carrier center is slow compared to the binding steps. There is really no experimental justification for this premise, and Sanders et al. (1984) have recently explored the consequences of discarding it. By the use of numerical methods they were able to fit a wide range of experimental data to computer-generated curves that assumed only variations in the rate constants for the association and dissociation reactions of the transport cycle. Their findings do not invalidate mechanistic interpretations drawn from earlier kinetic studies, such as proton wells and variable coupling stoichiometries, but they do render these interpretations somewhat suspect: the models that we have been discussing owe as much to their assumptions as to the experimental data.

Even for the best-studied porter, the $H^+-\beta$-galactoside symporter of *E. coli*, a molecular description of the reaction mechanism remains a distant prospect. Despite the intense efforts of half a dozen laboratories it is at present not certain whether the H^+/sugar stoichiometry is 1, 2, or variable; whether energy coupling by the proton current primarily increases J^o_{max}, alters the K_m for influx or for efflux, or modifies all these terms in ways that only a computer can express; whether protons and sugar bind to the porter in sequence or at random; whether slip reactions determine the accumulation ratio and, if so, what form they take; whether the functional carrier is the monomer, the dimer, or an oligomer; and whether $\Delta\tilde{\mu}_{H^+}$ shifts the equilibrium between two forms of this porter, one having a much higher substrate affinity than the other. Some of the apparent complexity of the galactoside porter may be due to variation of the kinetic parameters with the substrate. According to the recent work of Wright et al. (1985), the transport of lactose is especially complex, but the accumulation of β-galactosyl thiogalactoside can be described by a simple scheme: imposition of a protonic potential does not affect K_m but alters J_{max} by favoring the orientation of the unloaded binding site toward the exterior. The data would apparently be compatible with simultaneous translocation of protons and sugar, but do not exclude an alternative model (Lancaster, 1982) in which the entry of a proton reorients the sugar-binding site toward the exterior in the manner of a revolving door.

Resolution of the controversies over transport kinetics will not in itself carry us to the molecular level. The galactoside porter has been purified and reconstituted into liposomes in active form. It is known to span the membrane, its primary amino acid sequence has been deduced, and domains concerned with binding protons and substrate have been identified (Kaback, 1983; Wright et al., 1985). There are indications that histidine and sulfhydryl groups play a role in proton movements. But the findings do not yet convey a clear sense of how the lac porter functions as a working protein. For the foreseeable future, a circle with two arrows may have to do for a graphic summary of secondary energy coupling.

References

ADDISON, R., and SCARBOROUGH, G. A. (1982). Conformational changes of the *Neurospora* plasma membrane ATPase during the catalytic cycle. *Journal of Biological Chemistry* **257**:10421–10426.

AIDLEY, D. J. (1978). Op. cit., Chapter 9.

APPS, D. K. (1982). Proton-translocating ATPase of chromaffin granule membranes. *Federation Proceedings* **41**:2775–2780.

ARMSTRONG, C. M. (1981). Sodium channels and gating currents. *Physiological Reviews* **61**:644–683.

BAKKER, E. P. (1979). Op. cit., Chapter 3.

BENNETT, A. B., and SPANSWICK, R. M. (1983). Solubilization and reconstitution of an anion-sensitive H^+-ATPase from corn roots. *Journal of Membrane Biology* **75**:21–31.

BOWMAN, E. J., and BOWMAN, B. J. (1982). Op. cit., Chapter 9.

CHRISTENSEN, H. N. (1979). Exploiting amino acid structure to learn about membrane transport. *Advances in Enzymology* **49**:41–101.

CIDON, S., BEN-DAVID, H., and NELSON, N. (1983). ATP-driven proton fluxes across membranes of secretory organelles. *Journal of Biological Chemistry* **258**:11684–11688.

CONTI-TRONCONI, B. M., and RAFTERY, M. A. (1982). The nicotinic cholinergic receptor: Correlation of molecular structure with functional properties. *Annual Review of Biochemistry* **51**:491–530.

CRAIG, W. S. (1982). Op. cit., Chapter 9.

CRANE, R. K. (1977). Op. cit., Chapter 3.

CUPPOLETTI, J., and SEGEL, I. H. (1975). Kinetic analysis of active membrane transport systems: Equations for net velocity and isotope exchange. *Journal of Theoretical Biology* **53**:125–144.

DUAX, W. L., HAUPTMAN, H., WEEKS, C. M., and NORTON, D. A. (1972). Valinomycin crystal structure determination by direct methods. *Science* **176**:912–914.

DUPONT, Y. (1983). Is Ca^{2+}-ATPase a water pump? *FEBS Letters* **161**:14–20.

DUTTON, A., REES, E. D., and SINGER, S. J. (1976). An experiment eliminating the rotating carrier mechanisms for the active transport of Ca ion in sarcoplasmic reticulum membranes. *Proceedings of the National Academy of Sciences USA* **73**:1532–1536.

ECKERT, R., and RANDALL, D. (1983). Op. cit., Chapter 9.

EDDY, A. A. (1982). Op. cit., Chapter 9.

ENGELMAN, D. M., and ZACCAI, G. (1980). Bacteriorhodopsin is an inside-out protein. *Proceedings of the National Academy of Sciences USA* **77**:5894–5898.

FORGAC, M., and CANTLEY, L. (1984). Characterization of the ATP-dependent proton pump of clathrin-coated vesicles. *Journal of Biological Chemistry* **259**:8101–8106.

FUJIMURA, T., YAMATO, I., and ANRAKU, Y. (1983). Mechanism of glutamate transport in *E. coli* B. I. Proton-dependent and sodium ion-dependent binding of glutamate to a glutamate carrier in the cytoplasmic membrane. *Biochemistry* **22**:1954–1959.

GROSS, J., and MARMÉ, D. (1978). ATP-dependent Ca^{2+} uptake into plant membrane vesicles. *Proceedings of the National Academy of Sciences USA* **75**:1232–1236.

HAROLD, F. M. (1970). Op. cit., Chapter 3.

HAYDON, D. A., and HLADKY, S. B. (1972). Ion transport across thin lipid membranes: A critical discussion of mechanisms in selected systems. *Quarterly Review of Biophysics* **5**:187–282.

HEEFNER, D. L., and HAROLD, F. M. (1982). Op. cit., Chapter 5.

HENGGE, R., and BOOS, W. (1983). Op. cit., Chapter 5.

HIGGINS, C. F., HAAG, P. D., NIKAIDO, K., ARDESHIR, F., GARCIA, G., and FERRO-LUZZI AMES, G. (1982). Complete nucleotide sequence and identification of membrane components of the histidine transport operon of *S. typhimurium*. *Nature* **298**:723–727.

HILL, B. S., and HANKE, D. E. (1980). Properties of the chloride-ATPase from *Limonium* salt glands. *Journal of Membrane Biology* **51**:185–194.

HILLE, B. (1984). *Ionic Channels of Excitable Membranes*. Sinauer Associates, Sunderland, Mass.

JARDETZKY, O. (1966). Op. cit., Chapter 3.

JENCKS, W. P. (1980). The utilization of binding energy in coupled vectorial processes. *Advances in Enzymology* **51**:75–106.

JØRGENSEN, P. L. (1982). Op. cit., Chapter 9.

KABACK, H. R. (1983). Op. cit., Chapter 5.

KAKINUMA, Y., OHSUMI, Y., and ANRAKU, Y. (1981). Op. cit., Chapter 9.

KARLISH, S. J. D. (1980). Characterization of conformational changes in (Na, K) ATPase labeled with fluorescein at the active site. *Journal of Bioenergetics and Biomembranes* **12**:111–136.

KIMMICH, G., and RANDLES, J. (1980). Evidence for an intestinal Na^+/sugar transport coupling stoichiometry of 2.0. *Biochimica et Biophysica Acta* **596**:439–444.

KLINGENBERG, M. (1980). Op. cit., Chapter 7.

KLINGENBERG, M. (1981). Op. cit., Chapter 9.

KNAUF, P. A. (1979). Op. cit., Chapter 9.

KNOWLES, J. R. (1980). Enzyme-catalyzed phosphoryl transfer reactions. *Annual Review of Biochemistry* **49**:877–919.

KOMOR, E., SCHOBERT, C., and CHO, B.-H. (1983). The hexose uptake system of *Chlorella*: Is it a proton symport or a hydroxyl antiport system? *FEBS Letters* **156**:6–10.

KOMOR, E., and TANNER, W. (1974). The hexose-proton cotransport system of *Chorella*. pH-dependent change in K_m value and translocation constants of the uptake system. *Journal of General Physiology* **64**:568–581.

KOMOR, E., and TANNER, W. (1980). Op. cit., Chapter 9.

KRASNE, S., EISENMAN, G., and SZABO, G. (1971). Freezing and melting of lipid bilayers and the mode of action of nonactin, valinomycin and gramicidin. *Science* **174**:412–415.

KYTE, J. S. (1974). The reaction of sodium and potassium ion-activated adenosine triphosphatase with specific antibodies. *Journal of Biological Chemistry* **249**:3652–3660.

KYTE, J. S. (1981). Op. cit., Chapter 9.

LANCASTER, J. R. (1982). Op. cit., Chapter 5.

LATORRE, R., and MILLER, C. (1983). Conduction and selectivity in potassium channels. *Journal of Membrane Biology* **71**:11–30.

LEFEVRE, P. G. (1975). Op. cit., Chapter 9.

LEVER, J. E. (1984). A two-sodium ion/D-glucose symport mechanism: Membrane potential effects on phlorizin binding. *Biochemistry* **23**:4647–4702.

LOCKAU, W., and PFEFFER, S. (1983). ATP-dependent calcium transport in membrane vesicles of the cyanobacterium *Anabaena variabilis*. *Biochimica et Biophysica Acta* **733**:124–132.

MCLAUGHLIN, S., and EISENBERG, M. (1975). Antibiotics and membrane biology. *Annual Review of Biophysics and Bioengineering* **4**:335–336.

MITCHELL, P. (1979*b*). Op. cit., Chapter 3.

MITCHELL, P. (1981a). Op. cit., Chapter 3.

MOCZYDLOWSKI, E. G., and FORTES, P. A. G. (1981). Inhibition of sodium and potassium adenosine triphosphatase by 2', 3'-O-(2, 4, 6-trinitrocyclohexadienylidine) adenine nucleotides. *Journal of Biological Chemistry* **256**:2357–2366.

MOGI, T., and ANRAKU, Y. (1984). Mechanism of proline transport in *Escherichia coli* K12. *Journal of Biological Chemistry* **259**:7791–7796.

NAGLE, J. F., and MOROWITZ, H. J. (1978). Molecular mechanisms for proton transport in membranes. *Proceedings of the National Academy of Sciences USA* **75**:298–302.

NAGLE, J. F., and TRISTRAM-NAGLE, S. (1983). Hydrogen bonded chain mechanisms for proton conduction and proton pumping. *Journal of Membrane Biology* **74**:1–14.

NJUS, D., KNOTH, J., and ZALLAKIAN, M. (1981). Proton-linked transport in chromaffin granules. *Current Topics in Bioenergetics* **11**:107–147.

O'CONNOR, S. E., and GRISHAM, C. M. (1979). Manganese electron paramagnetic resonance studies of sheep kidney (Na^+ and K^+)-ATPase. Interactions of substrates and activators at a single Mn^{2+} binding site. *Biochemistry* **18**:2315–2323.

PARSEGIAN, A. (1969). Energy of an ion crossing a low dielectric membrane: solutions to four relevant electrostatic problems. *Nature* **221**:844–846.

PRESSMAN, B. C. (1976). Op. cit., Chapter 3.

RACKER, E., and HINKLE, P. C. (1974). Effect of temperature on the function of a proton pump. *Journal of Membrane Biology* **17**:181–188.

RAMOS, S., and KABACK, H. R. (1977). The relationship between the electrochemical proton gradient and active transport in *Escherichia coli* membrane vesicles. *Biochemistry* **16**:854–859.

REES-JONES, R., and AL-AWQATI, Q. (1984). Proton-translocating adenosinetriphosphatase in rough and smooth microsomes from rat liver. *Biochemistry* **23**:2236–2240.

RICKENBERG, H. V., COHEN, G. N., BUTTIN, G., and MONOD, J. (1956). La galactoside-permease d'*Escherichia coli*. Annales de'l Institut Pasteur **91**:829–857.

SANDERS, D. (1980). Op. cit., Chapter 9.

SANDERS, D., HANSEN, U.-P., GRADMANN, D., and SLAYMAN, C. L. (1984). Op. cit., Chapter 5.

SANDERS, D., SLAYMAN, C. L., and PALL, M. (1983). Op. cit., Chapter 9.

SINGER, S. J. (1974). The molecular organization of membranes. *Annual Review of Biochemistry* **43**:805–833.

SINGER, J. R. (1977). Thermodynamics, the structure of integral membrane proteins, and transport. *Journal of Supramolecular Structure* **6**:313–323.

STEINMETZ, P. R., and ANDERSON, O. S. (1982). Op. cit., Chapter 9.

STOECKENIUS, W., and BOGOMOLNI, R. A. (1982). Op. cit., Chapter 6.

TANFORD, C. (1982). Simple model for the chemical potential change of a transported ion in active transport. *Proceedings of the National Academy of Sciences USA* **79**:2882–2884.

TANFORD, C. (1983a). Mechanism of free energy coupling in active transport. *Annual Review of Biochemistry* **52**:379–409.

TANFORD, C. (1983b). Translocation pathway in the catalysis of active transport. *Proceedings of the National Academy of Sciences USA* **80**:3701–3705.

WEST, I. C. (1980). Op. cit., Chapter 5.

WIDDAS, W. F. (1980). Op. cit., Chapter 9.

WINKLER, H. H., and WILSON, T. H. (1966). The role of energy coupling in the transport of β-galactosides by *Escherichia coli*. *Journal of Biological Chemistry* **241**:2200–2211.

WRIGHT, J. K., DORNMAIR, K., MITAKU, S., MÖRÖY, T., NEUHAUS, J. M., SECKLER, R., VOGEL, H., WEIGEL, U., JÄHNIG, F., and OVERATH, P. (1985). Lactose: H$^+$ carrier of *Escherichia coli*: Kinetic mechanisms, purification and structure. *Annals of the New York Academy of Sciences*, in press.

11

The Major Organs of Movement

We seek to find nature one, a coherent unity. This gives to scientists their sense of mission, and let us acknowledge it, their aesthetic fulfillment: that every research carries the sense of drawing together the threads of the world into a patterned web.

Jacob Bronowski, *The Common Sense of Science*

• • •

How Muscles Contract
Calcium Ions Regulate Contraction
Undulipodia
Calcium Controls Movement and Behavior
Spasmonemes

Sessile plants and mobile animals epitomize two familiar and profoundly different ways of life. At a deeper level, however, the capacity to couple metabolic energy to the directed movement of bodies that are large by molecular standards is universal among eukaryotic organisms and cells. These movements come in a bewildering variety and range over the entire scale of natural sizes, but their diversity results largely from variations on a small number of molecular themes. It can hardly be an accident that energy transduction by means of actin, myosin, tubulin, and dynein and its regulation by calcium ions recur so persistently in discussions of cellular movements. These elements, together with some less familiar ones such as actin-binding proteins and spasmins, may be part of the ancestral endowment of all eukaryotic cells: more than a billion years have passed since the common ancestor of amoeba and humans flourished, but the molecular components of the motile machinery have preserved clear traces of their shared ancestry.

So far as we know, these particular arrangements have no counterpart in bacteria. Bacterial flagella are constructed of flagellin, which bears no homology to actin or tubulin, and they are energized by a proton circulation rather than by ATP (Chapter 5). Despite occasional claims that actin, myosin, tubulin, or molecules that may be ancestral to them are found in bacteria, the case has never been conclusively made. Moreover, bacteria apparently do not regulate cytoplasmic processes by monitoring the level of free calcium in the cytosol, a principle ubiquitous among eukaryotes high and low. For the present at least, we must rank this constellation of molecules among the criteria that distinguish the eukaryotic world from the prokaryotic one; and confess ignorance about its evolutionary origin.

The object of this chapter is twofold. Its first task is to survey the major classes of organs of motility: muscles, eukaryotic cilia and flagella, and spasmonemes. The second is to explore the molecular principles that underlie the transduction of chemical energy into mechanical work, since these are at present best displayed by the major organs of movement. Muscles, cilia, and flagella effect movement by forcing relatively rigid filaments to slide past one another, while spasmonemes probably contract by virtue of conformational changes in their constituent proteins. Additional principles of mechanochemical energy transduction, as yet ill-defined, will be considered in Chapter 12.

Most of what we know about the mechanisms of cellular movements stems historically and conceptually from the study of muscle. Contraction is the exemplar par excellence of the transduction of chemical energy into mechanical work. To be sure, muscle is a specialized tissue that appears relatively late in evolution, perhaps concurrently with the rise of the Metazoa at the beginning of the Cambrian epoch. But muscle is unique among motile systems in being composed predominantly of the small number of proteins that underlie its contraction. These are present in such regular order and at so high a concentration that electron microscopy and X-ray diffraction could interact with biochemistry to contribute important insights at all stages in the development of those techniques. The combination made muscle the paradigm for biological move-ment in general; any inquiry into its molecular basis must begin here.

How Muscles Contract

Sliding Filaments and the Cross-Bridge Cycle. When a striated muscle is stimulated with an electric pulse, it twitches: it shortens briefly by about a third and then returns to its original length. During the twitch the muscle may do work, perhaps by lifting a weight, and it will also produce heat. The energy released by the muscle is the sum of the heat of shortening and the work done (weight × distance). Efforts to identify the energy source for work and heat got under way in the 1920s and led eventually to the recognition that the immediate energy donor is ATP, produced either by respiration or by glycolysis. This simple answer was obscured for some years by the fact that creatine phosphate is much more abundant in muscle tissue than is ATP; therefore, if metabolism is blocked and the muscle is induced to contract, one observes only creatine phosphate breakdown while the ATP level holds constant. Not until 1962, when R. E. Davies introduced the use of dinitrofluorobenzene as an inhibitor of creatine kinase, was the obligatory relationship of ATP to contraction unambiguously established. Even so, the common view that creatine phosphate merely serves to buffer the ATP concentration requires revision: creatine phosphate is the form in which phosphoryl groups are translocated from the mitochondria to the contractile elements, and only there is ATP generated. Most of the heat evolved during a twitch can be attributed to the hydrolysis of ATP to ADP and P_i, but the time course of heat production is complex and includes contributions from other events, such as the release and reabsorption of calcium ions.

The chemical analysis of muscle began to make progress in the 1930s. Extraction of minced muscle with water removes salts and enzymes, leaving an insoluble residue of proteins. The pioneering research of J. T. Edsall, V. A. Engelhardt, A. Szent-Györgyi, and their associates defined the two major proteins that together make up some 80 percent of the insoluble fraction: myosin and actin. The molecular characterization of these and other participants in contraction (Table 11.1) still continues (Mannherz and Goody, 1976).

Myosin is insoluble in water but soluble in 0.6 M KCl. When V. A. Engelhardt and M. N. Ljubimova discovered in 1939 that myosin is an ATPase, they forged the first link between biological work and the emerging concept of ATP as the primary biological energy currency. Myosin is a large molecule and a very long one, composed of two heavy chains and four light ones. In the native molecule the chains are so intertwined as to generate a structure that somewhat resembles a double-headed golf club. Careful proteolytic dissection defined three functional domains (Figure 11.1a). Brief digestion of myosin with trypsin releases a large fragment (350 kdal) called heavy meromyosin, which contains both heads together with part of the tail. Papain, which cleaves the heavy meromyosin at

TABLE 11.1   Muscle, the Cast of Characters*

Component	Molecular mass (kdal)	Structure	Function
Myosin	470	Two heavy chains, four light, 160 nm long	ATPase; thick filaments
Actin	42	F actin is a linear polymer containing a variable number of G actin beads; monomer diameter, 5.5 nm	Major component of thin filaments
Tropomyosin	70	Two coiled subunits, 41 nm long, equal in length to 7 actin beads	Blocks binding of myosin to actin
Troponin	76	Troponin T; molecular size, 37	Binds tropomyosin and troponin C
		Troponin I; molecular size, 21	Controls inhibition of actomyosin
		Troponin C; molecular size, 18	Binds Ca^{2+}
α-Actinin	190	Two subunits, 30-nm rod	Component of Z line, anchors actin
β-Actinin	45		Determines length of thin filaments?

* Data for rabbit skeletal muscle, from Mannherz and Goody (1976). The contractile apparatus contains many additional proteins, often in small amounts, which play ill-defined roles in its function and organization.

another site, liberates the two heads individually (S_1 for subfragment 1; 120 kdal) as well as the tail fragment (S_2). The myosin heads, made up of the light chains plus portions of the heavy ones, are the business end of the molecule, containing both the binding site for actin and the catalytic site for ATP hydrolysis. The S_2 fragment is thought to represent a flexible hinge that links the heads to the tail proper. The various myosin fragments, being soluble in water, proved invaluable for exploring the structural and catalytic functions of both myosin and actin.

When the ionic strength of a myosin solution is reduced, the molecules aggregate spontaneously into a complex of remarkable structure. The long and relatively hydrophobic tails associate to form a stiff rod, with the heads clustered helically around each end; the polarity of the molecules in the aggregate reverses in the middle (Figure 11.1b). The structure of the complex closely resembles that of the thick filaments in which myosin resides in native muscle, as will be described shortly.

Actin, the other major protein of muscle, also comes in two forms. The

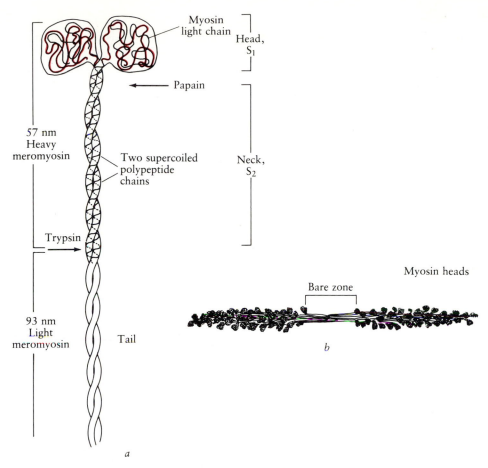

FIGURE 11.1   Myosin. (*a*) Topology and dissection by trypsin and papain. (After Lehninger, 1975, with permission of Worth Publishers.) (*b*) Myosin molecules aggregate spontaneously into a complex that resembles the thick filaments of muscle. (After Alberts et al., 1983, with permission of Garland Publishing.)

monomeric form, G actin, which contains one molecule of bound ATP, predominates at low ionic strength. At higher ionic strength, G actin polymerizes reversibly into helical filaments designated F actin. Concurrently, ATP is hydrolyzed to ADP which again remains bound. (The hydrolysis of ATP normally accompanies the polymerization of actin but is not required for polymerization to occur; some possible functions of this apparently wasteful step will be considered in the following chapter.) F actin is the species that is found in muscle. In situ, actin filaments take the form of a helical, double string of beads containing some 350 G actin monomers per micrometer; they have no overt enzymatic activities.

When solutions of myosin and actin are mixed in the absence of ATP, the two proteins interact, as shown by the sharp increase in viscosity; the adduct is called actomyosin. Subsequent addition of ATP has two effects. The viscosity

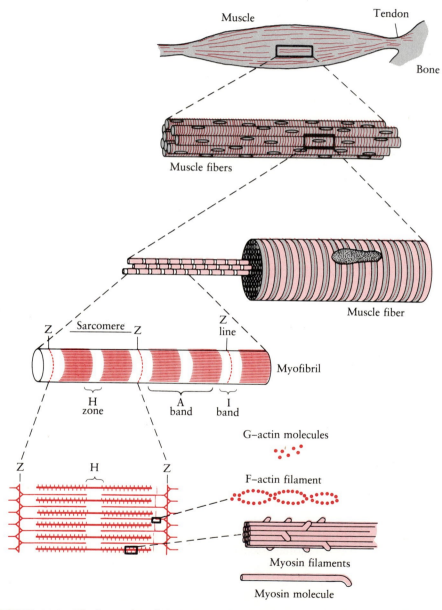

FIGURE 11.2   The hierarchy of skeletal muscle organization. (After Bloom and Fawcett, 1968, with permission of W. B. Saunders.)

falls, evidence that actomyosin has dissociated; at the same time ATP is hydrolyzed, but at a 200- to 500-fold greater rate than that due to myosin alone. In other words, actin enhances the ATPase activity of myosin. In a particularly dramatic experiment described in the 1940s by Albert Szent-Györgyi, a solution containing actin and myosin at high ionic strength was squirted into water through a fine needle. The proteins precipitated to form actomyosin threads, which could be made to shorten and develop tension by the addition of ATP. Evidently muscle contraction was the result of changes in the molecular structure of actomyosin induced by ATP. The full answer, it seemed, could not be far off.

Our entire conception of biological movement was to be transformed by the systematic application of X-ray diffraction and electron microscopy to the study of muscle, begun independently in about 1950 by Hugh E. Huxley and Andrew F. Huxley. But first something must be said about the anatomy of muscle (Figures 11.2 and 11.3). A fiber of vertebrate skeletal muscle is about one millimeter in diameter and several centimeters in length. It is a syncytium of cells containing several hundred nuclei enveloped by a common plasma membrane, the sarcolemma. Immersed in the communal cytoplasm are bundles of contractile fibrils, the myofibrils, each made up of a string of repeating units called sarcomeres. These sarcomeres, about 2 to 3 μm long and 1 μm in diameter, are the elementary units of contraction: muscle contracts because each individual sarcomere shortens. Sarcomeres are lined up in series and myofibrils line up in register; it is the latter order that produces the characteristic striations of skeletal muscle. To understand how muscles contract, we must discover how a sarcomere shortens.

It is not possible here to give a proper account of the methods used in these structural studies or even of the experimental results; the argument has been lucidly summarized by both Huxleys (Huxley, 1957; Huxley, 1969, 1976). The heart of the matter is that each sarcomere is packed with two kinds of filaments

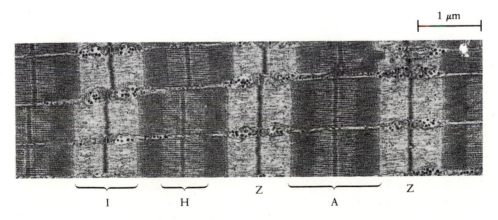

FIGURE 11.3 Electron micrograph of a longitudinal section of frog sartorius muscle, showing two whole and two half sarcomeres of three myofibrils. Zones designated I, H, and A bands and the Z line are labeled. (Electron micrograph courtesy of L. D. Peachey.)

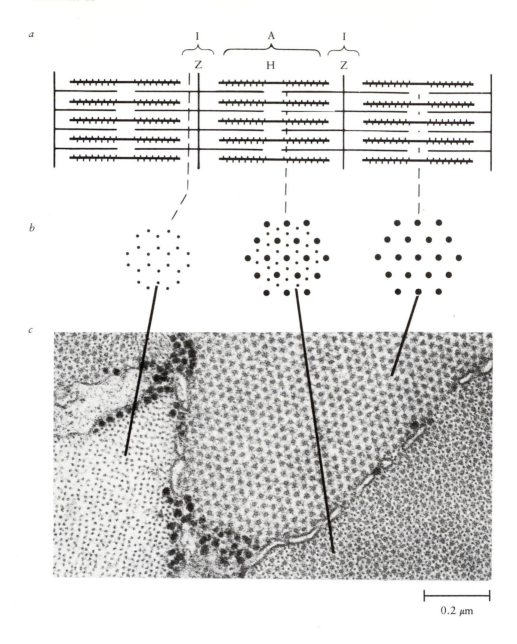

FIGURE 11.4 The physical basis of myofibril ultrastructure. (*a*) Diagram of three sarcomeres with thick and thin filaments, showing how their interdigitation gives rise to I, A, and H bands. (*b*) Imaginary sections through a sarcomere at various positions show thin filaments only (left), thick ones only (right), and both types (center). (*c*) Electron micrograph of a cross section in which the sarcomeres of adjacent myofibrils are out of register and can thus be matched with the corresponding profiles shown above. Frog sartorius muscle. (Electron micrograph courtesy of L. D. Peachey.)

(Figures 11.3 and 11.4). The thin filaments (diameter 8 nm), whose main constituent is actin, spring from the Z lines that define the boundaries of the sarcomere. Thick filaments (12 to 16 nm), mainly composed of myosin, lie in the middle of the sarcomere. The familiar striations of skeletal muscle identify zones that differ from one another with respect to their ultrastructure: a section through the I band on either side of the Z line reveals only thin filaments, a section through the H band, only thick filaments, while both filaments overlap in the remainder of the A band. The sliding-filament hypothesis of muscle contraction was proposed in 1954 by H. E. Huxley and J. Hanson and by A. F. Huxley and R. Niedergerke, primarily on the basis of the discovery that when a sarcomere shortens, neither the thin nor the thick filaments change in length; rather, they interdigitate. In a relaxed, extended muscle the zone of overlap is narrow, while in a fully contracted muscle it spans the entire sarcomere. The work done during muscular contraction was attributed to the development of active sliding forces between the thick and the thin filaments.

Both kinds of filaments exhibit structural polarity that corresponds to the direction of sliding. The polarity of the thin, or actin, filaments becomes manifest when they are "decorated" by binding heavy meromyosin (Figure 11.5); the

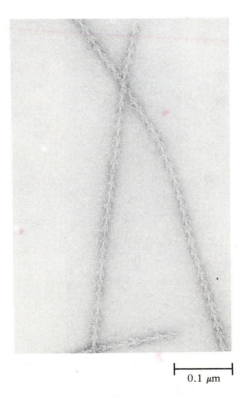

$$\vdash\!\!\!\!-\!\!\!\!\dashv$$
0.1 μm

FIGURE 11.5   Electron micrograph of rabbit muscle thin filaments decorated with myosin S_1. (Electron micrograph courtesy of K. A. Taylor and L. A. Amos.)

"arrowheads" point in the direction in which the filament tends to move. In situ, actin filaments are so arranged that their polarity in the left half of the sarcomere is opposite to that in the right half. The organization of the thick, or myosin, filaments is shown in Figure 11.6. Except for their bare midriff, the thick filaments are helically studded with projecting myosin heads; the polarity of each filament reverses at the midpoint. Direct visual evidence for the generation of sliding forces between the actin and myosin filaments comes from electron micrographs, most dramatically from those obtained recently using quick-frozen specimens. Figure 11.7 shows part of a sarcomere of insect flight muscle frozen after extraction with glycerol. The alternating thick and thin filaments run vertically through the section. They are connected by regular horizontal cross bridges, a state characteristic of muscles depleted of ATP (rigor). When ATP is

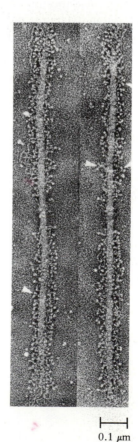

├──────┤
0.1 μm

FIGURE 11.6   Native thick filaments of rabbit muscle, showing the projecting myosin heads clustered at each end. (From Trinick and Elliott, 1979. Electron micrograph courtesy of A. Elliott.)

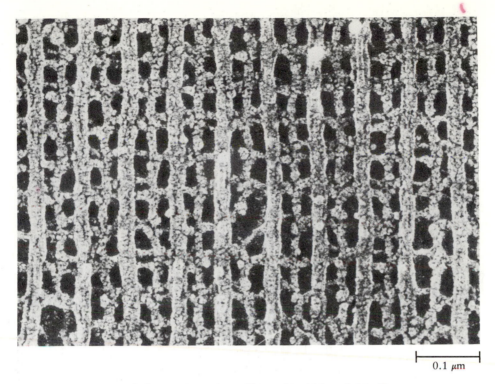

0.1 μm

FIGURE 11.7   Insect flight muscle in rigor. Alternating thick and thin filaments run vertically; horizontal cross bridges interconnect the two kinds of filaments. (From Heuser and Cooke, 1983, with permission from the *Journal of Molecular Biology*. Electron micrograph courtesy of R. Cooke.)

present, the cross bridges are dissociated and the filaments are disengaged. These cross bridges were identified with the myosin heads and the "cross-bridge theory" was born. Despite continuing controversy about all the details, its supremacy has never been seriously challenged.

The central thesis of the cross-bridge theory is that the myosin heads "walk" or "row" along the actin filaments, pulling them toward the center of the sarcomere (Figure 11.8). The physical basis of the sliding force is the cyclic formation and breakage of actomyosin links between the myosin heads and successive actin subunits. A myosin head first attaches to an actin monomer and then undergoes a vectorial change in configuration or orientation, such that the actin filament is pulled toward the middle. The cross-bridge then detaches, resumes its former configuration, and is ready to begin the next cycle by attaching to another actin bead. From the size of the myosin head one can estimate that the cycle shown in Figure 11.8 would involve incremental movements on the order of 10 nm, a value consistent with experimental data. It has also been calculated that the free energy available from the hydrolysis of a

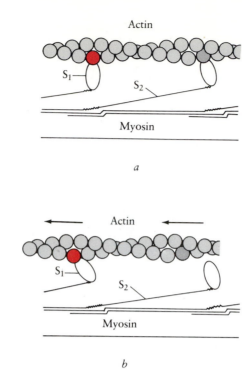

FIGURE 11.8   The cross-bridge cycle: how active changes in the angle of attachment of the cross bridge to actin filaments may produce sliding movements. (*a*) Left-hand bridge has just attached, right-hand bridge is already partly tilted. (*b*) Left-hand bridge has come to the end of its working stroke, right-hand bridge has detached and will probably not attach again until sliding movements bring an actin monomer into a favorable orientation. Note that bridges can act asynchronously since the subunit periodicities of actin and myosin differ. (After Huxley, 1976, with permission of Cold Spring Harbor Laboratory Press.)

molecule of ATP should suffice to support such a displacement against the full load a muscle can bear. The sarcomere as a whole shortens by multiple, repetitive, and independent cycles of the kind shown in Figure 11.8, each involving the movement of a single cross bridge by some 10 nm at the expense of one molecule of ATP. An important feature of this model is that actomyosin is formed transiently, and therefore ATP will be split only if the cycle goes to completion. Total ATP consumption will be determined by the rate of cross-bridge cycling, and the rate of energy release can be matched to the work being done (Huxley, 1957, 1974; Huxley, 1969; Tregear and Marston, 1979).

When a sarcomere shortens, its volume remains constant; the thick and thin filaments must then perforce lie farther apart than they did at rest. How does the cross bridge continue to maintain contact with its actin bead, on which the generation of force depends? Structural studies indicate that the myosin head is attached to the backbone of the thick filament by a flexible extension that can swing out from the thick filament as needed; a second flexible joint connects the head itself to the tail. Both joints are thought to correspond to protease-sensitive cleavage sites, where the molecular structure is relatively open. Between them, the two joints allow the head to maintain at all times the attitude required for contact (Figure 11.8).

We do not know precisely how the cross bridge generates the force that slides the two filaments past each other, but there is widespread agreement on the mechanical principles. Diffraction patterns and electron micrographs of insect muscle in the absence of ATP, that is, in a state of rigor, show the cross bridges attached to actin at an angle of about 45°. In relaxed muscle (ATP present, Ca^{2+} ions absent) the bridges are detached and stand away from the myosin backbone at right angles. A similar right-angle configuration is generated by analogues of ATP that bind to myosin but are not hydrolyzed. On the assumption that these two positions correspond to those of the normal cross-bridge cycle, as illustrated in Figure 11.8, we can account for the generation of the sliding force as follows. ATP causes the myosin head to detach from actin at the end of the working cycle and returns it to the vertical position. In this state it can bind to an actin subunit farther along the filament. The myosin head then tilts or rotates as a result of changes in its internal structure that are coupled to the hydrolysis of ATP, as suggested diagrammatically in Figure 11.9. The myosin head serves as a lever, connected to the backbone of the thick filament by an elastic linker that may correspond to the junction of the head with the tail (fragment S_2). The tilting

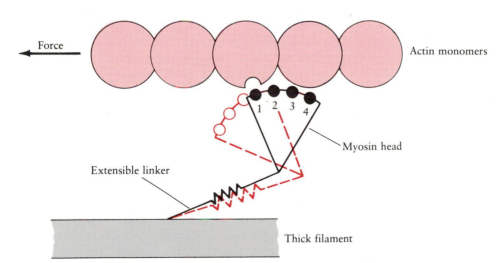

FIGURE 11.9   Hypothetical mechanism of cross-bridge tilting and force generation. The diagram shows a myosin head with four binding sites (black); these sites associate successively with a binding site on a particular actin bead. The myosin head tilts progressively (colored outline), which stretches the extensible linker. The tension pulls the actin filament to the left, causing it to slide past the myosin filament. (After Huxley, 1974.)

motion depicted in Figure 11.9 will stretch that linker, producing a force that tends to displace the actin filament to the left with respect to the myosin filament. Movement results, not from the activity of any one cross bridge but from hundreds of asynchronous cross bridges cycling on each thick filament (Huxley, 1969, 1976; Huxley, 1974; Tregear and Marston, 1979; Goody and Holmes, 1983).

Let it be clearly stated that this is a hypothesis, consistent with all the data available at present but not necessarily their only allowable interpretation. Students of muscle contraction are quick to point out that there is no assurance that the normal cross-bridge cycle involves the particular angles inferred from muscles in rigor and in relaxation or, indeed, that the cross bridge rotates at all in the normal cycle. There is no unequivocal evidence that the cross bridge contacts actin at its tip, as depicted in Figure 11.8. An early proposal that calls for association between actin and the hinge region of the myosin head has recently been revived. In the cycle shown in Figure 11.8, a single myosin head contacts a single actin bead. Recent data suggest, however, that each head makes contact with two neighboring actin beads, and it is quite possible that the cross-bridge cycle involves the concerted action of both heads found on each myosin molecule rather than a single one. These and other doubtful matters are fully considered in recent reviews by Pollack (1983), Goody and Holmes (1983), and Harrington and Rodgers (1984), who rightly insist that the case is far from closed. On the other hand, it would be a mistake to minimize the tremendous heuristic contribution that the cross-bridge theory has made. It remains the foundation for contemporary research on muscle and on all other actin- and myosin-based movements.

Energy Transduction.   How does the cross-bridge cycle harness the free energy of ATP hydrolysis to the generation of a vectorial sliding force? This central question takes us back to the biochemical realm. Since the determination of the intermediates and of the rate constants in biochemical mechanisms must be done in homogeneous solution, the system of choice is the interaction of heavy meromyosin (the S_1 myosin head) with F actin and with ATP. It will be recalled that ATP dissociates actomyosin. This can be construed as competition between ATP and actin for binding to the myosin head, although the binding sites for ATP and actin are not the same: ATP binds to the myosin head with great affinity, inducing an allosteric change that displaces the head from the actin monomer. (Control of this process by calcium ions will be outlined below.) About 1970 E. W. Taylor and R. W. Lymn, and also Tonomura, made the remarkable discovery that the displacing ATP is immediately hydrolyzed to ADP and P_i, but these products remain firmly bound at the catalytic site. There is no covalent phosphorylation of the myosin, yet energy is conserved in the sense that the hydrolysis of ATP to bound ADP and P_i is freely reversible; the free energy of that reaction is much smaller than that of ATP hydrolysis to free ADP and P_i. The myosin$\cdot$ADP$\cdot P_i$ complex recombines with actin and decays via a series of intermediates that can be detected by spectroscopic techniques; in the course of

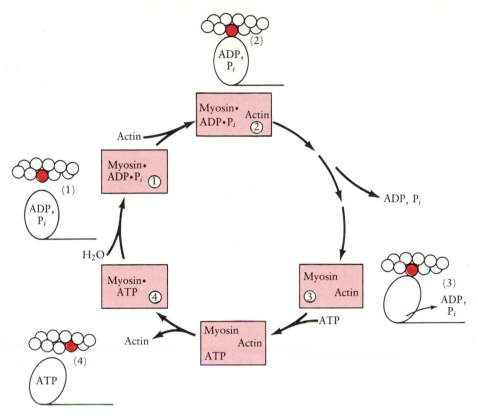

FIGURE 11.10   ATP hydrolysis by actomyosin, aligned with the cross-bridge cycle. (1) Myosin head with bound ADP and P_i. (2) Attachment of the myosin head to actin generates a 90° cross bridge. (3) Progressive tilting of the cross bridge with release of ADP and P_i. (4) Binding of ATP displaces the myosin head from actin; hydrolysis restores the 90° configuration.

these reactions, which terminate with actomyosin, P_i and ADP are released. These studies provided kinetic and biochemical evidence that actomyosin is a transitory entity that dissociates and re-forms with each turn of the ATPase cycle (Taylor, 1979).

In Figure 11.10 this sequence of enzymatic events has been aligned with the mechanical cross-bridge cycle. From the thermodynamic viewpoint, the complex of myosin with bound ADP and P_i (scheme 1) can be described as a state of high free energy that binds to an actin subunit, generating a 90° cross bridge (scheme 2). The newly formed cross bridge is likewise "energized," that is, it has a strong propensity to change its state for one of lower free energy. During the subsequent steps, the bulk of this free energy is utilized to slide the two filaments past one another. The end of the "power stroke" finds the myosin head still firmly attached to its actin bead, but in a tilted conformation, corresponding to that of the rigor cross bridge, while ADP and P_i have been liberated into solution (scheme 3). The arrival of a fresh molecule of ATP induces the release of the myosin head from actin (scheme 4). Hydrolysis of the ATP to bound ADP and P_i is accompanied by a return of the head to its original state, ready to form a $90°$ cross bridge with a *different* actin monomer and to start a new cross-bridge cycle. The combination with a new actin constitutes the vectorial component of the system.

The details are much more uncertain than the preceding confident description would suggest. The pathway of ATP hydrolysis is definitely more complex, and it is likely that the simple alternation between an attached state of the myosin head and a detached state is also oversimplified. (For some current controversies see Taylor, 1979; Eisenberg and Greene, 1980; Goody and Holmes, 1983.) The deeper issue is, just what does ATP do and how is energy transduced into work?

Overall, the driving force for the mechanical work performed between stages 2 and 3 is provided by the free energy of ATP hydrolysis. The most graphic and popular conception of the molecular mechanism regards the cross bridge as analogous to a loaded spring. The hydrolysis of ATP is thought to put the myosin head into a strained conformation, which attaches to actin in the $90°$ configuration; the strain relaxes as the head rotates with respect to the linker and pulls the actin bead along. But what do we mean by the metaphor of the loaded spring? A more abstract, but thermodynamically more informative description of the cycle puts the emphasis on the utilization of binding energies to effect mechanical work (Huxley, 1974; Jencks, 1980). The sole function of ATP is to detach the myosin head from actin at the end of the cross-bridge cycle. Energy input takes the form of the large favorable binding energy of ATP to myosin; this supplies the driving force required to overcome the tight association of myosin with actin. Bound ATP is immediately hydrolyzed to bound ADP and P_i, and the head rotates to the vertical position. This ensures that when the head next binds to actin, it binds to a subunit other than that from which it has just dissociated. But there is no need to regard myosin $\cdot$ ADP $\cdot P_i$ as a *strained* protein conformation. Rather, it is one that has a favorable free energy of binding to actin. The hypothetical positions 1, 2, 3, and 4, placed sequentially on the myosin head (Figure 11.9) represent states of progressively lower free energy (that is, tighter binding). Consequently, following attachment at M_1 the head tends to rotate successively through these positions, finishing up in the tilted configuration of the rigor cross bridge and at the bottom of a free-energy well. A third viewpoint regards the reorientation of the myosin head as an allosteric process, analogous to that which hemoglobin undergoes when it binds oxygen. Let us recall that the passage of the head through positions 1, 2, 3, and 4 is accompanied by a discharge of P_i and ADP. If the angle between the myosin head and the actin bead depends on the nucleotide bound at the catalytic site, then one can envisage this angle changing progressively as ATP is hydrolyzed and the products are released (Morales and Botts, 1979; Botts et al., 1984).

It will be obvious that the foregoing account does not consider specific molecular mechanisms but states general principles, analogous to those that describe the operation of primary transport systems and to the conformational-coupling mechanism of ATP synthesis by the F_1F_0 ATPase (Boyer, 1975; Jencks, 1980; Cross, 1981). In terms of the cross-bridge theory, conversion of the free energy of ATP hydrolysis into mechanical work depends on the successive placement of positions 1, 2, 3, and 4 on the myosin head and on the coupling rules that require the head to reorient when ATP is bound but forbid its passage through these positions unless it is associated with actin. The actual structural changes that accompany cross-bridge cycling remain unknown. There is much evidence in favor of the rotary motion discussed above, but changes in the internal structure of the head and even of the actin filament may well have a part to play. Some authorities maintain that the decisive structural change during contraction does not involve the myosin head at all, but a melting of the helical region of the S_2 filament into a shorter random coil (Ueno and Harrington, 1981).

Despite these uncertainties, the outlines of mechanochemical energy transduction by muscle are reasonably clear. The free energy of ATP hydrolysis is harnessed by means of changes in the structure of the molecules that make up the cross bridge. Mechanical work is not done when ATP binds to myosin, nor when it is hydrolyzed. Rather, free energy is somehow "stored" in the detached configuration of the myosin head and released with the performance of work as the head binds and tilts. Just what this means in molecular and geometric terms is for future research to answer.

Calcium Ions Regulate Contraction

The capacity to generate force is useful only to the extent that contraction and relaxation can be induced by physiological signals. Since the 1940s, beginning with the prescient work of L. V. Heilbrunn, various observations implicated calcium ions in the control of contraction. Early preparations of "actomyosin," for example, hydrolyzed ATP only in the presence of Ca^{2+}; synthetic actomyosin, prepared by mixing purified actin and myosin, does not require Ca^{2+}. Evidently, calcium ions are part of a regulatory mechanism rather than an element of contraction itself.

Just how this mechanism operates was clarified through the efforts of a generation of biochemists, beginning with S. Ebashi and A. Weber in the late 1950s (Murray and Weber, 1974; Gergely, 1976; Taylor, 1979; Perry, 1979; Adelstein and Eisenberg, 1980). The key discovery was that the thin filaments contain, in addition to actin, proteins that mediate the interaction of the thin filaments with calcium. This, in turn, regulates the ATPase activity of actomyosin. Briefly, tropomyosin (Table 11.1) is a filamentous protein that lies in the grooves of the F-actin chain, extending the length of seven actin beads (Figure 11.11a). Tropomyosin itself does not recognize calcium; that is the role of the trio of proteins called troponins, which are present to the extent of one troponin complex per actomyosin control unit. Several models of the control mechanism are under consideration, the most popular of which has long been the steric-

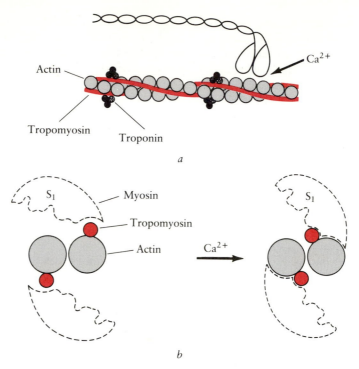

FIGURE 11.11   Control of energy transduction by calcium ions. (*a*) The biochemical machinery. The tropomyosin-troponin complex lies in the grooves of the thin filaments, extending over seven actin beads, and blocks actomyosin formation or activity; Ca^{2+} ions overcome the inhibition. (*b*) A current view of spatial relationships at the myosin-actin interface. In the absence of Ca^{2+}, tropomyosin is situated in the blocking position and prevents the association of the myosin head S_1 with actin. Binding of Ca^{2+} ions to troponin causes tropomyosin to shift to the active position and actomyosin formation proceeds. Drawn approximately to scale. (After Taylor and Amos, 1981, and Squire, 1981.)

blocking hypothesis. When Ca^{2+} ions are absent, the troponin-I subunit, in collaboration with troponin T and tropomyosin, is positioned such that it physically blocks the sites at which the myosin heads bind to the actin filament. When the level of ambient free calcium rises, from 10^{-7} M in resting muscle to about 10^{-5} M, troponin C binds calcium ions. A conformational change ensues in the troponin complex, which is transmitted to the tropomyosin and shifts the latter partway out of the groove. This exposes the binding sites on actin and allows the actomyosin cycle to proceed (Taylor and Amos, 1981; Figure 11.11*b*).

Steric blocking is probably part of the story, but not all of it. For instance, Chalovich and Eisenberg (1982) recently studied the binding of isolated myosin S_1 to calcium-regulated preparations containing actin together with tropomyosin and troponin. Binding was found to be independent of the presence of Ca^{2+} ions, but calcium greatly enhanced the ATPase activity of the adduct. Such observations are quite at odds with the elementary steric-blocking model; they suggest instead that tropomyosin-troponin inhibits a kinetic step in ATP hydrolysis, possibly the one that corresponds to the rotation of the cross-bridge angle, and that Ca^{2+} binding relieves that inhibition by an allosteric mechanism. But if so, why are thin and thick filaments free to slide apart in the relaxed state? The regulation of muscle contraction by Ca^{2+} ions no longer appears simple. But to a first approximation the salient point is still that a rise in the calcium concentration in the vicinity of the myosin head elicits cross-bridge cycling; conversely, removal of calcium blocks the cycle, cross bridges detach, and the thick and thin filaments are free to slide apart.

And what controls the calcium ion concentration? Muscle cells contain an extensive network of tubules and vesicles, called the sarcoplasmic reticulum. When muscle is minced, the sarcoplasmic reticulum breaks up into closed vesicles, which in the presence of ATP take up calcium ions most avidly. The mechanisms by which calcium ions are accumulated and also released have been thoroughly studied (MacLennan et al., 1976; Tada et al., 1978; Carafoli and Crompton, 1978). Indeed, the sarcoplasmic calcium ATPase is one of the best characterized primary transport systems (Chapter 9). As long as the muscle remains at rest, ATP-dependent accumulation of calcium by the sarcoplasmic reticulum maintains a low level of free calcium, about 10^{-7} M. Muscle fibers are enveloped by a specialized system of membranes derived from the sarcolemma, whose ramifications penetrate to the myofibrils and form a functional connection between the sarcoplasmic reticulum and the nerve that governs it. Arrival of an impulse at the neuromuscular junction depolarizes the sarcolemma. The depolarization spreads along these membranes into the interior of the fiber and ultimately triggers the release of calcium ions from the sarcoplasmic reticulum by mechanisms that are not well understood. The rise in cytosolic free Ca^{2+} ions elicits contraction. When nerve impulses cease, Ca^{2+} ions are reaccumulated by the sarcoplasmic reticulum and the relaxed state is restored.

Before concluding this section it is necessary to add that the preceding account pertains to the voluntary skeletal muscles of vertebrates. Other kinds of muscles, while relying on the same principles, often differ in important details. All muscles are controlled by calcium, but the tropomyosin-actin mechanism outlined above is not universal. In invertebrate muscle, Ca^{2+} regulation is often mediated through myosin itself. In the scallop, for instance, the activation of ATPase activity by actin is blocked as long as Ca^{2+} is absent; when the Ca^{2+} level rises, the ions bind to a specific light chain of myosin and relieve the inhibition. A second kind of myosin-based regulation is common in vertebrate smooth muscles and also in some nonmuscle cells. Here activation of the ATPase requires phosphorylation of a myosin light chain. Phosphorylation is mediated by a specific kinase, which is regulated by calcium ions. These muscles lack troponin; instead, a homologous protein called calmodulin (Chapter 13) monitors the cytosolic calcium level and is absolutely required for kinase activity. A third class of regulatory processes, whose physiological role is not clear, involves the phosphorylation of some myosins by protein kinases controlled by cyclic AMP (Adelstein and Eisenberg, 1980). These control mechanisms, and others as well, will turn up again when we come to consider other kinds of cellular movements.

Undulipodia

Sliding and Bending.   Cilia and flagella are motile appendages whose function is to generate relative motion between the cell and its liquid environment. In sedentary organisms, such as clams or the large protozoan *Stentor*, fields of cilia beat in unison to produce currents that carry food particles into the gullet. Analogous ciliary motions sweep the dust from our own lungs and transport the egg through the reproductive tract. Small organisms of many kinds propel themselves through fluid with the aid of cilia and flagella in search of food, light, mates, or safety. *Paramecium* and other ciliates swim by the coordinated beating of their cilia, many algae are endowed with one flagellum or more, and most biologists have at least a nodding acquaintance with the motion of sperm tails. The number, length, and beat patterns of cilia and flagella vary widely, and there is no unambiguous way of differentiating between the two kinds of appendages. We speak of cilia when they are short, numerous, and have a three-dimensional beating pattern, and of flagella when they are long, few in number, and undulate in a planar fashion. Their ultrastructure, however, is remarkably uniform in both arrangement and dimensions, as though a single design has been adapted to multiple purposes while being conserved in its essentials. Since this chapter is primarily concerned with the common principles of structure and energy transduction, I shall adopt the term "undulipodia" to designate eukaryotic cilia and flagella, as advocated by Margulis (1970). The term excludes bacterial flagella, whose structures and energetics are based on quite different principles (Chapter 5).

Undulipodia beat, that is, they propagate bends along their length, usually in the direction from base to tip. This sets the adjacent liquid in motion, and if the cell body is free to move, it propels it in one direction. In the case of flagella, whose beat is approximately planar, the cell body normally moves in the direction opposite to that of wave propagation. Ciliary movements are more complex and three-dimensional rather than planar. The stroke of a cilium consists of a planar effective stroke during which the shaft is almost straight, followed by a recovery stroke in the same plane (or sometimes with a sideward twist), as shown in Figure 11.12. In actuality, cilia are usually found in dense and regular arrays that do not permit the individual cilium to beat singly. Beating is coordinated, by mechanisms that are not well understood, to produce metachronal waves that pass over the cell surface somewhat like waves of bending pass over a field of grain blown by the wind. The hydrodynamic aspects of ciliary motions will not be discussed here; readers are referred to a comprehensive volume edited by Sleigh (1975) and to a lucid article by Holwill (1977).

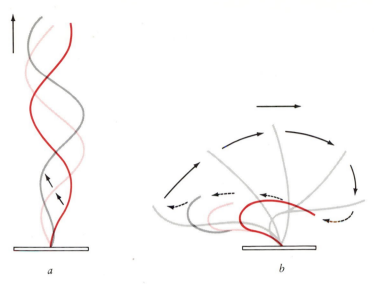

a   b

FIGURE 11.12   Patterns of motion in flagella and cilia. (*a*) Successive waves propagated toward the tip of a flagellum move water (arrows) and propel the cell body in the opposite direction. (*b*) The beat of a cilium consists of a straight-armed power stroke (light-gray cilia), followed by a curling return stroke (colored and dark-gray cilia); the latter may be in a third dimension, out of the plane of the paper. (From "How Cilia Move," by P. Satir. Copyright © 1974 by Scientific American., Inc. All rights reserved.)

The essential identity of all undulipodia, despite their diverse motions and wide range of lengths (5 to 1000 μm), becomes apparent under the electron microscope. The volume of the organelle, some 0.2 μm in diameter, is almost entirely filled by the motile machinery, called the axoneme, which sprouts from a complex basal body and extends to the tip. The limiting membrane is continuous with the plasma membrane so that the axoneme is topologically intracellular and in communication with the bulk cytoplasm. The great majority of axonemes are constructed from highly standardized elements according to a uniform pattern dubbed 9 + 2. Their architecture is shown schematically in Figure 11.13; details will be found in reviews by Holwill (1977) and Warner (1976, 1979). The most prominent elements are 11 microtubule doublets, which are built up of helical arrays of tubulin subunits. We shall return to microtubule structure in the following chapter. The cross section of an axoneme shows an outer ring of nine doublet microtubules, each composed of a complete subfiber A and an incomplete subfiber B. Attached to each subfiber A are paired arms, which contain an ATPase called dynein; the arms form intermittent cross bridges with the adjacent doublet. The bundle of nine doublets is held together by interdoublet links (nexin) and by a set of radial spokes that connect the outer ring of doublets with the central pair; this is why the ciliary membrane can be removed without damaging the axoneme's integrity. Note that the placement of each doublet is

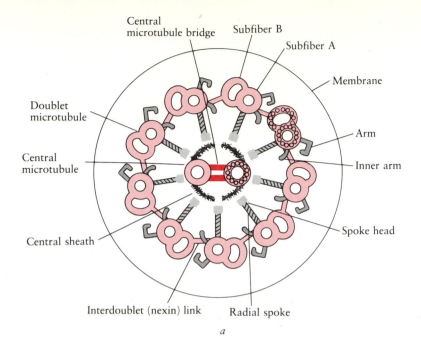

a

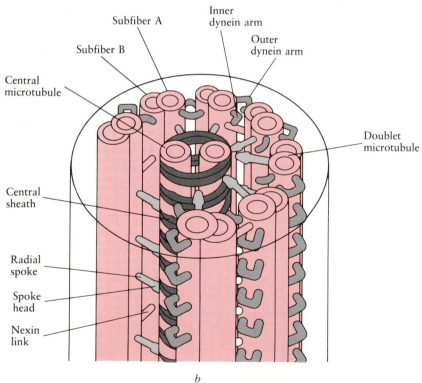

b

FIGURE 11.13   The anatomy of undulipodia. (*a*) Cross section, showing the arrangement of microtubules and linking structures. (From Holwill, 1977, with permission of Academic Press.) (*b*) Longitudinal perspective. (From "How Cilia Move," by P. Satir. Copyright © 1974 by Scientific American., Inc. All rights reserved.)

skewed with respect to the plane of the ring, and that all the dynein arms point in the same direction around the circumference.

The radial spokes, which are not made of tubulin, also spring from each subfiber A and point toward the center of the axoneme. Each spoke terminates in a distinct head that appears to touch the central sheath. Spokes come in groups of three (in some organisms, two), regularly spaced along the doublet; the spokes are apparently arranged helically within the axoneme matrix. Finally, there is a central structure of considerable complexity. The central "sheath" is thought to consist of parallel hoops projecting from the two central fibers and tilted with respect to their axis (Figure 11.13b). The two central tubules are not identical and may be twisted about each other. Clearly, undulipodia are mechanochemical devices of extraordinary intricacy, appreciably greater than that of muscle. Their structural complexity is mirrored by the resolution of solubilized axonemes into no fewer than 170 distinct proteins.

The ultimate goal is to explain how undulipodia beat, but this is well beyond our present capacity. A more immediate objective is to explain how axonemes bend, and substantial progress in this direction has been made. The essential first step was the demonstration by quantitative electron microscopy that the organelles operate by a sliding-filament mechanism (Satir, 1968, 1974). The nine doublets grow out of a fixed basal plate in the cell body. Therefore if the doublets do not change in length but merely slide past one another, then doublets on the inside of a bend should protrude farther into the tip than doublets on the outside of the bend. The expected displacement was calculable from the geometry and checked by using the radial spokes as a register for adjacent doublets (Figure 11.14). These experiments make it clear that microtubules do not contract and must perforce slide.

A dramatic visual confirmation of the sliding-filament hypothesis was supplied by Summers and Gibbons (1971), who worked with sperm tail "models" whose outer membrane had been removed by gentle treatment with nonionic detergents. This leaves the axoneme exposed but intact; on addition of ATP beating resumes and the models propel themselves through the medium. This procedure, which has been successfully applied to a variety of undulipodia, demonstrates that in every case the immediate energy donor is the magnesium salt of ATP. Now, when the naked axonemes were first treated briefly with trypsin, the effect of ATP was quite different: the axonemes did not beat but rather telescoped apart into structures up to seven times their original length (Figure 11.15). Evidently trypsinization destroys a component required to convert active sliding into bending and makes it possible to study active sliding by itself. The nature of the trypsin-sensitive entity is not entirely clear. Trypsinization destroys both the radial spokes and the nexin interdoublet links; it seems clear that both are normally required to convert sliding into bending (see below). Mutants of *Chlamydomonas* whose flagella lack spokes cannot beat, but ATP makes them slide apart. On the other hand, axonemes from *Tetrahymena* cilia come apart in response to ATP (endogenous proteases may be responsible), yet they possess intact spokes (Warner, 1979).

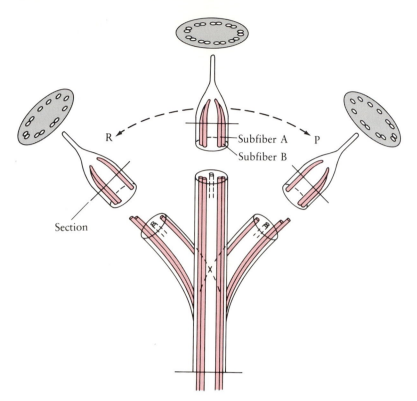

FIGURE 11.14 Evidence for a sliding-filament mechanism in undulipodia. If doublets do not change in length but slide past one another, doublets on the inside of a bend should protrude farther into the tip than those on the outside of the bend. Cross sections at the tip accord with this expectation. P, power stroke; R, return stroke. (After Satir, 1974, and Eckert and Randall, 1983.)

By analogy with muscle, one expects the sliding force to be generated between adjacent doublets, and there is strong evidence that the instruments thereof are the two dynein arms (Figure 11.13). The original evidence came once again from the work of Gibbons and his associates, who showed that extraction of sea urchin sperm tails with 0.5 M KCl selectively removed the outer row of arms and diminished the frequency of beating in the presence of ATP; the shape of the bending wave, however, was unchanged. In terms of the sliding filament model, this was interpreted to mean that the rate of sliding is a function of the number of arms working. Indeed, readdition of the extracts to the depleted axonemes allowed the arms to rebind and restored the normal beating frequency. The extracts contain an ATPase, now referred to as dynein 1, which is almost certainly the enzyme responsible for the bulk of flagellar energy conversion (Gibbons and Gibbons, 1973). Similar but not identical dynein ATPases are found in the flagella of *Chlamydomonas* and the cilia of *Tetrahymena* and

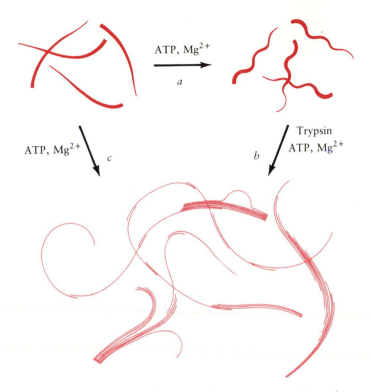

FIGURE 11.15 (a) When isolated cilia or flagella are stripped of their plasma membrane and reactivated by the addition of ATP and Mg^{2+}, they normally resume beating. (b) When reactivated flagella are lightly trypsinized, the doublet microtubules slide apart. (c) Cilia from *Tetrahymena* slide apart even without prior proteolysis on addition of ATP and Mg^{2+}. (After Warner, 1979, with permission of Academic Press.)

Paramecium, where they perform an analogous function (Gibbons, 1975; Blum and Hines, 1979; Warner, 1979; Warner and Mitchell, 1980; Gibbons, 1981). Some of the most striking evidence comes from work with mutants of *Chlamydomonas*: mutant axonemes that lack arms do not beat, contain no dynein ATPase, and do not slide apart (Huang et al., 1979).

Now, relative sliding cannot of itself induce bending; it will only give rise to forces of compression or of tension within the axoneme. To cause bending, one requires struts set at right angles, strong enough to withstand the axial thrust. This is almost certainly the function of the radial spokes, and the ultrastructural studies of Satir, Warner, and their colleagues provide considerable insight into the mechanism by which local bending is produced. Briefly, longitudinal sections of axonemes show that in straight stretches the spoke heads are free, while in bent ones they are attached to the central sheath projections. It appears, then, that at the leading edge of a propagating bend the spokes are caused to attach, while at the trailing edge they detach. That the radial spokes convert sliding into bending seems to be confirmed by the isolation of spokeless mutants of *Chlamydomonas* that retain the capacity to generate sliding forces but not to bend. On the other hand, the finding that other spokeless mutants can initiate bends of an abnormal pattern indicates that we still do not understand the function of spokes. This admission is doubly true of the two central fibers; it is believed that these have to do with coordination of the spoke cycle to produce local, propagated bending on alternate sides of the axoneme, but just how this is accomplished is not yet clear.

This survey of undulipodial mechanics leaves the reader (and the author) with a sense of inordinate complexity, but this impression may be misleading. Biologists have described flagella and cilia that do not conform to the full 9 + 2 pattern, lacking spokes or the central fibers, but that bend nonetheless. A recent note (Prensier et al., 1980) reports a flagellum from a parasitic protozoan that is fully functional even though it contains only three doublet microtubules and lacks spokes, central fibers, and sheath. Evidently, there exist relatively simple arrangements that support the basic mechanochemical functions of undulipodia, even if they are not as sophisticated as the deluxe model, and it is a pity that these simpler devices have received so little attention.

Dynein Arms and Energy Transduction.   From the viewpoint of bioenergetics, the central problem is how the dynein arms transduce the free energy of ATP hydrolysis into sliding forces. In outline, the answer appears to be as follows. The dynein arms form intermittent cross bridges between subfiber A and the adjacent subfiber B, thereby exerting a vectorial force on the latter. Repetitive formation and detachment of the bridges causes the arm to "walk" along subfiber B, pushing the latter toward the tip in increments of about 2 nm (Sale and Satir, 1977; Warner, 1979). There is good reason to believe that the outer and inner arms, despite differences in structure and composition, are functionally equivalent in that either set can support sliding; one must therefore expect both to be capable of forming cross bridges. The first prediction of this model is that the arms can exist in two states, attached and detached. Thin sections usually show the arms free, but conditions have recently been devised that preserve the cross bridges in an attached state, apparently analogous to rigor in muscle. For example, when cilia of mussel gills are extracted with Triton, most of the arms are found attached to subfiber B; subsequent addition of ATP and magnesium ions causes detachment of the arms (Figure 11.16).

The cross-bridge cycle appears to be broadly analogous to that by which myosin heads generate sliding forces between thick and thin filaments of muscle. In several organisms ATP was shown to detach the dynein arms even in the presence of vanadate, which blocks the hydrolysis of ATP. When hydrolysis is allowed to proceed, the dynein arms pass through a complex sequence of configurations. These were interpreted as indicating that ATP binding dissociates the cross bridges while its hydrolysis restores the attached rigor state (Sale and

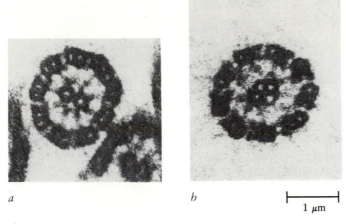

a   *b*   ⊢————————⊣
                          1 μm

FIGURE 11.16   Axonemes from mussel gill cilia in cross section, showing the standard pattern of nine outer doublet microtubules and two single ones in the center. (*a*) With the dynein arms attached. (*b*) With the dynein arms detached. Arrows indicate typical dynein arms. (From Satir et al., 1980, with permission of Claitor's Publishing Division. Electron micrograph courtesy of P. Satir.)

Gibbons, 1979; Satir et al., 1981; Goodenough and Heuser, 1982; Avolio et al., 1984). A formal cross-bridge cycle analogous to that for muscle is presented in Figure 11.17.

Thanks to recent advances in the techniques of specimen preparation, the morphological study of dynein arms has been pressed far beyond the simple conception of projecting linkers (Figure 11.18*a*). The outer dynein arms of *Chlamydomonas* and *Tetrahymena* consist of three globular heads connected by slender and flexible stalks to a rootlike base. The root anchors the "bouquet" to subfiber A, while the heads interact with subfiber B in the ATP-dependent reaction to produce sliding force (Goodenough and Heuser, 1982, 1985; Johnson and Wall, 1983; Avolio et al., 1984). The general construction of dynein arms resembles that of myosin with its multiple heads and long, narrow stem. The cross-bridge cycle involves a reorganization of the entire structure. One of several possible interpretations is depicted in Figure 11.18*b*, but readers should keep in mind that other investigators obtained somewhat different images and offer alternative views of what transpires during the cycle. What is clear is that dynein arms are sophisticated devices whose subtle motions are but crudely represented by the formal cross-bridge cycle.

The biochemical complexity of dynein arms mirrors the anatomical one. Gibbons and his associates isolated what appear to be the intact outer arms of sea urchin sperm flagella. The complex, called dynein 1 (> 1250 kdal), contains at least three large subunits (> 300 kdal) and three smaller ones, including a latent ATPase activity. Smaller preparations of dynein 1 presumably contain partially degraded outer arms. Harsher extraction procedures remove a second ATPase,

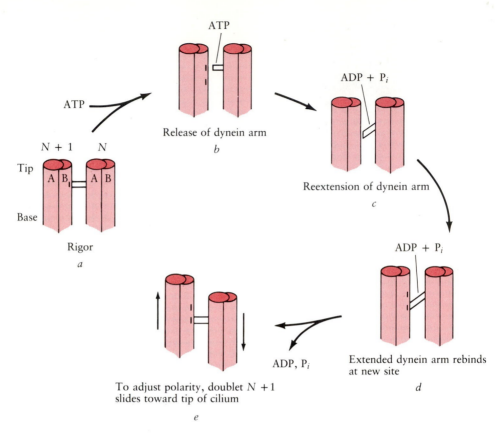

FIGURE 11.17   A hypothetical dynein cross-bridge cycle. (*a*) In the absence of ATP and at the end of the working cycle, the dynein arm is attached at 90°. (*b*) Addition of ATP causes the arm to detach and shorten. (*c*) As ATP is hydrolyzed to bound ADP and P$_i$, the arm extends in a tilted configuration. (*d*) The arm binds to subfiber B at a new site and reorients (*e*), returning to the rigor configuration while pushing subfiber B toward the tip. The stage at which ADP and P$_i$ are released is speculative. (After Satir et al., 1981, with permission of Alan R. Liss.)

dynein 2, whose location in the axoneme is unknown (Gibbons, 1981). These two dyneins do not simply correspond to outer and inner arms. Genetic studies with *Chlamydomonas* (summarized by Luck, 1984) indicate that mutants devoid of outer arms lack a set of 13 polypeptides including two distinct dynein ATPases (12S and 18S); mutants devoid of inner arms are deficient in a different set of 10 polypeptides, including two further dynein ATPases (11S and 12.5S). The finding that each arm contains two ATPases has recently been extended to sea urchin sperm flagella, but the significance of these multiple ATPases is obscure; presumably, each corresponds to one of the globular dynein heads.

Dynein ATPases are defined as a distinct family of energy-transducing

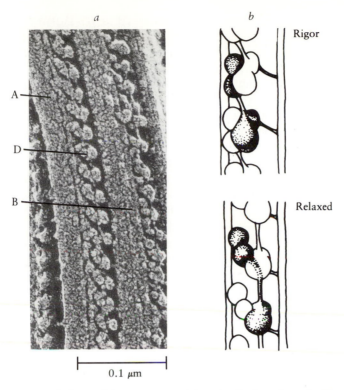

FIGURE 11.18   Morphology of dynein arms. (*a*) Longitudinal section of a *Tetrahymena* axoneme, with ATP present. The multiheaded dynein arms (D) are attached to subfiber A (A) but detached from subfiber B. (*b*) An interpretation of configurational transitions during the cross-bridge cycle. (After Goodenough and Heuser, 1985, with permission of the *Journal of Molecular Biology*. Electron micrograph courtesy of U. Goodenough.)

enzymes, specifically associated with microtubule-based sliding filament mechanisms. They are inhibited by vanadate ions and by EHNA, erythro-9-[3-(2-hydroxynonyl)]-adenine, apparently a specific reagent but not a very potent one. Known dyneins characteristically contain at least one subunit of very large molecular mass, 300 to 600 kdal. They bear no obvious chemical or structural resemblance to myosins, but catalyze ATP hydrolysis in a formally analogous manner. Kinetic studies (Johnson, 1983) suggest that rapid binding of ATP to dynein induces its dissociation from the adjacent tubulin subfiber; this is followed by hydrolysis to ADP and P_i, which initially remain bound. These observations fit neatly into a cross-bridge cycle (Figure 11.17), analogous to that for myosin. But our understanding of energy transduction by dynein ATPases is still quite rudimentary, and it remains to be seen how deep the parallels go.

Calcium Controls Movement and Behavior

The regulation of undulipodial beating is no less elaborate than that of muscle contraction and once again involves calcium ions, but the mechanisms are quite different. In muscle, calcium ions control the energy-transducing step itself; in undulipodia, regulation is removed from the primary mechanism of force generation, affecting instead the rate of beating and the direction of the power stroke. Two kinds of control are beginning to emerge. In clam gill cilia and sea urchin sperm tails a sudden rise in the cytosolic calcium level arrests beating. In *Paramecium* cilia, *Chlamydomonas* flagella, and many other systems a rise in the cytosolic calcium level alters the pattern of beating. We shall examine the latter mechanism in a little detail because it provides an elegant illustration of a general principle: the role of calcium ions as the link between sensory perception and the teleonomic control of movement and behavior.

Let us begin with the avoiding reaction that H. S. Jennings described in 1905 in his classic treatise on the behavior of lower organisms. When *Paramecium* encounters an obstacle or a chemical stimulus such as sodium ions, it backs up briefly and then resumes forward motion in a slightly different direction (Figure 11.19*a*). This pattern results from a temporary reversal of the direction of ciliary beating over the entire cell surface. That the direction of the power stroke is a function of the concentration of calcium ions, probably in the basal region of the cilia, was dramatically demonstrated by Naitoh and Kaneko (1972). Organisms whose membranes had been disrupted by gentle extraction with detergents continued to swim nicely when supplied with ATP and Mg^{2+}: they swam forward when the ambient calcium concentration was below $10^{-8} M$, backward when it was raised above $10^{-6} M$. The cytosolic calcium level in turn is controlled by the plasma membrane, as documented in a remarkable series of electrophysiological studies by R. Eckert, C. Kung, and their associates (Eckert, 1972; Kung et al., 1975; Kung and Saimi, 1982).

The electric potential across the *Paramecium* plasma membrane at rest is about -40 mV. This is an ionic diffusion potential, determined primarily by the potassium concentration gradient. Like other eukaryotes, *Paramecium* maintains a low level of cytosolic calcium by pumping the ion out, presumably by means of a calcium ATPase. There is thus a substantial electrochemical gradient favoring the diffusion of Ca^{2+} inward. Now, mechanical and electric stimuli alter the membrane potential in a manner clearly related to motile behavior. Gently tapping the anterior of *Paramecium* with a small stylus causes it to back up and also generates an action potential: an intracellular microelectrode records transient depolarization of the membrane, sometimes an overshoot to positive readings, followed by recovery (Figure 11.20). The action potential was traced to an influx of calcium ions in amounts sufficient to raise the level within the ciliary matrix from 10^{-7} to $4 \times 10^{-5} M$. Calcium influx is thermodynamically passive and mediated by voltage-sensitive channels. The sequence of events in the avoidance reaction comes about as follows (Figure 11.19 *b*). Anterior stimulation

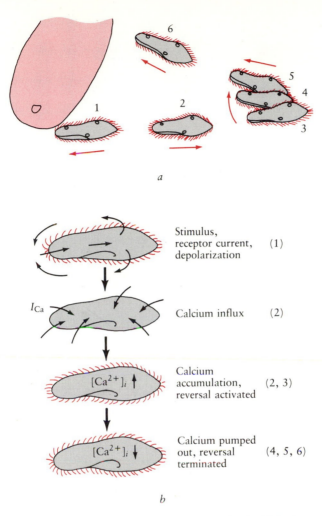

FIGURE 11.19   The avoiding reaction in *Paramecium*. (*a*) As the biologist sees it. The large oval is a solid object, the numbers refer to successive positions occupied by the animal, and the arrows show the direction of movement. (After Jennings, 1905, with permission of Indiana University Press.) (*b*) As the electrophysiologist interprets it. The ionic events described in the text are keyed to the behavioral response. (After Eckert, 1972. Copyright 1972 by the American Association for the Advancement of Science.)

causes a local depolarization of the membrane, initiated by special touch receptors (receptor current). The depolarization spreads over the entire cell surface, including the ciliary membrane, where most of the voltage-sensitive calcium channels are located. The latter open in response to depolarization, calcium rushes in, and the cilia reverse in response to the elevated calcium level.

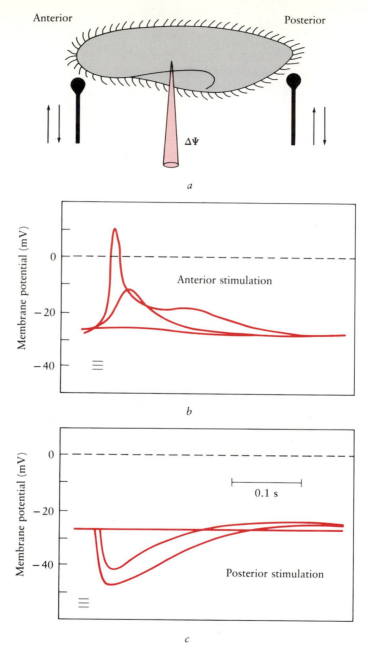

FIGURE 11.20   The action potential in *Paramecium*. (*a*) The anterior or posterior of the
organism is stimulated by tapping with a small stylus, while an intracellular electrode
monitors the membrane potential. (*b*) Time course of the action potential. Anterior
stimulation with taps of increasing force produces transient depolarization and eventually a
frank action potential, both due to Ca^{2+} influx. (*c*) Stimulation of the posterior produces
hyperpolarization due to K^+ efflux. The lower traces in (*b*) and (*c*) show the intensity and
duration of each stimulus. (After Eckert, 1972. Copyright 1972 by the American Association
for the Advancement of Science.)

But the opening of the channels is transient: within a few milliseconds they recover their initial low conductance, Ca^{2+} is pumped out, and forward motion resumes. Just how the calcium level determines the direction of the power stroke is not known.

Stimulation of the posterior end causes *Paramecium* to swim forward faster, as if to escape a possible threat. This response is correlated with hyperpolarization of the membrane (Figure 11.20) and with an increase in the rate of ciliary beating. Hyperpolarization was traced to a transient increase in the conductance of K^+ channels located at the cell's rear (enhanced K^+ efflux, hence a more negative potential). How this stimulates the cilia to beat faster is not clear, but again calcium may be the determining factor; perhaps, as the membrane hyperpolarizes, its calcium conductance falls. This may result in a lower steady-state level of free calcium within the ciliary cytosol and in accelerated beating (Eckert et al., 1976).

It may seem to some readers that I have dwelt to excess on the behavioral responses of an obscure organism, but there are good reasons for this. First, *Paramecium* lends itself to genetic analysis as well as to electrophysiology and is thus a prime candidate for the analysis of transmembrane signaling at the molecular level. Second, the control of locomotor behavior by calcium currents is widespread among ciliates, flagellated algae, and the specialized cells of higher organisms (for examples see Eckert, 1972; Schmidt and Eckert, 1976; Hyams and Borisy, 1978; Bessen et al., 1980); even amoeboid movement may be controlled by local calcium fluxes. Calcium need not always come from outside: often it is an intracellular calcium reservoir that discharges in response to an environmental stimulus. Finally, it is not a long step from these responses to the ever-deepening involvement of calcium ions in the control of cytoplasmic streaming, exocytosis, cell division, and other motile phenomena that are the subject of the next chapter. There is nothing trivial about what happens when *Paramecium* bumps into a bit of debris.

Spasmonemes

The waters of the local duck pond may be slimy but are sure to yield a rich harvest of ciliated and flagellated protists. With luck, there will be a specimen of *Vorticella*: a tiny funnel with a fringe of cilia, attached to the substratum by a stalk, not unlike a long-stemmed wineglass. Usually the stalk is extended while the organism feeds, but tap the dish and the stalk instantly collapses; *Vorticella* hunkers down and will not extend again until the danger is past. The contractile organelles of protozoa, called spasmonemes, have intrigued microscopists for a century. What justifies their inclusion in the present chapter, despite their restricted distribution, is that spasmonemes seem to embody molecular principles of movement quite unlike those of muscles and cilia and a different energy source as well.

In 1958, L. Levine and H. Hoffmann-Berling discovered the peculiar relationship between spasmoneme contraction and calcium (Hoffmann-Berling, 1958). They extracted *Vorticella* with glycerol, which disrupts the organization of the cell but leaves the contractile apparatus intact. Addition of micromolar levels of calcium induced the glycerinated stalks to contract; removal of calcium by chelation caused them to extend again, and the cycle could be repeated indefinitely in the absence of any obvious energy source. A little later the problem was taken up again by W. B. Amos, T. Weis-Fogh, and their associates, and it is to their systematic studies that we owe most of our insight into the way that changes in intracellular calcium concentration drive spasmoneme contraction (Amos, 1975; Amos et al., 1975; Routledge et al., 1976).

The contractile spasmoneme runs longitudinally within the stalk. When it contracts, it does not just thicken but is thrown into a curve, a helix, or a zigzag. This happens because the organelle is stiffened on one side by extracellular fibers that resist longitudinal compression. In the common *Vorticella* the spasmoneme is some 50 μm long and 1 μm in diameter. However, there exist large colonial vorticellids, the most spectacular of which is *Zoothamnium* (Figure 11.21). This consists of thousands of individuals, each connected by a small stalk to the communal spasmoneme trunk, which may be 1 mm in length and 30 to 40 μm across. The spasmoneme proper can be surgically removed from the trunk and studied in vitro. Calcium ions induce it to contract at a remarkable rate: in vivo the trunk is fully contracted within 8 ms, and even the isolated organelle shortens 15 times faster than the fastest striated muscle. Extension is much slower, a matter of several seconds.

Glycerinated stalks or excised spasmonemes are very sensitive to the Ca^{2+} concentration: fully contracted at $10^{-6} M$, extended at $10^{-8} M$. ATP is not required and metabolic poisons are ineffective, suggesting right from the beginning (Hoffmann-Berling, 1958) that calcium ions themselves must provide the energy for contraction. This is by no means impossible: as the calcium concentration changes, the chemical potential of Ca^{2+} varies as well according to the relationship

$$\Delta\mu_{Ca} = RT \ln \frac{[Ca^{2+}]_{upper}}{[Ca^{2+}]_{lower}}$$

The range from 10^{-6} to $10^{-8} M$ involves a change in chemical potential of 2.4 kcal per mole of Ca^{2+}, or 10 kJ. Now, the work output of a spasmoneme was measured at about 11 joules per kilogram (fresh weight). To do that much work, the organelle would have to bind 11 $J/10^4$ J/mol, or one millimole of calcium per kilogram, quite a modest amount, and it would also have to be capable of varying the internal calcium concentration over that range. Actual measurements made with an electron microprobe indicated that when isolated spasmonemes were shifted from 10^{-8} to $10^{-6} M$ Ca^{2+}, they bound some nine millimoles of Ca^{2+} per kilogram, nearly 10 times the minimum required to account for their

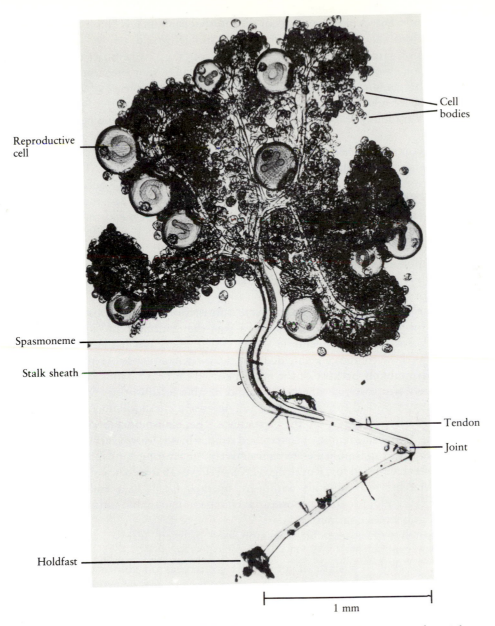

Cell bodies

Reproductive cell

Spasmoneme

Stalk sheath

Tendon

Joint

Holdfast

1 mm

FIGURE 11.21   A mature colony of *Zoothamnium geniculata* fixed in a state of partial contraction. (From Amos, 1975, with permission of Raven Press. Electron micrograph courtesy of W. B. Amos.)

work output. Direct measurements of the internal calcium level have not yet been reported, but the required concentration change would be well within the capacities of sarcoplasmic reticulum. Stalks contain a system of membranous sacs that, on cytochemical evidence, sequester calcium. One presumes, then, that contraction is induced and also energized by the sudden release of calcium ions from these vesicles in response to a disturbance in the neighborhood. As the calcium is gradually reaccumulated, the stalk extends once again and life returns to normal.

The contractile elements themselves are represented by bundles of longitudinal fibers, closely appressed to the calcium-binding reticulum. Their protein composition has been described by Routledge et al. (1976; Routledge, 1978). Actin is absent and so is tubulin; the major proteins found in spasmonemes of all species examined are a pair of very similar molecules, called spasmins (about 20 kdal). Spasmins bind calcium with high affinity, as shown by the effects of Ca^{2+} on the proteins' electrophoretic behavior; about two calcium ions are bound per protein monomer. Superficially at least, spasmins resemble troponin C and calmodulin; whether they do in fact belong to that same family of calcium-binding proteins remains to be seen.

The molecular basis of spasmoneme contraction also remains to be established. Hoffmann-Berling's original idea was that mutual repulsion of electric charges keeps the spasmoneme in an extended state. When the protein binds Ca^{2+} ions these charges are shielded and the protein collapses. This mechanism now seems unlikely because of the high specificity for Ca^{2+} ions. If contraction were purely electrostatic, other ions should be able to substitute. Measurements of changes in optical birefringence argue against a sliding-filament mechanism and also call into question the hypothesis that spasmoneme contraction is analogous to the shortening of a stretched rubber band. Moreover, when calcium is removed, the spasmonemes extend actively, generating a pushing force that needs to be accounted for. Routledge (1978) believes that calcium alters the angle of bonding between subunits within an ordered fibril; in a coiled or folded filament this could produce a substantial change in length and ultimately explain how spasmonemes contract.

Organelles that resemble spasmonemes in their ultrastructure are not uncommon among protozoa. The large ciliate *Stentor*, for instance, which can contract quickly to a sixth of its normal size, relies on contractile structures called myonemes, which are certainly homologous to spasmonemes. But it now appears that organelles of this type may also be widespread among plant and animal cells. The basal bodies of many cilia and flagella bear striated rootlets of unknown function; the rootlets of certain algal flagella contract in response to calcium and contain proteins that resemble spasmins (Salisbury and Floyd, 1978; Salisbury et al., 1984). Nothing is yet known of the mechanism by which rootlets contract, but it seems likely that the study of the obscure vorticellids pointed the way to the discovery of a new and ubiquitous principle of biochemical mechanics.

References

ADELSTEIN, R. S., and EISENBERG, A. (1980). Regulation and kinetics of the actin-myosin-ATP interaction. *Annual Review of Biochemistry* **49**:921–956.

ALBERTS, B., BRAY, D., LEWIS, J., RAFF, M., ROBERTS, K., and WATSON, J. D. (1983). *Molecular Biology of the Cell.* Garland, New York.

AMOS, W. B. (1975). Contraction and calcium binding in the vorticellid ciliates, in *Molecules and Cell Movement,* S. Inoué and R. E. Stephens, Eds. Raven Press, New York.

AMOS, W. B., ROUTLEDGE, L. M., WEIS-FOGH, T., and YEW, F. F. (1975). The spasmoneme and calcium-dependent contraction in connection with specific calcium-binding proteins. *Symposia of the Society for Experimental Biology* **30**:273–301.

AVOLIO, J., LEBDUSKA, S., and SATIR, P. (1984). Dynein arm substructure and the orientation of arm-microtubule attachment. *Journal of Molecular Biology* **173**:389–401.

BESSEN, M., FAY, R. B., and WITMAN, G. (1980). Calcium control of waveform in isolated flagellar axonemes of *Chlamydomonas. Journal of Cell Biology* **86**:446–455.

BLOOM, W., and FAWCETT, D. W. (1968). *A Textbook of Histology,* 9th ed. Saunders, Philadelphia.

BLUM, J. J., and HINES, M. (1979). Biophysics of flagellar motility. *Quarterly Review of Biophysics* **12**:103–180.

BOTTS, R., TAKASHI, R., TORGERSON, P., HOZUMI, T., MUHLRAD, A., MORNET, D., and MORALES, M. F. (1984). On the mechanism of energy transduction in myosin subfragment-1. *Proceedings of the National Academy of Sciences USA* **81**:2060–2064.

BOYER, P. D. (1975). Op. cit., Chapter 7.

CARAFOLI, E., and CROMPTON, M. (1978). Op. cit., Chapter 9.

CHALOVICH, J. M., and EISENBERG, E. (1982). Inhibition of actomyosin ATPase by troponin-tropomyosin without blocking the binding of myosin to actin. *Journal of Biological Chemistry* **257**:2432–2437.

CROSS, R. L. (1981). Op. cit., Chapter 7.

ECKERT, R. (1972). Bioelectric control of ciliary activity. *Science* **176**:473–481.

ECKERT, R., NAITOH, Y., and MACHEMER, H. (1976). Calcium in the bioelectric and motor functions of *Paramecium. Symposia of the Society for Experimental Biology* **30**:233–255.

ECKERT, R., and RANDALL, D. (1983). Op. cit., Chapter 9.

EISENBERG, E., and GREENE, L. E. (1980). The relation of muscle biochemistry to muscle physiology. *Annual Review of Physiology* **42**:293–309.

GERGELY, J. (1976). Troponin-tropomyosin-dependent regulation of muscle contraction by calcium, in *Cell Motility,* R. Goldman, T. Pollard, and J. Rosembaum, Eds. Cold Spring Harbor Laboratory Press, New York. Book A, pp. 137–149.

GIBBONS, I. R. (1975). The molecular basis of flagellar motility in sea-urchin spermatozoa, in *Molecules and Cell Movements,* S. Inoué and R. E. Stephens, Eds. Raven Press, New York, pp. 207–231.

GIBBONS, I. R. (1981). Cilia and flagella of eukaryotes. *Journal of Cell Biology* **91**:1075–1245.

GIBBONS, B. H., and GIBBONS, I. R. (1973). The effect of partial extraction of dynein arms on the movement of reactivated sea-urchin sperm. *Journal of Cell Science* **13**:337–357.

GOODENOUGH, U. W., and HEUSER, J. E. (1982). Substructure of the outer dynein arm. *Journal of Cell Biology* **95**:798–815.

GOODENOUGH, U. W., and HEUSER, J. E. (1985). Structural comparison of purified dynein proteins with *in situ* dynein arms. *Journal of Molecular Biology* **180**:1083–1118.

GOODY, R. S., and HOLMES, K. C. (1983). Cross-bridges and the mechanism of muscle contraction. *Biochimica et Biophysica Acta* **726**:13–39.

HARRINGTON, W. F., and RODGERS, M. E. (1984). Myosins. *Annual Review of Biochemistry* **53**:35–73.

HEUSER, J. E., and COOKE, R. (1983). Actin-myosin interactions visualized by the quick-freeze, deep-etch replica technique. *Journal of Molecular Biology* **169**:97–122.

HOFFMANN-BERLING, H. (1958). Der Mechanismus eines neuen, von der Muskel-Kontraktion verschiedenen Kontraktionszyklus. *Biochimica et Biophysica Acta* **27**:247–255.

HOLWILL, M. E. J. (1977). Some biophysical aspects of ciliary and flagellar motility. *Advances in Microbial Physiology* **16**:1–48.

HUANG, B., PIPERNO, G., and LUCK, D. J. (1979). Paralyzed flagellar mutants of *Chlamydomonas reinhardtii* defective for axonemal doublet arms. *Journal of Biological Chemistry* **254**:3091–3099.

HUXLEY, A. F. (1957). Muscle structure and theories of contraction. *Progress in Biophysical Chemistry* **7**:255–318.

HUXLEY, A. F. (1974). Muscular contraction. *Journal of Physiology* **243**:1–43.

HUXLEY, H. E. (1969). The mechanism of muscular contraction. *Science* **164**:1356–1366.

HUXLEY, H. E. (1976). The relevance of studies on muscle to problems of cell motility, in *Cell Motility*, R. Goldman, T. Pollard, and J. Rosenbaum, Eds. Cold Spring Harbor Laboratory Press, New York, Book A, pp. 115–126.

HYAMS, J. S., and BORISY, G. G. (1978). Isolated flagellar apparatus of *Chlamydomonas*: Characterization of forward swimming and alteration of waveform and reversal of motion by calcium ions *in vitro*. *Journal of Cell Science* **33**:235–253.

JENCKS, W. P. (1980). Op. cit., Chapter 10.

JENNINGS, H. S. (1905). *Behavior of the Lower Organisms*. Indiana University Press, Bloomington, Reprinted 1976.

JOHNSON, K. A. (1983). The pathway of ATP hydrolysis by dynein: Kinetics of a presteady state phosphate burst. *Journal of Biological Chemistry* **258**:13823–13825.

JOHNSON, K. A., and WALL, J. S. (1983). Structure and molecular weight of the dynein ATPase. *Journal of Cell Biology* **96**:669–678.

KUNG, C., CHANG, S. Y., SATOW, Y., HOUTEN, Y., and HANSMA, H. (1975). Genetic dissection of behavior in *Paramecium*. *Science* **188**:898–904.

KUNG, C., and SAIMI, Y. (1982). The physiological basis of taxes in *Paramecium*. *Annual Review of Physiology* **44**:519–534.

LEHNINGER, A. L. (1982). Op. cit., Chapter 2.

LUCK, D. J. L. (1984). Genetic and biochemical dissection of the eukaryotic flagellum. *Journal of Cell Biology* **98**:789–794.

MACLENNAN, D. H., STEWART, P. S., ZUBRZYCKA, E., and HOLLAND, P. C. (1976). Composition, structure and biosynthesis of sarcoplasmic reticulum, in *Cell Motility*, R. Goldman, T. Pollard, and J. Rosenbaum, Eds. Cold Spring Harbor Laboratory Press, New York, Book A, pp. 153–163.

MANNHERZ, H. G., and GOODY, R. S. (1976). Proteins of contractile systems. *Annual Review of Biochemistry* **45**:427–465.

MARGULIS, L. (1970). Op. cit., Chapter 6.

MORALES, M. F., and BOTTS, J. (1979). On the molecular basis for chemomechanical energy transduction in muscle. *Proceedings of the National Academy of Sciences USA* **76**:3857–3859.

MURRAY, J. M., and WEBER, A. (1974). The cooperative action of muscle proteins. *Scientific American* 230(2):58–71.

NAITOH, Y., and KANEKO, H. (1972). ATP-Mg-reactivated triton-extracted models of *Paramecium*: Modification of ciliary movement by calcium ions. *Science* 176:523–524.

PERRY, S. V. (1979). The regulation of contractile activity in muscle. *Biochemical Society Transactions* 7:593–617.

POLLACK, G. M. (1983). The cross-bridge theory. *Physiological Reviews* 63: 1049–1113.

PRENSIER, G., VIVIER, M., GOLDSTEIN, S., and SCHREVEL, J. (1980). Motile flagellum with a "3 + 0" ultrastructure. *Science* 207:1493–1494.

ROUTLEDGE, L. M. (1978). Calcium-binding proteins in the vorticellid spasmoneme. *Journal of Cell Biology* 77:358–370.

ROUTLEDGE, L. M., AMOS, W. B., YEW, F. F., and WEIS-FOGH, T. (1976). New calcium-binding contractile proteins, in *Cell Motility*, R. Goldman, T. Pollard, and J. Rosenbaum, Eds. Cold Spring Harbor Laboratory Press, New York, Book A, pp. 93–113.

SALE, W. S., and GIBBONS, I. R. (1979). Study of the mechanism of vanadate inhibition of the dynein cross-bridge cycle in sea urchin sperm flagella. *Journal of Cell Biology* 82:291–298.

SALE, W. S., and SATIR, P. (1977). Direction of active sliding of microtubules in *Tetrahymena* cilia. *Proceedings of the National Academy of Sciences USA* 74:2045–2049.

SALISBURY, J. L., BARON, A., SUREK, B., and MELKONIAN, M. (1984). Striated flagellar roots: Isolation and partial characterization of a calcium-modulated contractile organelle. *Journal of Cell Biology* 99:962–970.

SALISBURY, J. L., and FLOYD, G. L. (1978). Calcium-induced contraction of the rhizoplast of a quadriflagellate green alga. *Science* 202:975–976.

SATIR, P. (1968). Studies on cilia. III: Further studies on the cilium tip and a "sliding filament" model of ciliary motility. *Journal of Cell Biology* 39:77–94.

SATIR, P. (1974). How cilia move. *Scientific American* 231(4):44–52.

SATIR, P., WAIS-STEIDER, J., and AVOLIO, J., (1980). The mechanomorphology of microtubule sliding, in *38th Annual Proceedings of Electron Microscopy Society of America*, G. W. Bailey, Ed. Claitor's, Baton Rouge, La.

SATIR, P., WAIS-STEIDER, J., LEBDUSKA, S., NASR, A., and AVOLIO, J. (1981). The mechanochemical cycle of the dynein arm. *Cell Motility* 1:303–327.

SCHMIDT, J. A., and ECKERT, R. (1976). Calcium couples flagellar reversal to photostimulation in *Chlamydomonas reinhardtii*. *Nature* 262:713–715.

SLEIGH, M. A., Ed. (1975). *Cilia and Flagella*. Academic, New York.

SQUIRE, J. (1981). Muscle regulation: A decade of the steric blocking model. *Nature* 291:614–615.

SUMMERS, E., and GIBBONS, I. R. (1971). Adenosine triphosphate-induced sliding of tubules in trypsin-treated flagella of sea-urchin sperm. *Proceedings of the National Academy of Sciences USA* 68:3092–3096.

TADA, M., YAMAMOTO, Y., and TONOMURA, Y. (1978). Molecular mechanism of sarcoplasmic calcium transport. *Physiological Reviews* 58:1–79.

TAYLOR, E. W. (1979). Mechanism of actomyosin ATPase and the problem of muscle contraction. *CRC Critical Reviews in Biochemistry* 6:103–164.

TAYLOR, K. A., and AMOS, L. A. (1981). A new model for the geometry of the binding of myosin crossbridges to muscle thin filaments. *Journal of Molecular Biology* 147:297–324.

TREGEAR, R. T., and MARSTON, S. B. (1979). The crossbridge theory. *Annual Review of Physiology* **41**:723–736.

TRINICK, J., and ELLIOTT, A. (1979). Electron microscope studies of thick filaments from vertebrate skeletal muscle. *Journal of Molecular Biology* **131**:133–136.

UENO, H., and HARRINGTON, W. F. (1981). Conformational transition in the myosin hinge upon activation of muscle. *Proceedings of the National Academy of Sciences USA* **78**:6101–6105.

WARNER, F. D. (1976). Cross-bridge mechanisms in ciliary motility: The sliding-bending conversion, in *Cell Motility*, R. Goldman, T. Pollard, and J. Rosenbaum, Eds. Cold Spring Harbor Laboratory Press, New York, Book C, pp. 891–914.

WARNER, F. D. (1979). Cilia and flagella: Microtubule sliding and regulated motion, in *Microtubules*, K. Roberts and J. S. Hyams, Eds. Academic, New York.

WARNER, F. D., and MITCHELL, D. R. (1980). Dynein: The mechanochemical coupling adenosine triphosphatase of microtubule-based sliding filament mechanisms. *International Review of Cytology* **66**:1–43.

Filaments, Tubules, and Vesicles: Topics in Cellular Motility

Seek simplicity, and then distrust it.

A. N. Whitehead

In den Einzelheiten steckt der Teufel.
(The devil lurks in the details.)

German proverb

• • •

Whatever Became of Cytoplasm?
Biochemical Dynamics of the Cytoskeleton
Amoeboid Movements
Cytoplasmic Streaming
Cytoplasmic Transport of Vesicles
Cell Division

The most obvious difference between prokaryotic and eukaryotic cells is one of size: the volume of a bacterial cell is on the order of $1 \mu m^3$, that of even a small amoeba is a thousand times larger. To put it another way, a bacterial cell is small enough for diffusion to distribute ions and metabolites while macromolecules can be produced at, or very near, their site of function. The greater size of eukaryotic cells entails the need to transport molecules and large bodies over relatively long distances, from micrometers to centimeters and more in special cases. Perhaps it would be better to turn this point around: the invention of novel mechanisms for the movement of cells and their parts may have been what made eukaryotic cells functionally workable. We do not know what fraction of their energy budget eukaryotic cells expend on movement (this will in any case vary widely between cell types), but even a casual listing of cellular movements suffices to show their importance: amoeboid locomotion, mitosis and cell cleavage, cytoplasmic streaming, the flow of vesicles to sites of growth or activity, endocytosis, and exocytosis. . . . It is no exaggeration to insist that the utilization of metabolic energy for the directional transport of subcellular structures is a sine qua non of eukaryotic life.

The province of this chapter thus includes a large slice of cell biology and a burgeoning literature that mocks an author's delusions of adequacy. To survey cell motility comprehensively in a chapter is impossible; I have therefore elected to discuss a few examples of cellular motility in a little detail, with emphasis on the shifting perception of the mechanisms by which eukaryotic cells convert energy into mechanical work.

Whatever Became of Cytoplasm?

Biochemists, Erwin Chargaff once wrote, are peculiar people: they would take a fine Swiss watch, grind it up in a mortar, and then meticulously examine the debris in the hope of learning how the watch worked. There is truth enough in the taunt to discomfit a member of the tribe even today; but molecular rambunctiousness has been much gentled of late by the lessons of the electron microscopists. Every thin section of a eukaryotic cell displays a labyrinth of organelles, compartments, and filaments: it is obvious that the higher levels of biological order are easily destroyed and are likely to be overlooked unless precautions are taken to preserve at least part of the local organization. The electron microscope transformed our conception of cellular architecture and later laid the foundations for an understanding of how cells move, grow, direct their interior traffic, and establish each its particular form.

The concept of "protoplasm" as the seat of life processes evolved in about the middle of the nineteenth century. Most of the early cytologists appear to have envisaged it as a viscous fluid, but some were impressed by the contractile properties of protoplasm and by its immiscibility with water, features that suggested a degree of structural organization. In time, as nuclei and other inclusions were recognized, protoplasm was replaced by cytoplasm and karyo-plasm (the cytoplasm of the nucleus), and modern microscopes revealed a growing array of intracellular organelles: mitochondria, chloroplasts, lysosomes, endoplasmic reticulum, Golgi bodies, and a miscellany of tubules and vacuoles. Organelles often occupy remarkably fixed positions, and many communicate with one another and with the plasma membrane via mobile vesicles. By 1960 the concept of cytoplasm as a simple fluid in which the various organelles are suspended and enzymes and solutes are dissolved had begun to give way to an ever more structured conception of the cell's interior. The history of this development has been surveyed in a collection of essays that celebrate the twenty-fifth anniversary of the *Journal of Cell Biology* (Gall et al., 1981).

Microtubules.   In the early 1960s, thanks particularly to the introduction of glutaraldehyde as a fixative, cytologists began to describe a novel but widespread class of intracellular elements: long, slender filaments, about 25 nm in diameter, and apparently of tubular construction, which became known as microtubules. In a prophetic review of the subject, to which he had himself made many of the most notable contributions, Keith Porter (1966) clearly spelled out the basic features of microtubule structure; recognized that microtubules are universal constituents of eukaryotic cells, essentially identical whether they come from cilia, flagella, mitotic spindles, or locations in the cytoplasm; and pointed out that they seemed to function both in the transport of matter within the cytoplasm and in determining cell shape and organization. He even realized that micro-tubules in vivo must include functional constituents beyond those visible in the electron micrographs. Subsequent reviews of the microtubule literature include articles by Hepler and Palevitz (1974), Stephens and Edds (1976), Gunning and Hardham (1982), a book by Dustin (1984), and an invaluable compendium edited by Roberts and Hyams (1979).

The torrent of publications over the past two decades has abundantly confirmed Porter's insight that microtubules are universal elements of the cellular cytoskeleton. Unlike axonemal microtubules, cytoplasmic microtubules take the form of individual hollow cylinders composed of 13 parallel strands. In cross section one can generally make out 13 protomers, each 5 nm across, arranged in a ring; these are the ends of the protofilaments. The elementary subunit, or protomer (5 nm, 110 kdal), is itself a duplex consisting of one molecule each of the closely related α- and β-tubulins, plus two molecules of GTP whose significance will be discussed below. Tubulins, whatever their source, are much alike but not identical; a given cell may manufacture several distinct tubulins for particular purposes.

When tubulin is isolated from cells, it carries along several additional proteins known collectively as microtubule-associated proteins, or MAPs. These are certainly not adventitious contaminants; rather, they correspond to the fuzzy material that generally surrounds microtubules seen in section and forms links with adjacent microtubules or organelles. Tubulin also has specific binding sites for certain drugs, including the alkaloids colchicine and vinblastine and a series of synthetic benzimidazole derivatives such as nocodazole; these are playing a major role in research on microtubule assembly and function.

Cytoplasmic microtubules may extend the length of a cell and assume a variety of organizational patterns that are obviously related to the shape of the cell in which they reside. In many plant cells a belt of microtubules runs just beneath the plasma membrane. The early observation that cell wall fibrils parallel these microtubules suggested that the arrangement of microtubules somehow specifies the organization of microfibrils in the wall, and there is now abundant evidence to support this interpretation (Chapter 14). In protozoa, bundles and regular arrays of microtubules traverse anatomical features such as the axopodia of heliozoans or the feeding structures of ciliates, not to mention the axonemes of cilia and flagella. The microtubules of animal cells make up a complex scaffold that radiates from the vicinity of the nucleus to all regions of the cell (Figure 12.1*a*), and they compose the mitotic spindles that are a hallmark of all eukaryotic cells. It is very important to remember that some microtubular structures are permanent (axonemes, for instance); others are transient and may assemble or disassemble within minutes in response to physiological needs, temperature changes, or drugs (mitotic spindles and cytoplasmic tubules).

Microtubules are intrinsically polarized structures, and their assembly and disassembly are likewise directional. In the living cell, microtubule assembly is initiated at specialized structures called microtubule-organizing centers. The cytoplasmic microtubules of animal cells spring from centrioles located just outside the nucleus (Figure 12.1*a*). The basal bodies, or kinetosomes, from which cilia and flagella sprout are microtubule-organizing centers of well-defined ultrastructure based on a short, cylindrical array of microtubules. Centrioles are also constructed in this manner, but very often the loci from which cytoplasmic microtubules grow appear structurally quite amorphous. Just how microtubule-organizing centers function is still unknown.

Microfilaments.   Hard upon the discovery of microtubules came that of a second ubiquitous class of structural elements, the microfilaments. These take the form of double-stranded helical filaments, 6 to 7 nm in diameter, whose chief component is actin. Microfilaments occur in a variety of patterns ranging from microfilament bundles (as in the stress fibers that are conspicuous features of well-spread animal cells in culture) to a loose meshwork of short filaments, often subjacent to the plasma membrane of motile cells. When the microfilament system is visualized as a whole, it appears as a continuous three-dimensional sheath enclosing the bulk of the cytoplasm together with its microtubular scaffold. Thin actin filaments also ramify throughout the cytoplasm, especially in animal cells, where they are often interwoven with thick actin cables (Figure 12.1*b* and *c*). The growth of knowledge about microfilaments and actin can be traced through a succession of reviews by Pollard and Weihing (1974), Hepler and Palevitz (1974), Clarke and Spudich (1977), Korn (1978), Pollard (1981), and Tilney (1983).

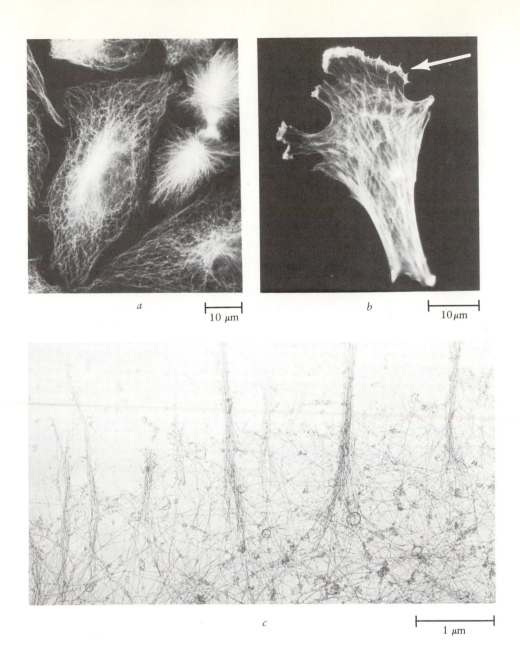

a   ⊢ 10 μm ⊣   *b*   ⊢ 10 μm ⊣

c   ⊢ 1 μm ⊣

FIGURE 12.1   Glimpses of cellular architecture. (*a*) Fibroblasts (3T3 cells) in culture, stained with fluorescent antibody to show the cytoplasmic microtubule network. The cell in the center is dividing. (From Brinkley et al., 1981; reproduced from the *Journal of Cell Biology* by copyright permission of Rockefeller University. Electron micrograph courtesy of B. R. Brinkley.) (*b*) Chick heart fibroblast stained with fluorescent antibodies to display stress fibers and microfilaments. Note the cell's leading edge (arrow). (From Small et al., 1982, with permission of Cold Spring Harbor Laboratory Press. Electron micrograph courtesy of J. V. Small.) (*c*) Leading edge of chick heart fibroblast after detergent extraction, showing arrangement of actin filaments. The prominent spikes are filopodia. (Electron micrograph courtesy of M. Schliwa.)

The morphology of microfilaments, resembling that of the thin filaments of muscle, first suggested that they might be composed of actin. Evidence for this proposition was obtained by Ishikawa et al. (1969), who applied to tissue culture cells H. E. Huxley's technique of "decorating" thin filaments of muscle with heavy meromyosin; it will be recalled that this procedure generates a characteristic arrowhead pattern (Figure 11.5). The method has been widely applied to animal and plant cells, both to identify actin filaments and to determine their polarity; in microfilaments anchored to the plasma membrane, for instance, the arrowheads usually point away from the membrane (Tilney, 1983).

Actin is a major cell constituent, comprising as much as 15 percent of the total protein of amoebas, fibroblasts, and neurons; it is quite abundant even in many plant cells. Its structure has been very well conserved, as suggested initially by the capacity of actin from any source to bind heavy meromyosin from striated muscle in the sterically specific fashion. All actins are alike in molecular size (near 42 kdal) and in amino acid composition, and all contain one molecule of bound ADP per monomer. Determination of the amino acid sequence of various actins has confirmed their conservatism; for example, 94 percent of the sequence of *Acanthamoeba* actin is identical with that of muscle (Korn, 1978). All this suggests that the interaction of actin with myosin in striated muscle is the climax of a far more ancient partnership. But there is too much actin in most cells and too little myosin for movement on the muscle model to be the sole biological function of actin. Rather, like tubulin, actin is part of a dynamic scaffold that helps to shape the cell in which it resides. The actin scaffold is again subject to rapid assembly and disassembly, particularly during cellular locomotion; in some cells as much as half the actin complement may be unpolymerized. Several fungal metabolites exert their effects by interfering with either the assembly or the disassembly of actin filaments, including the cytochalasins and phalloidins, and these have become valuable tools in the drive to unravel the tangled biology of these structures.

The Cytoplasmic Matrix.   Filamentous cellular elements may be composed of proteins other than actin and tubulin. Myosin is very widespread, probably universal, having been isolated from lower eukaryotes, animal cells, and various plants, but it is usually a minor constituent, less than 1 percent of total cell protein, and thick filaments corresponding to those of muscle are not conspicuous. At least passing mention should here be made of the "intermediate filaments," a heterogeneous class of filaments about 10 nm in diameter composed of proteins related to keratin (Goldman et al., 1979; Lazarides, 1980). They are found in a variety of animal cells, often constituting a basket that surrounds the nucleus.

With the discovery of cytoplasmic filaments, cells were seen to possess internal scaffolding on which mechanical force could be exerted and which helped confer on them their characteristic form. In recent years, however, advances in the techniques of sample preparation and the development of the high-voltage electron microscope have carried the solidification of the cytoplasm much further. Porter and his colleagues (Wolosewick and Porter, 1979; Porter and Tucker, 1981; Schliwa and van Blerkom, 1981; Porter et al., 1983) discovered that the spaces between the relatively gross microtubule and microfilament bundles are occupied by a tracery of slender and ephemeral fibrils, 2 to 3 nm in diameter, which they call microtrabeculae after the spongy regions in bone (Figure 12.2). Other investigators, employing alternative techniques to preserve the details of cytoplasmic organization (Heuser and Kirchner, 1980; Small et al., 1982), find somewhat different structures, notably a ubiquitous tangle of fine actin filaments (Figure 12.3), which may or may not correspond to Porter's microtrabecular lattice. Resolution of the dispute about the chemical nature of this lattice is of the utmost importance to cell physiology, for these filaments apparently connect various cytoplasmic elements—microtubules and microfilaments, plasma membrane and organelles—into a gelatinous network endowed with contractile properties. Some organelles, chiefly mitochondria and chloroplasts, retain a degree of independence, but in most respects the entire cytoplasmic matrix behaves as an integrated structure that responds as a unit to various stimuli.

What, then, of the cytosol, with its familiar complement of soluble enzymes, ions, and metabolites? There are innumerable open spaces within the microtrabecular lattice, room and to spare for the diffusion of small molecules. But one increasingly wonders about those "soluble" enzymes of classical biochemistry.

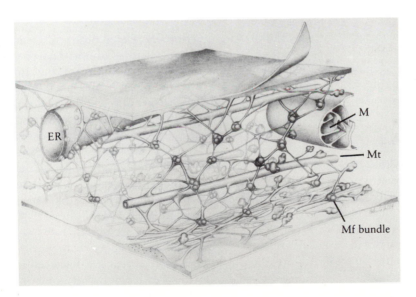

FIGURE 12.2   Model of the microtrabecular lattice, some 300,000 times its actual size, showing its relationship to other cytoplasmic elements. The microtrabeculae suspend the endoplasmic reticulum (ER), mitochondria (M), microtubules (Mt), and microfilament (Mf) bundles; polyribosome clusters are located at junctions of the lattice. (From Porter et al., 1983, with permission of Alan R. Liss. Photograph courtesy of K. R. Porter.)

0.1 μm

FIGURE 12.3   The structure of cytoplasm. Cytoskeleton from the leading edge of a fibroblast showing details of the actin meshwork. (From Small et al., 1982, with permission of Cold Spring Harbor Laboratory. Electron micrograph courtesy of J. V. Small.)

For one thing, many enzymes that are functionally related occur in clusters: fatty acid synthase, the pathway of aromatic amino acid synthesis, and the pyruvic dehydrogenase complex, among others. The function of enzyme clusters is presumably to ensure the correct channeling of diffusible metabolites, and probably also to speed their transit (Welch, 1977). But even beyond this, many "soluble" enzymes are, in the living cell, bound more or less firmly to structural elements: in muscle, phosphofructokinase and other glycolytic enzymes are bound to actin filaments, in erythrocytes to the plasma membrane. Experiments with whole cells that have been stratified by centrifugation suggests that few if any proteins are free in the cell sap. Even water is not exempt: as much as a third of the cell's water exhibits an ordered structure, somewhat like that of ice, within which much of cellular biochemistry is confined (Kell, 1979; Fulton, 1982; Clegg, 1982). Potassium ions, Na^+, and small metabolites are mobile in what remains of the cytosol, but calcium ions do not diffuse freely, a feature of obvious importance in understanding their role as messengers for localized cell functions (Chapter 13).

What has happened, then, is that over the years our perception of the cytoplasm has changed. We no longer see it as a fluid that permits random interactions among molecules large and small, but as a highly structured matrix that has about it the air of a well-managed bureaucracy: everything that is not forbidden becomes compulsory, and disorder signals pathology. It would be difficult to exaggerate the importance of this discovery: it is the foundation upon which we must build models of cellular motility and, more important, a realistic conception of biological order.

Biochemical Dynamics of the Cytoskeleton

The cytoskeleton is a dynamic structure whose form and organization can alter rapidly in response to either normal or abnormal stimuli. Amoeboid movements entail continuous, cyclic changes in the organization of cellular actin. When animal cells are chilled, their cytoplasmic microtubules disassemble in a vectorial manner beginning at the periphery; when they are rewarmed, microtubules grow back from the cell center adjacent to the nucleus. A more radical reorganization accompanies the division of animal cells: cytoplasmic microtubules disassemble and the mitotic spindle takes their place; stress fibers vanish, locomotion ceases, the cell rounds up, and during cytokinesis a transient contractile belt of microfilaments cleaves the cytoplasm. Evidently the cytoskeleton as it exists at any moment is the static expression of a dynamic steady state in which protomers of actin, tubulin, and perhaps others enter and leave their filamentous polymers, and these in turn pass in and out of more highly organized structures. Since the ordered construction and transformation of the cytoskeleton enters into all the modes of motility discussed in this chapter, an outline of the biochemical basis of these operations is pertinent.

Polarized Filaments.   As mentioned above, both microfilaments and microtubules are polarized structures whose ends are functionally different. In the case of microfilaments the polarity is plain to be seen when the filament is decorated with heavy meromyosin (Figure 11.5). The polarity of microtubules is not so readily documented, but here also appropriate techniques have now been devised (Euteneuer and McIntosh, 1981). Monomers can add to and dissociate from either end, as suggested in Figure 12.4a. Since the two ends are not equivalent, the rates of net elongation or shortening may be different at the two ends. However, if the exchange tends toward equilibrium, then at any particular concentration of monomers one would expect both ends to be either growing or diminishing. The reason is that the equilibrium constant, the ratio of the rate constants for the "on" and "off" reactions, must be the same at both ends. Moreover, at a particular concentration of monomers called the critical concentration, neither elongation nor shortening should occur and the filament should be stable, in equilibrium with the monomers.

In fact, as A. Wegner (1976) was the first to recognize, the assembly of microfilaments and microtubules is not an equilibrium process and obeys different laws. Actin is a case in point; G-actin monomers polymerize spontaneously to F actin, and they do so by adding preferentially to the "barbed" end, while dissociation occurs preferentially from the "pointed" end. (These designations refer to the flights of the arrowheads seen on meromyosin binding, Figure

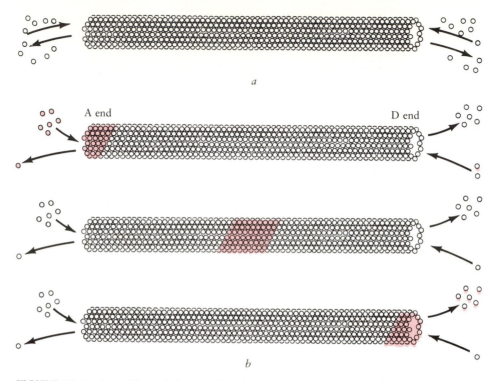

FIGURE 12.4   Assembly and disassembly of microtubules. (*a*) At equilibrium, monomers combine with and dissociate from both ends. (*b*) GTP hydrolysis supports a steady state, such that addition of monomers is favored at the assembly end (A) and dissociation at the disassembly end (D). As a result, monomers pass clear through the tubule before being discharged (treadmilling). (After Margolis and Wilson, 1981, with permission of *Nature*, Macmillan Journals.)

11.5.) Similarly, tubulin monomers add preferentially to one end of the microtubule and come off the other, as indicated in Figure 12.4*b*. It follows that the critical concentration is different at the two ends: over a particular range of monomer concentrations a filament may elongate at one end and shorten at the other, while the overall length remains constant. One of the consequences is the phenomenon called treadmilling (Figure 12.4*b*): an individual monomer incorporated at the assembly end may pass clear through the filament before dissociating from the disassembly end. The rate of subunit flux through the polymer depends greatly on the conditions of incubation: in some circumstances, rates as high as 50 μm/h have been observed (Cote and Borisy, 1981; Margolis and Wilson, 1981).

   The thermodynamic basis of this remarkable behavior is that the polymerization of G actin to F actin and of tubulin into microtubules is accompanied by an irreversible step, the hydrolysis of a nucleotide triphosphate. G actin contains a molecule of bound ATP, which is hydrolyzed to bound ADP concurrently with the polymerization step. ATP hydrolysis is not required for polymerization to occur: G actin loaded with a nonhydrolyzable analogue polymerizes perfectly well, but it does so symmetrically. By the same token, tubulin contains two molecules of bound GTP, one of which is converted into GDP during polymerization. Tubulin bearing a nonhydrolyzable analogue can polymerize but cannot establish a difference in critical concentration between the two ends (Wegner, 1976; Margolis, 1981). It appears, then, that cells expend ATP not in order to polymerize monomers into filaments but to do so in a polarized, vectorial manner.

To the biologist, the critical question is whether treadmilling occurs in vivo, and that is at present not known. Margolis et al. (1978) based a model of mitosis on this supposition and more recently suggested ingenious ways in which treadmilling may contribute to other instances of intracellular transport (Margolis and Wilson, 1981). These ideas remain speculative so long as there is no evidence for the occurrence of treadmilling in the living cell. A very appealing alternative hypothesis has been put forward by Kirchner (1980). He argues that most, if not all, cellular actin and tubulin filaments are not free but are attached to specific sites such that the attached end is blocked. For such filaments a flux of subunits is not possible, but the expenditure of ATP and GTP during polymerization nevertheless serves a useful purpose. Filaments anchored at the disassembly end, as microtubules seem to be, cannot shorten but will elongate from the assembly end until they come into pseudoequilibrium with the monomer concentration in the cytoplasm. At this concentration, any filament with a free disassembly end must depolymerize. The net result of the disparity between the critical concentrations at the two ends will be the elimination of unattached filaments and the suppression of spontaneous polymerization. In terms of this hypothesis, the object of GTP and ATP expenditure is not treadmilling per se but the stabilization of the cytoskeleton.

From Filament to Assembly.   In living cells, microtubules and microfilaments do not function individually but as members of a larger assembly such as the mitotic spindle or an actin meshwork. The transformation of single filaments into structures of higher order, which may moreover be attached to the plasma membrane or other loci, involves a complex set of interactions that are just beginning to be understood.

The multiple states of cellular actin, which are at present under intense investigation, illustrate the problems (pertinent reviews include Korn, 1982; Weeds, 1982; Geiger, 1983; Tilney, 1983). Briefly, actin filaments assemble (as explained above) by the preferential addition of monomers to the barbed end. This is frequently the end attached to a membrane; how monomers find access to it is not very clear. In any event, on the basis of in vitro experiments a variety of proteins have been reported to modulate the elongation of actin filaments (Figure

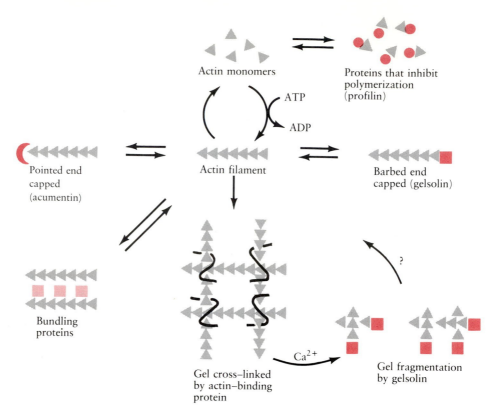

Actin monomers

Proteins that inhibit polymerization (profilin)

ATP

ADP

Pointed end capped (acumentin)

Actin filament

Barbed end capped (gelsolin)

Bundling proteins

Gel cross–linked by actin–binding protein

Ca²⁺

Gel fragmentation by gelsolin

?

FIGURE 12.5   Organization of actin into filaments and networks. A schematic diagram to illustrate some of the roles that actin-binding proteins play in the assembly and disassembly of actin-based structures. (After Stossel, 1983, with permission of Alan R. Liss.)

12.5). Many cells contain profilins, proteins that bind to G actin and thus stabilize the pool of unpolymerized actin. Other proteins are thought to "cap" F-actin filaments, either at the barbed end (inhibiting elongation) or at the pointed end (inhibiting disassembly), while some have the capacity to nucleate filament formation. Conversion of filaments into structures of higher order involves proteins that mediate the formation of parallel alignments (bundling proteins) or of cross-linked gels (gelation or branching proteins). The converse process, the solation of actin gels, is the function of severing proteins that remove cross-links, shorten filaments, and sometimes cap filament ends. Finally, still other proteins link actin bundles or filaments to the plasma membrane. In most cases it has not been determined whether energy input is required.

Actin-binding proteins can be regarded as a tool kit on which cells draw to manipulate that basic structural material, actin. Obviously, actin-binding proteins are not autonomous but operate under the control of physiological signals, many of which are probably still unknown (Chapter 13). It is clear, however, that calcium ions play a crucial regulatory role. For example, actin gels generally form only at low levels of free Ca^{2+}, 10^{-7} M or less; higher concentrations of Ca^{2+} (10^{-6} M) disrupt the gels, either by breaking cross-links or by activating Ca^{2+}-sensitive severing proteins. At the same time Ca^{2+} ions favor the formation of myosin filaments, which requires the phosphorylation of myosin light chains by a calmodulin-activated protein kinase (Adelstein and Eisenberg, 1980), and as in muscle, Ca^{2+} ions activate contraction. The local pH also appears to be a significant control point. The regulatory proteins that respond to these signals are a diversified family, and the details of regulation and information flow differ significantly from one cell to another. We shall examine some of them in their physiological context below.

Amoeboid Movements

In popular parlance the amoeba stands for life at a primitive level, unformed yet pregnant with evolutionary potential. In reality, an amoeba is a well-coordinated carnivore, undoubtedly the product of long adaptation, whose mode of locomotion is not nearly so clumsy as it looks. At the least, it must be accounted an evolutionary success: amoeboid movements are common among microorganisms, have been retained by many tissue cells of higher animals (our own leukocytes, fibroblasts, platelets, and embryonic cells, for example), and even survive in the plant world.

The first indication that amoeboid movement and muscle contraction belong to the same universe of discourse came in 1952. A. G. Loewy was studying the spectacular protoplasmic streaming movements of the large acellular slime mold *Physarum polycephalum*. When he added ATP to extracts of these organisms, the viscosity decreased sharply. The preparation hydrolyzed the ATP, and when it was gone, the viscosity recovered. Loewy correctly inferred that his extracts contained a substance analogous to actomyosin that could undergo cyclic changes in structure by interaction with ATP and thus do mechanical work at the expense of the free energy of ATP hydrolysis. Pure actin and myosin from *Physarum* were first isolated in the 1960s by Hatano and his associates in Japan and have since been obtained from a great variety of nonmuscle cells (see the reviews by Pollard and Weihing, 1974; Clarke and Spudich, 1977; Korn, 1978; Taylor and Condeelis, 1979; Pollard, 1981).

The fundamental working hypothesis that underlies research on amoeboid movements and on cellular motility in general assumes that myosins generate shearing forces by interaction with actin filaments, as suggested in Figure 12.6. Either the actin or the myosin may be attached to other structures, such as organelles or the plasma membrane, so that this general mechanism could in principle generate a variety of cytoplasmic movements. With this model as our point of departure we shall now examine selected cases of amoeboid movement in an effort to relate what the biochemist finds on gels to what the biologist sees under the microscope.

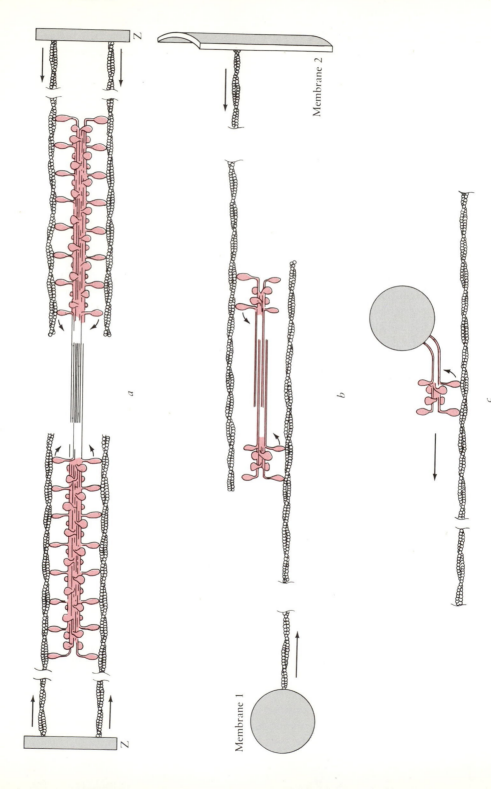

FIGURE 12.6   Sliding forces generated by actomyosin can in principle support a variety of movements. (*a*) Contraction, as in skeletal muscle. (*b*) Transport of a vesicle along an actin filament. (*c*) A vesicle "walking" along an actin filament. (Inspired by a drawing by J. A. Spudich.)

Movement of the Giant Amoeba.   In a giant amoeba on the move (Figure 12.7*a*), cytoplasm streams visibly out of the tail or uroid into the advancing pseudopods. A century of morphological and biophysical studies demonstrated that movement is accompanied by cyclic transformation of the physical state of the cytoplasm from relatively solid "ectoplasm" to the more fluid "endoplasm." The pioneers of amoebal physiology, such as S. O. Mast in the 1920s, interpreted this change in terms of the conversion of a gel into a sol and proposed that contraction of the ectoplasm in the uroid exerts pressure on the solated endoplasm in the cell's interior, pushing it outward to generate a pseudopod. The alternative view, championed especially by R. D. Allen (Allen, 1973; R. D. Allen and N. S. Allen, 1978), localizes the contractions that generate motive force at the tips of advancing pseudopods: as the endoplasm at the front is converted back into ectoplasm it contracts, pulling more endoplasm forward into the contraction zone. (We may crudely liken Mast's view to extruding toothpaste from the tube by squeezing the bottom, Allen's to extracting it by applying a vacuum to the top.) Allen argues his case forcefully, but the site of contraction remains a matter on which investigators differ, and recent observations (see below) suggest that both regions must be active.

Whether the cytoplasm is pushed out of the tail or pulled into the pseudopod, the overall process of movement entails two things: a directional circulation of protoplasm within the cell and the reversible transformation of ectoplasm into endoplasm in the uroid and then back into ectoplasm at the pseudopod tip. The relatively fluid endoplasm occupies the cell's interior, while the rigid and contractile ectoplasm forms a sleeve around the periphery (Figure 12.7*b*). Since the chief structural elements of amoebal cytoplasm are actin filaments, the molecular basis of movement is thought to reside in two actin-based processes: contraction proper, involving the generation of sliding forces between actin and myosin filaments, and the reversible interconversion of actin gel and actin sol (Taylor, 1976; Taylor and Condeelis, 1979). At the pseudopod front, where the flow everts, actin changes from the sol state into a gel; the gel corresponds to the meshwork of contractile filaments situated just beneath the plasma membrane. This region exhibits strain birefringence, suggesting that actin is partly cross-linked by local interactions with actin-binding proteins or with myosin. Toward the amoeba's rear gel turns into sol: binding proteins dissociate from actin and the meshwork dissolves, producing a more fluid structure.

The vectorial circulation of protoplasm from uroid to pseudopod and back again is a necessary feature of amoeboid movement but is not sufficient by itself. In order for the organism as a whole to move forward it must push or pull on the substrate: mechanical tension must be transmitted to particular loci on the plasma membrane and across the membrane to the substrate. By analogy with fibroblasts, in which this aspect of movement is better defined than in amoebas, one expects actin filaments to make transient connections with the plasma membrane at specialized sites that correspond to foci of adhesion. A clear and intuitively comprehensible explanation of the amoeba's distinctive crawl is not yet available, but it appears to bear some resemblance to the movements of a

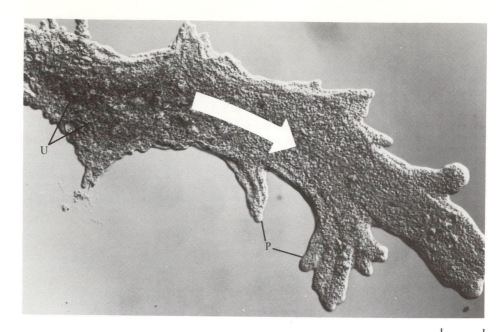

100 μm

a

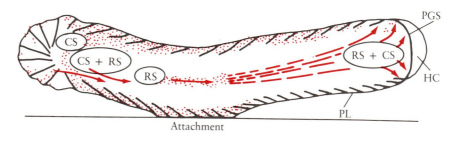

b

FIGURE 12.7   Amoeboid movement. (*a*) A giant amoeba, *Chaos carolinensis*. Note the uroid (U), pseudopods extending in the direction of movement (P), and the flow of cytoplasm (arrow). (Courtesy of R. D. Allen.) (*b*) Schematic representation of structure and contractility in various regions of an amoeba. As the endoplasm approaches the tip of the advancing pseudopod, it gains in structure and contractility. The front peels away from the plasma membrane (PL) to form the plasmagel sheet (PGS) with a clear hyaline cap (HC) beyond. Ectoplasm forms at the tip of the pseudopod by conversion of actin from the relaxed state (RS) to the contractile state (CS). In the ectoplasmic sleeve, actin and myosin interact to produce contraction. The ectoplasm remains in the contractile state and is reconverted to endoplasm in the uroid. (From Taylor, 1976, with permission.)

tank equipped with caterpillar tracks. The cell, however, is much the more flexible vehicle: the local signals that control the flow of protoplasm and its coupling to the plasma membrane can start these processes at almost any point on the cell's surface, thereby initiating a new pseudopod.

Recent research on amoebal movement has centered on the interconversion of the various states of actin in vitro and its relationship to contraction proper, matters in which the plasma membrane plays no direct role. This was first demonstrated two decades ago when R. D. Allen found that the cytoplasm of an amoeba confined in a capillary tube remained motile even after disruption of the plasma membrane. A most dramatic illustration was provided by Taylor et al. (1973), who disrupted single cells in a special buffer containing ATP and just the right amount of Ca^{2+} ($7 \times 10^{-7} M$; "flare solution"): the cytoplasmic droplet, devoid of membranes, erupts into loops of cytoplasm that stream for a period of minutes (Figure 12.8). One can thus speak of "motile extracts" that preserve, at least for a time, their underlying contractile structure. The extensive work on this

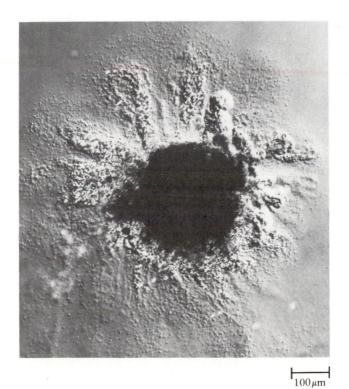

$\vdash\!\!\!\!\!\!\!\dashv$
$100\,\mu m$

FIGURE 12.8   Movements in a naked cytoplasmic droplet from a giant amoeba. When placed in an appropriate buffer, the droplet throws out loops that stream vigorously. (From Taylor et al., 1973, by copyright permission of Rockefeller University. Electron micrograph courtesy of D. L. Taylor.)

system, carried out chiefly by D. L. Taylor and his associates, must be dealt with here in summary fashion (Taylor, 1976; Taylor and Condeelis, 1979; Hellewell and Taylor, 1979; Taylor et al., 1980a, 1980b). In essence these studies showed that the extracts may exist in one of several alternative physical states depending on the temperature, the pH, and the concentration of free Ca^{2+} ions and of ATP. Buffered extracts (pH 7) prepared in the absence of either calcium ions or ATP are fluid, exhibit no streaming, and lack F-actin filaments and myosin aggregates. Addition of $Mg \cdot ATP$, followed by warming to $25°C$, transforms the extract into a gel containing a mesh of F-actin filaments. These ionic conditions correspond to those that relax the cytoplasm of living amoebas. The presence of Ca^{2+} ions ($10^{-6} M$) causes the gel to solate, despite the continued presence of numerous actin filaments, and if myosin is also present, the fluid contracts and streams. Injection of Ca^{2+} ions into living cells elicits local contraction.

The interpretation of these observations is that gelation results from the cross-linking of F-actin filaments by association with one or more actin-binding proteins; it does not require myosin. In the gel state, interaction of actin with myosin is blocked or ineffective. Calcium ions have three effects. In the absence of myosin they solate the gel; if myosin is present, Ca^{2+} ions also stimulate the aggregation of myosin into thick filaments and permit actomyosin cross-bridge cycling. These three interactions are the pillars of the "solation-contraction hypothesis" of amoebal locomotion (Figure 12.9). In Taylor's view, the gel of cross-linked actin serves the cell as a cytoskeleton (amoebas have few microtubules), determines its shape, and also transmits tension from regions of contraction to adhesion foci on the plasma membrane. Contraction is due to sliding movements effected by myosin and ATP in the conventional manner, but these can only take place after the cross-links of the actin gel have been locally dissociated; contractions then pull together portions of cytoplasm that remain internally cross-linked, resulting in mechanical work. Coherent movements require two additional features about which little is known: devices that transmit contractile force to the substratum and a meaningful spatial pattern of regulatory signals.

The movements of a living amoeba are far from random, though they may appear so to a casual observer; the efficiency with which hungry amoebas clear their dish of motile ciliates is ample testimony to their dexterity. It is not yet clear how environmental signals control either contractions or the interconversion of actin states and determine their direction in space. However, substantial evidence from extracts as well as living cells implicates calcium ions. As early as 1928, H. Pollack observed that injection of the calcium-chelating dye alizarin inhibited pseudopod extension; as the amoeba attempted to do so a "shower of red granules" appeared in the cytoplasm, evidence for the release of free calcium ions. The same conclusion was drawn by Taylor et al. (1980a) from experiments in which aequorin luminescence was used to monitor free calcium ions: during movement the calcium concentration rose above $7 \times 10^{-7} M$ in the uroid as well as locally in the pseudopods. The cytoplasmic calcium level can also be manipulated by local injection of Ca^{2+} or of calcium chelators, which elicit

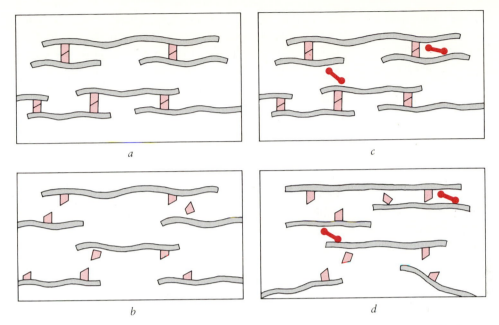

FIGURE 12.9   Proposed role of actin sol-gel transitions in amoeboid movement. The bars represent actin, the dumbbells myosin, and the blocks gelation factors. (*a*) In the absence of myosin. At pH 6.8 and low Ca^{2+} concentration gelation occurs, represented here as the cross-linking of actin filaments. (*b*) If the pH is raised above 7 or if the Ca^{2+} concentration is raised above 7.6×10^{-7} M, solation occurs, represented here as the dissociation of cross-links from each other and from actin, leaving free actin filaments. (*c*) With myosin present. At pH 6.8 and low Ca^{2+} gelation occurs as above, and contraction is inhibited because the cross-links prevent sliding. (*d*) Raising the pH or the Ca^{2+} concentration dissociates the cross-links; the actin filaments are free to interact with myosin and contraction ensues. (After Hellewell and Taylor, 1979, with copyright permission of Rockefeller University.)

contraction and relaxation, respectively. Finally, Nuccitelli et al. (1977) used their vibrating probe (Chapter 13) to monitor the flow of electric current into amoebal cells during movement; they observed a steady current of Ca^{2+} ions entering the uroid, as well as current pulses at the beginning of pseudopod formation. Such observations suggest at least that calcium levels and fluxes are major elements in the control of contractile activity (and incidentally point to the uroid as a major site of contraction); recent observations make it likely that the local pH is a second regulatory factor. But they fall short of bridging the gap between the interactions of known proteins and the purposeful competence of a hungry amoeba on the prowl.

Dynamics of Actin and Myosin during Movement.   For the sake of clarity the preceding discussion was sharply centered on the familiar giant amoeba, but there is no doubt that similar events underlie the motions of amoebas in general, of the slime mold *Physarum*, and of many amoeboid animal cells. In fact, when one turns to the molecular cast of the play and to its deployment in space, most of our information comes from smaller cells that are more amenable to mass culture, and particularly from animal cells.

Macrophages, members of the immune system of higher animals, will serve as an example (Figure 12.10). T. P. Stossel and his coworkers isolated an actin-binding protein from these cells that connects F-actin filaments by forming perpendicular cross-links between them (Figure 12.5). In consequence, the actin sol turns into a stiff gel of filaments knotted together into a three-dimensional network, with actin-binding protein at the vertices. A second protein, called gelsolin, counteracts the effects of actin-binding protein, causing solation when Ca^{2+} ions are present. The molecular basis of gelsolin action is not yet certain. Apparently it does not displace actin-binding protein but intercalates into the filaments by binding tightly to actin monomers; this fragments the network into segments whose barbed ends are capped by gelsolin-Ca^{2+}, but leaves the vertices intact. Removal of Ca^{2+} causes gelsolin to dissociate and actin-binding protein regains the upper hand, restoring the gel state. (When a gradient of Ca^{2+} ion concentration is imposed on an in vitro system containing actin, actin-binding protein, gelsolin, and myosin, contraction of the gel becomes vectorial: the direction of contraction is from the high end of the Ca^{2+} gradient to the low

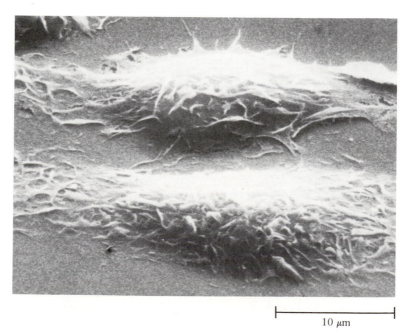

10 μm

FIGURE 12.10   Migrating macrophages from mouse peritoneal fluid. Note the ruffled surface and edges; fingerlike projections along the edge of the cytoplasmic sheet form adhesion sites with the substrate. (Electron micrograph courtesy of K. Leung.)

end.) A third protein, called acumentin, caps the pointed ends of actin filaments; its biological role is less obvious than that of the other two. Cytochalasin and phalloidin toxins mimic the actin-binding proteins; the inhibitory effects of these substances on many kinds of amoeboid movement are probably to be ascribed to interference with gel-sol transformations (Yin et al., 1980; Stossel et al., 1982; Stossel, 1983).

The molecular events described above, together with others yet to be discovered, must be the chemical basis of the changes in cytoskeletal structure that accompany movement. Pertinent observations come particularly from studies on fibroblasts in culture. These cells move slowly and are best tracked with the aid of time-lapse films, but they are thin and lend themselves to various cytological procedures that display the organization of the cytoskeleton. Moving fibroblasts are flattened and polygonal; the leading edge is represented by a thin, flat pseudopod called a lammellopodium, terminating in a ruffled fringe (Figure 12.1*b* and *c*). This is the cell's main locomotive organ. The ruffles make contact with the substratum, forming specialized adhesion plaques, and the entire cell is then pulled forward by about 10 μm at a time. The ruffles, together with long and slender projections called filopodia, also serve as sense organs: two cells whose ruffles touch cease moving toward each other.

These movements are again presumed to involve both actin and myosin and can be correlated with dramatic changes in the organization of actin filaments (Lazarides and Revel, 1979). In well-spread, relatively quiescent cells one sees prominent "stress fibers," long bundles of microfilaments that run just below the plasma membrane and make contact with it at the adhesion plaques. These stress fibers, 0.1 to 0.2 μm in diameter, define the cell's axis and obviously play a structural role. But in highly motile cells, particularly in such mobile regions as ruffles, actin is present in the form of a mesh of cross-linked filaments just beneath and in contact with the plasma membrane (Figure 12.3). The transformation of one arrangement into the other is a most remarkable process. As a rounded, motile cell stops moving and begins to flatten, actin filaments in the vicinity of the nucleus proceed to organize themselves into a regular order so reminiscent of Buckminster Fuller's architecture that Lazarides and Revel called it a geodome (Figure 12.11). Additional actin filaments assemble onto the vertices of the geodome, initiating the bundles of filaments. Ultimately, over the course of several hours, the geodome disassembles, leaving the bulk of the cell's actin in the form of the stress fibers diagnostic of quiescent cells. But should circumstances change, the fibers can disassemble within minutes to regenerate the actin meshwork of motile cells.

Myosin, like actin, is subject to changes of state in relation to cellular motility. Animal cells and amoebas generally contain small bipolar myosin aggregates, consisting of 10 to 20 myosin molecules arranged tail to tail (compared to about 500 myosin molecules in a thick filament of skeletal muscle). These myosin aggregates are not permanent but form only when myosin has been phosphorylated by a Ca^{2+}-dependent kinase. Note that actin and myosin respond oppositely to a rise in the ambient Ca^{2+} level: the signal that causes

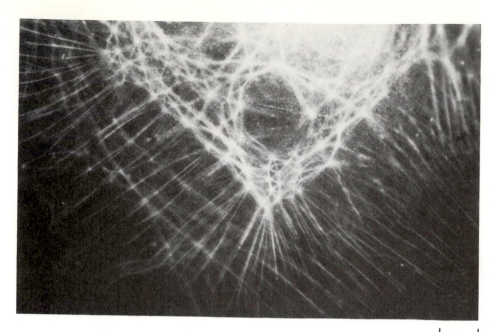

10 µm

FIGURE 12.11 The "geodome," a regular network of actin filament bundles, is an intermediate stage in the conversion of the diffuse actin mesh of an actively moving cell into the filament bundles of a spread-out, quiescent one. This three-dimensional network surrounds the nucleus, and its vertices provide organization sites for actin filament bundles that project toward the periphery of the cell. (Electron micrograph courtesy of E. Lazarides.)

myosin to aggregate into filaments also induces actin gels to solate, in accord with expectations from the solation-contraction hypothesis (Figure 12.9).

By the application of immunofluorescence, actin, myosin, and some ancillary constituents of the contractile machinery have been localized within their anatomical framework. In both amoebas and macrophages, proteins that cross-link actin filaments into a gel have been found at the cell margins. Another actin-binding protein prominent in animal cells is α-actinin. It may be recalled (Chapter 11) that this is a constituent of the Z lines in striated muscle, where it probably serves to anchor the actin filaments. In fibroblasts, α-actinin has been found in a number of foci: in the adhesion plaques, where stress fibers make contact with the substratum via the plasma membrane; in regions where actin filaments contact the membrane; and regularly arrayed along the filament bundles themselves, which seem almost to be divided into primitive sarcomeres. The impression that α-actinin identifies sites where actin filaments become organized is reinforced by its occurrence at the vertices of the cellular geodome as well.

The chief regulatory signal that controls fibroblast motility may be the cytosolic Ca^{2+} level, mediated by mechanisms reminiscent of smooth rather than skeletal muscle (Adelstein and Eisenberg, 1980). Tropomyosin has been identified in fibroblasts, but it appears to be concentrated in the filament bundles of quiescent cells rather than in the mobile actin meshwork of the ruffles; its role would appear to be a structural rather than a dynamic one. Troponins are probably absent and the effects of Ca^{2+} ions appear to be mediated by calmodulin. One regulatory mechanism, which is also found in smooth muscle, involves the control of myosin light-chain kinase; this enzyme is known to be situated along the microfilament bundles (Holzapfel et al., 1983). It must be added that the control of fibroblast motility is far from being fully understood and is not to be glibly summarized.

Yet some kind of summary is in order here, if only to emphasize once again the larger unity that embraces muscle as well as nonmuscle motility and the distinctive qualities of each. In striated muscle, thick and thin filaments form an exceedingly regular array that makes up the bulk of each cell; these organs are designed for speed and precise control but need perform only a single kind of motion whose direction is ordained by the structure of the sarcomere. Contraction of striated muscle is regulated by calcium ions via troponin and tropomyosin, but alternative regulatory mechanisms exist in smooth muscle. The movements of an amoeba or of a fibroblast also depend on the active sliding of actin microfilaments past myosin bundles, with ATP as the primary energy donor, but these cells lack the structural order that make the workings of muscle intuitively comprehensible. Moreover, cellular movements are far more diverse than those of muscle and closely intertwined with the changing shape of the cell. Regulatory signals affect not only the rate and extent of filament sliding but also the state of organization of actin and myosin filaments, their association with the plasma membrane, and the spatial direction of the assemblage as a whole. Hence, no doubt, the profusion of regulatory proteins whose diversity from one organism to another contrasts with the relative uniformity of actin and myosin. Calcium ions are once again the best-known signal, exerting their effects on at least three levels: actomyosin cycling, the formation of myosin bundles, and the polymerization of actin into filaments and nets. The multiplicity of interactions obscures the unifying molecular principles; perhaps for that reason one is left with the uneasy feeling that our capacity to explain amoeboid movement may have outstripped our understanding of it.

Cytoplasmic Streaming

The bulk flow of protoplasm from one part of a cell to another provides some of the most impressive instances of cellular motility, and doubly so when it occurs in a member of the plant world. The cytoplasm of many mycelial fungi flows toward the growing tip and back again in discrete streams, carrying nuclei and other organelles like so much flotsam; the acellular slime molds exhibit a kind of vigorous motion called shuttle streaming. But the most spectacular example is surely the ceaseless rotation of the cytoplasm within the cells of some higher plants and of green algae such as *Chara* or *Nitella*. Cyclosis has mystified biologists ever since Corti first described it in 1774 and is by no means fully understood today. Words cannot do it justice, and I would urge a reader who has not yet seen cyclosis to lay aside this book and search the nearest stream (or biological supply house catalog) for a live specimen.

Nitella is a sizable plant but structurally very simple (Figure 12.12a). Internodal cells, the object of choice for the investigation of streaming, may be several centimeters in length. The bulk of their volume is occupied by the central vacuole, leaving a thin layer of living protoplasm some 200 μm wide between plasma membrane and tonoplast. The outer part of the protoplasmic layer is stationary and contains a solid zone of chloroplasts arranged in regular files; this is referred to as the ectoplasm. Between this and the tonoplast is the mobile layer, the endoplasm, which rotates continuously throughout the life of each cell at a speed of 40 to 120 μm/s. In the long, cylindrical internodal cells the stream flows axially along one-half of the cell, around the end, and back along the other half-cylinder (Figure 12.12). The flow is everywhere parallel to the files of chloroplasts, and as the latter wind helically around the axis, so does the stream. Flow is made manifest by organelles and other particles carried in the endoplasm and even in the adjacent vacuole; the rate of particle movement is the basis for calculating the rate of streaming.

What is the purpose of this continuous motion, with its attendant energy expenditure? Hope and Walker (1975) estimate that the mechanical work amounts to 2 nanowatts (nW) per cell, a small fraction of the 1 to 2 microwatts (μW) available from respiration; they argue that its function is probably the transport of salts within the cells and up to the growing tip of the plant. Apical cells of *Nitella*, where most of the growth takes place, maintain the requisite turgor by accumulating chloride. The rate of chloride accumulation is nearly tenfold greater than the flux of chloride across the plasma membrane; this suggests that much of the chloride comes from neighboring cells, passing across the node by conductive channels called plasmodesmata. In general, while diffusion is adequate to account for the distribution of solutes in small cells (10 μm or less), in large cells cytoplasmic streaming serves as a mechanism of bulk transport.

In the 1950s N. Kamiya and his associates in Osaka carried out elegant studies on the physical basis of cytoplasmic streaming (summarized by Kamiya, 1976, 1981), including measurements of the rate of streaming at various depths. These led to the conclusion that streaming is due to a shearing force generated at the interface between the stationary ectoplasm and the moving endoplasm. Moreover, they noted a connection between those stationary files of chloroplasts that parallel the direction of flow and the generation of force: if, for example, the chloroplasts were locally dislodged, active streaming ceased until they began to regenerate. Subsequently E. Kamitsubo and also R. Nagai and L. I. Rebhun discovered linear bundles of fibrils that run along that interface, linking the chloroplasts, and proposed that they are the physical basis of force production. It is noteworthy that few microtubules are present in this region; instead, the chief constituent of the fibrils was shown to be actin, arranged in the form of microfilament cables (Figure 12.13). The polarity of these actin bundles was

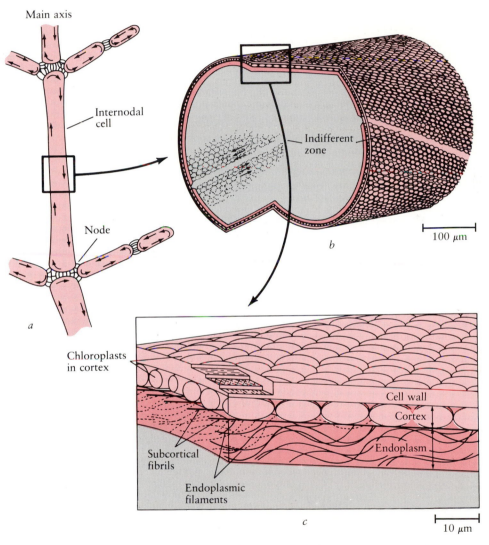

FIGURE 12.12   Cytoplasmic streaming in the giant-celled algae. (*a*) Part of the plant, showing the directions of streaming in internodal cells and laterals. (From Hope and Walker, 1975, with permission of Cambridge University Press.) (*b*) Cross section of *Nitella* internodal cell, showing the cortex with chloroplasts in relation to the huge vacuole. (*c*) Spatial relationships between chloroplast files, subcortical fibrils, and endoplasm (with endoplasmic fibrils). (*b* and *c* after Allen, 1974, with copyright permission of Rockefeller University.)

determined by allowing permeabilized cells to bind heavy meromyosin. The arrowheads point in the direction opposite to that of streaming; this is what one would expect if streaming were generated by myosinlike molecules in the stream, working along stationary actin filaments (Palevitz and Hepler, 1975; Kersey et al., 1976). Together with the isolation of myosin from *Nitella* (Kato and Tonomura, 1977) these studies are responsible for the growing consensus that streaming is due to interactions between actin and myosin that are in principle analogous to those that take place in muscle (N. S. Allen and R. D. Allen, 1978; R. D. Allen and N. S. Allen, 1978; Kamiya, 1981).

Important as this conclusion unquestionably is, it falls short of a mechanistic explanation of cytoplasmic streaming. As matters now stand, two views have begun to crystallize. The first, which would probably command a plurality of the vote, puts the emphasis on myosinlike molecules attached to endoplasmic organelles of unknown identity. Some years ago, Williamson (1975) described experiments in which *Chara* internodal cells were perfused so as to wash away both the vacuolar contents and the bulk of the endoplasm; the behavior of the remaining microfilaments and of organelles immobilized adjacent to them was

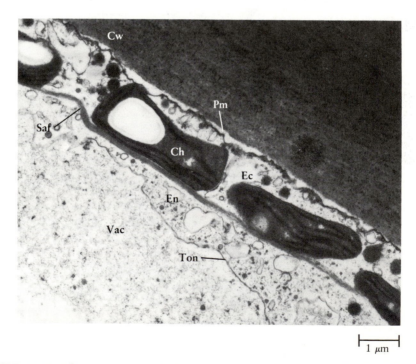

1 µm

FIGURE 12.13   Electron micrograph showing a portion of an internodal cell of *Chara* sp. The image includes a chloroplast (Ch), a portion of the cell wall (Cw), and the tonoplast (Ton) limiting the vacuole (Vac). The subcortical actin fibrils (Saf) are prominent at the interface between ectoplasm (Ec) and streaming endoplasm (En). A plasma membrane (Pm) separates the ectoplasm from the cell wall. It is evident that the fibril is made up of numerous filaments. (Electron micrograph courtesy of J. Pickett-Heaps.)

then monitored. Williamson observed that when the cells were perfused with solutions containing ATP, Mg^{2+}, and less than 0.1 μM Ca^{2+}, streaming resumed and the quiescent organelles began to travel; in time they detached from the bundles and streaming ceased. Excess Ca^{2+} and also cytochalasin B stopped both movements; pyrophosphate failed to support movements but caused the organelles to detach. All this is consistent with organelles "walking" along actin tracks with the aid of a myosin-ATPase engine (Figure 12.6c). A paper by Nagai and Hayama (1979) described what may be the ultrastructural basis of this interaction, in the form of knobs (myosin?) that link those endoplasmic organelles to the actin bundles. Various inhibitors that would be expected to interfere with such a process have been injected into the cells and shown to stop streaming (Nothnagel et al., 1982).

The myosin-dependent translocation of particles along actin tracks has now been demonstrated directly in a remarkable study by Sheetz and Spudich (1983). These investigators cut open internodal cells of *Nitella*, washing away the cytoplasm and exposing the actin cables. They then added fluorescent beads (1 to 10 μm in diameter) covalently coated with heavy meromyosin from muscle and watched the beads traveling along the cable at 0.5 to 10 $\mu m/s$. Movement was dependent on the presence of ATP and of functionally competent myosin but did not require Ca^{2+} ions; the polarity was the same as that of streaming. These findings are consistent with the notion that organelle transport is primary and drags the endoplasmic fluid along.

The minority view is defended by N. S. Allen (Allen, 1974; N. S. Allen and R. D. Allen, 1978), who discovered masses of thin microfilaments (actin?) branching from the subcortical actin bundles into the endoplasm itself (Figure 12.12c). In the living cell these filaments appear to undulate, and one can calculate that if these undulations were active, they would generate ample force to drive the fluid endoplasm, much as cilia set the adjacent liquid in motion (Chapter 11). But do they undulate actively? Other investigators question this, suspecting that the fringe of free endoplasmic actin is an artifact generated by fraying of the subcortical fibrils.

This, perhaps, is as far as the noncombatant observer can safely venture. But he might point out that both conceptions have loose ends. One would wish to know the meaning of the fact that entry of Ca^{2+} ions halts cyclosis; in muscle at least, actomyosin cycling is stimulated by Ca^{2+}. And why does cytochalasin B inhibit cyclosis? This popular inhibitor is now thought to "cap" the ends of F actin filaments and thereby interfere with sol-gel transitions; this would seem to point toward parallels between cytoplasmic streaming and amoeboid movement that have no place in either model as it stands. Hydrodynamic calculations (Nothnagel and Webb, 1982) are not in good accord with the proposal that endoplasm flows in response to the drag exerted by moving organelles; interactions between mobile myosin and a gel of actin extending into the endoplasm would fit the data better. One wonders whether Allen's filaments may be an aspect of Porter's cytomatrix, and soon one comes to ponder the parable of the blind men trying to make out the elephant.

Cytoplasmic Transport of Vesicles

Until quite recently, biological transport referred chiefly to the movement of molecules across cell membranes, but with the growing appreciation of cellular architecture a complementary field of research has arisen that is concerned with transport within the cytoplasm itself. This is a major facet of cell physiology, for we can regard all of growth and development as a sequence of transport processes by which constituents of the medium are assimilated and carried to particular destinations in the cell's matrix. Intracellular transport has already been exemplified by gross movements of the entire cytoplasm; this section will focus on the traffic in vesicles, a mode of containerized transport that seems to be another universal feature of eukaryotic cells.

Apical Growth.   Fungi provide splendid examples of vectorial transport, thanks to their habit of apical growth. Mycelial fungi live in slender tubes that may elongate rapidly under favorable circumstances: 50 μm/min is not uncommon for *Neurospora crassa*. Much of the mycelial biomass is contributed by the cell wall, whose growth is largely confined to the terminal 10 μm of the hypha. This is not to say that the apex is the sole site of metabolic activity: macromolecule synthesis takes place all along a young hypha, and many fungi are capable of translocating nutrients and metabolites over distances of many centimeters. But deposition of cell wall at the apex documents that fungal growth is highly polarized and begs the question of how this is accomplished.

The elaborate apparatus that underlies apical growth was discovered about 1970. The apex of a growing hypha is filled with a dense mass of membrane-bound vesicles, whose abundance decreases sharply with distance from the tip. These vesicles are thought to contain enzymes and precursors required for the assembly of cell wall and plasma membrane (Figure 12.14a). They are produced well behind the apex in the Golgi region (or its functional equivalent; many fungi lack a classical Golgi apparatus), move to the apex, and fuse locally with the plasma membrane (Bartnicki-Garcia, 1973; Grove, 1978; Gooday, 1983). From the dimensions of the apical vesicles one can estimate that to account for the growth of *Neurospora* at 40 μm/min, 37,500 vesicles must fuse with the tip every minute and from their abundance that they must move out from their site of production at a rate of 12.5 μm/min (Gooday and Trinci, 1980). At least two kinds of vesicles can be distinguished cytologically, and there are probably more. One group should correspond to the chitosomes, particles isolated from broken cells that contain a latent form of chitin synthase; others may carry membrane lipids, and in certain algae at least, Golgi vesicles carry complete prefabricated units of cell wall. Be this as it may, the hyphal apex represents a highly ordered apparatus for the production, translocation, and exocytosis of precursor vesicles.

How is this unidirectional transport accomplished? The answer is not known, but the inquiry naturally centers on the possible involvement of microtubules and microfilaments. Until recently research has been hampered by difficulties in preserving either structure for electron microscopy, but these are

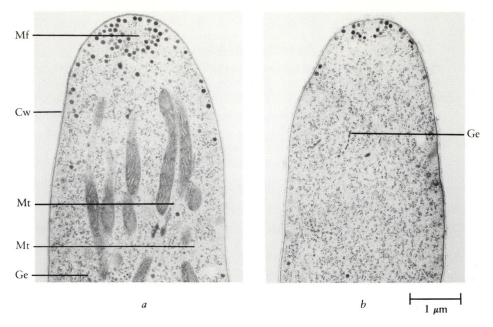

FIGURE 12.14   Organization of the apical region of hyphae of *Fusarium acuminatum*. (*a*) Normal hypha. Cw, cell wall; Ge, Golgi equivalent. Note the numerous microtubules (Mt) and microfilaments (Mf) within the apical cluster of vesicles. (*b*) After 10-minute exposure to a drug that inhibits the assembly of microtubules, few microtubules are present. The number of vesicles and microfilaments near the cell apex is reduced, and mitochondria have been displaced to a subapical position. (From Howard and Aist, 1980, with copyright permission of Rockefeller University. Electron micrographs courtesy of R. J. Howard.)

now being overcome. In a recent study, Howard and Aist (1980; Howard, 1981) demonstrated microtubules oriented parallel to the axis of *Fusarium* hyphae; mitochondria and vesicles were aligned with these microtubules, apparently en route to the apex. Exposure of the cell for as little as 10 minutes to certain benzimidazole drugs that depolymerize fungal microtubules drastically altered the cytology (Figure 12.14*b*): microtubules disappeared, and mitochondria as well as apical vesicles were displaced to the rear. The authors suggest, as had many previous investigators (Hepler and Palevitz, 1974; Heath and Heath, 1978; Hyams and Stebbings, 1979), that microtubules are involved in the translocation of cellular vesicles and organelles. But it is quite uncertain whether microtubules generate motile force, as they do in flagellar axonemes, or define transport channels while the work is done by some other agency. Equally doubtful is the role of actomyosin: actin filaments are generally inconspicuous in fungi (although Howard and Aist noted their presence among the apical vesicles), and the effects of cytochalasin are variable.

The transport of vesicles to the tip of growing hyphae is but one example of the orderly traffic in vesicles that occurs in the great majority of eukaryotic cells. Phagocytosis by amoebas or leukocytes, receptor-mediated endocytosis of lipo-proteins or hormones, the export of proteins by secretory cells, and transport of materials along pollen tubes and nerve axons all supply additional illustrations of the flow of both proteins and membranes in a directional manner to particular destinations. Anatomical and physiological details vary greatly from one organism to the next, but these appear to be variations on a limited number of common themes that can be phrased in the form of questions. How do cells recognize external macromolecules and larger bodies, often with a high degree of specificity? How do the various kinds of endocytic vesicles bud off the plasma membrane? Conversely, what is the molecular mechanism of exocytosis, and how is this process controlled? How are vesicles addressed, sorted, and transported from one cellular locus to another? What is the role of microtubules, microfilaments, and the Golgi apparatus in all this? How do cells manage the flow of membrane lipid that the traffic in vesicles entails? How do regulatory signals from the genome and the environment impinge on these interwoven patterns of flow? The answers to these questions are incomplete at best, and a comprehensive discussion is beyond the scope of a chapter on cellular motility. (Recent pertinent reviews include Pearse and Bretscher, 1981; Besterman and Low, 1983; Steinman et al., 1983.) We shall focus here on a single aspect of this field, the coupling of metabolic energy to the directional translocation of vesicles.

Erythrophores.   Some of the most illuminating observations have been made by K. R. Porter and his colleagues with a seemingly unlikely object, the pigment cells of fish epidermis (Luby and Porter, 1980; Stearns and Ochs, 1982; Beckerle and Porter, 1983; reviews by Porter et al., 1983, and Schliwa, 1984). Briefly, the erythrophores of the squirrel fish are disk-shaped cells, about 40 μm in diameter, packed with small granules containing a red carotenoid pigment. These granules alternate between two states, dispersed throughout the cell or aggregated into a dense central mass (Figure 12.15). Aggregation of the pigment in response to a stimulus (a nerve impulse, or exposure to epinephrine) is complete within two to three seconds; dispersal is a much slower process, with the pigment granules moving out of the aggregate in a saltatory, "irresolute" manner.

In the dispersed state, the pigment granules are arrayed in linear files radiating from the cell center. These coincide with a cytoskeleton consisting chiefly of radial microtubules, visible in Figure 12.15. Actin filaments are scarce

FIGURE 12.15   Pigment migration in the erythrophore of the squirrel fish, *Holocentrus*. (*a*) With pigment dispersed. The granules are arranged in radial files along microtubules that extend from the cell's center. (*b*) With pigment granules aggregated. Numerous microtubules are visible, as is the smooth endoplasmic reticulum. N, nucleus; C, cell center; M, mitochondria; I, central pigment mass; II, a middle zone lacking pigment but containing the microtrabecular lattice; Ser, smooth endoplasmic reticulum. (Electron micrographs courtesy of K. R. Porter.)

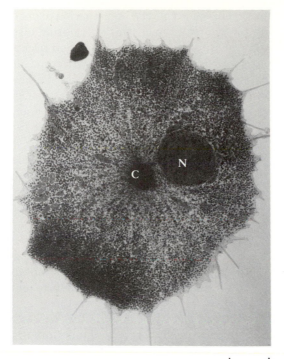

a

b

and mostly confined to the cell periphery. Aggregating pigment granules travel in straight lines along these microtubule "tracks"; during dispersal each granule retraces its path, returning to the position that it had occupied originally. The granules do not move individually. Rather, they are embedded in a cytoplasmic matrix that expands or contracts as a whole, carrying the pigment granules along. This matrix does not move the mitochondria, nucleus, microtubules, or the vesicular endoplasmic reticulum that ramifies throughout the cell; it appears to be a separate cytoplasmic structure whose function is to move the pigment granules along a scaffolding provided by the microtubules.

Microtubules confer direction on the motions of the lattice, but what supplies the motive power? The answer is not yet known, but the complexity of the process is becoming apparent. When the erythrophores are depleted of metabolic energy, the pigment granules aggregate in the cell center; they disperse when metabolism resumes. Apparently pigment dispersal and aggregation are distinct kinds of motion: the former requires ATP, while the latter does not. Actin filaments are scarce in erythrophores, and inhibitors of actomyosin ATPase or of actin reorganization did not inhibit pigment migration. Dynein is a more promising candidate: both aggregation and dispersal were inhibited by vanadate, a popular inhibitor of dynein. However, EHNA, erythro-9-[3-(2-hydroxynonyl)]-adenine, perhaps a more specific dynein inhibitor, prevented dispersal while leaving aggregation untouched. Cells deprived of microtubules by drug treatment no longer supported radial movements of the pigment granules, but undirected saltatory movements continued. Dynein or something like it may play a role in saltatory motions and in dispersal, but Beckerle and Porter (1983) believe that the contraction and expansion of the cytoplasmic lattice are effected neither by dynein nor by actomyosin but by a third energy-transducing system that may be analogous to spasmin. In support of this suggestion they cite the findings that ATP is not required for aggregation, and that aggregation is triggered by Ca^{2+} ions at 5×10^{-6} M while lower concentrations elicit dispersal (compare Chapter 9). ATP may then be needed not for expansion of the cytoplasmic lattice per se but to support the uptake of Ca^{2+} ions by the endoplasmic reticulum. If this provocative hypothesis is corroborated by future work, it will constitute a major advance, for it promises to clarify a variety of baffling cellular movements, including mitosis.

Axonal Transport.   Cytoplasmic transport reaches its apogee with the movement of substances through vertebrate nerves. Neurons keep their nuclei in the cell body; proteins and other macromolecules produced there must be transported down the length of the axon, which may extend as much as a meter or more. The transport of substances through axons was discovered in 1948 by Paul Weiss, who applied a constriction to exposed nerves and noticed that the segment distal to the ligature degenerated while that proximal to the ligature turned swollen and congested. Subsequent research led to the recognition that there are several kinds of axonal transport. Slow transport, or axoplasmic flow, refers to the translocation of much of the cytoplasm down the axon at a rate of about 1 mm per day (approximately equal to the rate of chromosome movement during mitosis). Major constituents so moved include tubulin, actin, and neurofilament protein. Axoplasmic flow requires continued synthesis of new protein and probably serves to replenish and renew the axoplasm of mature axons. Much more spectacular is the fast, or axonal, transport of a variety of proteins and other substances at rates from 100 to 700 mm per day. Virtually all this material is in the form of membrane-bound vesicles: elements of the endoplasmic reticulum, some mitochondria, and vesicles containing acetylcholinesterase or neurotransmitters destined for the synaptic terminal. There is also a retrograde flow of vesicles back to the cell body, including membrane material to be recharged with fresh transmitter. The vesicular nature of the freight and its discontinuous (saltatory) motions are characteristic features that axonal transport shares with other varieties of cytoplasmic transport.

Axonal transport has been extensively studied (Grafstein and Forman, 1980; Schwartz, 1980; Schliwa, 1984), yet its mechanism remains as elusive as that of other cytoplasmic movements. Longitudinal microtubules and neurofilaments (composed of a keratinlike protein) are prominent in axons and are often linked to vesicles by distinct cross bridges. This, together with repeated observations showing that axonal transport ceases when the microtubules are depolymerized with colchicine, supports the general belief that microtubules play an important role in the transport process. But just what that role is remains unclear. One possibility that has been discussed for years is that microtubules provide the motive force, with the help of dynein ATPase arms. This is by no means a farfetched notion, particularly since Haimo et al. (1979) found that microtubules assembled in vitro from brain tubulin bind purified *Chlamydomonas* dynein, forming sidearms with regular periodicity. Cytoplasmic dynein has been isolated from sea urchin eggs and may prove to be a general cell constituent. On the other hand, several investigators have reported that cytoplasmic movements continue in the absence of microtubules. For instance, Brady et al. (1980) found that exposure of axons to high concentrations of $CaCl_2$ depolarizes microtubules without inhibiting fast transport. Another difficulty is the uniform polarity of axonal microtubules: it is hard to see how unidirectional microtubules could energize both anterograde and retrograde vesicle transport (Heidemann et al., 1981). On balance, it seems most likely that axonal microtubules provide mechanical support and define transport pathways but are not the source of mechanical power.[1]

[1] I cannot resist mentioning another spectacular instance of this kind, from the work of Edds (1975) with the heliozoan *Echinosphaerium*. The body of this common protozoan is studded with needlelike axopodia, supported by a central axoneme of microtubules; small particles can be seen moving in saltatory fashion within the axopodium. Edds passed a fine glass needle through the cell, raising the cell membrane together with its underlying cortex and creating an artificial axopodium. Saltatory movement of particles was observed in these artificial appendages, despite the absence of any microtubules.

A circumstantial, but not entirely persuasive, case can be made that ATP and actomyosin are ultimately behind axoplasmic transport. That organelle movement requires ATP is all but certain (Adams, 1982), but actomyosin is implicated chiefly by the observation that injection of certain actin-binding proteins into neurons inhibits axonal transport. Cytochalasins and phalloidin often inhibit cytoplasmic transport process, but not always; the meaning is unclear. There are suggestions that a cytoplasmic matrix, analogous to that of the erythrophore, may be the key to the mystery (Stearns, 1982), but the mystery has yet to be unlocked.

Cell Division

A true nucleus, with discrete chromosomes and a mitotic apparatus to ensure their equal distribution at division, is a hallmark of the eukaryotic cell. Cell division is so familiar that it requires a conscious effort to recall the many kinds of precisely controlled movements needed for its execution. In animal cells mitosis begins with the breakdown of cytoplasmic microtubules and the duplication of the centrioles, which lie outside the nucleus, approximately in the cell center. The daughter centrioles then travel around to opposite sides of the nucleus and proceed to organize a basket of microtubules that becomes the mitotic spindle. As the tubules elongate from the spindle poles, the chromosomes condense into portable packets, the nuclear membrane breaks down, and some of the spindle fibers make contact with each chromosome at a specialized region, the kinetochore. Opinion is divided whether in the living cell the kinetochore draws to itself spindle fibers that emanate from the poles or is itself a microtubule-organizing center from which fibers grow toward the poles. In any case there are two kinds of spindle fibers, those that attach to chromosomes and those that do not (Figure 12.16).

Through movements described as jerky or irresolute the chromosomes, each now connected to both poles by microtubule bundles, collect at the spindle's equator (metaphase). It appears that each chromosome is pulled in both directions and comes to rest at the point where the forces balance. Here the sister chromatids separate, each starting the journey toward the pole to which its kinetochore is linked at a leisurely speed of about 1 μm/min (anaphase). In the meantime the cell has begun to divide, a distinct process referred to as cytokinesis. The cytoplasm and plasma membrane constrict in the plane of the equatorial plate; the spindle, even before its task is completed, begins to disassemble. Eventually just a narrow isthmus filled with a remnant of the spindle links the daughter cells until this also is pinched off and discarded. We cannot here survey the diverse forms that mitosis takes in different creatures (for recent reviews see Dustin, 1984, and McIntosh, 1979) but will consider the functioning of the mitotic apparatus as a specialized machine, precisely controlled in space and time, that performs a particular task of intracellular transport.

It should be clear from the outset that cytokinesis and mitosis are based on different structures and probably depend on different molecular mechanisms. Constriction of the cell is effected by the contractile ring, a band of microfilaments that assembles beneath the plasma membrane, defines the cleavage furrow, and operates in the manner of a purse string (Schroeder, 1976). There is good evidence that the ring contracts by an actomyosin mechanism that slides the microfilaments into an ever narrowing circle. For example, both actin and myosin accumulate in the contractile ring as it forms, and cytokinesis is blocked by cytochalasin B and by antibodies to myosin. Cytokinesis continues in cells that have been carefully permeabilized with detergents provided the experimenter supplies ATP and the proper ionic milieu (Schroeder, 1976, 1981; Aubin et al., 1979, 1981; Cande et al., 1981; Kiehart et al., 1982). Schroeder has measured the tension generated by the contractile ring, about 2.5×10^{-3} dyne, and found that it remains constant for the duration of cytokinesis. Moreover, the ring maintains a constant width and thickness even as its circumference shrinks, evidence that controlled disassembly accompanies contraction. In principle at

1 µm

FIGURE 12.16   Metaphase spindle of a PtK cell as observed by high-voltage electron microscopy. (Electron micrograph courtesy of J. R. McIntosh.)

least, cytokinesis appears to be akin to the amoeboid movements discussed earlier in this chapter.

The volume of information about the mitotic spindle is incomparably greater, yet the mechanism by which chromosomes are transported remains quite mysterious. What has clearly emerged from a century of research is the central role of the spindle itself in effecting movement: mitosis is a prime example of microtubule-linked motility (McIntosh, 1979; Inoué, 1981). The chromosomes themselves are entirely passive; each behaves as though it were tethered to the pole by a string that exerts on it a force directed toward the pole, from before metaphase through anaphase (Nicklas, 1975). Concurrently the spindle itself lengthens, increasing the distance between the poles. These appear to be independent motions that in some organisms occur sequentially rather than concurrently and probably involve distinct mechanisms. The crux of the problem is the nature of the forces produced and their regulation in space and time.

Since in the course of mitosis the spindle itself undergoes both rearrangement and elongation, McIntosh et al. (1969) suggested more than a decade ago that the spindle may generate force by a sliding-filament mechanism akin to that seen in flagellar axonemes. Evidence for sliding movements has come from studies on spindle organization, particularly from the reconstruction of the entire spindle from serial sections (McIntosh et al., 1979; McIntosh, 1979). This approach has been most successful with the small, tight spindles of diatoms, but recently has been extended to other algae as well. As Figure 12.17 clearly shows, the mitotic spindle of *Diatoma vulgare* consists of two half-spindles that interdigitate at the equator to form a zone of overlap. As the spindle lengthens, the zone of overlap narrows or vanishes, a strong visual indication that elongation results from a sliding interaction between two sets of fibers of opposite polarity. In some cases one can see links between the fibers, which may represent cross bridges that mediate the sliding. Note, however, that not all mitotic spindles visibly slide apart and that sliding is, in any event, not sufficient to account quantitatively for spindle elongation: the individual fibers also grow in length, a process that may itself exert mechanical force.

Spindle growth cannot explain how the chromosomes are moved poleward, a process accompanied by shortening of those microtubules that are attached to the kinetochores. Not only chromosomes but also granules and inclusions within the spindle move toward the poles, as though the entire spindle matrix flowed slowly poleward. The forces involved in this transport are vanishingly small. Nicklas (1975) estimated that a single actin filament of muscle, or one ciliary doublet microtubule, would generate many times the requisite power: splitting 20 molecules of ATP would suffice to support the mechanical work of moving an average chromosome to the spindle pole. A satisfactory model of chromosome movement continues to elude us; let us glance at some of the models presently under discussion, if only to define the nature of the problem.

1. The detection in the mitotic spindle of actin, myosin, and calmodulin gave rise to the hypothesis that the spindle microtubules serve as a scaffold that guides

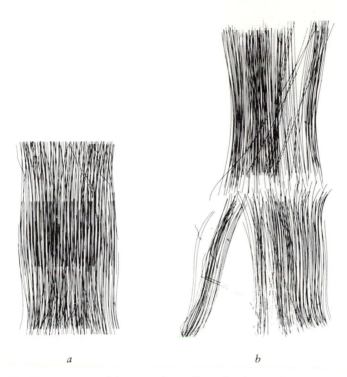

a   *b*

FIGURE 12.17   Organization of the central spindle in the diatom *Diatoma*, reconstructed from serial transverse sections. (*a*) Metaphase spindle, composed of two overlapping half-spindles. (*b*) Telophase spindle. The half-spindles have moved apart and only the longest microtubules still interdigitate; there is also a slight increase in average microtubule length. Both spindles have been shortened relative to their width by a factor of 2 to improve clarity. (From McIntosh et al., 1979, with copyright permission of Rockefeller University. Photograph courtesy of J. R. McIntosh.)

(and limits) chromosome movement, while the work is performed by an actomyosin system. In recent years this view has been largely abandoned because the mitotic apparatus and the contractile ring exhibit clearly different responses to experimental perturbation, even in the same cell. I mentioned above that the contractile ring accumulates actin and myosin during its formation and is disrupted by reagents that interfere with actin or myosin. By contrast, in the same cell there was no accumulation of extra actin or myosin in the mitotic spindle (Aubin et al., 1979, 1981). By use of permeabilized cells in which chromosome separation continues, Cande et al. (1981) were able to show that cytochalasins, phalloidin, and myosin S_1 (all of which blocked cytokinesis) had no effect on chromosome movements; these were, however, retarded by conditions and reagents that inhibit microtubule polymerization. It is hard to avoid the conclusion that neither actomyosin-based filament sliding nor actin sol-gel transitions are responsible for chromosome transport. The presence of actin and functionally related proteins in the spindle would then be fortuitous, reflecting their abundance in cytoplasm generally. Calmodulin, however, accumulates in the spindle during assembly (Welsh et al., 1978) and may be functional there.

2.   Sliding movements between the microtubules of undulipodia are energized by dynein ATPase (Chapter 11). A flurry of excitement was therefore occasioned by the discovery that the addition of dynein plus ATP to permeabilized cells stimulated chromosome movements and that the process was inhibited by vanadate; spindle elongation was also inhibited by the dynein inhibitor EHNA, but chromosome movements were unaffected (Cande, 1982). Cytoplasmic dynein has been isolated from sea urchin eggs; it forms cross-links with microtubules that are dissociated on the addition of ATP, and its activity is regulated by Ca^{2+} ions and calmodulin (Hollenbeck et al., 1984; Hisanaga and Pratt, 1984). Cytoplasmic dynein may well be derived from the mitotic apparatus, but attempts to establish this have not been fully successful. For the present, the role of dynein in mitosis and other microtubule-linked transport processes remains moot.

3.   Could it be that microtubule polymerization and depolymerization itself is a force-generating process? This idea was first developed by Inoué and Sato (see Inoué, 1981), who argued that the spindle fibers are in a state of dynamic equilibrium with tubulin monomers in the cytoplasm. Addition of monomers to the ends of fibers would cause the spindle to elongate; conversely, depolymeriza-tion of the kinetochore fibers would shorten them and might concurrently pull the attached chromosome toward the poles. Models of this kind all require fancy mechanical arrangements to maintain tension on the fiber while tubulin monomers are added or subtracted and are closely linked to our understanding of microtubule growth.

It will be recalled that tubulin monomers add to one end of the microtubule and dissociate from the other, with the concurrent hydrolysis of GTP to GDP; microtubules are not in equilibrium but in a steady state, characterized by a vectorial flow of monomers through each tubule from the assembly end to the disassembly end (treadmilling; Figure 12.4). Margolis et al. (1978; Margolis and Wilson, 1981) based an intriguing model of mitosis on the premise that treadmilling occurs in vivo and serves as a force-generating mechanism (Figure 12.18). Assume that all the microtubules in each half-spindle are parallel, that subunits assemble onto the fiber at the end distal to the pole (in the case of chromosomal fibers, subunits would add in the vicinity of the kinetochore), and that disassembly takes place at the poles. Each half-spindle thus grows in length and slides apart thanks to interactions (presumably ATP-linked) between fibers of opposite polarity in the zone of overlap. Metaphase reflects a balanced state: the symmetrical structure of the mitotic apparatus is maintained by the internal tension of microtubule sliding working against the linkage between sister chromatids. At the beginning of anaphase the link between the sister chromatids is cleaved and kinetochore fibers cease to grow. Disassembly at the poles continues, however, which allows the chromosomes to move closer to the poles

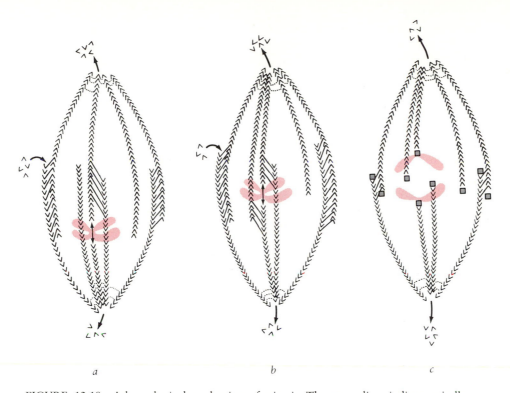

a   *b*   *c*

FIGURE 12.18   A hypothetical mechanism of mitosis. The arrow lines indicate spindle microtubules; the arrows point in the direction of monomer flow in the microtubules due to steady-state assembly and disassembly. Note that all microtubules in one half-spindle are parallel. The kinetochore is represented by a line through the chromosome.
(*a*) Prometaphase. Microtubules assemble and elongate; antiparallel microtubule sliding at this stage causes chromosomes to migrate toward the metaphase plate. (*b*) Metaphase. Microtubules are at steady state; the structure of the mitotic apparatus is maintained by the internal tension of antiparallel microtubule sliding that pulls on sister chromatid linkages. (*c*) Anaphase is characterized by the cleavage of sister chromatid links and a simultaneous blockage of microtubule assembly at the kinetochore (blocks), while microtubule sliding continues. (After Margolis et al., 1978, with permission of *Nature*, Macmillan Journals.)

even while the spindle as a whole lengthens thanks to the continued elongation of the nonchromosomal fibers.

This model also is not fully satisfactory. A sophisticated mechanical device must be postulated to explain how kinetochore fibers can disassemble at the poles while maintaining mechanical tension. A more serious difficulty arises from microtubule polarity. The model is consistent with the demonstration (Euteneuer and McIntosh, 1981) that all the microtubules in a given half-spindle have the same polarity. But it achieves this consistency by postulating that kinetochore fibers elongate by the addition of monomers to the fiber's kinetochore end. Numerous in vitro studies have shown that kinetochores can function as microtubule-organizing centers, but the microtubules so nucleated grow by addition of monomers to their distal end, so that their polarity is opposite to that of microtubules that emanate from the poles. The polarized growth of microtubules and sliding forces between the antiparallel halves of the spindle do seem to be part of the mechanism of mitosis, but it seems unlikely that these same forces also move the chromosomes to the poles.

4.   What other mechanisms might be called on to move the chromosomes? There is the cytoplasmic matrix, and in the opinion of Pickett-Heaps et al. (1982, 1984), this is at present the most promising candidate. In their view the spindle matrix is contractile and chromosomes are attached to this matrix via the kinetochore fibers. The role of these and other microtubules in mitosis is analogous to that of the microtubule cytoskeleton in the erythrophore: they give spatial coherence to the contractile lattice but are neither energized tracks nor passive traction cables. The two half-spindles are to be pictured as mutually repulsive contractile lattices whose workings are coordinated in time. A striking corroboration of this hypothesis is the discovery that when certain diatoms are treated with inhibitors of energy metabolism, the chromosomes come to rest at the poles (Pickett-Heaps and Spurck, 1982). The result is analogous to the aggregation of pigment granules in energy-depleted erythrophores and may well be a hint that a spasminlike contractile system underlies the anaphase movement of chromosomes.

Pickett-Heaps et al. (1982) entitled their review "Rethinking Mitosis" and offer pointed critiques of all the current models, not exempting their own. The deficiencies go beyond particular inconsistencies or dubious postulates: the deeper problem is that none of the models does justice to the complexity of the process. Mitosis now appears to involve no fewer than three distinct kinds of movement: resolute chromosome movement to the poles (ATP not required), saltatory oscillation of the chromosomes (ATP required), and spindle elongation (ATP required). Mitosis is not a simple process and will evidently not be accommodated by a unitary, elementary explanation. Perhaps other instances of cellular motility should also be reconsidered with an attitude of thoroughgoing skepticism. Reflection on the hypotheses that have been proposed to account for amoeboid movement, cytoplasmic streaming, and intracellular transport leaves one with the suspicion that actomyosin may be necessary but not sufficient, and that the models at hand, though hardly simple, do not yet make sufficient allowance for biological complexity.

References

ADAMS, R. J. (1982). Organelle movement in axons depends on ATP. *Nature* **297**:327–329.
ADELSTEIN. R. S., and EISENBERG, A. (1980). Op. cit., Chapter 10.
ALLEN, N. S. (1974). Endoplasmic filaments generate the motive force for rotational streaming in *Nitella*. *Journal of Cell Biology* **63**:270–287.

ALLEN, N. S., and ALLEN, R. D. (1978). Cytoplasmic streaming in green plants. *Annual Review of Biophysics and Bioengineering* 7:497–526.

ALLEN, R. D. (1973). Biophysical aspects of pseudopodium formation and retraction, in *The Biology of Amoeba*, K. Jeon, Ed. Academic, New York, pp. 201–247.

ALLEN, R. D., and ALLEN, N. S. (1978). Cytoplasmic streaming in amoeboid movement. *Annual Review of Biophysics and Bioengineering* 7:469–495.

AMOS, L. A. (1979). Structure of microtubules, in *Microtubules*, K. Roberts and J. S. Hyams, Eds. Academic, New York, pp. 1–64.

AUBIN, J. E., OSBORN, M., and WEBER, K. (1981). Inhibition of cytokinesis and altered contractile ring morphology induced by cytochalasins in synchronized PtK_2 cells. *Experimental Cell Research* 136:63–79.

AUBIN, J. E., WEBER, K., and OSBORN, M. (1979). Analysis of actin and microfilament-associated proteins in the mitotic spindle and cleavage furrow of PtK_2 cells by immunofluorescence microscopy. *Experimental Cell Research* 124:93–109.

BARTNICKI-GARCIA, S. (1973). Fundamental aspects of hyphal morphogenesis. *Symposia of the Society for General Microbiology* 23:245–267.

BECKERLE, M. C., and PORTER, K. R. (1983). Analysis of the role of microtubules and actin in erythrophore intracellular motility. *Journal of Cell Biology* 96:354–362.

BESTERMAN, J. M., and LOW, R. B. (1983). Endocytosis: A review of mechanisms and plasma membrane dynamics. *Biochemical Journal* 210:1–13.

BRADY, S. T., CROTHERS, S. D., NOSAL, C., and MCCLURE, W. O. (1980). Fast axonal transport in the presence of high Ca^{2+}: Evidence that microtubules are not required. *Proceedings of the National Academy of Sciences USA* 77:5909–5913.

BRINKLEY, B. R., COX, S. M., PEPPER, A., WIBLE, L., BRENNER, S. L., and PARDUE, R. L. (1981). Tubulin assembly sites and the organization of cytoplasmic microtubules in cultured mammalian cells. *Journal of Cell Biology* 90:554–562.

CANDE, W. Z. (1982). Inhibition of spindle elongation in permeabilized mitotic cells by erythro-9-[3-(2-hydroxynonyl)]adenine. *Nature* 295:700–701.

CANDE, W. Z., MCDONALD, K., and MEEUSEN, R. L. (1981). A permeabilized cell model for studying cell division: A comparison of anaphase chromosome movement and cleavage furrow constriction in lysed PtK_1 cells. *Journal of Cell Biology* 88:618–629.

CLARKE, M., and SPUDICH, J. A. (1977). Nonmuscle contractile proteins: The role of actin and myosin in cell motility and shape determination. *Annual Review of Biochemistry* 46:797–822.

CLEGG, J. S. (1982). Interrelationships between water and cell metabolism in *Artemia* cysts. IX: Evidence for organization of soluble cytoplasmic enzymes. *Cold Spring Harbor Symposia on Quantitative Biology* 46:23–37.

COTE, R. H., and BORISY, G. G. (1981). Head-to-tail polymerization of microtubules *in vitro*. *Journal of Molecular Biology* 150:577–602.

DUSTIN, P. (1984). *Microtubules*. 2d ed. Springer-Verlag, Berlin.

EDDS, K. T. (1975). Motility in *Echinosphaerium nucleofilum*. An analysis of particle motions in the axopodia and a direct test of the involvement of the axoneme. *Journal of Cell Biology* 66:145–155.

EUTENEUER, U., and MCINTOSH, J. R. (1981). Structured polarity of kinetochore microtubules in PtK_1 cells. *Journal of Cell Biology* 89:338–345.

FULTON, A. B. (1982). How crowded is the cytoplasm? *Cell* **30**:345–347.

GALL, J. G., PORTER, K. R., and SIEKEVITZ, P., Eds. (1981). Discovery in cell biology. *Journal of Cell Biology* **91**:1s–306s.

GEIGER, B. (1983). Membrane-cytoskeleton interactions. *Biochimica et Biophysica Acta* **737**:305–341.

GOLDMAN, R. D., MILSTED, A., SCHLOSS, J. A., STARGER, J., and YERNA, M. J. (1979). Cytoplasmic fibers in mammalian cells: Cytoskeletal and contractile elements. *Annual Review of Physiology* **41**:703–722.

GOODAY, G. W. (1983). The hyphal tip, in *Fungal Differentiation: A Contemporary Synthesis*, J. E. Smith, Ed. Marcel Dekker, New York, pp. 315–356.

GOODAY, G. W., and TRINCI, A. P. J. (1980). Wall structure and biosynthesis in fungi. *Symposia of the Society for General Microbiology* **30**:207–251.

GRAFSTEIN, B., and FORMAN, D. S. (1980). Intracellular transport in neurons. *Physiological Reviews* **60**:1167–1283.

GROVE, S. N. (1978). The cytology of hyphal tip growth, in *The Filamentous Fungi*, vol. III; *Developmental Mycology*, J. E. Smith and D. E. Berry, Eds. Wiley, New York, pp. 28–50.

GUNNING, B. S., and HARDHAM, A. R. (1982). Microtubules. *Annual Review of Plant Physiology* **33**:651–698.

HAIMO, L. T., TELZER, B. R., and ROSENBAUM, J. L. (1979). Dynein binds to and crossbridges cytoplasmic microtubules. *Proceedings of the National Academy of Sciences USA* **76**:5759–5763.

HEATH, I. B., and HEATH, M. C. (1978). Microtubules and organelle movements in the rust fungus *Urolmyces phaseoli var. vignae. Cytobiologie* **16**:393–411.

HEIDEMANN, S. R., LANDERS, J. M., and HAMBORG, M. A. (1981). Polarity orientation of axonal microtubules. *Journal of Cell Biology* **91**:661–665.

HELLEWELL, S. B., and TAYLOR, D. L. (1979). The contractile basis of amoeboid movement. VI: The solation-contraction coupling hypothesis. *Journal of Cell Biology* **83**:633–648.

HEPLER, P. K., and PALEVITZ, B. A. (1974). Microtubules and microfilaments. *Annual Review of Plant Physiology* **25**:309–362.

HEUSER, J. E., and KIRCHNER, M. W. (1980). Filament organization revealed in platinum replicas of freeze-dried cytoskeletons. *Journal of Cell Biology* **86**:212–234.

HISANAGA, S., and PRATT, M. M. (1984). Calmodulin interaction with cytoplasmic and flagellar dynein: Calcium-dependent binding and stimulation of adenosinetriphosphatase activity. *Biochemistry* **23**:3032–3037.

HOLLENBECK, P. J., SUPRYNOWICZ, F., and CANDE, W. Z. (1984). Cytoplasmic dynein-like ATPase cross-links microtubules in an ATP-sensitive manner. *Journal of Cell Biology* **99**:1251-1258.

HOLZAPFEL, G., WEHLAND, J., and WEBER, K. (1983). Calcium control of actin-myosin based contraction in triton models of mouse 3T3 fibroblasts is mediated by the myosin light chain kinase (MLCK)-calmodulin complex. *Experimental Cell Research* **148**:117–126.

HOPE, B. A., and WALKER, N. A. (1975). *The Physiology of Giant Algal Cells*. Cambridge University Press, London.

HOWARD, R. J. (1981). Ultrastructural analysis of hyphal tip cell growth in fungi. *Journal of Cell Science* **48**:89–103

HOWARD, R. J., and AIST, J. R. (1980). Cytoplasmic microtubules and fungal morphogenesis: Ultrastructural effects of methyl benzimidazole-2-ylcarbamate determined by freeze-substitution of hyphal tip cells. *Journal of Cell Biology* **87**:55–64.

HYAMS, J. S., and STEBBINGS, H. (1979). In *Microtubules*, K. Roberts and J. S. Hyams, Eds., Academic Press, New York, pp. 487–530.

INOUÉ, S. (1981). Cell division and the mitotic spindle. *Journal of Cell Biology* **91**:131S–147S.

ISHIKAWA, H., BISHOFF, R., and HOLTZER, H. (1969). Formation of arrowhead complexes with heavy meromyosin in a variety of cells. *Journal of Cell Biology* **43**:312–328.

KAMIYA, N. (1976). Molecular basis of motility, introductory remarks, in *International Cell Biology, 1976–1977,* B. R. Brinkley and K. R. Porter, Eds. Rockefeller University Press, New York, pp. 361–366.

KAMIYA, N. (1981). Physical and chemical basis of cytoplasmic streaming. *Annual Review of Plant Physiology* **32**:205–236.

KATO, T., and TONOMURA, Y. (1977). Identification of myosin in *Nitella flexilis. Journal of Biochemistry* **82**:777–782.

KELL, D. B. (1979). Op. cit., Chapter 3.

KERSEY, Y. M., HEPLER, P. K., PALEVITZ, B. A., and WESSELS, N. K. (1976). Polarity of actin filaments in Characean algae. *Proceedings of the National Academy of Sciences USA* **73**:165–167.

KIEHART, D. P., MABUCHI, I., and INOUÉ, S. (1982). Evidence that myosin does not contribute to force production in chromosome movement. *Journal of Cell Biology* **94**:165–178.

KIRCHNER, M. W. (1980). Implications of treadmilling for the stability and polarity of actin and tubulin polymers *in vivo. Journal of Cell Biology* **86**:330–334.

KORN, E. D. (1978). Biochemistry of actomyosin-dependent cell motility (a review). *Proceedings of the National Academy of Sciences USA* **75**:588–599.

KORN, E. D. (1982). Actin polymerization and its regulation by proteins from nonmuscle cells. *Physiological Reviews* **62**:672–737.

LAZARIDES, E. (1980). Intermediate filaments as mechanical integrators of cellular space. *Nature* **283**:249–256.

LAZARIDES, E., and REVEL, J. P. (1979). The molecular basis of cell movement. *Scientific American* **240**(5):100–113.

LUBY, K. J., and PORTER, K. R. (1980). The control of pigment migration in isolated erythrophores of *Holocentrus ascensionis* (Osbeck). I: Energy requirements. *Cell* **21**:13–23.

MARGOLIS, R. L. (1981). Role of GTP hydrolysis in microtubule treadmilling and assembly. *Proceedings of the National Academy of Sciences USA* **78**:1586–1590.

MARGOLIS, R. L., and WILSON, L. (1981). Microtubule treadmills, possible molecular machinery. *Nature* **293**:705–711.

MARGOLIS, R. L., WILSON, L., and KIEFER, B. I. (1978). Mitotic mechanisms based on intrinsic microtubule behavior. *Nature* **272**:450–452.

MCINTOSH, J. R. (1979). Cell division, in *Microtubules*, K. Roberts and J. S. Hyams, Eds. Academic, New York, pp. 381–442.

MCINTOSH, J. R., HEPLER, P. K., and VAN WIE, D. G. (1969). Models for mitosis. *Nature* **224**:659–663.

MCINTOSH, J. R., MCDONALD, K. L., EDWARDS, M. K., and ROSS, B. M. (1979). Three-dimensional structure of the central mitotic spindle of *Diatoma vulgare. Journal of Cell Biology* **83**:428–442.

NAGAI, R., and HAYAMA, T. (1979). Ultrastructure of the endoplasmic factor responsible for cytoplasmic streaming in *Chara* internodal cells. *Journal of Cell Science* **36**:121–136.

NICKLAS, R. B. (1975). Chromosome movement: Current models and experiments on living cells, in *Molecules and Movement,* S. Inoué and R. E. Stephens, Eds. Raven Press, New York, pp. 97–116.

NOTHNAGEL, E. A., SANGER, J. W., and WEBB, W. W. (1982). Effects of exogenous proteins on cytoplasmic streaming in perfused *Chara* cells. *Journal of Cell Biology* **93**:735–742.

NOTHNAGEL, E. A., and WEBB, W. W. (1982). Hydrodynamic models of viscous coupling between mobile myosin and endoplasm in characean algae. *Journal of Cell Biology* **94**:444–454.

NUCCITELLI, R., POO, M.-M., and JAFFE, L. F. (1977). Relations between amoeboid movement and membrane-controlled electrical currents. *Journal of General Physiology* **69**:743–763.

PALEVITZ, B. A., and HEPLER, P. K. (1975). Identification of actin *in situ* at the ectoplasm-endoplasm interface of *Nitella*: Microfilament-chloroplast association. *Journal of Cell Biology* **65**:29–38.

PEARSE, B. M., and BRETSCHER, M. S. (1981). Membrane recycling by coated vesicles. *Annual Review of Biochemistry* **50**:85–101.

PICKETT-HEAPS, J. D., and SPURCK, T. P. (1982). Studies on kinetochore function in mitosis. II: The effects of metabolic inhibitors on mitosis and cytokinesis in the diatom *Hantzschia amphioxys. European Journal of Cell Biology* **28**:83–91.

PICKETT-HEAPS, J. D., SPURCK, T., and TIPPIT, D. (1984). Chromosome motion and the spindle matrix. *Journal of Cell Biology* **99**:137s–143s.

PICKETT-HEAPS, J. D., TIPPIT, D. H., and PORTER, K. R. (1982). Rethinking mitosis. *Cell* **29**:729–744.

POLLARD, T. D. (1981). Cytoplasmic contractile proteins. *Journal of Cell Biology* **91**:156s–165s.

POLLARD, T. D., and WEIHING, R. R. (1974). Actin and myosin in cell movements. *CRC Critical Reviews in Biochemistry* **2**:1–65.

PORTER, K. R. (1966). Cytoplasmic microtubules and their function, in: *Principles of Biomolecular Organization, Ciba Foundation Symposium,* G. E. W. Wostenholme and M. O'Connor, Eds. Little, Brown, Boston, pp. 308–345.

PORTER, K. R., BECKERLE, M., and MCNIVEN, M. (1983). The cytoplasmic matrix, in *Spatial Organization of Eukaryotic Cells,* J. R. McIntosh, Ed. *Modern Cell Biology* **2**:259–302; Alan R. Liss, New York.

PORTER, K. R., and TUCKER, J. B. (1981). The ground substance of the living cell. *Scientific American* **244**(3):56–67.

ROBERTS, K., and HYAMS, J. S., Eds. (1979). *Microtubules.* Academic, New York.

SCHLIWA, M. (1984). Mechanisms of intracellular organelle transport. *Cell and Muscle Motility* J. W. Shay, Ed. **5**:1–81.

SCHLIWA, M., and VAN BLERKOM, J. (1981). Structural interaction of cytoskeletal components. *Journal of Cell Biology* **90**:222–235.

SCHROEDER, T. E. (1976). Actin in dividing cells: Evidence for its role in cleavage but not in mitosis, in *Cell Motility*, R. Goldman, T. Pollard, and J. Rosenbaum, Eds. Cold Spring Harbor Laboratory, New York, pp. 265–277.

SCHROEDER, T. E. (1981). The origin of cleavage forces in dividing eggs. *Experimental Cell Research* **134**:231–240.

SCHWARTZ, J. H. (1980). The transport of substances in nerve cells. *Scientific American* **242**(4):152–171.

SHEETZ, M. P., and SPUDICH, J. A. (1983). Movement of myosin-coated fluorescent beads on actin cables *in vitro*. *Nature* **303**:31–35.

SMALL, J. V., RINNERTHALER, G., and HINSSEN, H. (1982). Organization of actin meshworks in cultured cells. *Cold Spring Harbor Symposia on Quantitative Biology* **46**:599–611.

STEARNS, M. E. (1982). High voltage electron microscopy studies of axoplasmic transport in neurons: A possible regulatory role for divalent cations. *Journal of Cell Biology* **92**:765–776.

STEARNS, M. E., and OCHS, R. L. (1982). A functional *in vitro* model for studies of intracellular motility in digitonin-permeabilized erythrophores. *Journal of Cell Biology* **94**:727–739.

STEINMAN, R. M., MELLMAN, I. S., MULLER, W. A., and COHN, Z. A. (1983). Endocytosis and the recycling of plasma membrane. *Journal of Cell Biology* **96**:1–23.

STEPHENS, R. E., and EDDS, K. T. (1976). Microtubules: Structure, chemistry and function. *Physiological Reviews* **56**:709–777.

STOSSEL, T. P. (1983). The spatial organization of cortical cytoplasm in macrophages, in *Spatial Organization of Eukaryotic Cells*, J. R. McIntosh, Ed. *Modern Cell Biology* **2**:203–223; Alan R. Liss, New York.

STOSSEL, T. P., HARTWIG, J. H., YIN, H. L., ZANER, K. S., and STENDAHL, O. I. (1982). Actin gelation and the structure of cortical cytoplasm. *Cold Spring Harbor Symposia on Quantitative Biology* **46**:569–578.

TAYLOR, D. L. (1976). Motile model systems of amoeboid movement, in *Cell Motility*, R. Goldman, T. Pollard, and J. Rosenbaum, Eds. Cold Spring Harbor Laboratory, New York, pp. 797–821.

TAYLOR, D. L., and CONDEELIS, J. S. (1979). Cytoplasmic structure and contractility in amoeboid cells. *International Review of Cytology* **56**:57–144.

TAYLOR, D. L., CONDEELIS, J. S., MOORE, P. L., and ALLEN, R. D. (1973). The contractile basis of amoeboid movement. I: The chemical control of motility in isolated cytoplasm. *Journal of Cell Biology* **59**:378–394.

TAYLOR, D. L., BLINKS, J. R., and REYNOLDS, G. (1980*a*). Contractile basis of amoeboid movement. VIII: Aequorin luminescence during amoeboid movement, endocytosis and capping. *Journal of Cell Biology* **86**:599–607.

TAYLOR, D. L., WANG, Y.-L., and HEIPLE, J. M. (1980*b*). Contractile basis of amoeboid movement. VII: The distribution of fluorescently labeled actin in living amoebas. *Journal of Cell Biology* **86**:590–598.

TILNEY, L. G. (1983). Interactions between actin filaments and membranes give spatial organization to cells, in *Spatial Organization of Eukaryotic Cells*, J. R. McIntosh,

Ed. *Modern Cell Biology* 2:163–199; Alan R. Liss, New York.

WEEDS, A. (1982). Actin-binding proteins: Regulators of cell architecture and motility. *Nature* 296:811–816.

WEGNER, A. (1976). Head to tail polymerization of actin. *Journal of Molecular Biology* 108:139–150.

WELCH, G. R. (1977). On the role of organized multienzyme systems in cellular metabolism: A general synthesis. *Progress in Biophysics and Molecular Biology* 32:101–191.

WELSH, M. J., DEDMAN, J. R., BRINKLEY, B. R., and MEANS, A. R. (1978). Calcium-dependent regulator protein: Localization in mitotic apparatus of eukaryotic cells. *Proceedings of the National Academy of Sciences USA* 75:1467–1481.

WILLIAMSON, R. E. (1975). Cytoplasmic streaming in *Chara*: A cell model activated by ATP and inhibited by cytochalasin B. *Journal of Cell Science* 17:655–668.

WOLOSEWICK, J. J., and PORTER, K. R. (1979). The microtrabecular lattice of the cytoplasmic ground substance: artifact or reality? *Journal of Cell Biology* 82:114–139.

YIN, H.-L., ZANER, K. S., and STOSSEL, T. P. (1980). Ca^{2+} control of actin gelation. Interaction of gelsolin with actin filaments and regulation of actin gelation. *Journal of Biological Chemistry* 255:9494–9500.

13

Signals for Communication and Control

The isomorphism of entropy and information establishes a link between the two forms of power: the power to do and the power to direct what is done.

François Jacob, *The Logic of Life*

• • •

The Flow of Information
Directional Movement
Informational Transactions in Eukaryotic Cells
Developmental Signals

The purposeful behavior of living organisms is often their most conspicuous aspect, the one that immediately distinguishes the quick from the dead. The capacity to regulate the internal environment and to respond to cues from the external one in a manner that enhances the likelihood of survival and reproduction is shared by all living things. Jacques Monod incorporated it into the very definition of living things as objects endowed with a project or purpose (Monod, 1971).

There is no fundamental difference between regulatory and sensory communications, even though one mode looks inward and the other outward. Both require a cell to recognize a change in some significant parameter and to respond appropriately. Moreover, the mechanisms that link stimulus to response are much the same whether the stimulus is external or internal and whether the response is intended to restore the original state or to take advantage of the new one. The primary stimulus does not normally elicit the response directly; instead, the stimulus elicits the production of what physiologists call a signal, a specialized molecule, ion, or chemical grouping whose function is to convey information and thus to incite the cell to action. Cyclic AMP and Ca^{2+} ions are familiar examples of intracellular signals; hormones and neurotransmitters perform analogous roles in the bloodstream of higher animals.[1]

It is intuitively evident that discrimination between stimuli, the generation of intracellular signals, and the selection of the appropriate response cannot be thermodynamically spontaneous. They are in fact instances of a singular class of work, informational work, whose teleonomic purpose is the establishment of structural and functional order. The object of this chapter is to consider the management of regulatory and sensory information as useful work, a province on the map of bioenergetics. But before we do so, a bird's eye view of informational transactions in general may help to set biological signals in perspective and to clarify the special meaning that biologists assign to this workaday term.

The Flow of Information

Information and Order. Information is an elusive but fundamental concept whose relationship to other facets of energetics was explored in Chapter 1. Suffice it to recall here that entropy is a measure of disorder, randomness, and exhaustion. The quintessential characteristic of life is the converse of the above, biological order: regularity, predictability, and the capacity for action. At the

---

[1] In everyday speech, signal and stimulus are nearly synonymous. For the purposes of this chapter, stimulus refers to a change in the internal or external environment to which the cell responds, while signal refers to ions or molecules that convey information in a symbolic manner (for example, Ca^{2+} or cyclic AMP). Whether a hormone is a stimulus or a signal depends on whether one's viewpoint is the target cell or the whole animal.

molecular level, order is expressed in the rigidly determinate association of atoms into particular molecules (amino acids, proteins, lipids, nucleic acids), which constitute but a small selection of the myriads that might have been formed from these atoms. At the cellular level, order is expressed in the regularity and reproducibility of cellular organization, in every aspect of form and behavior. These things entail work of the kinds considered in earlier chapters: biosynthesis, transport, movement. But the establishment of order in itself entails an added measure of work, which is a function of the number of ways in which the constituents of a system can be arranged. To impose order, that is, to reduce the entropy of living matter by selecting one particular arrangement, one requires information (Equations 1.21 and 1.22). Information, as Maxwell's demon learned to his cost, is a charge on the energy budget; indeed, one may loosely think of information as a special kind of energy required for the work of establishing order.

Biosynthetic reactions illustrate the connection between energy, entropy, and information in a straightforward manner. When chloroplasts fix CO_2 and deposit the product as starch, simple precursors (CO_2 and H_2O) are transformed into a complex, specific, and highly determinate compound with the aid of light energy and a set of catalysts. Part of the light energy goes toward making a product whose entropy is less than that of its precursors: free energy is conserved, so to speak, in the form of organization. But the input of light energy is not sufficient, as any sufferer from sunburn can attest. If energy is to be converted into organization, its flow must be guided through channels that are in themselves highly ordered. That is where the enzymes of photosynthesis come in, and the architecture of the chloroplast in general. Enzymes owe their catalytic powers to their capacity to bind the reactants, imposing spatial order on the anarchic motions of free molecules; this capacity in turn is a reflection of the ordered, determinate structure of protein molecules. Chloroplasts embody in their own organization the information required to convert the energy of incident light into the displacement of the $NADP^+$/NADPH ratio from equilibrium and ultimately to generate starch, with minimal waste of materials and energy.

But what, precisely, is meant by information? By now the reader will have divined that it is easier to process information than to define it, and that this word is once again a metaphor whose meanings shift with the boundaries of professional specialties. To the the physicist, it is a numerical quantity that measures the uncertainty in the outcome of an experiment. Biologists use the term in the opposite (or rather, complementary) sense as a measure of regularity and predictability. In molecular parlance, information is closely allied to the chemist's notion of complexity and specificity. But the word also points to ideas that fall outside the vocabulary of physical scientists: to utility, responsiveness, purpose, and evolutionary descent. For the needs of this chapter, I shall stipulate a definition of information complementary to that of energy: the capacity to establish order or, in Jacob's happy phrase, the power to direct what is done (Jacob, 1973).

Information, as a concept, is even more abstract than energy. But living things never handle information as such; they deal with molecules whose structures embody greater or lesser degrees of order, specificity, and therefore information. Like the chemical factory to which it is often likened, a cell is designed to manufacture and manipulate chemical substances. But the products derive meaning only from their organization, which duplicates that of the original cell. Every aspect of cellular activity or architecture that contributes to this purpose is engaged in information processing; the flow of information is nearly coterminous with life itself.

For the purpose of delimiting a chapter on biological information, the foregoing is at once too little and too much. We shall be concerned here with those processes whose chief or even sole function is the transmission of information. These are of two kinds (Figure 13.1). The central information-processing machinery of every cell and organism is the replication, transcription, and translation of the genetic repository by mechanisms designed to ensure linear correspondence between a segment of DNA and its RNA or protein transform. I

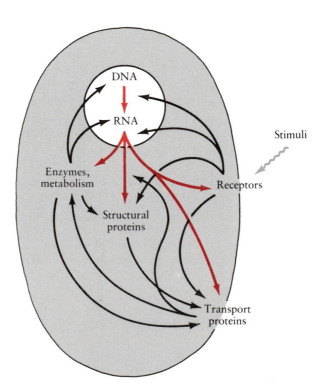

FIGURE 13.1.   Two kinds of biological information. The central dogma describes the flow of messages from DNA to protein (colored arrows). All biological activities, including transcription and translation, are controlled by a multitude of signals from within the cell and from the external environment (black arrows), which weave the genome and the remainder of the cell into a functional whole.

shall designate this mode of information transmission a message. The hormone that controls metabolism and the light that elicits directional movement also convey information, but of a different kind from that which issues from the genome. The mechanisms involved here are collectively called signals. Signals do not specify molecular structures but regulate the rates of chemical and physical processes. (From the informational viewpoint it might not be inappropriate to describe even an enzyme as a special kind of signal.) Signals control all vital activities and therefore hold a central place in bioenergetics. Both messages and signals select a useful response from among the multitude of possible responses; and work must by done by organized components of the living machinery in order to garner the information and make use of it.

Genetic Information: Messages. That DNA makes RNA, which in turn makes protein, has been called the central dogma of molecular biology. It is not a statement about the transformation of matter but about the transmission of information, and this is the aspect that concerns us here: just what kind of information does the genome supply, and how is it related to cellular functions?

Few readers will need to be reminded that DNA is the repository of genetic information in all cells and organisms (the fact that in certain viruses RNA plays that role need not detain us). The information is encoded in the linear sequence of purine and pyrimidine bases, like the dots and dashes of Morse code or the marks on a punched tape. Information is transmitted by transcription into RNA, whose primary sequence of purine and pyrimidine bases corresponds point by point to that of the DNA template. Some RNA transcripts are physiologically useful in their own right, particularly transfer RNAs and the various kinds of ribosomal RNAs. The majority of the transcripts are messenger RNAs, whose function is to be translated into protein. The latter is an exceedingly complex process, mediated by transfer RNAs and ribosomes, and by no means fully understood. However, in the end the linear sequence of amino acids corresponds precisely to the sequence of bases in the messenger RNA. A triplet of three successive bases, called a codon, specifies each amino acid according to the universal set of equivalences known as the genetic code. Most of the 64 possible codons stand for one amino acid or another, but a few stand for the initiation of translation or for its termination.

The special fitness of DNA and RNA to be carriers of information derives from the enormous number of combinations that can be generated by four symbols in a sequence of reasonable length. A standard-sized protein, 300 amino acids long, will be coded by a stretch of nucleic acid containing 900 bases. These 900 bases can in principle be arranged in 4^{900}, or about 10^{540}, different ways, an inconceivably large number. The selection of a single sequence for use implies that 1800 bits of determinacy have been built into the cell's genetic library (Chapter 1). This information is not precisely energy, but in conjunction with the appropriate machinery it has the power to direct what protein is made.

The genome of *E. coli* is about 4×10^6 base pairs in length, sufficient to code for the approximately 2500 proteins that this organism can produce. Many of these proteins are themselves part of the molecular machinery of transcription and translation (ribosomes alone make up more than a third of the mass of rapidly growing cells), while others ensure the accuracy of macromolecule production by detecting and correcting the inevitable errors. But proteins do not contribute in any way to the information stored in the genome. The object of the entire exercise is to ensure that the base sequence of the RNA transcript faithfully mirrors that of the corresponding gene and is in turn accurately reproduced in the amino acid sequence of the protein.

A cell, as I have said more than once, is an integrated hierarchy of structures and processes, "not a bundle of parts but an organization of parts, of parts fitting one with the other" (Thompson, 1917). The network of switches and dials required to coordinate the output of multiple structural genes and to relate it to cues from the environment is built into the genome itself in the form of regulatory genes. These constitute a higher level of information processing (more precisely, a hierarchy of such levels) that determines which of several alternative programs will be expressed and with what intensity; they are concerned, so to speak, with the qualification of the instructions engraved in DNA. In *E. coli* regulatory genes probably make up no more than a tenth of the total information, but in eukaryotic cells the proportion is likely to be much higher.

The meaning of the metaphor that regulatory genes qualify the genome is best explained by a simple example that also illustrates the interplay of messages and signals in the expression of genetic information. *Escherichia coli* utilizes lactose with the aid of an inducible enzyme, β-galactosidase, together with a specific porter. The two proteins are produced above basal level only when lactose is present, and their induction is repressed by glucose, which is the cell's preferred carbon source. The communications network that underlies this purposeful organization is shown diagrammatically in Figure 13.2. Three structural genes, coding for β-galactosidase, the lactose-H^+ symporter, and a third enzyme (a transacetylase, which we shall ignore) form a continuous sequence that is transcribed as a unit, called an operon. The operon is controlled by a stretch of DNA just upstream from the structural gene, the operator-promoter region, which includes the binding site for RNA polymerase. In the absence of lactose the operator is blocked by a cytoplasmic repressor protein (specified by a separate gene located elsewhere in the genome), which prevents the initiation of transcription. When lactose is present, the operon is opened to transcription; the inducer is not lactose itself but allolactose, a minor product of transglycosylation catalyzed by the basal level of β-galactosidase. Allolactose binds to the repressor protein and effects an allosteric change in the latter's configuration. This results in release of the repressor protein from the operator site, transcription commences, and the information encoded in the structural genes is expressed through the production of the corresponding proteins.

Glucose exerts its overriding influence on the inducibility of this operon and of some others as follows. Glucose deficiency leads to an increase in the cell's content of cyclic AMP. (The mechanism by which the cells sense starvation and translate it into enhanced adenylate cyclase activity is still uncertain.) Cyclic

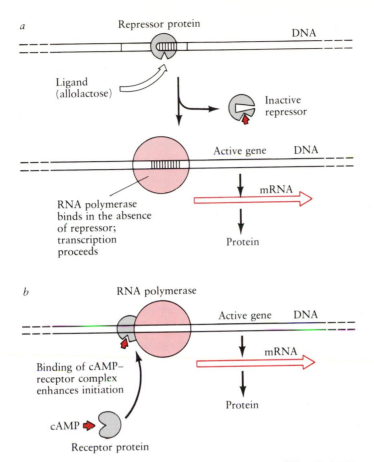

FIGURE 13.2.   Transcriptional control of the lactose operon of *E. coli*. (*a*) Negative control. When the repressor protein binds allolactose, it dissociates from its binding site on DNA, allowing transcription to proceed. (*b*) Positive control. The complex of cyclic AMP with its receptor protein binds to a site on the lactose operon, enhancing the initiation of transcription. (After Alberts et al., 1983, with permission of Garland Publishing.)

AMP combines with a cytoplasmic binding protein, and the complex in turn associates with a genomic site close to the operator-promoter locus and enhances the rate of transcript initiation. The utility of this stratagem is obvious: it is designed to minimize the expenditure of energy on the production of catabolic enzymes so long as glucose is available while ensuring rapid induction of the lactose operon as soon as glucose is exhausted. The genes that code for the repressor protein and the cyclic AMP receptor protein stand for the "when" and "but" that qualify the basic message of the lactose genes; their expression is under the control of signals from the cytoplasm and from the environment.

Bacteria display numerous variations on this theme of transcriptional control by negative and positive effector proteins. There must be many more in eukaryotic cells, whose genome is thought to be regulated by hundreds of such proteins. Eukaryotes also have regulatory devices that operate at later stages of information transfer: RNA processing following transcription, RNA export from the nucleus, and translation itself are all potential sites for the control of gene expression. These matters are beyond our scope (see *The Molecular Biology of the Cell*, by B. Alberts et al., 1983, for a lucid and contemporary introduction), but their logical status is important to the present argument. The genome is self-contained with respect to the primary structures of biological molecules: the information it embodies is both necessary and sufficient. But all these molecules, including those that regulate gene expression, must be manufactured by the remainder of the cell; orders as to what segment of the genetic information is to be read out come generally (perhaps always) from outside the genome. In one sense, the genome controls the cytoplasm; in another, the converse is true. The logic of biological information management is circular. Signals from the cytoplasm or from the environment select which of the programs encoded in DNA will be played, but the genome determines what signals the cell can read and how it interprets those signals.

And yet there is a profound difference between the information that the genome supplies and that which it receives. Messages, information that specifies the primary structure of macromolecules, flow from DNA into RNA and thence unidirectionally into proteins. The essence of the central dogma is that messages do not flow back from the protein level to that of nucleic acids. (The fact that in certain viruses RNA serves as the genetic repository and DNA is an intermediate in virus replication does not contravene this thesis.) Proteins serve as signals to control the output of the genome, but they do not alter its determinacy content. This conclusion is crucial to a comprehension of evolution by natural selection. There is no mechanism by which the genome can respond teleonomically to information from the external world. No matter how pressing the need, no organism can alter its genetic constitution to favor its survival in the face of environmental change. Modifications to the genome can occur only by random mutations, passed through the filter of natural selection. Even in this narrow passage the opportunities for change are circumscribed by the burden of earlier genetic decisions, embedded deep in the genome beyond amendment, recall, or repeal.

From Gene to Function.   The purpose of genetic messages is to transmit information that will eventually be expressed as a physiological process. Since the genes determine the structure of macromolecules they specify functions as well, but the relationship is less direct than it appears (Stent, 1978).

In daily life we readily distinguish two levels of meaning in any message: the explicit meaning, which is conveyed directly by the symbols of which the message is composed; and the implicit one, which depends on the context. The question, "What are you doing here?" is straightforward enough on the explicit level, but its full meaning (and the response it evokes) depends heavily on who is addressing whom under what circumstances. In like manner, when we ask how a cell's genetic complement specifies any one of the physiological functions discussed in the preceding chapters, we seek the meaning, so to speak, of a portion of the genome.

The explicit meaning of genetic information is plain: it consists of the primary sequences of amino acids in proteins and of nucleotides in transfer RNA, ribosomal RNA, and regions such as operators that control gene expression. But the biological connotations of this information are far wider, for they include the ripple of consequences that are implicit in the message. The sequence of amino acids ensures that a protein will fold into a particular configuration. Once that has happened, parts of the resulting structure may have the power to catalyze a particular chemical reaction; the principles of chemical catalysis are not explicitly spelled out in the sequence of nucleotides in DNA but arise implicitly from correct folding of the chain and sometimes from the mutual association of subunits. Should the enzyme in question be the F_1 headpiece of an ATPase, a higher level of contextual meaning appears when the F_1 portion associates with an F_0 segment embedded in a membrane, acquiring thereby the capacity to translocate protons. Association with other proteins in a bacterial plasma membrane may call forth a still higher level of meaning, which we call oxidative phosphorylation. Is oxidative phosphorylation genetically determined? Yes, but there is no one segment of the genome that can be identified as specifying that function. It is a function that arises by the interaction of independent elements within a particular structural context. Should our proton-translocating ATPase find itself in the plasma membrane of a streptococcal cell, its functional meaning would be quite significantly altered.

Genes do determine biological functions, in the sense that functions result automatically from the execution of the genetic instructions, but functions are specified implicitly rather than explicitly. The broader the process under consideration, the more remote it is likely to be from what is explicitly spelled out in any genetic message, and the larger the contextual element. We shall return to this subject in Chapter 14 when we consider the genesis of cellular forms. It should come as no surprise that morphogenesis is but obliquely related to the macromolecular structures specified by the genome.

Signals.   Besides the genetic messages that issue from the genome alone a network of signals extends through all levels of cellular operations, weaving genome, cytoplasm, and various membrane-bound compartments into a functional unity. Genetic messages, like written communications, consist of conventional symbols in linear array. Signals, by contrast, come in a kaleidoscopic variety and lack the universal equivalences of the genetic code. But signals also can be regarded as symbols from a metabolic or sensory code (Tomkins, 1975), that varies somewhat from one class of organisms to another. Hormones, neurotransmitters, cyclic nucleotides, and Ca^{2+} ions are not metabolic intermediates, energy carriers, or catalysts. They are strictly regulatory substances, controlling the rates of chemical reactions and other processes without directly participating in any essential way in these processes. Signals convey information in a symbolic manner and thereby incite the cell to perform the actions for which its macromolecular constituents are designed. The intrinsic information content of a signal is small, and it derives meaning only from being channeled through a system of specific receptors and effectors that the cell has encoded in its genome and produced in advance. In a nutshell, messages instruct whereas signals select.

One of the hazards in generalizing about biological communications is their great diversity. Light, temperature changes, mechanical disturbance, and chemical substances from ions to proteins act as stimuli, to which cells respond variously by directional movement, shifts in metabolic pathways, the discharge of a product, or perhaps by a change in form or growth habit. The flow of information from sensory stimulus to response almost always involves intracellular signals, of which cyclic AMP and Ca^{2+} ions are the most familiar. We shall consider examples of sensory information processing in later sections of this chapter. But stimuli do not always come from outside the plasma membrane. Communications within the cell are exemplified by the regulatory molecules that coordinate the metabolic web (Chapter 2), by the homeostatic control of the internal pH, and by the regulation of gene expression through cytoplasmic proteins; these also are signals. The genome itself is often a source of signals, such as the caps at the ends of messenger RNA molecules that have to do with RNA processing and the leader sequences that address certain proteins to the plasma membrane for insertion or export. In many organisms the very passage of time monitored by the biological clock induces cells to respond with some particular action, elicited by signals of unknown nature.

In order for a cell to extract useful information from a stimulus, the latter must generally be processed. The stages are particularly plain in the case of sensory perceptions. The primary stimulus, a change in some external parameter, must be recognized by a receptor specific enough to distinguish one potential stimulus from another: galactose, not glucose; blue light, not red. This intelligence must next be transduced into a form compatible with cellular operations, just as energy from the environment must be processed before it can serve the uses of life. Finally, there is an effector that executes the action incited by the signal. Signal processing does not always conform to this pattern, but recognition, transmission, and execution are usually distinguishable aspects of information flow that bring a measure of uniformity to the cacophony of biological communications.

Informational Work.   Whether the information takes the form of messages or of signals, work must be done (Chapter 1). To be precise, a gain of one bit of information requires the cell to perform $kT \ln 2$ units of work, where k is Boltzmann's constant and T is the absolute temperature (at 27°C, one bit is equivalent to 3×10^{-21} joule). The same amount of work is required for each binary decision built into the genome. How do cells pay for this sort of work?

A system at equilibrium can neither gather information nor respond to it.

Signal processing entails the consumption of some preexisting gradient of potential energy, just as other kinds of work do. For example, certain hormones trigger a flux of Ca^{2+} ions into the cell, dissipating the calcium gradient that the cell created by doing work. The gradients that support information processing are the same that support other kinds of work: the ATP-ADP couple, other compounds of high group transfer potential, and the electrochemical potential of Ca^{2+} and other ions. Signals convey information by exerting kinetic control over the dissipation of thermodynamic gradients. It is obviously advantageous to keep such episodes brief and of limited extent; but some energy dissipation is an unavoidable price that cells must pay for information. The cost of synthesizing receptors and transducers is no less essential, since it is these macromolecules that give meaning to the signal. In the case of genetic messages, biosynthesis accounts for the lion's share of the energy cost.

It is sometimes said that information makes up but a small fraction of the cell's energy budget. This assertion stems from attempts to express the cost of biological order in units of energy, as follows.[2] It will be recalled (Chapter 1) that the information content of a cell is extraordinarily high, some 10^9 bits per bacterial cell. This is equivalent to $10^9 \times kT \ln 2 = 3 \times 10^{-12}$ joule per cell, and given that there are 2×10^{12} cells in one gram of dry bacteria, to six joules per gram of cells. Bacterial growth yields are on the order of 10 grams of cells per mole of ATP (Chapter 5), or one gram per 0.1 mole. Taking the free energy of ATP hydrolysis to be 32 kJ/mol, we need 3200 joules to grow one gram of bacteria. Of these only six joules, 0.2 percent, are the cost of order—a bargain by any standard. But the calculation is misleading, for it assumes that the energetics of information processing can be separated from those of matter. This is not so: the informational macromolecules alone, DNA, RNA, and protein, make up as much as half of the mass of a bacterial cell, and their biosynthesis consumes two-thirds of the assignable ATP production (Table 5.2). If we add to this the cost of protein phosphorylation, maintaining calcium gradients, and other items, we reach the conclusion that information processing must be a major expense. This should come as no surprise: if biological order is as close as we can come to the essence of life (Chapter 1), information management must be the central purpose for which living things mobilize energy.

The viewpoint of molecular biology is presently so overpowering that it blinds many scientists to the fundamental duality of biological information management. Messages and signals are complementary: the former govern the structure of the living machinery, the latter its workings, and it is their mutual interactions that produce a living cell. Molecular genetics is too vast a topic to be usefully treated in this book, but the control of cellular activities by means of signals is an integral part of bioenergetics. In the remainder of this chapter we shall look more closely at some of the molecular mechanisms by which prokaryotic and eukaryotic cells monitor changes in their internal or external

[2] I am indebted to Dr. H. J. Morowitz for his help with this calculation.

milieu and make functional use of this intelligence. The examples were selected to illustrate the expenditure of metabolic energy on informational work and the sophisticated adaptations by which cells maximize the benefits so purchased.

Directional Movement

Bacterial Chemotaxis.   Most bacterial perceptions and responses are designed for homeostasis, the regulation of metabolism, and for the rapid adjustment of the cell's enzyme complement to whatever nutrients the environment has to offer. The induction of the lactose operon (Figure 13.2), the control of RNA production by guanosine polyphosphates, and the relationship between cell turgor and potassium transport (Chapter 5) illustrate the use of signals for regulatory purposes. In addition, many bacteria recognize sensory stimuli and respond with directional movement. Bacterial chemotaxis can be regarded as a simple model for behavior in general; and even on this first rung of the evolutionary ladder we find conduct that suggests memory, choice, judgment, and learning. There are many recent reviews of this field; particularly useful ones are those of Koshland (1977, 1979, 1981), Macnab (1979), Boyd and Simon (1982), and Taylor (1983).

That bacteria migrate toward sources of oxygen and nutrients, and away from noxious substances has been known since the late nineteenth century, but these matters did not arouse widespread interest until the 1960s, when Julius Adler demonstrated that bacterial chemotaxis could be studied with the methods of molecular genetics (Adler, 1966). Adler showed that attractants and repellents need not be metabolized; that each class of chemoeffectors has its specific receptor; and that one can isolate mutants that are either defective in their response to individual attractants or incapable of chemotaxis in general. Today we recognize that bacterial chemotaxis is a simple but astonishingly sophisticated mode of behavior. *Escherichia coli* has about three dozen distinct receptors, most of them located either in the plasma membrane or in the periplasmic space. Binding of the chemoeffector to its receptor generates a signal that is processed through a common pathway involving another nine proteins. The output is conveyed to the flagellar motor and controls the manner of swimming (Figure 13.3). Bacteria respond effectively to gradients as shallow as one part in 10^4 over the length of a cell. They distinctly prefer some attractants to others and readily learn to ignore a constant stimulus while remaining attentive to those that vary. We understand in principle how bacteria sense gradients and how the motor responses bring about directed movement. What is still intriguingly unclear is how the information gathered by the receptors is transmitted to the motor and how it alters the latter's rotation.

As explained in Chapter 5, the propulsive force for bacterial swimming is supplied by a rotary motor at the flagellar base, whose energy source is an inward current of protons. When the motor turns counterclockwise, the flagella coalesce into a bundle and the cell swims smoothly. Periodically the sense of rotation

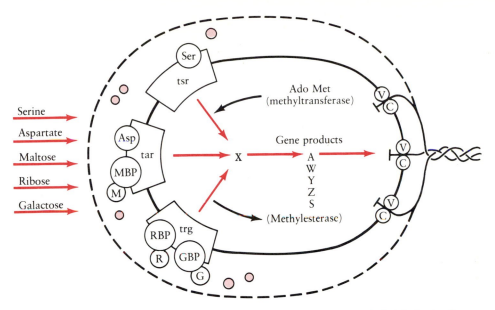

FIGURE 13.3.   Information flow in bacterial chemotaxis. A variety of chemical stimulants, only a few of which are listed, pass through pores in the outer membrane and bind to specific receptors in the plasma membrane or in the periplasmic space (colored circles). Association with the transducer proteins (trg, tsr, or tar) generates a signal, X, whose nature is unknown. The intensity of signal output is modulated by methylation and demethylation of the transducer proteins. The signal is processed further with the aid of cytoplasmic proteins (A, W, Y, Z, S) whose function is unknown. It then impinges on the flagellar motor via gene products V and C to modulate the frequency of flagellar reversal. (After Koshland, 1981, and Taylor, 1983, with copyright permission of Annual Reviews.)

undergoes spontaneous reversal; the flagellar bundle flies apart and the cell tumbles. When another spontaneous reversal restores counterclockwise rotation, the cell resumes forward swimming in a new direction. Chemotactic effectors somehow alter the *frequency* of tumbling: attractants suppress it, while repellents enhance it (weakly). As a consequence when the bacterium "senses" that it is going in a favorable direction, tumbling is diminished and the cell is more likely to stay the course. Conversely, when it is going in an unfavorable direction it is more likely to tumble and, perhaps, choose a better course next time. "By taking giant steps in the right direction and small steps in the wrong direction, it biases its walk very effectively in the direction which aids its survival" (Koshland, 1977).

How does the cell "know" whether it is moving in a favorable direction or an unfavorable one? In principle there are two ways to sense a gradient. The cell might "compare" the instantaneous extent of receptor occupancy at its head and at its tail, a formidable analytical problem since *E. coli* is only 2 μm long yet attains an accuracy of one part in 10,000. Alternatively, a moving cell might compare the extent of receptor occupancy at the present instance with that of a little while earlier. The latter is in fact what *E. coli* does. The cell's "memory span," on the order of 1 minute, is long enough to let it cover 20 to 100 body lengths between "measurements." By this trick the analytical problem is reduced to measuring one part in 100 to 1000, still a very respectable achievement. The bacterial memory span is short by comparison with ours, but optimal for its purpose: long enough to be useful in guiding the cell's migration, yet not so long that the cell is confused by information rendered irrelevant by a subsequent change in the direction of travel.

Signal Recognition and Transduction.   Of all the elements of the sensory response, the surface receptors are the best known. They number about 30; 20 recognize attractants including serine, aspartate, glucose, maltose, ribose, galactose, and N-acetyl glucosamine; repellents include leucine, weak acids, and certain metal ions. *Escherichia coli* also recognizes oxygen as a powerful attractant, but by another pathway, which involves the respiratory chain (see below). Chemoreceptor proteins are remarkably diverse, and many serve more than one function. Ribose, galactose, and some other sugars are recognized by the periplasmic binding proteins that also take part in the active transport of these sugars (Chapter 5); the sugar-binder complex then links up with a separate signal-transducing protein. Glucose is recognized by enzyme II of the phosphotransferase system (Chapter 5). Serine and aspartate bind directly to the integral membrane proteins that also mediate signal transduction, as will be described shortly.

Much of the capacity of *E. coli* to integrate and evaluate sensory information has its roots in the competition between ligands for receptors or between receptor-ligand complexes for sites on the transducer (Figure 13.4). For instance, ribose and galactose bind to separate cytoplasmic receptor proteins, but these two share a common transducer, which has a higher affinity for the ribose-binding protein. So long as ribose is present, the cell will not "see" galactose since all the sites are occupied, and it will not waste energy responding to a galactose gradient. But it can still sense the gradient of a nitrogen source, aspartate for example, which feeds into a separate transducer. The objective of *E. coli* is to migrate toward a source of nutrients, usually decaying organic matter. It need not be able to detect all organic substances individually, just a judiciously selected sample, and this is the case.

The next stage of information processing is effected by the transducer proteins: they make up the comparator with its "memory" that allows the cell to detect a gradient and to relay this information to the flagellar motor. By the ingenious application of genetic and biochemical approaches, this central function has been assigned to a set of integral proteins that span the plasma membrane (Figures 13.3 and 13.4). Domains exposed to the exterior bind the chemical stimulants, either directly (serine, aspartate) or with the aid of an adaptor (the periplasmic binding proteins for ribose, galactose, and maltose). Binding of these ligands somehow alters a functional property of the protein or of

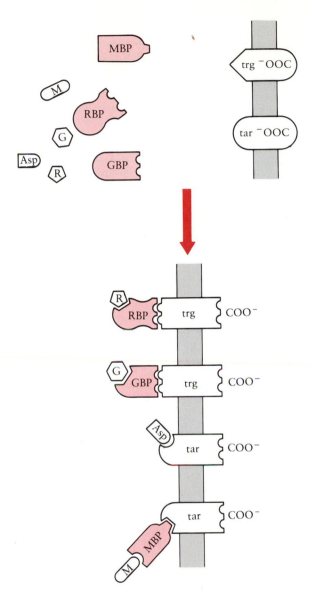

FIGURE 13.4.   Receptors and transducers. Ribose (R), galactose (G), and maltose (M) each bind to periplasmic receptor proteins (RBP, GBP, MBP) and induce conformational changes in the latter. The RBP and GBP then bind to the trg transmembrane protein, eliciting production of a cytoplasmic signal. The MBP binds to the tar protein, which also binds aspartate directly; both associations again elicit production of the same signal. Ligand binding to these transducer proteins is accompanied by conformational changes that alter the susceptibility of multiple glutamic acid residues to methylation and demethylation. This is indicated symbolically by showing interior carboxyl groups before ligand binding and exterior carboxyls afterward. (From Koshland, 1981; reproduced from the *Annual Review of Biochemistry*, vol. 50, with permission of Annual Reviews.)

the domain that projects into the cytoplasm, possibly a catalytic activity, and generates a signal that is passed to the flagellar motor. Signal transduction is modulated by methylation of the transducer protein, a matter to which we shall return later. Figure 13.4 shows two of the four transducer proteins known to exist in *E. coli*. The *tar*-gene product binds aspartate directly and also binds the complex of maltose with its periplasmic protein. The *trg*-gene product accepts ribose and galactose, each in the form of a protein complex. Serine binds directly to a third transducer, specified by the *tsr* gene. All transducer proteins appear to generate the same signal since effectors that bind to each add up algebraically to produce a unified response.

At this point we encounter a large gap in our understanding of sensory processing in bacteria: we do not know the identity of the signal generated by the transducer and therefore cannot say what functional property of the transducer is modified by binding the chemoeffector. One possibility is that the transducer is linked to the motor by a cascade of conformational transitions in adjacent proteins, a molecular domino effect, but there is good reason to discount such mechanical coupling. A few years ago it was proposed that chemical stimulants modulate the flux of Ca^{2+} ions across the plasma membrane, much as in the avoidance reaction of *Paramecium*, but the balance of the evidence runs against this view. A more attractive hypothesis states that the signal consists of an abrupt change in the membrane potential, since we know that the protonic potential drives the flagellar motor. Fluctuations in the membrane potential and in the protonic potential do affect the tumbling fequency, and some gradients are apparently sensed in this manner; oxygen is a notable instance (Taylor, 1983). But at least in *E. coli*, the signal generated by the receptors for sugars and amino acids seems to be something other than a parameter of the protonic potential. On balance, the most plausible hypothesis is that the binding of the chemical effector modifies a catalytic function of the transducer protein and thereby alters the concentration of a cytoplasmic metabolite that is the signal proper. This notion obviously owes a debt to the manner in which cyclic AMP links stimuli to responses in eukaryotic cells. Indeed, Black *et al.* (1980) reported that attractants elicit a transient accumulation of cyclic GMP, but it now seems unlikely that this is in fact the elusive cytoplasmic signal.

The nature of the tumble regulator, or switch, that alters the sense of rotation of the flagellar motor in response to the signal is as uncertain as the signal itself. A gene that codes for this protein has been identified, however. Furthermore, it is clear that certain physiological stimuli, including oxygen gradients, bypass the signal-processing apparatus outlined above to impinge directly on the tumble regulator. It seems likely that the nature of this switch will remain obscure as long as we do not understand how the proton flux drives the motor.

The preceding account dealt with *E. coli,* and with a selected set of stimuli at that, because these have received the most extensive study. But it should not be taken for granted that the manner of information processing is the same for all stimuli and for all bacteria. We saw above that oxygen taxis in *E. coli* is probably mediated via the protonic potential. This is not an exceptional case but a major alternative pathway of signal transduction. For instance, the protonic potential appears to carry the signal in many phototactic bacteria. But halobacteria, which are strongly phototactic, may not use the protonic potential as a mediator; their signal is in dispute (Baryshev et al., 1983; Taylor, 1983). Spirochetes are another class of organisms in which fluctuations of the membrane potential play a major role in signal transmission (Goulbourne and Greenberg, 1983); there are even indications that a calcium flux may be involved. It may well turn out that the growing diversity of bacterial transport mechanisms is matched by their variety of sensory pathways.

Explaining Behavior.   Regardless of the chemical identity of the chemotactic signal, it is already possible to rationalize most of the behavioral responses of bacteria with the aid of a simple set of postulates; these make up what Koshland (1977) designated the response regulator model (Figure 13.5). The model assumes that the frequency of tumbling is a function of the level of some cytoplasmic substance, the signal or response regulator R. There is a single pool of R in each cell, and the rates of formation and decomposition of R are variously influenced by different stimuli. The time-dependent characteristics of the pool of R constitute what was earlier called the cell's memory. For illustrative purposes, Figure 13.5 assumes that an increase in the R level suppresses tumbling; until we know the nature of R, this is an arbitrary assumption. On this premise, a cell moving up an attractant gradient will experience an increase in R that lasts until decomposition catches up. The signal is transient, and if the cell encounters no reinforcing stimulus, tumbling soon resumes. A higher attractant level will perpetuate the elevated level of R and the suppression of tumbling, making it more likely that the cell will persist in its course of swimming.

Among the observations that support this model, some of the most impressive concern the behavior of nonchemotactic mutants. One class of mutants never tumbles, but can be made to do so by a strong repellent stimulus; in these the basal steady-state level of R may be above the threshold that suppresses tumbling. A second class of mutants tumbles constantly, presumably because the steady-state level of R is far below the threshold; an extra-strong attractant stimulus raises the level of R and affords the mutant a period of smooth swimming.

We have not yet mentioned adaptation, an important aspect of bacterial behavior and of sensory physiology in general. Addition of ribose to a suspension of *E. coli* instantaneously suppresses tumbling. However, after a little while tumbling resumes despite the continued presence of ribose. Such adaptation to an unvarying stimulus is a necessary feature of all sensory systems, whose purpose is normally to detect change against a constant background. The capacity to adapt enables the organism to sense a second stimulus on top of the first—for instance, a gradient of serine, which, being a nitrogen source as well as a carbon source,

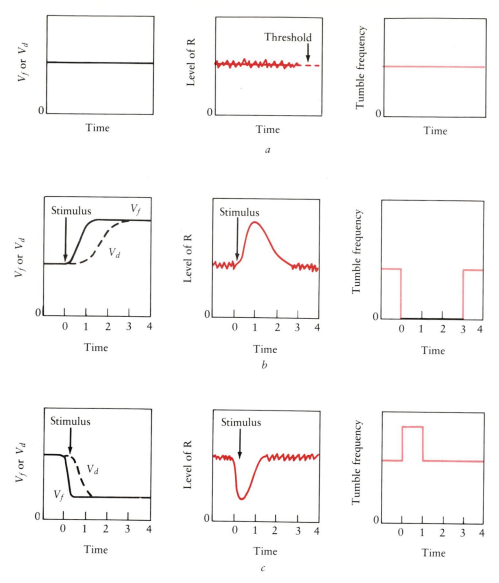

FIGURE 13.5.   Elements of the response regulator model of bacterial behavior. Panels show the variation over time of the formation and decomposition of the response regulator R (V_f and V_d), the level of R, and the frequency of tumbling. (*a*) In the absence of a gradient, $V_f = V_d$. The level of the response regulator remains essentially constant and so does the cell's frequency of tumbling. (*b*) An increase in attractant concentration causes V_f to spurt ahead of V_d. There is a transient increase in the level of R and a transient decrease in the frequency of tumbling. (*c*) A decrease in the attractant concentration causes V_d to exceed V_f. There is a transient decrease in the level of R, accompanied by a transient increase in the frequency of tumbling. (After Koshland, 1977. Copyright 1977 by the American Association for the Advancement of Science.)

would be even more useful than ribose. Adaptation to persistent stimuli is a familiar feature of more advanced organisms as well. Where would we be if we were forever conscious of the feel of our clothes or of meaningless background noise? To make practical use of a sensory apparatus it is necessary to set it back to zero when the stimulus becomes constant.

Bacteria accomplish this by methylation of the transducer proteins. Almost 20 years ago, Adler and his colleagues found that a minimum level of methionine was required for the chemotactic response. Later it turned out that methionine was not needed for tumbling or for motility but for controlled behavior; and that it serves as a precursor for the production of S-adenosyl methionine. The latter donates methyl groups to the transducer proteins, which bear multiple methyl-accepting sites. The degree of methylation is a function of two enzymes, a methyltransferase and a demethylase, both of which are specified by genes that belong to the chemotaxis cluster.

Apparently binding of attractants to the transducer proteins elicits a conformational change that results in the exposure of glutamate residues to the cytoplasm; these become methylated, reducing the signal output. Conversely, repellents alter the conformation of the transducer proteins in such a fashion as to lower the degree of methylation and increase signal output. A cell in the presence of a constant level of an attractant tumbles as frequently as one that has never seen the stimulus. An observer could not tell them apart by their behavior, but the level of transducer methylation would be distinctly higher in the cell exposed to the attractant than in the uninitiated one. It should be added here that not all adaptive responses in bacteria are mediated by methylation. Adaptation to oxygen, for example, does not involve methylation; its mechanism is unknown.

Some aspects of bacterial chemotaxis may prove to be peculiar to the prokaryotic level of cellular organization: temporal sensing, for example, would be most useful to very small cells, and there is evidence that the larger eukaryotic cells can detect spatial gradients. Nevertheless, the pathways of signal generation and processing by prokaryotic and eukaryotic cells have many fundamental features in common.

1.   The range of external stimuli to which a cell can respond is a function of its receptor complement.

2.   The input from many receptors may be integrated by processing through a common pathway to generate a unified signal.

3.   Stimuli compete at several levels, including the receptors, transducers, and effector. These interactions produce a scale of values such that some stimuli carry more weight than others.

4.   Higher levels of behavioral control are achieved by adaptation. Methylation as a mechanism of adaptation has so far been documented only in bacteria but may be more widely distributed.

5.   Several molecular features are common if not universal. One is the stimulus-induced association of mobile proteins. Another is the use of two antagonistic enzymes to modulate the level of activity: methyl transferase and demethylase, protein kinase and protein phosphatase.

6.   The work of information processing is done at the level of signal transduction. This usually entails amplification of the signal over many orders of magnitude. In bacteria we still do not understand the molecular basis of signal transduction, and therefore our comprehension of information flow as a special kind of work remains seriously deficient. By contrast, in many instances of information processing by eukaryotic cells this aspect is particularly well understood.

Informational Transactions in Eukaryotic Cells

Eukaryotic cells are generally both larger than prokaryotic ones and more sophisticated, in the sense that they display a much wider repertoire of perceptions and responses. Chemotaxis, tropisms, induced secretion and contraction, immune reactions, and the triggering of development or differentiation are but some of the instances in which an external stimulus elicits a programmed, predictable response. Each involves a unique set of receptors and may call on any of the cell's work functions. A measure of uniformity nevertheless prevails at the level of signal transduction, which rings the changes upon a manageable number of recurring themes; ion currents, cyclic nucleotides, and protein phosphorylation are as central to eukaryotic signal transduction as ATP and the protonic potential are to energy conversions.

The Many Faces of Cyclic AMP.   A particularly fruitful avenue of research opened in the late 1950s with the studies of E. W. Sutherland and his colleagues on the mechanism by which epinephrine (adrenaline) stimulates the breakdown of glycogen by the liver, part of the animal's preparation for either fight or flight. In the intact tissue epinephrine activates glycogen phosphorylase kinase, which in turn phosphorylates glycogen phosphorylase and initiates glycogen utilization. Epinephrine had no effect in cell-free extracts, but incubation of membrane fragments with epinephrine and ATP produced a factor that could activate phosphorylase kinase. The substance was quickly identified as 3′, 5′–cyclic AMP. It is neither a metabolic intermediate nor an energy carrier but serves as an intracellular signal (Sutherland, 1972).

Figure 13.6 outlines the flow of information for the mobilization of glycogen by skeletal muscle, a case that is particularly well worked out. Epinephrine binds to a receptor protein at the outer surface of the plasma membrane. This activates adenylate cyclase, the enzyme that converts ATP into cyclic AMP, causing the cAMP level in the cytoplasm to rise by two orders of magnitude. Cyclic AMP in turn activates a particular protein kinase, cAMP-dependent protein kinase,

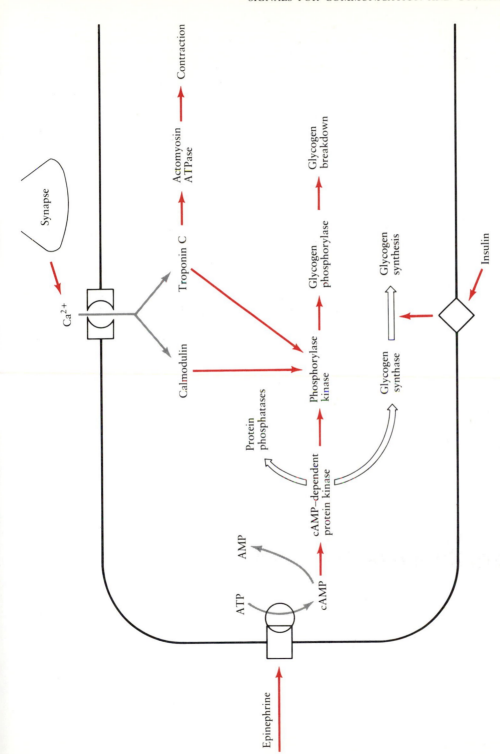

FIGURE 13.6. Pathways of information flow for the mobilization of glycogen by skeletal muscle. Binding of epinephrine to its receptor specifically stimulates cAMP synthesis, which activates cAMP-dependent protein kinase. Phosphorylation activates phosphorylase kinase but inactivates other enzymes of glycogen metabolism. The effects of Ca^{2+} ions are synergistic with those of cAMP. Colored arrows, stimulatory signals; open arrows, inhibitory signals.

which is to be distinguished from other protein kinases that occur in the muscle. The cAMP-dependent protein kinase also phosphorylates a number of other proteins, including glycogen phosphorylase kinase, glycogen synthase, and an inhibitor of protein phosphatase. Some enzymes are activated by phosphorylation, others are inhibited; in the present example the net result is that glycogen breakdown is stimulated while glycogen synthesis is inhibited. The hydrolytic enzyme cAMP phosphodiesterase quickly returns the cAMP level to baseline; it operates in concert with the phosphatases that act on the various metabolic enzymes themselves. Thanks to the interplay of the many enzymes that make up this corner of the metabolic web, cAMP does not simply act as an on/off switch: glycogen metabolism is precisely modulated to track the level of epinephrine in the circulation, which in turn reflects the physiological demand for metabolic energy. The control of glycogen metabolism remains the archetype of cAMP-mediated coupling between stimulus and response and serves as a model for many analogous instances that are less fully explored. Various aspects of this extremely active field have been reviewed by Greengard, 1978; Krebs and Beavo, 1979; Chock et al., 1980; Rodbell, 1980; Ross and Gilman, 1980; Limbird, 1981; Rasmussen, 1981; Cohen, 1982; and Flockhart and Corbin, 1982.

The activation of adenylate cyclase by epinephrine is a somewhat circuitous process. Epinephrine binds tightly and specifically to a receptor protein that is mobile in the fluid phase of the plasma membrane. Having bound the hormone,

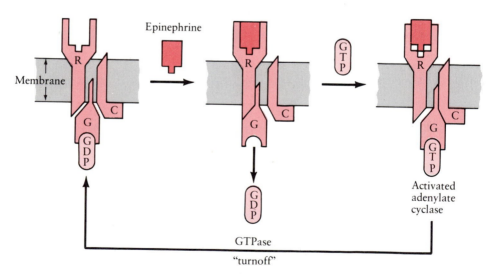

FIGURE 13.7. A proposed mechanism for the activation of adenylate cyclase by hormones. The hormone receptor R, the GTP-binding protein G, and the catalytic subunit C are independent proteins, mobile in the cytoplasmic membrane. Once R has bound epinephrine, it activates G; activated G binds GTP and then activates adenylate cyclase. The enzyme is deactivated by the hydrolysis of GTP. (After Limbird, 1981; reprinted with permission of the Biochemical Society, London.)

the receptor activates the cyclase with the aid of a third protein, the GTP-binding or G protein (Figure 13.7). Sustained activation does not require the formation of a ternary complex, but a succession of binary ones. The charged receptor collides with the G protein, inducing a conformational change that allows the G protein to bind cytoplasmic GTP, and collision of the G(GTP) protein with the catalytic element triggers cyclase activity in the latter. The G protein has a catalytic action of its own: it hydrolyzes GTP and is consequently self-inactivating. Sustained activation of the adenylate cyclase requires the continued presence of epinephrine. Note that the pathway of signal transmission traverses the plasma membrane and depends on a train of conformational changes to carry the communication.

Cyclic AMP in turn acts on a protein kinase, and once again the molecular mechanism is indirect and hinges on the specific association of protein subunits (Figure 13.8). The protein kinase consists of two kinds of subunits, one catalytic and the other regulatory. The complex, containing two subunits of each sort, is inactive. When cAMP binds to the regulatory subunit, the complex dissociates,

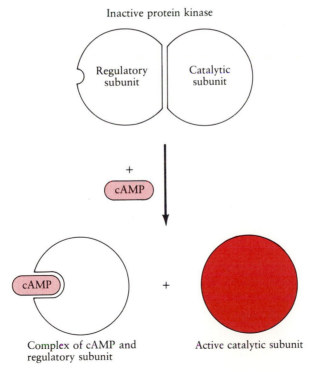

FIGURE 13.8. The activation of cAMP-dependent protein kinase. Binding of cAMP to the regulatory subunit induces a conformational change, causing this subunit to dissociate from the complex with liberation of the active catalytic subunit. (After Alberts et al., 1983, with permission of Garland Publishing.)

liberating the active catalytic subunit. As the cytoplasmic cAMP level declines, the bound nucleotide dissociates from the regulatory subunit; ligand-free molecules reassociate with the catalytic element, and protein phosphorylation ceases.

The enzyme that mobilizes glycogen is glycogen phosphorylase, but the key to its control is glycogen phosphorylase kinase; Figure 13.9 shows the elements of which this protein is constructed. There are four kinds of subunits, each present in two copies. Subunit γ bears the catalytic site. Subunits α and β are both phosphorylated by cAMP-dependent protein kinase, but only the phosphorylation of β activates the enzyme; the functional significance of α phosphorylation is not known. Subunit δ is the calcium-binding protein calmodulin, which will be discussed below. Phosphorylase kinase is open to activation by phosphoryl transfer only when Ca^{2+} ions are bound to the δ subunit. It is known that calmodulin undergoes a large change in conformation when it binds Ca^{2+}; presumably this change is transmitted to the β subunit as well as to the catalytic γ subunit. From a functional viewpoint, the result is that glycogen phosphorylase kinase is under dual control by both cAMP and Ca^{2+} ions. The physiological utility of this arrangement is plain, since Ca^{2+} ions also elicit muscle contraction, which consumes the ATP supplied by glycogenolysis.

What has been said above does not yet do justice to the sophistication of the

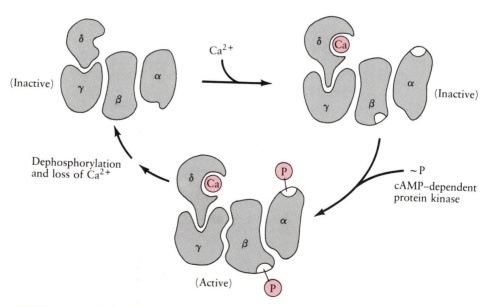

FIGURE 13.9.   A schematic mechanism for the activation of glycogen phosphorylase kinase. Binding of Ca^{2+} to the δ subunit induces one set of conformational changes, phosphorylation of the β subunit another, producing active enzyme. (After Alberts et al., 1983, with permission of Garland Publishing.)

control exercised over glycogen phosphorylase kinase. In addition to the sites available to cAMP-dependent protein kinase there are others, accessible to different kinases. These have their own signals, one of which may be insulin. In this manner diverse stimuli, in addition to epinephrine and Ca^{2+} ions, have input into determining the rate of energy mobilization.

The activating effects on glycogen phosphorylase kinase are coordinated with inhibitory effects at two additional control points. Cyclic AMP–dependent protein kinase also phosphorylates glycogen synthase, causing the latter enzyme to be *inactivated;* here again Ca^{2+} ions act synergistically with cAMP. Both signals also impinge on protein phosphatases in a manner too complex to be described here, reducing the rate of dephosphorylation. The overall result of these multiple sites of action is to sharpen the response to epinephrine: within a fraction of a second, glycogen utilization begins to compensate for the ATP and creatine phosphate consumed by muscle contraction.

At first sight the molecular machinery outlined above seems needlessly elaborate, but its complexity yields benefits in terms of speed, amplification, and modulation of both signal and response.

1. The level of circulating epinephrine is so low that only a few molecules are likely to bind to any one muscle fibril. These suffice to elicit a maximum response thanks to two steps that greatly amplify the signal. Each receptor molecule occupied by epinephrine triggers the formation of numerous molecules of cAMP, and each catalytic subunit of protein kinase mediates the transfer of multiple phosphoryl groups to effector proteins. Overall, the initial stimulus is amplified by some six orders of magnitude.

2. It is no less imperative to attenuate the signal as soon as it ceases to be useful. Attenuation is accomplished in three stages: by cAMP phosphodiesterase, by the GTPase activity of the G protein, and by a set of protein phosphatases. As a consequence, unless epinephrine is continually present the cAMP signal is transient: the cytoplasmic level varies from 10^{-7} to 10^{-5} M and returns to baseline within seconds.

3. In keeping with the need for rapid response, the steady-state level of cAMP is set by the balance between relatively high rates of both synthesis and degradation.

4. Each stage of signal transmission is open to modulation by alternative signals, some stimulatory and others inhibitory. This is particularly true at the effector level. Many functional proteins can undergo phosphorylation at several sites, controlled by separate kinases and phosphatases that respond to disparate signals.

In the course of the past two decades, cAMP has been identified as the intracellular signal in many instances of stimulus-response coupling, especially those mediated by peptide hormones and by biogenic amines. Each target cell possesses a particular set of receptors and effectors. In adipocytes, for instance, epinephrine and glucagon initiate the mobilization of fat, whereas in liver they act on glycogen breakdown. But the pathway of signal transmission is essentially the same as that described above: the hormone stimulates adenylate cyclase and the cAMP produced elicits the response through the agency of one or several cAMP-dependent protein kinases. There must be many such pathways, for cAMP is a universal constituent of animal cells. So far as one knows, all cAMP effects in eukaryotic cells involve cAMP-dependent protein kinases; but many of the functions modulated in this manner remain to be identified.

This is a matter of considerable urgency, for it is now clear that cAMP and the entire pathway of signal transmission sketched above are widespread (possibly universal) characteristics of eukaryotic cells. Yeast, *Neurospora,* and *Dictyostelium* are among the lower eukaryotes known to possess adenylate cyclase, cAMP phosphodiesterase, and cAMP-dependent protein kinase. In some cases it has been shown that the cyclase includes a GTP-binding protein of the conventional kind and that the regulatory subunit of the kinase can substitute for its counterpart from animal cells (Pall, 1981; Casperson et al., 1983; Majerfeld et al., 1984). In general we do not know what essential functions the cAMP system controls, but a broad hint comes from recent genetic studies with yeast. Matsumoto et al. (1982) isolated mutants that were deficient in adenylate cyclase and permeable to cAMP; in the absence of the nucleotide growth of the cells was arrested at a stage prior to budding, suggesting that cAMP is required to initiate a new division cycle. By using suppressor mutations it could even be shown that this function of cAMP, whatever it may prove to be, is once again mediated through cAMP-dependent protein kinase. The next step, and it is a hard one, is to identify the physiological substrate or substrates of that kinase; extrapolation from work to be touched on later points to elements of the cytoskeleton.

The Importance of Being Phosphorylated.   One who surveys the frenzied diggings of sensory and regulatory physiology cannot fail to be impressed by the numerous instances in which some physiological activity is controlled by phosphorylation of the effector protein or of an ancillary polypeptide. The list of individual proteins known to undergo phosphorylation lengthens almost daily; some examples are collected in Table 13.1, which lays no claim to completeness. In many cases, phosphorylation demonstrably either increases or decreases functional activity; in others this remains a surmise. Further covalent modifications include methylation, acylation, ADP-ribosylation, and adenylation, and these are likewise of regulatory significance. But there is no doubt that phosphorylation is the premier mechanism for the regulation of protein activity at the molecular level, at least in eukaryotic cells. Phosphorylated proteins have been found in bacteria as well, but this regulatory device seems less dominant there than it is in eukaryotes (Greengard, 1978; Krebs and Beavo, 1979; Cohen, 1982; Flockhart and Corbin, 1982).

The modulation of activity by phosphorylation can be very delicate because many proteins expose multiple sites to protein kinases. The acceptors are usually

TABLE 13.1   A Sampler of Protein Kinases and Their Substrates*

Protein kinase	Physiological substrate	Possible substrates
cAMP-dependent	Phosphorylase kinase Glycogen synthase Hormone-sensitive lipase Troponin I	Histones Fructose bisphosphatase Phosphofructokinase S-6 ribosomal protein K^+ channel
cGMP-dependent	?	Ribosomal proteins Glycogen synthase cAMP-dependent protein kinase Phosphorylase kinase

Ca^{2+}-dependent
  Phosphorylase kinase
  Myosin light-chain
    kinase
	Glycogen phosphorylase
	Myosin
		Several other enzymes

Protein kinase C	Proteins involved in secretion	Many cellular proteins

Other kinases
  pp60^{src}
  pp60^{sarc}
  EGF receptor
  PDH kinase
  ?
	?
	?
	EGF receptor (?)
	Pyruvate dehydrogenase
	Rhodopsin
	Na^+-K^+ ATPase
	Light-harvesting complex
	  of chloroplasts
		Cytoskeletal proteins,
		  inositol lipids, vinculin

Polypeptide-dependent
  kinase
	?	Na^+-K^+ ATPase,
		  Ca^{2+} ATPase

* From Flockhart and Corbin (1982) and elsewhere.

serine or threonine residues, rarely tyrosine. A well-known case of multisite phosphorylation is the glycogen synthase of skeletal muscle, in which five distinct protein kinases phosphorylate seven widely spaced serine residues. Cyclic AMP–dependent protein kinase attacks sites called 1a, 1b, and 2; phosphorylase kinase, site 2. Three further kinases place phosphoryl groups at sites 3a, 3b, and 3c; 2; and 5, respectively. All seven sites can be found to be phosphorylated in living muscle. The kinetic effects of the various modulations are complex. All reduce catalytic activity, full phosphorylation most of all, but the more significant consequences of phosphorylation may be the altered affinity of the enzyme for its substrate, UDP glucose, and changes in its response to Ca^{2+} and other ions. As Cohen (1982) puts it, multisite phosphorylation is a simple mechanism that markedly increases the regulatory potential of enzymes. Phosphorylation at one site may amplify or antagonize the effects of phosphorylation at other sites or it may alter the rates at which the latter sites are phosphorylated or dephosphorylated. The rate at which any given site is modified may in turn be modulated by substrates, activators, or inhibitors; which thus amplify or suppress the effects of the covalent reaction. Phosphorylation and dephosphorylation should not be regarded as a mechanism for turning an enzyme on or off but rather for tuning it across a range of forms, each of which responds in its own way to substrates and to regulatory molecules.

It must be emphasized that of all the protein kinases in skeletal muscle, only one is cAMP-dependent. In general it is technically quite easy to detect proteins that are subject to phosphorylation: when eukaryotic cells are briefly exposed to $^{32}P_i$ and the proteins are then separated by two-dimensional electrophoresis, one finds hundreds of radioactive spots. It is much harder to identify individual protein kinases, let alone the natural substrate of each kinase. Nevertheless, the number of individually recognizable kinases grows steadily (Table 13.1). Besides cAMP-dependent protein kinase, many cell types contain one that is activated by cyclic GMP. A number of protein kinases are activated by Ca^{2+} ions; a protein kinase from reticulocytes is activated by double-stranded RNA; and most animal cells contain protein kinase C, an enzyme that requires Ca^{2+} and is markedly activated by diacylglycerol, which comes from the breakdown of inositol phospholipids (Flockhart and Corbin, 1982; Nishizuka, 1984). A recent addition to the roster is a protein kinase from Ehrlich ascites cells that is activated by a variety of small polypeptides (Racker et al., 1984). Most kinases act on serine or threonine residues, but at present there is intense interest in the rare tyrosine-specific protein kinases, for these seem to play a critical role in the control of cellular growth (see below). The insulin receptor at the surface of animal cells is a tyrosine-specific protein kinase, the receptor for epidermal growth factor another.

Some of these protein kinases phosphorylate many cellular proteins, others may attack just a few, possibly a single one. No one knows how many different kinases to expect in any given cell, nor how many different signals impinge on them. But it is hard to avoid the impression that we are just beginning to explore the variety of protein kinases and that most cellular functions are affected by their attentions, directly or indirectly.

Calcium Ions as Cytoplasmic Signals.   L. V. Heilbrunn was one of the first, and surely the most exuberant, proponent of calcium as a regulator of cellular activities. Legend has him proclaiming "Kalzium macht alles," calcium does it all, nearly 40 years ago. Events have vindicated his prescience: calcium ions do not control everything, but they certainly serve as ubiquitous regulatory signals in the physiology of eukaryotic cells.

The avoiding reaction of *Paramecium*, which was described in another context in Chapter 11, is a well-studied example. It will be recalled that on collision with an obstacle, the organism backs up due to a temporary reversal of the ciliary beat and then resumes forward motion in a new direction. The manner in which a calcium flux controls the cells' behavior was deduced from the changes in membrane potential recorded when the anterior end of an impaled cell was prodded with a small stylus (Figure 11.20). The first response is a partial depolarization, due to calcium influx via ion channels located at the anterior end and controlled by touch receptors. When this initial depolarization exceeds a certain threshold, it is immediately followed by an action potential, a massive depolarization of the cell due to the opening of a separate set of voltage-sensitive calcium channels located in the ciliary membranes. The calcium channels close spontaneously within a few milliseconds, and the resting potential is restored by K^+ efflux through a set of K^+ channels located on the body surface. The calcium ions that enter the cilia during the brief period when the channels are open are the signal that controls the pattern of swimming. Cilia are small, and Ca^{2+} influx is sufficient to raise the Ca^{2+} concentration in the ciliary lumen from the basal level, about $10^{-7}\ M$, to $4 \times 10^{-5}\ M$. At the latter concentration, Ca^{2+} ions reverse the direction of ciliary beating by a mechanism that is still unknown. As soon as calcium influx ceases, the basal level is restored as Ca^{2+} ions in the cilia are either pumped out or sequestered in intracellular compartments (Eckert, 1972; Kung et al., 1975; Kung and Saimi, 1982).

Paramecium displays a second behavioral response that likewise depends on a Ca^{2+} signal, the explosive discharge of missiles called trichocysts that seem to play a role in prey capture. Trichocysts are housed in silos at the base of the flagella, and their ejection can be triggered artificially by calcium ionophores. Curiously, mechanical stimuli that make the cell back up do not cause it to fire its trichocysts; and certain chemical substances that trigger trichocyst discharge have no effect on swimming behavior. Since the entire organism is a single cell, one wonders how *Paramecium* keeps ciliary calcium segregated from the calcium pool that controls the trichocysts. This example illustrates an aspect of the Ca^{2+} signal that is of general significance. Unlike Na^+ or K^+, Ca^{2+} ions do not readily diffuse through the cytoplasm. (This was first discovered when $^{45}Ca^{2+}$ injected into nerve axons was seen to remain close to the site of injection.) In some cells the immobility of Ca^{2+} ions may be due to association with anionic macromolecules or with specific calcium-binding proteins, in others to the sequestration of Ca^{2+} by cytoplasmic organelles. These processes make Ca^{2+} ions a suitable signal for the control of highly localized responses, such as movement or exocytosis.

It will be recalled from Chapter 9 that all cells expel Ca^{2+} ions, maintaining the cytoplasmic free Ca^{2+} level around $10^{-7}\ M$ and generating a sharp inward calcium gradient between the extracellular fluid and the cytosol. There are now dozens, possibly hundreds, of cases in which a stimulus is coupled to a response via a transient flux of Ca^{2+} ions that serves as a signal or trigger. Sometimes the Ca^{2+} ions come from the medium, as in the case of *Paramecium*, but the ion may

TABLE 13.2   Cellular Functions Controlled by a Calcium Signal

Organism	Stimulus	Response
Ciliated and flagellated microorganisms	Collision with obstacle	Avoidance behavior
Paramecium	Chemicals, prey	Trichocyst discharge
Green algae, *Blastocladiella*	Osmotic shock, current injection	Osmotic adjustment
Mosses, *Neurospora, Achlya*	Ca^{2+} ionophores, cytokinin	Branching
Eggs of brown algae	Ion gradients, unidirectional light	Polarized outgrowth
Eggs of sea urchin or fish	Fertilization, Ca^{2+} entry	Activation of development
Skeletal muscle	Nerve impulse	Ca^{2+} release from sarcoplasmic reticulum, contraction
Synaptic junction	Depolarization	Exocytosis of neurotransmitters
Macrophages, neutrophils	Immune effectors	Exocytosis of histamine, activation of bactericidal activities
Fish scale erythrophores	Epinephrine	Pigment aggregation
Cultured mammalian cells	Ca^{2+} injection	Disassembly of cytoskeleton
Insect cells in culture	Ca^{2+} injection	Closure of gap junction channels

also be released from intracellular storage vesicles. Table 13.2 lists selected examples; it is not intended to be exhaustive but to document how widely this pathway of signal processing is spread throughout the eukaryotic world, from protoctists to higher plants and to our own muscles and nerves. Many examples come from developmental biology, a subject to which we shall return in a later section of this chapter. Numerous recent reviews cover aspects of the bustling field of calcium research (Carafoli and Crompton, 1978; Kretsinger, 1980; Cheung, 1980, 1982; Rasmussen, 1981; Means et al., 1982; Rasmussen and Barrett, 1984.)

From the chemical viewpoint Ca^{2+} ions and cAMP have nothing in common, but their informational roles are much alike. Primary stimuli, which may be either physical or chemical, impinge on selective, membrane-bound receptors that control a calcium channel. The best-characterized calcium channels are voltage-sensitive, opening in response to depolarization; ciliary calcium channels are a case in point. It seems likely, however, that other calcium channels can respond to physical stimuli (light, pressure) or to chemical activators. In any event calcium responses are rapid, transient, and attain a high degree of amplification since each channel passes some 10^6 ions per second. The signal is quickly attenuated by removal of the Ca^{2+} ions from the cytosol. Finally the cytosolic Ca^{2+} level is monitored by selective calcium binding proteins, which transmit the signal to one or more effectors.

In most cases the receptor for cytosolic Ca^{2+} ions belongs to a ubiquitous class of small proteins called calmodulins. The calmodulins are all very similar in primary structure and bind four Ca^{2+} ions per monomer with dissociation constants on the order of 1 μM. This is an appropriate affinity for a protein whose function is to register a rise in the concentration of free Ca^{2+} ions from a basal level of 10^{-7} M to a peak around 10^{-5} M. Calmodulins have been isolated from fungi, protozoa, higher plants, and many kinds of animal cells, where they constitute as much as 1 percent of the cytoplasmic protein complement. Whether calmodulin or a homologous protein occurs in prokaryotes is in dispute. I think calmodulins are a diagnostic feature of the eukaryotic cell and absent from eubacteria; it would be interesting to look for them in archaebacteria.

Calmodulin was discovered in 1970 as an activator of cAMP phosphodiesterase. Since then it has been shown to be loosely or tightly associated with a wide variety of proteins whose activities are modulated by Ca^{2+} ions (Figure 13.10): several metabolic pathways, the synthesis and degradation of cAMP, contractile activities, transport systems, and cytoskeletal proteins can all respond to global or localized changes in the concentration of free Ca^{2+} ions with the aid of calmodulin. Sometimes the molecular mechanism is understood, at least in principle. Examples are glycogen phosphorylase kinase, one of whose subunits is calmodulin (Figure 13.9), and the initiation of contraction in skeletal muscle, which involves troponin C, a protein closely related to calmodulin. In other cases we have only the outlines of a biochemical explanation. For instance, the self-assembly of tubulin into microtubules requires the participation of an ancillary structural protein called Tau. At low Ca^{2+} levels Tau is free to associate with tubulin, and microtubule assembly can proceed. Higher levels of Ca^{2+} ions prevent assembly and promote disassembly, apparently by the following mechanism: Ca^{2+} ions bind to calmodulin and the latter complex binds Tau, rendering it unavailable. The ability of Tau to promote microtubule assembly is subject to further modulation by phosphorylation of Tau (Kakiuchi and Sobue, 1983; Lindwall and Cole, 1984). Future research will no doubt reveal a diversity of such couplings. The one common principle that can be discerned at this time is that Ca^{2+} binding alters the three-dimensional configuration of calmodulin; and this in turn impinges on the configuration and catalytic activities of proteins that are associated with calmodulin.

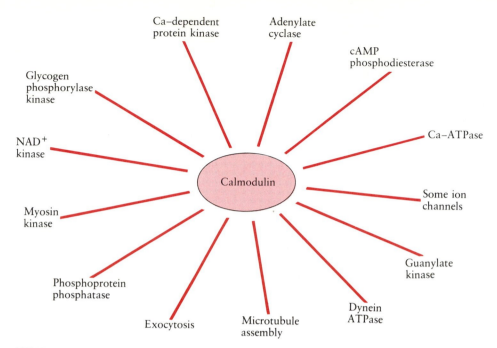

FIGURE 13.10.   Calmodulin has been implicated in the control of various enzymes and functions, some of which are indicated here. (After Cheung, 1980. Copyright by the American Association for the Advancement of Science.)

Intersecting Signals: Reinforcement and Antagonism.   A decade ago it was widely believed that some cellular activities are controlled by cAMP and others by Ca^{2+} ions. It will be clear from the foregoing that this is far too simple and stereotyped a view of cellular signaling. Glycogen metabolism (Figure 13.6) is but one of many responses that are subject to dual control by both cAMP and Ca^{2+} ions; they are what Rasmussen (1981) calls synarchic signals (the adjective derives from ancient Greece, where certain messages were borne by two heralds, sometimes with part of the message entrusted to each). Another example of dual control is the Ca^{2+}-activated K^+ channel of snail neurons, which responds to Ca^{2+} only after phosphorylation by a cAMP-dependent kinase; in some neurons phosphorylation elicits an increase in conductance, in others a decrease (Kandel and Schwartz, 1982).

Figure 13.11 illustrates in simplified form some of the interactions involved in the response of animal cells to a series of hormones, neurotransmitters, and growth promoters; the discharge of serotonin from blood platelets in response to thrombin is a case in point (Michell, 1975; Nishizuka, 1984; Berridge and Irvine, 1984; Rasmussen and Barrett, 1984). It has been recognized for some time that the pathway of information flow involves both Ca^{2+} ions and a novel class of signaling molecules derived from the breakdown of inositol-containing mem-

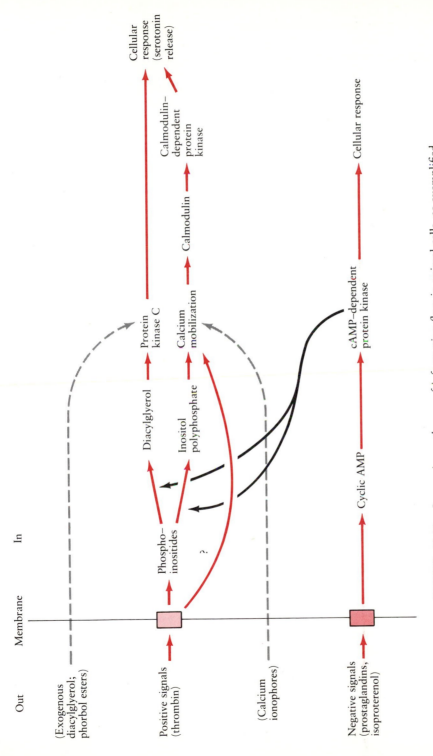

FIGURE 13.11.   Interacting pathways of information flow in animal cells, as exemplified by the discharge of serotonin from blood platelets in response to thrombin. Binding of thrombin to a surface receptor elicits the breakdown of phosphoinositides to diacylglycerol and inositol polyphosphate; the former activates protein kinase C, the latter induces Ca^{2+} mobilization. Prostaglandins and drugs that stimulate cAMP production antagonize the effects of thrombin. Colored arrows, stimulatory signals; black arrows, inhibitory signals; dashed arrows, artificial stimuli. (After Nishizuka, 1984, with permission of *Nature*, Macmillan Journals.)

brane lipids. It now appears that interaction of the primary stimulus (such as thrombin) with a surface receptor triggers the breakdown of polyphosphoinositides, generating inositol polyphosphate and 1,2-diacylglycerol. Each of these products serves as a signal: the former elicits the release of calcium ions from intracellular storage sites, while the latter activates protein kinase C by enhancing its affinity for calcium ions. The two signals can be differentially elicited by both natural and artificial stimuli, including calcium ionophores, exogenous diacylglycerol, and the tumor-promoting phorbol esters. In the cell the two signals reinforce each other: serotonin discharge entails the phosphorylation of at least two proteins, one reaction mediated by protein kinase C and the other by a kinase activated by Ca^{2+} plus calmodulin.

The effects of thrombin are antagonized by a variety of pharmacological agents, including prostaglandins and the drug isoproterenol. These bind to a different set of surface receptors and elicit the production of cAMP, which, by mechanisms that are not entirely clear, causes the attenuation of the thrombin-induced signals. Agents that release cyclic GMP also have this effect. The preceding pathway, with variations, appears to be involved in a great diversity of physiological processes, including the activation of eggs, the initiation of cell growth by mitogens and growth factors, and the transformation of cells infected with tumor viruses. It is instructive to ponder the multiplicity of energy-consuming reactions required to sustain this network of signals, not to mention the outlay of energy represented by genetic messages and the biosynthesis of catalytic proteins.

It would be all too easy to multiply examples of signal-mediated responses from all reaches of cell physiology or to burrow deeper into the molecular mechanisms by which these signals transmit information. But let us for a moment stand back from the trees to look at the wood. Cyclic AMP and Ca^{2+} ions play central roles in eukaryotic communications, and it is not likely that many additional symbols of such wide currency will be discovered. But these two channels of transmission interact, sometimes to reinforce each other and sometimes to diminish their output. One or both channels may be further modulated by additional signals, some of which are known while others remain conjectural. Cyclic GMP, for instance, is certainly an informational molecule whose roles converge with those of Ca^{2+}, cAMP, and inositol polyphosphates, though we do not yet understand the nature of this interplay. The growing number of protein kinases and other enzymes that catalyze the covalent modification of proteins hints at signaling molecules as yet unsuspected. Beyond that, we have more than a few examples of regulation and stimulus-response coupling that are mediated by the cytoplasmic pH (Nuccitelli and Deamer, 1982); will protons prove to be another common ionic signal?

Genetic messages flow through linear channels, but signals make webs, and each such web has an individual character. The molecular switches that transmit information, such as Ca^{2+}-calmodulin or diacylglycerol–protein kinase C, are more or less universal elements that recur with minor variations. But the physiological meaning that the signals convey has little to do with the chemical nature of these switches. It depends, instead on the connections that the elements make. Information flow is a function of the design of the wiring diagram, an entity that is greater than the sum of its strands and knots.

Developmental Signals

Few manifestations of life are as dramatic or have remained so obstinately mysterious as those collected under the loose rubric of development. Even the meaning of the term is in dispute. For present purposes, development will encompass all situations in which a constellation of cellular characteristics changes, more or less permanently, in a temporally and spatially defined sequence, resulting in a new cell that has a different morphology, behavior, and an altered set of specific macromolecules (Fulton, 1977). The maturation of a fertilized egg into an embryo, the malignant transformation of a somatic cell, and the intricate maneuvers that turn a water mold into a bag full of motile zoospores all entail the conversion of an entire pattern of cellular organization into another in response to some primary stimulus. To a point, it is useful to regard such radical transmutations as extreme examples of stimulus-response coupling, in which a transient signal is locked into place by stable changes in the pattern of gene expression.

Ionic Triggers of Development. It is remarkable how often developmental processes are initiated not by mysterious commands from the nucleus but by a flux of ions into the cytoplasm, either natural or artificial. Examples can be quoted from fungi and protists, ferns, mosses, algae and higher plants, animal cells in culture, eggs, and embryos. In most cases the ion in question is Ca^{2+}, which may enter the cytosol either from the environment or from an internal reservoir, but protons, Na^+ ions, even K^+ have been implicated more than once (see also Chapter 11). In no instance do we understand the chain of causality that links, say, a transient rise in the cytosolic Ca^{2+} level to the initiation of cell division minutes or hours later, but it is evidently an important aspect of information flow during development.

The activation of sea urchin eggs is a well-studied case in point. Normally an egg is activated by fertilization: soon after the entry of the sperm its respiratory rate rises, protein synthesis commences, and in time the egg begins to cleave, ultimately producing the many-celled embryo. But these developmental events do not necessarily require sperm entry. Many eggs can be activated without fertilization, either mechanically or chemically; some cleave several times before development is aborted, and frog eggs go on to produce normal offspring by parthenogenesis. What lies behind the initiation of all this activity is a sudden rise in the cytosolic Ca^{2+} level. This is normally elicited by the sperm, which physically carries Ca^{2+} ions into the egg cytosol, but Ca^{2+} influx from the medium following a prick by a fine needle is also sufficient to start the process. Either way, the first burst of Ca^{2+} triggers the release of more Ca^{2+} ions from intracellular stores, so that the level of Ca^{2+} ions in the cytosol shoots up from 10^{-7} to 10^{-5} M within a minute. In the transparent eggs of certain tropical fish one can follow Ca^{2+} release visually by monitoring light emitted when Ca^{2+} ions interact with the photoprotein aequorin. A ring-shaped wave of free Ca^{2+} ions traverses the entire egg, beginning from the point of sperm entry and subsiding again even as the egg begins to waken (Epel, 1977, 1978; Jaffe and Nuccitelli, 1977; Jaffe, 1979; Vacquier, 1981; Whitaker and Steinhardt, 1982).

The brief pulse of cytosolic Ca^{2+} is soon followed by other events, some of which serve as signals to elicit still later responses (Figure 13.12). An early effect is the exocytosis of cortical granules; another is the rise in cytosolic pH, apparently by Na^+-H^+ antiport. In sea urchin eggs (but not in other kinds of eggs) both the calcium signal and the pH rise appear to be required for rapid protein synthesis to commence. In some eggs phosphorylation of a particular ribosomal protein is also required for translation to begin; it remains to be seen how general this proves to be. Dormant messenger RNA is put into service and microtubules and microfilaments undergo reorganization before DNA synthesis and genome replication commence. There is no reason to believe that the initial Ca^{2+} flux impinges on the genome directly. More probably it is but the first in a cascade of signals and responses, flowing through multiple channels, some of which eventually reach those genes that are programmed to respond to activation. And we have only the most general notion, biased by analogy and preconceptions, of what this information flow looks like.

The role of Ca^{2+} ions in egg activation is well known, but it is by no means the only example of a developmental pathway initiated by ion movements. Recall, for example (Chapter 9), that many mitogens act on cultured animal cells by triggering sodium influx. In the cellular slime mold *Dictyostelium*, cAMP plus a factor of unknown structure called DIF induces the amoebas to differentiate into prestalk cells. There is some evidence that DIF inhibits a proton pump, and the consequent acidification of the cytoplasm favors the path that leads to stalk cells. By contrast, cAMP plus ammonia (and other reagents that alkalinize the cytoplasm) induces the amoebas to differentiate into prespore cells (Gross et al., 1983; Jamieson et al., 1984). Besides, the literature is replete with descriptions of interesting developmental effects elicited by an artificial flux of Ca^{2+} ions or of protons. The calcium ionophore A23187 causes branching in several fungi and in moss; the germination of certain spores; activation of a variety of eggs; and the differentiation of lymphocytes and myeloid tumor cells. Proton conductors induce branching in *Achlya* and stimulate the germination of certain fungal spores. In other cases, ion movements appear to disorganize a fine-tuned developmental sequence. Sometimes the polarity of development is influenced by an external electric field, perhaps because the field modulates ion flow across the plasma membrane (Jaffe and Nuccitelli, 1977; Jaffe, 1979, 1981; for recent examples see Chen and Jaffe, 1979; Saunders and Hepler, 1982; Hesketh et al., 1983; McNally et al., 1983).

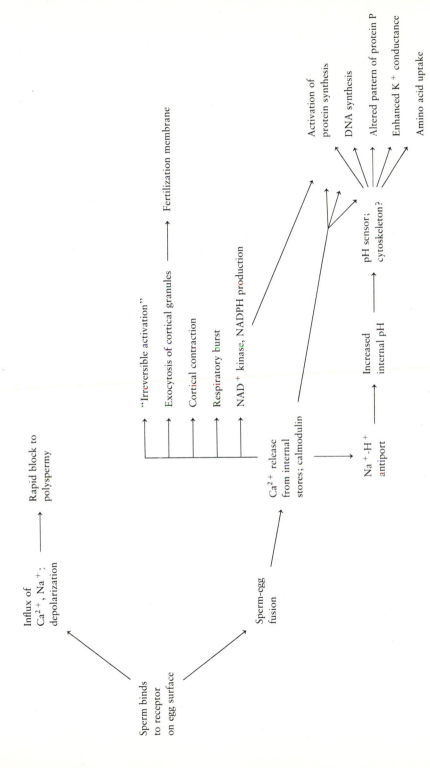

FIGURE 13.12. Sequence of events and the flow of information during the activation of a fertilized sea urchin egg. Ca²⁺ ions and the cytoplasmic pH both play important roles in this complex network of signals. (After Epel, 1978, and elsewhere, with permission of Academic Press.)

The hypothesis offered to explain such findings is generally that the artificial ion flux either mimics a *natural* one or overwhelms some intracellular ionic gradient. These explanations may well be correct, but an attitude of constructive skepticism is surely warranted whenever an argument rests primarily on ionophores and artificially induced ion fluxes. For just this reason, particular interest attaches to the study of natural ionic currents in relation to growth, development, and differentiation.

Transcellular Ion Currents in Development.   The recent surge in research on transcellular ion currents provides strong reinforcement for the hypothesis that ion fluxes are natural and significant strands in the web of developmental communications. Briefly, we have learned that many cells and organisms segregate ion pumps from ion leaks, generating transcellular ion currents that are visibly correlated with morphology and function. Furthermore, the pattern of ion flow changes in the course of development in such a way as to suggest that ion fluxes often help determine when and where growth or morphogenesis will occur. The idea is far from novel, having been promulgated by that forgotten pioneer of electrobiology, Elmer Lund (1947). But its resurrection and present influence are largely due to the work of L. F. Jaffe and his students, to whom we also owe the instrumentation required to measure minute extracellular electric fields and currents.

The paradigms in this field originated in studies on the developing zygotes of the marine brown algae *Fucus* and *Pelvetia* (Jaffe et al., 1974; Jaffe and Nuccitelli, 1977; Jaffe, 1979, 1981; Nuccitelli, 1982). During the first day after fertilization the egg, spherical to begin with, develops a protuberance, which becomes progressively more prominent (Figure 13.13). The first division, about 24 hours after fertilization, produces two daughter cells that differ in form, composition, and developmental fate: the protuberance, or rhizoid cell, is destined to grow into the holdfast, while the thallus cell will give rise to the frond. At the early stages of development the rhizoid is the growing part, as the embryo's first object is to secure firm lodging on a rock. What makes this system so useful is that unlike animal eggs, algal zygotes have no preformed axis of symmetry. The site and direction of outgrowth are determined by any one of a variety of external gradients: unidirectional light, imposed electric fields, gradients of pH or of certain ions. The usual laboratory procedure is to illuminate the eggs from one side; the rhizoid develops from the shaded half. The various stimuli are thought to produce a localized change in the plasma membrane that confers polarity on the embryo; the axis is labile at first but becomes fixed even before the first cell division. The process by which zygotes of *Pelvetia* and *Fucus* establish their axis of polarity has come to be seen as a prototype for cellular differentiation in general. Jaffe proposed that isotropic cells, such as eggs or spores, first develop a polarized pattern of electric currents that is then permanently imprinted on the structure of the protoplasm.

With the development of an ultrasensitive device called the vibrating probe, it became feasible to explore the pattern of electric current around a single egg

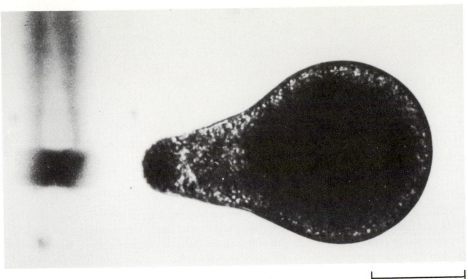

50 μm

FIGURE 13.13.   Growing embryo of the marine alga *Pelvetia* at the two-cell stage, with the vibrating probe in measuring position. (From Nuccitelli and Jaffe, 1974. Photograph courtesy of L. F. Jaffe.)

and to subject Jaffe's hypothesis to experimental scrutiny. It is now established that shortly after fertilization, the egg begins to drive an electric current through itself such that positive charges enter the presumptive rhizoid and leave from elsewhere on the egg's surface (Figures 13.13 and 13.14). Current begins to flow as early as 30 minutes after fertilization, preceding any visible signs of polarization such as the local fusion of Golgi vesicles with the plasma membrane. The pattern is unstable at first but becomes progressively better defined as polarization proceeds; the site of stable current entry predicts the site of outgrowth. Experimental manipulations that alter the current pattern, such as shining a second beam of light onto the egg, also alter the locus of outgrowth in a corresponding manner (Nuccitelli and Jaffe, 1974; Nuccitelli, 1978). We can therefore assert that the current precedes and predicts the position of outgrowth and may well be causally involved in localizing it.

The ionic basis of the transcellular electric current is complex, but what matters for the polarization of growth is probably that portion which reflects the entry of calcium ions into the tip of the rhizoid (Robinson and Jaffe, 1975). The most persuasive evidence that this calcium influx determines the position of outgrowth comes from the work of Robinson and Cone (1980), who grew *Pelvetia* embryos in a gradient of the calcium ionophore A23187 and found that the majority of the rhizoids emerged on the side corresponding to maximum calcium influx. The polarization of outgrowth by applied electric fields, first studied by Lund, can also be rationalized in terms of the stimulation of calcium movements. All this is consistent with the hypothesis that the zygote, isotropic to

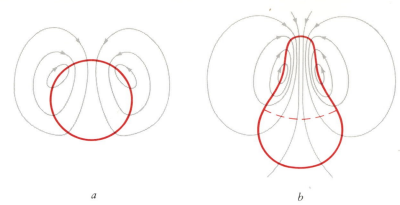

a   *b*

FIGURE 13.14.   The current pattern of the *Pelvetia* egg as mapped by use of the vibrating probe. (*a*) Fertilized egg just before outgrowth begins; the egg is still spherical, but the current pattern predicts that outgrowth will proceed from the upper surface. (*b*) Two-celled embryo. Current enters the rhizoid cell and leaves from the thallus cell. (After Nuccitelli, 1978, and Jaffe, 1979, with permission of Academic Press.)

begin with, utilizes the spatial information conveyed by the unilateral light stimulus to segregate calcium pumps from calcium leaks and that the position of the latter determines the locus where the rhizoid subsequently forms.

   Transcellular ion currents are by no means confined to eggs; fungi are at present the least specialized organisms to have been explored in some detail (Kropf et al., 1983, 1984; Gow et al., 1984). Growing hyphae of the water mold *Achlya* drive longitudinal proton currents through themselves. These can be monitored either electrically or by measurement of the pH profile along the length of a hypha (Figure 13.15). Protons are expelled from the hyphal trunk, probably by a H^+-translocating ATPase, and reenter the growing tip by symport with amino acids. We may describe the pattern of current flow as a spatially extended chemiosmotic system, with proton pumps and proton leaks separated by as much as 300 μm. Remarkably, a new point of proton entry precedes the emergence of a branch and predicts its locus. This proton entry may be instrumental in localizing the new branch, for the application of proton-conducting ionphores to growing hyphae elicits branching (Harold and Harold, 1985). These data suggest that the inward limb of the proton current, generated by the spatial segregation of proton pumps from proton leaks, may serve both as a driving force for amino acid transport and as a signal that helps to polarize tip growth.

   The mechanisms by which a local ion flux may localize the site of growth or development are far from clear; two general alternatives, by no means mutually exclusive, are under consideration. In Jaffe's view (Jaffe et al., 1974; Jaffe, 1979, 1981) the localized calcium influx generates an electric field across the cytoplasm, which in turn exerts a directional driving force on all charged particles within its

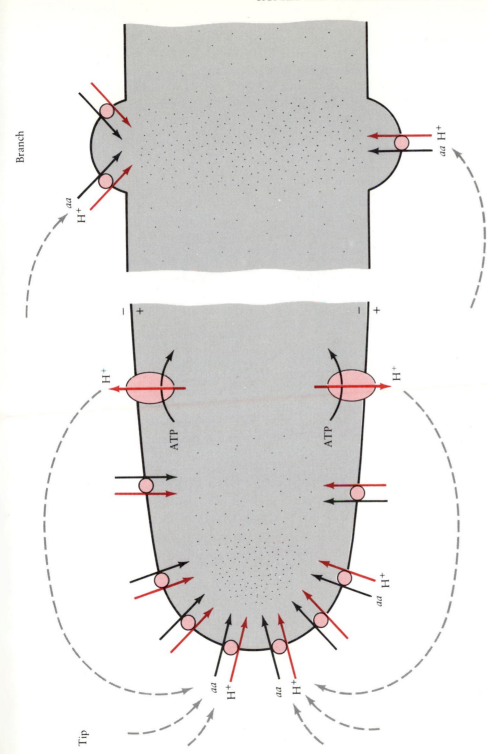

FIGURE 13.15.   Hypothesis to explain the generation of a transcellular proton current by the filamentous water mold *Achlya*. Protons are expelled by a proton-translocating ATPase whose distribution is not specified. They return to the cytoplasm via H^+–amino acid symporters, which are localized preferentially at the hyphal tip and at branch points. The cytoplasmic gradient of protons and/or the transcellular electric field, tip positive (stippling), may convey spatial information. (After Kropf et al., 1984, with copyright permission of Rockefeller University.)

range. Polarization could result, for example, from the electrophoretic redistribution of membrane proteins. The alternative hypothesis, which I am inclined to favor, proposes that the localized ion flux serves as an informational signal, perhaps exerting its effects by communicating with the cytoskeleton (Quatrano, 1978; Kropf et al., 1983). Calcium ions have been implicated in the regulation of contraction, in the assembly of actin filaments and microtubules, in vesicle fusion, and in the activation of protein kinases; any or all of these may implement the instruction "grow here." We shall return to this issue in Chapter 14.

Insight into the genesis and functions of transcellular ion currents is still limited, and there is no rigorous proof that they play the developmental roles attributed to them in the preceding discussion. However, it is clear that spatial segregation of ion pumps from ion leaks is a widespread, possibly universal feature of eukaryotic cells and organisms. Figure 13.16 presents a selection of organisms in which a transcellular ion current is plainly correlated with the spatial distribution of functions and foci of development. As a rule, the entry of current, energetically downhill, marks the region of maximum growth or activity. Representatives of six fungal classes are known to drive transcellular proton currents into the growth tip, presumably because this is the locus of nutrient uptake. Higher up in the plant world, there is good evidence that protons enter barley roots just behind the growing tip; individual root hairs are also electrically polarized, the current entering the tip (Weisenseel et al., 1979). Germinating pollen grains of the lily drive an ion current through themselves. In this case protons carry the outward current from the grain while potassium ions flow into the tip of the elongating pollen tube, and there is reason to believe that calcium entry into the tip, albeit a minor component of the current, may be the critical element in polarizing growth. An amoeba that is moving and changing shape drives a steady calcium current into the uroid and out of the extending pseudopodia. Examples from the animal world are multiplying: frog eggs, chicken embryos, muscle cells, and regenerating limbs are electrically polarized, and in some cases there is evidence for localized calcium influx.

FIGURE 13.16.   A miscellany of transcellular electric currents in growing or developing organisms. The left-hand column shows three microbial systems. Current enters the growing tips of the rhizoids of *Blastocladiella emersonii,* just behind the tips of *Achlya bisexualis* hyphae, and into the uroid of *Amoeba proteus.* The middle column shows examples from plants. Current enters the elongating zone of barley roots, the tips of the root hairs, the rhizoid of fucoid embryos, and the germ tube of germinating lily pollen. The right-hand column depicts animal systems. Current enters the animal pole of the *Xenopus* oocyte, traverses the follicle complex of the silkworm oocyte in a rather complex pattern and leaves the primitive streak of the early chick embryo. (After Jaffe, 1979, 1981, and Nuccitelli, 1982, with permission of Raven Press, New York, and the Royal Society, London.)

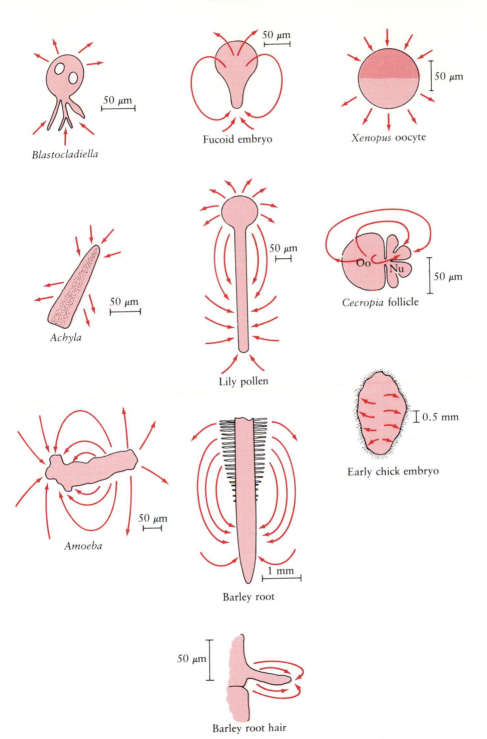

Blastocladiella

Fucoid embryo

Xenopus oocyte

Achyla

Lily pollen

Cecropia follicle

Amoeba

Barley root

Early chick embryo

Barley root hair

This very profusion suggests that there is no special relationship between transcellular currents and growth or development. In my view, the heart of the matter is that many (if not all) transport systems occupy definite cellular locations. The majority are also electrogenic, for thermodynamic reasons discussed in the preceding chapters, and may therefore be detected by the vibrating probe. Local electrogenic fluxes of Ca^{2+} and other ions are particularly prominent, for they serve as signals that control local work functions: exocytosis, contraction, protein phosphorylation, and cytoskeletal rearrangements, which are involved both in physiological responses and in growth or development. In most instances information is probably conveyed chemically by such molecular symbols as Ca^{2+} ions or protons; but the electric field generated by the transcellular ionic flux may also be physiologically significant (Chapter 14). This is a subject whose basic principles are just beginning to emerge. Only by studying each case on its own terms can we hope to disentangle the extent to which localized ion currents are a cause of cellular differentiation, a consequence thereof, or both.

Switching between Patterns.   The inclusive sense in which development is used in this chapter underscores not the differential expression of genetic information but the conversion of one pattern of cellular organization into another. Whether the initial impetus comes from the genome or from some extracellular source, development calls for reorganization of the network of messages and signals that integrates cellular functions. In general we know little about such drastic switches, but there are model systems that afford at least glimpses of what is involved.

Some of the most instructive instances come from cancer research. When certain avian or mammalian cells in culture are infected with Rous sarcoma virus, they are not killed but undergo "transformation" to a malignant phenotype: growth of the cells continues without restraint, instead of halting as soon as a confluent layer has formed. Transformed cells also differ from normal ones in shape, metabolic pattern, and cytoskeleton structure. All these changes can be blamed on a single viral gene, which codes for a particular protein kinase. This kinase is not cAMP-dependent; it phosphorylates certain tyrosine residues, whereas most protein kinases phosphorylate serine or threonine. A very similar kinase is also present in normal cells, but transformation results in a tenfold increase in the level of phosphotyrosine. The virus-specific kinase has several natural substrates, one of which is a protein called vinculin. Vinculin is localized at sites where microfilament bundles make contact with the plasma membrane; these are also the sites of formation of adhesion plaques, which mediate the cell's attachment to its substratum. It appears, then, that the phenotype of transformed cells results at least in part from interference with cytoskeletal functions by the abnormal phosphorylation of vinculin (Erikson et al., 1980; Hynes, 1982). The viral enzyme may alter the pattern of vinculin phosphorylation or merely increase its extent. Recent findings suggest that protein kinase C and phosphoinositide turnover are once again involved, so that the network of signals is sure to be more complex than outlined above (see Berridge and Irvine, 1984; Schliwa et al., 1984). But the salient point remains that the introduction of a new protein kinase suffices to convert the cell from one functional order into another.

Changes in the pattern of phosphorylation may also underlie another switch in growth habit, this time in the reverse direction. Chinese hamster ovary (CHO) cells grow in culture with a fibroblastlike morphology, forming highly ordered colonies of spindle-shaped cells. Certain lines derived from the primary explants are described as "spontaneously transformed," exhibiting the changes in cell shape, growth habit, and colony morphology that characterize malignant cells. Hsie and Puck (1971) found that all these changes could be reversed by growing the transformed cells in the presence of the cAMP analogue dibutyryl cAMP. It now appears that the analogue inhibits cAMP phosphodiesterase, allowing the cytoplasmic cAMP level to rise, and the restoration of the normal growth habit is due at least in part to changes in the phosphorylation of cytoskeletal proteins mediated by a cAMP-dependent protein kinase (Hsie and Puck, 1971; Pastan and Willingham, 1978; Nielson and Puck, 1980; Lockwood et al., 1982; Hsie, 1982). Mutants refractory to reverse transformation have been isolated and shown to contain a defective protein kinase. Reverse transformation is a phenotypic change, dependent on the continued presence of dibutyryl cAMP, but it is not difficult to imagine permanent, developmental changes arising from genetically controlled alterations in the metabolism of cAMP or in the pattern of protein phosphorylation.

One should not conclude from the preceding examples that a special link exists between the kind of global changes that are here designated as developmental and cAMP or protein phosphorylation. Examples could have been chosen to illustrate a role for protein acetylation in developmental control, and for DNA methylation. The prevalence of protein phosphorylation simply reflects its central place in the regulation of cellular activities.

Stable changes in the pattern of gene expression are a feature of all but a few developmental processes; to many developmental biologists the modulation of gene expression is the very essence of differentiation. Are cyclic nucleotides, protein phosphorylation, and ionic signals also instrumental in regulating the output of messages by the genome? It must again be emphasized that cytoplasmic signals impinge on the genome in many ways. Ionic triggers of development were discussed above, and the literature of developmental biology holds many instances in which cytoplasmic cues determine which set of genes is to be transcribed. In the case of bacteria, the molecular mechanisms by which cytoplasmic proteins and signal molecules govern the operation of the genome are known in great detail (Figure 13.2), and this provides models for research into the control of gene expression in eukaryotes. Many students believe that covalent modification of macromolecule structure by methylation (DNA) or by phosphorylation (histones and other nuclear proteins) holds the key to the differential activation of some genes (or blocks of genes) and the repression of others; and a variety of observations implicate cAMP in the control of eukaryotic gene transcription (Williams et al., 1980; Miles et al., 1981; Chrapkiewicz et al., 1982).

It seems increasingly probable that the same kinds of signals that govern cellular physiology also guide the workings of the genome, but not enough is known of the regulation of gene expression in eukaryotes to pursue this topic here.

Development entails not only a shift in the pattern of gene expression but the transformation of one network of cellular communications into another. There must be as many such networks as there are developing organisms; each is to some extent unique, and if we would understand how a particular path of development comes about, we must discover its particular set of interdigitating messages and signals, its circuitry of commands to let this and hinder that. In many cases we know a few strands of the communications web, but for no single developing cell or organism could we sketch the web in its entirety, not even in broad outline. Cellular physiology and molecular genetics are separate disciplines whose practitioners rarely speak each other's language; developmental biology might make better progress if more researchers were multilingual.

References

ADLER, J. (1966). Chemotaxis in bacteria. *Science* **153**:708–716.

ALBERTS, B., BRAY, D., LEWIS, J., RAFF, M., ROBERTS, K., and WATSON, J. D. (1983). Op. cit., Chapter 11.

BARYSHEV, V. A., GLAGOLEV, A. N., and SKULACHEV, V. P. (1983). The interrelation of phototaxis, membrane potential and K^+/Na^+ gradient in *Halobacterium halobium*. *Journal of General Microbiology* **129**:367–373.

BERRIDGE, M. J., and IRVINE, R. F. (1984). Inositol trisphosphate, a novel second messenger in cellular signal transduction. *Nature* **312**:315–321.

BLACK, R. A., HOBSON, A. C., and ADLER, J. (1980). Involvement of cyclic GMP in intracellular signaling in the chemotactic response of *Escherichia coli*. *Proceedings of the National Academy of Sciences USA* **77**:3879–3883.

BOYD, A., and SIMON, M. (1982). Bacterial chemotaxis. *Annual Review of Physiology* **44**:501–517.

CARAFOLI, E., and CROMPTON, M. (1978). Op. cit., Chapter 9.

CASPERSON, F., WALKER, N., BRASIER, A., and BOURNE, H. R. (1983). A guanine nucleotide-sensitive adenylate cyclase in the yeast *Saccharomyces cerevisae*. *Journal of Biological Chemistry* **258**:7911–7914.

CHEN, T. H., and JAFFE, L. F. (1979). Forced calcium entry and polarized growth of *Funaria* spores. *Planta* **144**:401–406.

CHEUNG, W. Y. (1980). Op. cit., Chapter 9.

CHEUNG, W. Y. (1982). Calmodulin. *Scientific American* **246**(6):62–70.

CHRAPKIEWICZ, N. B., BEALE, E. G., and GRANNER, D. K. (1982). Induction of the messenger ribonucleic acid coding for phosphoenolpyruvate carboxykinase in H4-II-E cells. *Journal of Biological Chemistry* **257**:14428–14432.

CHOCK, P. B., RHEE, S. G., and STADTMAN, E. R. (1980). Interconvertible enzyme cascades in cellular regulation. *Annual Review of Biochemistry* **49**:813–843.

COHEN, P. (1982). The role of protein phosphorylation in neural and hormonal control of cellular activity. *Nature* **296**:613–619.

ECKERT, R. (1972). Op. cit., Chapter 11.

EPEL, D. (1977). The program of fertilization. *Scientific American* **237**(5):129–138.

EPEL, D. (1978). Mechanisms of activation of sperm and egg during fertilization of sea urchin gametes. *Current Topics in Developmental Biology* **12**:185–246.

ERIKSON, R. L., PURCHIO, A. F., ERIKSON, E., COLLETT, M. S., and BRUGGE, J. S. (1980). Molecular events in cells transformed by Rous Sarcoma Virus. *Journal of Cell Biology* **87**:319–325.

FLOCKHART, D. A., and CORBIN, J. D. (1982). Regulatory mechanisms in the control of protein kinases. *CRC Critical Reviews in Biochemistry* **12**:133–186.

FULTON, C. (1977). Cell differentiation in *Naegleria gruberi*. *Annual Review of Microbiology* **31**:597–629.

GOULBOURNE, E. A., JR., and GREENBERG, E. P. (1983). A voltage clamp inhibits chemotaxis of *Spirochaeta aurantia*. *Journal of Bacteriology* **153**:916–920.

GOW, N. A. R., KROPF, D. L., and HAROLD, F. M. (1984). Proton currents and pH gradients along growing hyphae of the water mould *Achlya bisexualis*. *Journal of General Microbiology* **130**:2967–2974.

GREENGARD, P. (1978). Phosphorylated proteins as physiological effectors. *Science* **199**:146–152.

GROSS, J. D., BRADBURY, J., KAY, R. R., and PEACEY, M. J. (1983). Intracellular pH and the control of cell differentiation in *Dictyostelium discoideum*. *Nature* **303**:244–245.

HAROLD, R. L., and HAROLD, F. M. (1985). Ionophores and inhibitors modulate branching in *Achlya bisexualis*. *Journal of General Microbiology*, in press.

HESKETH, T. R., SMITH, G. A., MOORE, J. P., TAYLOR, M. V., and METCALFE, J. C. (1983). Free cytoplasmic calcium concentration and the mitogenic stimulation of lymphocytes. *Journal of Biological Chemistry* **258**:4876–4882.

HSIE, A. W. (1982). Reverse transformation of Chinese hamster ovary cells by cyclic AMP and hormones, in *Cell Growth*, C. Nicolini, Ed. Plenum, New York, pp. 557–573.

HSIE, A. W., and PUCK, T. T. (1971). Morphological transformation of Chinese hamster cells by dibutyryl adenosine cyclic 3′, 5′-monophosphate and testosterone. *Proceedings of the National Academy of Sciences USA* **68**:358–361.

HYNES, R. (1982). Phosphorylation of vinculin by pp60^{src}: What might it mean? *Cell* **28**:437–438.

JACOB, F. (1973). *The Logic of Life: A History of Heredity*. Pantheon, New York.

JAFFE, L. F. (1979). Control of development by ionic currents, in *Membrane Transduction Mechanisms*, R. A. Cone and J. E. Dowling, Eds. Raven Press, New York, pp. 199–231.

JAFFE, L. F. (1981). The role of ionic currents in establishing developmental patterns. *Philosophical Transactions of the Royal Society of London* **B295**:553–566.

JAFFE, L. F., and NUCCITELLI, R. (1977). Electrical controls of development. *Annual Review of Biophysics and Bioengineering* **6**:445–476.

JAFFE, L. F., ROBINSON, K. R., and NUCCITELLI, R. (1974). Local cation entry and self-electrophoresis as an intracellular localization mechanism. *Annals of the New York Academy of Science* **238**:372–389.

JAMIESON, G. A., JR., FRAZIER, W. A., and SCHLESINGER, P. H. (1984). Transient increase in intracellular pH during *Dictyostelium* differentiation. *Journal of Cell Biology* 99:1883–1887.

KAKIUCHI, S., and SOBUE, K. (1983). Control of the cytoskeleton by calmodulin and calmodulin-binding proteins. *Trends in Biochemical Sciences* 8:59–62.

KANDEL, E. R., and SCHWARTZ, J. E. (1982). Molecular biology of learning. *Science* 218:433–443.

KOSHLAND, D. E., JR. (1977). A response regulator model in a simple sensory system. *Science* 196:1055–1063.

KOSHLAND, D. E., JR. (1979). A model regulatory system: Bacterial chemotaxis. *Physiological Reviews* 59:811–862.

KOSHLAND, D. E., JR. (1981). Biochemistry of sensing and adaptation in a simple bacterial system. *Annual Review of Biochemistry* 50:765–782.

KREBS, E. G., and BEAVO, J. A. (1979). Phosphorylation-dephosphorylation of enzymes. *Annual Review of Biochemistry* 48:923–959.

KRETSINGER, R. H. (1980). Structure and evolution of Ca-mediated proteins. *CRC Critical Reviews in Biochemistry* 8:119–174.

KROPF, D. L., CALDWELL, J. H., GOW, N. A. R., and HAROLD, F. M. (1984). Transcellular ion currents in the water mold *Achlya*: Proton/amino acid symport as a mechanism of current entry. *Journal of Cell Biology* 99:486–496.

KROPF, D. L., LUPA, M. D. A., CALDWELL, J. H., and HAROLD, F. M. (1983). Cell polarity: Endogenous ion currents precede and predict branching in the water mold *Achlya*. *Science* 220:1385–1387.

KUNG, C., CHANG, S. Y., SATOW, Y., HOUTEN, Y., and HANSMA, H. (1975). Op. cit., Chapter 11.

KUNG, C., and SAIMI, Y. (1982). Op. cit., Chapter 11.

LIMBIRD, L. L. (1981). Activation and attenuation of adenylate cyclase. *Biochemical Journal* 195:1–13.

LINDWALL, G., and COLE, R. D. (1984). Phosphorylation affects the ability of Tau to promote microtubule assembly. *Journal of Biological Chemistry* 259:5301–5305.

LOCKWOOD, A. H., TRIVETTE, D. D., and PENDERGAST, M. (1982). Molecular events in cAMP-mediated reverse transformation. *Cold Spring Harbor Symposia on Quantitative Biology* 46:909–919.

LUND, E. (1947). *Bioelectric Fields and Growth.* University of Texas Press, Austin.

MACNAB, R. (1979). Op. cit., Chapter 5.

MAJERFELD, I. H., LEICHTLING, B. H., MELIGENI, J. A., SPITZ, E., and RICKENBERG, H. V. (1984). A cytosolic cyclic AMP-dependent protein kinase in *Dictyostelium discoideum*. *Journal of Biological Chemistry* 259:654–661.

MATSUMOTO, K., UNO, I., OSHIMA, Y., and ISHIKAWA, T. (1982). Isolation and characterization of yeast mutants deficient in adenylate cyclase and cAMP-dependent protein kinase. *Proceedings of the National Academy of Sciences USA* 79:2355–2359.

MCNALLY, J. G., COWAN, J. D., and SWIFT, H. (1983). The effects of the ionophore A23187 on pattern formation in the alga *Micrasterias*. *Developmental Biology* 97:137–145.

MEANS, A. R., TASH, J. S., and CHAFOULEAS, J. G. (1982). Physiological implications of the presence, distribution and regulation of calmodulin in eukaryotic cells. *Physiological Reviews* 62:1–39.

MICHELL, R. H. (1975). Inositol phospholipids and cell surface receptor function. *Biochimica et Biophysica Acta* 415:81–147.

MILES, M. F., HUNG, P., and JUNGMANN, R. A. (1981). Cyclic AMP regulation of lactate dehydrogenase. *Journal of Biological Chemistry* 256:12545–12552.

MONOD, J. (1971). Op. cit., Chapter 1.

NIELSON, S. E., and PUCK, T. T. (1980). Deposition of fibronectin in the course of reverse transformation of Chinese hamster ovary cells by cyclic AMP. *Proceedings of the National Academy of Sciences USA* **77**:985–989.

NISHIZUKA, Y. (1984). The role of protein kinase C in cell surface signal transduction and tumor promotion. *Nature* **308**:693–698.

NUCCITELLI, R. (1978). Öoplasmic segregation and secretion in the *Pelvetia* egg is accompanied by a membrane-generated electrical current. *Developmental Biology* **62**:13–33.

NUCCITELLI, R. (1982). Transcellular ion currents: Signals and effectors of cell polarity. *Modern Cell Biology* **2**:451–481.

NUCCITELLI, R., and DEAMER, D. W., Eds. (1982). *Intracellular pH: Its Measurement, Regulation and Utilization in Cellular Functions.* Alan R. Liss, New York.

NUCCITELLI, R., and JAFFE, L. F. (1974). Spontaneous current pulses through developing fucoid eggs. *Proceedings of the National Academy of Sciences USA* **71**:4855–4859.

PALL, M. L. (1981). Adenosine 3′,5′-phosphate in fungi. *Microbiological Reviews* **45**:462–480.

PASTAN, I., and WILLINGHAM, M. (1978). Cellular transformation and the morphological phenotype of transformed cells. *Nature* **286**:645–650.

QUATRANO, R. S. (1978). Development of cell polarity. *Annual Review of Plant Physiology* **29**:487–510.

RACKER, E., ABDEL-GHANY, M., SHERRILL, K., RIEGLER, C., and BLAIR, E. A. (1984). New protein kinase from plasma membrane of Ehrlich ascites tumor cells activated by natural polypeptides. *Proceedings of the National Academy of Sciences USA* **81**:4250–4254.

RASMUSSEN, H. (1981). Calcium and cAMP as synarchic messengers. Wiley, New York.

RASMUSSEN, H., and BARRETT, P. Q. (1984). Calcium messenger system: An integrated view. *Physiological Reviews* **64**:938–984.

ROBINSON, K. R., and CONE, R. (1980). Polarization of fucoid eggs by a calcium ionophore gradient. *Science* **207**:77–78.

ROBINSON, K. R., and JAFFE, L. F. (1975). Polarizing fucoid eggs drive a calcium current through themselves. *Science* **187**:70–72.

RODBELL, M. (1980). The role of hormone receptors and GTP-regulatory proteins in membrane transduction. *Nature* **284**:17–22.

ROSS, E. M., and GILMAN, A. G. (1980). Biochemical properties of hormone sensitive adenylate cyclase. *Annual Review of Biochemistry* **49**:533–564.

SAUNDERS, M. J., and HEPLER, P. K. (1982). Calcium ionophore A23187 stimulates cytokinin-like mitosis in *Funaria. Science* **217**:943–954.

SCHLIWA, M., NAKAMURA, T., PORTER, K. R., and EUTENEUER, U. (1984). A tumor promoter induces rapid and coordinated reorganization of actin and vinculin in cultured cells. *Journal of Cell Biology* **99**:1045–1059.

STENT, G. S. (1978). Genes and the embryo, in *Paradoxes of Progress.* Freeman, New York.

SUTHERLAND, E. W. (1972). Studies on the mechanism of hormone action. *Science* **117**:401–408.

TAYLOR, B. L. (1983). Role of proton motive force in sensory transduction in bacteria. *Annual Review of Microbiology* **37**:551–573.

THOMPSON, D. W. (1917). *On Growth and Form,* abridged ed., J. T. Bonner, Ed. (1961). Cambridge University Press, London.

TOMKINS, G. M. (1975). The metabolic code. *Science* **189**:760–763.

VACQUIER, V. D. (1981). Dynamic changes of the egg cortex. *Developmental Biology* **84**:1–26.

WEISENSEEL, M. H., DORN, A., and JAFFE, L. F. (1979). Natural H$^+$ currents traverse growing roots and root hairs of barley (*Hordeum vulgare*). *Plant Physiology* **64**:512–518.

WHITAKER, M. J., and STEINHARDT, R. A. (1982). Ionic regulation of egg activation. *Quarterly Review of Biophysics* **15**:593–666.

WILLIAMS, J. G., TSANG, A. S., and MAHBUBANI, H. (1980). A change in the rate of transcription of a eukaryotic gene in response to cyclic AMP. *Proceedings of the National Academy of Sciences USA* **77**:7171–7175.

CHAPTER

14

Morphogenesis and Biological Order

The form . . . of any portion of matter, whether it be living or dead, and the changes of form which are apparent in its movements and in its growth, may in all cases alike be described as due to the action of force. In short, the form of an object is a "diagram of forces" in this sense, at least, that from it we can judge or deduce the forces that are acting or have acted upon it.

D'Arcy Thompson, *On Growth and Form*

• • •

Self-Assembly
Shaping a Bacterial Cell
Cytoskeleton and Form in Eukaryotic Cells
Localizing Morphogenetic Work
Inheritance of Spatial Order
Epilogue

At first sight, a chapter on morphogenesis may seem out of place in a book on bioenergetics, but in fact these two subjects intersect on several levels. The term "morphogenesis" denotes the generation of recognizable forms during growth and development, and this immediately implies the performance of physiological work: biosynthesis, the processing of genetic information, and the transport of building blocks large and small to their destination all require energy and come within the purview of bioenergetics. A deeper and more interesting nexus between bioenergetics and all aspects of developmental biology arises from the concept of biological order (Chapter 1). A cell or an organism constitutes a dynamic arrangement of molecules in space whose maintenance, like that of a flame or an eddy, depends on a continuous flux of matter and energy. Growth of a cell implies not merely the accretion of more substance but the enlargement of the pattern, and all developmental changes involve the progressive transformation of one spatial pattern into another. The orderly placement of molecules in living matter has sometimes been likened to that in a crystal, but this analogy is fundamentally erroneous. The shapes of cells and the organization of their constituents do not represent states of minimum free energy, as a crystal does (though this is certainly one element in the generation of some biological forms). Rather, in the course of their growth and development, organisms exert force and perform work to impose on themselves forms "of their own choosing." Morphogenesis is thus a prime illustration of the truism that living things convert energy into organization.

To be sure, morphogenesis is but one theme in the great symphony of biological order. In Chapter 1 we concluded, following Rupert Riedl (1978), that order in biology measures the degree of predictability, or determinacy. Predictability is so fundamental a feature of life that it is quite impossible to think about biology without taking it for granted. We expect the molecules that make up the living world to constitute a recurring and minute sample of those that are chemically possible. We know that their transformations in metabolism are confined to a handful of narrow channels, each closely regulated. We are not surprised to find that one specimen of *Amoeba proteus* or of elephant is in essence identical with any other; were it not so we could not classify living things, or even recognize them. We accept that these patterns have remained the same for very long periods of time, hundreds of million years in some cases, yet have given rise to novel patterns by descent with modification. And we confidently attribute both change and persistence to the extreme fidelity with which genetic information is transmitted and expressed. Each of these marvels is an instance of biological order, which extends beyond the individual organisms to the structure of biological communities and human societies, and each level must be understood in terms that include energy, work, and information.

This chapter makes no attempt to discuss biological order in general, but will examine morphogenesis as a concrete illustration of the manner in which biological order arises. In keeping with the scope of this book, the discussion will center on single cells and unicellular organisms, for their growth and development represent the integration of the diverse energy-linked processes described in the preceding chapters into an autonomous whole and constitute the culmination of cellular bioenergetics. However, even at the cellular level morphogenesis remains an area of darkness punctuated by patches of light that dazzle as much as they illuminate. The fundamental problem is that there are so many kinds of organisms, a veritable riot of living forms. In no case do we clearly understand how cellular morphology and organization arise, and no single organism can be taken as representative of morphogenesis in general. By the same token, there is certainly no unitary molecular mechanism of morphogenesis in the same sense that there is a single mechanism of oxidative phosphorylation. Seeking to understand cells as dynamic, integrated entities is quite different from analyzing their component parts and conceptually more difficult, but this must be one goal of those who profess an interest in biological order.

The object of this chapter is to display some of the principles that underlie the generation of biological forms, with the understanding that every cell and organism is likely to represent a unique set of variations on these general themes. We can travel only a little way along this path, for we understand next to nothing about global organization above the molecular level. So it is only fair to warn the reader that the following discussion relies more heavily on my personal selection and interpretation of the facts than did the previous chapters, and that it is conditioned by two judgments I believe to be sound, but that are not universally shared. The first is simply that the spatial order we perceive in structural form is but the static expression of the forces that engendered this form (Thompson, 1917; Weiss, 1967). It should therefore be illuminating to examine morphogenesis from the viewpoint of bioenergetics as spatially localized work. The second judgment is that cellular forms, being the expression of dynamic processes, are implicitly determined by the cell's genetic endowment but are not explicitly specified there; morphogenesis during growth and development results from the integrated workings of the whole cell, rather than from the execution of specific orders issued by the genome.

Self-Assembly

The biochemical approach to morphogenesis originated not with chick embryos and sea urchin eggs but in protein chemistry. The first general principle of molecular morphogenesis was spawned by research on the way denatured enzymes spontaneously fold back into the native, active form and on the recombination of purified viral proteins and nucleic acids to produce infectious particles. The principle of self-assembly states that complex biological structures often arise spontaneously by the association of their constituent macromolecules. Given the proper pH, ionic environment, and temperature, self-assembly is often thermodynamically favored and requires no further input of energy or information. In such cases, transcription and translation of the genetic messages that specify molecular structure are sufficient to account for the conversion of energy into spatial organization.

The classic instance of self-assembly is the formation of infectious tobacco mosaic virions when the capsid proteins and TMV-RNA are mixed in solution. The work of H. Frankel-Conrat, R. C. Williams, D. L. D. Caspar, and A. Klug on this process holds a place in most textbooks of biochemistry and a brief reminder must suffice here. In the native virus particle, 300 nm in length, 2130 identical protein subunits are wound helically about a single strand of RNA, enclosing it in a cylindrical shell. The purified constituents reassociate to yield particles having the correct form, dimensions, and biological activity; the length, for instance, is determined by that of the RNA core and the diameter by the polypeptide subunits. There is nothing random about the pathway of virion assembly. As shown in Figure 14.1, protein monomers first associate into disks,

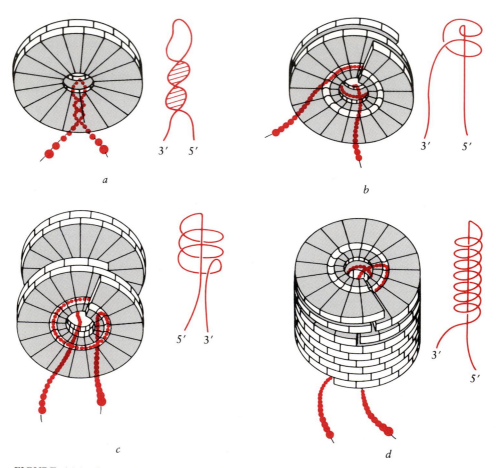

FIGURE 14.1   Stages in the self-assembly of tobacco mosaic virus. (*a*) Intercalation of the initiation region of RNA into a disk of capsid monomers. (*b*) Binding of RNA and dislocation of the disk into a "lock washer." (*c*) A second disk adds on top of the first. (*d*) Elongation of the virion. (From "The Assembly of a Virus," by P. J. G. Butler and A. Klug. Copyright 1978 by Scientific American, Inc. All rights reserved.)

each containing 34 of the subunits arranged in two tiers. Nucleation of a virus particle begins with the insertion of a special loop of RNA into the central hole of the protein disk. As the RNA binds there, the disk is dislocated into the helical form called a lock washer, which is the start of the capsid helix. Additional disks rapidly join the complex, each threaded in turn on a loop of RNA, where it snaps into the lock washer configuration and clamps the RNA into place until the particle is complete (Butler and Klug, 1978).

How is this possible? Formation of a virus particle certainly entails a substantial increase in order and a concomitant decrease of entropy since the individual macromolecules take up defined positions in space. Does not the spontaneous occurrence of this process violate the second law? The answer is no, for the increase in entropy of the solvent more than makes up for the loss of entropy by the virus particle. The reason is that the capsid proteins join up along relatively hydrophobic surfaces, thereby excluding water from the interface between them. This entails an increase in entropy, for water molecules in the vicinity of a hydrophobic surface are constrained into relatively ordered structures resembling ice. The gain in solvent entropy comes from the liberation of these bound water molecules, and it in turn stabilizes the bond between the protein subunits. Specific ionic and hydrogen bonds also have a role to play in ensuring a close fit. But what promotes the association of macromolecules into a stable complex, in this and in most other instances of self-assembly, is not mutual adhesion so much as the exclusion of water that results from the adhesion. No external energy source is required because the increase in entropy suffices to "pay" for the work done.

Formation of the virus particle also seems to entail a gratuitous gain in information, or determinacy. Where does the information come from that specifies the characteristic shape of tobacco mosaic virus? This is a genuine paradox, from which we draw two lessons that will be useful later. The first is that one cannot estimate how much information a given morphogenetic process requires unless one knows the pathway of construction. Determinacy is, as the word plainly suggests, a measure of the likelihood of finding a given atom in a particular place. A model builder, told to make up a TMV particle from a kit of atoms, would require extensive instructions to know where each atom goes. This task would be eased by allowing the builder to start with peptide and nucleotide units and would be made quite manageable by providing the primary sequences of RNA and capsid proteins together with the folding rules, since these suffice to determine the forms of the macromolecules. The second lesson is a reminder that genetic messages convey both explicit and implicit information (Stent, 1978; Chapter 13). The RNA coding sequence determines only the amino acid sequence of the capsid protein and its own replication; we cannot identify any part of it with specification of the shape of the virus. Yet selection, acting at the level of the virion, will certainly have favored sequences compatible with the efficient packaging of the protein into the infectious virus under the conditions that prevail in the host plant. The implicit, teleonomic meaning of the message is to make virus, even though all it explicitly calls for is protein and RNA.

Formation of the TMV virus illustrates another feature that is characteristic of self-assembly in general: events occur in a particular sequence, which is much more rapid than random association, effectively irreversible, and avoids the accumulation of abortive complexes. The biological significance of this feature is particularly apparent in another well-known example of viral morphogenesis, the production of the bacteriophage T4. Some 60 proteins collaborate to build this sophisticated creature, which resembles a miniature moon lander, by a sequence of steps shown in Figure 14.2 in simplified form. Most, but not all, depend on self-assembly. Nearly 20 years of research, recently summarized by Wood (1979) and King (1980), have shown that the sequence reflects not the temporal appearance of the various proteins but the rule that governs their association: no two proteins interact with each other unless another pair of proteins has previously interacted. For example, tail tube subunits show no inclination to polymerize in the absence of the base plate; on addition of base plates to a suspension of tail tube monomers, the latter polymerize onto particular sites on the base plates and (for reasons not yet clear) stop when the tube has attained its final length of 100 nm. By the same token, sheath subunits do not aggregate by themselves, nor do they polymerize onto base plates; only the addition of base plates that already bear the tail tube initiates the addition of sheath monomers. The mechanism is as follows. The monomers are produced in a conformation that has no tendency to polymerize. Binding of the first tail tube subunit to a base plate switches it into an "active" form that can now bind further tube subunits, each of which is switched in turn (King, 1980). The utility of this coupling rule in ensuring the production of infectious virus particles will be obvious, but its significance is far broader, for it certainly orders the construction of many intracellular organelles as well. The entire process of T4 production is a mine of mechanisms that may participate in the initiation, termination, and regulation of intracellular self-assembly.

That self-assembly is a major mechanism of cellular morphogenesis is now beyond question. Ribosomes can be dissociated into their constituent proteins and RNA; when these are incubated together under the proper conditions, they reconstitute the large and the small subunits, according to a clear assembly sequence. Another example of much current interest is the spontaneous assembly of microtubules from the purified tubulin monomers. Evidently, all the structural information required to build a microtubule is contained in the tubulins; associated proteins, GTP, calcium ions, and protons may nucleate and regulate the assembly of microtubules, but they do not determine their form. Bacterial flagella, the clathrin cage of coated vesicles, collagen fibrils, the dynein arms of cilia, and filaments of myosin and actin likewise owe their morphology to their constituent proteins. It is worth noting that in extending the concept of self-assembly to intracellular organelles we slightly alter its meaning. Our interest is not so much in the dispensability of energy input as in the conclusion that no structural information needs to be supplied beyond that contained in the primary sequences of amino acids and nucleotides.

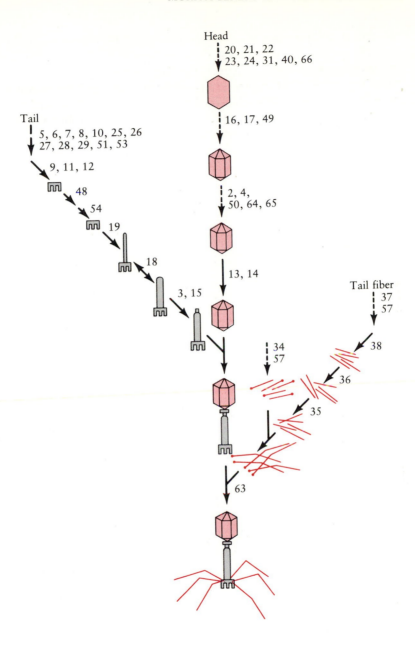

FIGURE 14.2   A partial sequence of gene-controlled steps in the assembly of the bacteriophage T4. Heads, tails, and tail fibers assemble separately and subsequently join to generate the complete particle. Numbers designate the gene whose product participates in each step; most of the steps consist of the self-assembly of macromolecules, but a few represent enzyme-catalyzed reactions. (From Wood, 1979, with permission of Academic Press.)

Many cellular elements are membranous, and it is surely of great significance that formation of the phospholipid bilayer is also a kind of self-assembly: phospholipids in water organize themselves spontaneously into bilayers, for these represent a configuration of lowest free energy. The process exemplifies what Tanford (1978) has called the hydrophobic effect. As is well known, the two ends of a phospholipid molecule have opposite thermodynamic preferences: the polar head groups seek to associate with water, while the acyl chains prefer a nonaqueous medium. These opposing tendencies are satisfied by structures that place the acyl chains in the middle, shielded from contact with water, and expose the polar groups at the surface. The simplest structure of this kind is a micelle (Figure 14.3a), but phospholipids do not form stable micelles. The reason is that

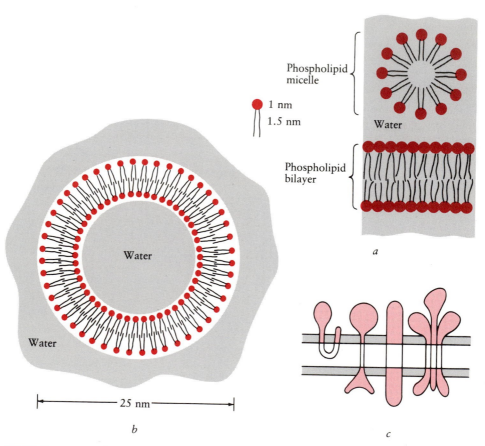

FIGURE 14.3   Biological membranes as self-assembling entities. (a) Phospholipid micelle and segment of a bilayer vesicle. (b) Bilayers spontaneously form closed vesicles. (c) Membrane proteins. Hydrophilic portions in contact with water are colored, hydrophobic portions embedded in the bilayer are blank. (After Alberts et al., 1983, and Tanford, 1978, with permission of Garland Publishing and the American Association for the Advancement of Science.)

unlike soap or most detergents, phospholipids have two acyl chains per polar head. As a consequence, the head groups are too small to cover the surface and water molecules make contact with the acyl chains, destabilizing the structure. A configuration with lower free energy, having a more favorable surface-to-volume ratio, is the closed vesicle in which the bilayer surrounds an internal aqueous cavity (Figure 14.3b). It should be pointed out that phospholipid molecules have no affinity for each other: they are squeezed together solely by the exclusion of water. Therefore while proteins self-assemble into rigid structures joined by specific ionic or hydrogen bonds, phospholipid membranes are typically fluid. The phospholipid bilayer vesicle, self-sealing, topologically closed and impermeable to ions and polar molecules, is one of the fundamental biological forms; it might stand as a symbol for bioenergetics much as the double helix has come to epitomize molecular biology.

Real biological membranes are studded with proteins, some peripheral but most of them integrally set into the lipid phase; many span the bilayer, exposing portions of their anatomy to one or both aqueous phases. It is now a relatively routine matter to extract such proteins by means of detergents, preserving their active configuration, and then to reconstitute functional enzymes or transport carriers into phospholipid vesicles. Evidently the structure of membrane proteins is such as to render their insertion into membranes thermodynamically favorable, requiring neither energy nor additional information. This inference is consistent with what we know of the structure of membrane proteins. Some have a hydrophobic tail that anchors the polar portion to the membrane; proteins that span the membrane usually have a hydrophobic midriff, which is buried in the bilayer, while the remainder protrudes into the aqueous phases on either side, like the heads of a rivet (Figure 14.3c). The manner of self-association between membrane proteins and hydrophobic environments testifies again to the role of purely thermodynamic forces in the generation of biological order and form (Tanford, 1978).

Might we not at least in principle hope to account for all biological forms and structures in terms of self-assembly? We need not quibble over the finer points of definition, for what one wants to know is whether cellular forms are uniquely determined by the primary structures of gene products, and that is no longer credible. Biological membranes illustrate the point in a straightforward fashion. The chemical structures of membrane proteins are such as to favor their insertion into the bilayer but cannot, unaided, ensure the vectorial orientation in the membrane, which is mandatory for their physiological activities; to that end we require the apparatus of vectorial biosynthesis with its expenditure of energy. It is even less reasonable to expect primary protein structures alone to explain how the various membranous organelles of a cell each acquire their own protein complement, or how regions of a single membrane may come to differ. These processes of sorting and distribution involve cytoplasmic transport guided by the cytoskeleton. And it now appears that this pervasive cellular framework is also not to be explained as the result of self-assembly alone. Helical structures such as microtubules and microfilaments can be packed into a variety of patterns, all of which are consistent with the bonding rules for globular proteins (DeRosier et al., 1980). It follows that the diverse forms assumed by the cytoskeleton in cells cannot be specified by the primary structures of their constituent proteins but must be fashioned by processes that operate at the cellular, global level. Specific dovetailing of macromolecular building blocks is certainly an element in most morphogenetic processes, but self-assembly by itself is not a sufficient foundation for reflection on the genesis of biological forms.

Shaping a Bacterial Cell

Bacteria typically consist of a single compartment bounded by the plasma membrane, and their forms are elementary ones: the coccus, the rod with hemispherical caps, and the helix. Bacteria presumably display the principles of cellular morphogenesis more plainly than larger cells do, but even at this most basic level of cellular organization the problems are formidable, and we are just beginning to discern how bacteria make themselves.

The shape of a bacterium is almost always maintained by the cell wall, which lies external to the plasma membrane. We know this from the fact that mechanical disruption of bacteria produces wall fragments that retain the shape of the cell from which they came, while enzymatic digestion of the wall results in spherical, deformable, and osmotically fragile protoplasts or spheroplasts. The chemical structure of bacterial cell walls is inordinately complex and cannot be reviewed here (see Tipper and Wright, 1979). Suffice it to say that the component that confers rigidity and mechanical strength on the wall is a polysaccharide unique to eubacteria, called peptidoglycan or murein. Its structure consists of linear chains of alternating units of N-acetylglucosamine and N-acetylmuramic acid, cross-linked by short polypeptide chains some of whose constituent amino acids do not occur in proteins. Although the individual polysaccharide chains are only about 100 units long, the structure as a whole is so well knit together by covalent cross-links that one can isolate the entire peptidoglycan layer of a cell as a single sacculus. Other components of the cell wall, such as the teichoic acids, contribute to the shape of the wall, but the peptidoglycan layer is probably the most important element and certainly the best known.

The wall maintains the cell's shape, but what determines the shape the wall grows into? It seems quite clear that this shape is not directly determined by the chemical structures of cell wall polysaccharides, in the sense that a protein's amino acid sequence determines the form into which it folds: the cell wall is not a self-assembling structure (Daneo-Moore and Shockman, 1977). Rather, the cell must impose form on the wall as it grows, before it is fixed by cross-links, as the potter molds the clay before it sets. In principle this might be done with the aid of a protein scaffold that fits together in a way determined by the structure of the constituent proteins; the hexagonal head of T4 bacteriophage is known to be shaped in this manner on a matrix of proteins that do not remain in the finished head (King, 1980). This hypothesis has never been rigorously falsified but has nevertheless lost favor, largely because evidence for the existence of bacterial scaffold proteins has not been forthcoming (Henning, 1975). Instead, microbiologists are coming to see the shape of the wall as a consequence of the spatial and kinetic constraints that govern the activities of the enzymes responsible for wall biosynthesis.

This conception emerges most compellingly from studies with the gram-positive bacterium *Streptococcus faecium*, which displays the simplest mode of wall growth and therefore deserves to be examined in some detail. Streptococci characteristically grow and divide into two equal, spheroidal daughters (Figure 14.4). The thick wall is typically gram-positive in composition, with peptidogly-

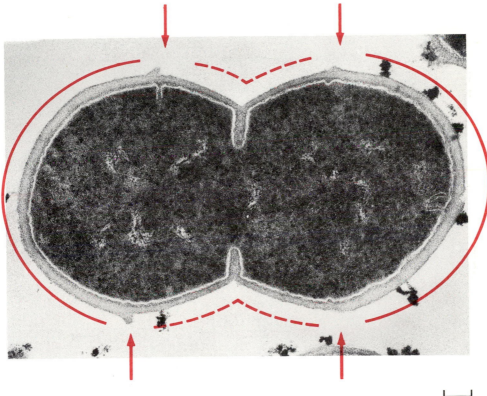

0.1 µm

FIGURE 14.4   A dividing cell of *Streptococcus faecium*, showing the septum and the zone of equatorial growth. Newly externalized peripheral wall (dashed line) is markedly thinner than the old wall (solid line), which was allowed to thicken during a period of starvation in order to magnify the contrast. Old wall has been segregated to the poles by growth of the new wall. Two new wall bands (arrows) mark the juncture between old and new wall and predict the site of septum formation when the daughter cells divide in turn. (From Higgins et al., 1971, with permission of the American Society for Microbiology. Electron micrograph courtesy of M. L. Higgins.)

can making up half the mass and teichoic acids another quarter. The equator of each cell is marked by a conspicuous raised band, which predicts the site at which the septum will form when the cell next divides. Some 20 years ago, R. M. Cole and J. J. Hahn discovered, by a pioneering application of immunofluorescence, that wall growth in streptococci is confined to the cell's equator, that is, to the region of the growing septum (Figure 14.4). In retrospect it is clear that their important discovery owes something to the choice of organism. In streptococci the cell wall, once formed, is stable and does not turn over; most other bacteria continuously remodel their walls, and restricted growth zones, albeit common and perhaps universal, are not so easily demonstrable.

The sequence of events in the growth and division of a streptococcal cell appears to be as follows (Shockman et al., 1974; Daneo-Moore and Shockman,

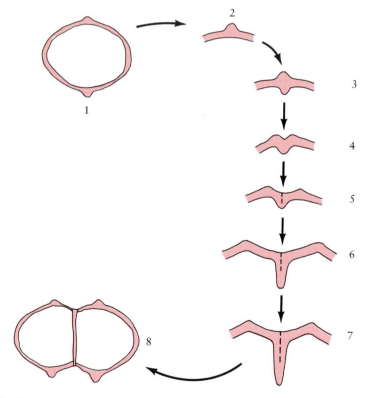

FIGURE 14.5   Sequence of events in surface enlargement in *Streptococcus faecium*. The cycle begins with the activation of a preexistent site of wall synthesis located beneath the wall band (1). The nascent septum grows down into the cytoplasm, while a notch in the wall marks the beginning of its duplication. As the septum elongates, new wall is externalized in response to turgor pressure, forming an equatorial zone of newly made peripheral wall (2 to 6). When surface expansion halts the septum continues to elongate (7), finally dividing the original cell into two compartments (8). (After Shockman et al., 1974, with permission of the New York Academy of Science.)

1977). Wall synthesis is initiated just beneath the wall band, forming a nascent septum that grows down into the cytoplasm (Figure 14.5). Both the wall band and the septum are duplex in structure. Very early a notch develops in the wall band, splitting it in two; the split continues into the septum, which peels apart into two leaflets. The septum is externalized even as it grows, giving rise to new peripheral walls, which progressively push apart the two daughter wall bands. As the growth zone widens, the cell assumes the characteristic shape of a diplococcus (Figure 14.4). Each daughter-cell-to-be has its own equatorial wall band, which marks the junction between the old wall (laid down in the preceding division cycle) and the new wall, and the daughters are joined along the septal annulus, as yet incomplete. The new wall is first deposited as septum, but as it is externalized the wall undergoes progressive thickening until it attains the normal width. Finally, when the surface has expanded to the extent required to make two cells, expansion ceases; the septum closes and the cells can go on to separate.

Synthesis of the cell wall calls for the incorporation of precursors manufactured in the cytosol into a structure external to the plasma membrane, but the cell's osmotic integrity must be maintained at all times. The solution to this dilemma is a biosynthetic pathway that spans the membrane: peptidoglycan elements, each consisting of an acetylglucosamine-acetulmuramic acid dimer plus a polypeptide chain, pass across the membrane in nascent form bound to a special lipid carrier molecule. Insertion of a new unit into the existing wall takes place at a site prepared by a special hydrolase that cleaves the disaccharide backbone (Tipper and Wright, 1979). Teichoic acids arise by an analogous pathway, and their insertion into the existing wall is closely coupled to that of peptidoglycan. These enzymatic reactions, whose biochemistry is very well defined, must be so arranged in space as to generate the morphological sequence outlined in the preceding paragraph. According to Shockman and his colleagues, insertion of new wall elements occurs chiefly in a narrow zone beneath the wall band, which becomes the leading edge of the growing septum. Outgrowth of the cell wall from this region can then be mathematically modeled in terms of the rates of three concurrent processes: the rate of linear extension of the leading edge of the septum, the rate at which the two leaflets of the cross wall peel apart and become externalized, and the rate of wall thickening, which is not restricted to the septal zone but occurs all over the surface. Each wall band would be capable of synthesizing an area of wall corresponding to the new poles that arise there, and the form of these poles could be determined by the kinetic and regulatory parameters governing the activity of the enzymes that lay down the new wall.

The model developed above accounts for many aspects of growth and division but does not explain the unique shape of the emerging cells. This was tacitly credited to the unspecified kinetic parameters of the enzymes that assemble the cell wall. Thanks to the recent contributions of A. L. Koch we can propose a more explicit hypothesis. The shapes of bacterial cells can be understood on the simple assumption that the cell wall expands in response to tension created by turgor pressure. The cell yields to this force by localized synthesis of additional wall; the various bacteria assume different forms because wall growth takes place in different regions of their surface. Streptococci are a particularly simple case; growth is confined to the septum, and the hemispherical shapes of the new poles arise in effect by surface tension as newly externalized plastic wall seeks the configuration of minimum surface area for the volume it encloses (Koch et al., 1981; Koch, 1983). Koch and his colleagues thus return, with evident relish, to an idea discussed at length by D'Arcy Thompson (1917) but which has enjoyed little support in modern times.

The argument is mathematical in nature, and I wish to do no more than indicate the gist of it. A growing streptococcus maintains a substantial turgor pressure, on the order of 20 atmospheres, which is transmitted to the rigid wall and constitutes a force tending to expand its surface area. Now, Koch noted that soap bubbles blown on a rigid ring assume a shape very similar to that of the half-cells shown in Figure 14.4. Calculations were made for *S. faecium* on the assumptions that (1) the zone of growth is very narrow and immediately adjacent to the ingrowing septum, (2) the septal material comes under tension just before it becomes part of the external wall, (3) the septal wall is twice as thick as the adjacent external wall, and (4) the hydrostatic pressure and wall growth activity are constant. If the growing wall responds to tension in such a way as to minimize its surface energy, as a soap bubble does, the calculations predict a shape that agrees well with actual measurements (Koch et al., 1981). But how can the rigid cell wall respond to tension as though it were a fluid film? The authors argue on thermodynamic grounds that a cell wall must initially be assembled in an environment free of strain and can be subjected to tension only after new monomers have formed at least two covalent bonds with the framework of the wall. This attachment takes place at the leading edge of the septum. As the leaflet splits, just before the new wall is externalized, it comes under tension. Local bond scission is favored because the tension reduces the activation energy of the hydrolytic reaction, accelerating it by many orders of magnitude. As bonds split in the old wall new units are pressed into the rent, taking a share of the load (Figure 14.6). (Readers of sea tales may recall that sailors repair a vessel holed below the waterline in an analogous manner: canvas is placed over the hole from the outside and hydrostatic pressure forces the patch into place.) In the process of insertion the wall is rendered temporarily plastic, so that its form comes to reflect tension as though it were fluid (Koch et al., 1981).

In subsequent papers, Koch and his colleagues extended their approach to the spherical staphylococci; to rod-shaped bacteria, whose cylindrical portions can be attributed to zones in which growth is diffuse and accompanied by turnover of old wall; and to the tips of fungal hyphae (Koch et al., 1982; Koch, 1983). This is not the place to query details, but it is necessary to point out major postulates that for the present are unverified. The surface stress theory demands that new wall units be attached only in unstressed regions of the old wall and that hydrolases split only bonds that are protected by new units ready to bear a load. It also requires close regulation of the turgor pressure, small fluctuations of which may be the signal that initiates division. Nevertheless, there can be no

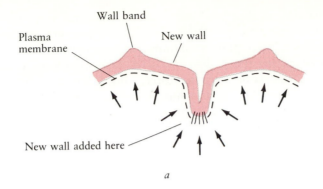

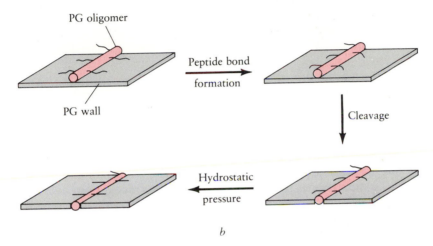

FIGURE 14.6   The surface stress theory of bacterial morphogenesis. (*a*) The driving force for surface expansion is turgor pressure, which is uniform at all points. Surface expands only by the addition of new wall material to the septum; newly externalized wall is plastic and its form is effectively governed by surface tension. (*b*) Peptidoglycan (PG) oligomers are first attached via their peptide cross bridges to the existent, stressed wall. Such linkages, if formed at random, would leave the newly inserted unit loosely attached and not in an extended conformation. It has been proposed that hydrolytic or transferase enzymes act only on bonds under stress and only in a region surrounded by newly formed linkages. Following cleavage, the newly inserted unit will then be pulled into the wall. As the new unit comes to bear part of the stress due to hydrostatic pressure, it assumes an extended conformation and allows the wall to expand. (After Koch et al., 1981, with permission of the Society for General Microbiology.)

doubt of the explanatory and heuristic powers of the surface stress theory. And this microbiologist, for one, takes pleasure in the belief that three centuries after van Leeuwenhoek first described the shapes of bacteria, we are beginning to understand the forces that produce those shapes.

If the surface stress theory is fundamentally correct, we must conclude that the conversion of energy into cellular organization proceeds in a manner quite unlike the self-assembly of preformed units into a predetermined shape. Instead, bacteria underscore the thesis that biological form is the static expression of dynamic forces. The scalar driving force, turgor pressure, is given spatial direction by channeling at the growing septum, much as the shape of a pot thrown on the wheel recalls the balance between the centrifugal force and the localized restraint exerted by the potter's hand. Moreover, the growing cell does not draw on the genome for direct instructions that specify its shape. Instead, new structures are laid down on structures inherited from the preceding generation, in this case the immortal wall bands, which provide an elementary spatial memory. Both features are common in cellular morphogenesis and will shortly reappear in other guises.

Cytoskeleton and Form in Eukaryotic Cells

Bacteria come in a handful of elemental shapes, but the forms of single-celled eukaryotic organisms are legion. Fungal hyphae, yeast, amoebas, flagellated algae, and ciliated protozoa obviously represent distinct patterns of organization, each at a high level of structural order. The unity that underlies the morphological diversity of eukaryotic cells is nevertheless quite apparent, thanks to the use of standard parts in their construction. Chloroplasts and chromosomes, Golgi bodies and ribosomes, microtubules and nuclear membrane pores are all examples of standard parts that differ in detail from one organism to another yet remain "in essence" the same. It seems reasonable to infer that the morphological features by which we recognize particular kinds of eukaryotic cells generally result from comparatively minor variations in the intensity, timing, or localization of processes that are in themselves widely distributed. Self-assembly plays a large part in molding molecules and subcellular organelles, but there is good reason to believe that the cell as a whole is shaped by different forces and that the chief instrument by which eukaryotic cells shape themselves is the cytoskeleton.

Microtubules Guide Cellulose Deposition.   That the arrangement of microtubules in cells is correlated with their shape was recognized from the beginning, and Porter (1966) already envisaged that the relationship may be a causal one. This is exceedingly difficult to prove, but the literature of the past two decades reports many examples in which the local assembly of microtubules precedes and predicts subsequent morphogenetic events, suggesting that microtubules are an important element in the localization of these events (Hepler and Palevitz, 1974; Stephens and Edds, 1976; Raff, 1979; Tucker, 1979; Gunning and Hardham, 1982; Dustin, 1984). Some of the most persuasive examples come from the cells of plants, which, like those of fungi and bacteria, are enclosed by rigid walls that maintain the cell's shape. In many cases one observes a clear correlation between the formation of a microtubular array and the subsequent deposition of cell wall.

For example, during the cycle of growth and division in many plant cells, a transitory band of microtubules appears about the cell's midriff, lying parallel to the plane of the plasma membrane. It appears before prophase (hence the term "preprophase band"), is gone by the time mitosis is complete, and its location predicts with remarkable precision the place where the cell plate will eventually form and divide the cell in two. Another example comes from the differentiation of the guard cells of stomata (these are the pores that mediate gas exchange across the surface of a leaf). The pore arises by the division of a precursor cell into two parallel, elongated daughter cells. Each daughter proceeds to lay down a thickened cell wall along their common border, and retraction of the wall forms the opening of the pore. Deposition of the stomatal walls is preceded by the assembly of a localized and transient population of microtubules whose position identifies the locus of future wall synthesis, while their orientation predicts that of the cellulose fibrils in that wall.

In some instances the link between microtubule assembly and the subsequent deposition of an ordered cell wall can be dissociated and its causal nature probed by experiment. For example, protoplasts of the unicellular alga *Mougeotia* can regenerate their cell wall and, in doing so, revert from a spherical shape to the cylindrical one of the normal organism. Marchant and Hines (1978) have described the sequence of events. A network of cortical microtubules can be seen in the protoplasts, and when wall regeneration begins, these become progressively linked to the plasma membrane. Concurrently cellulose microfibrils are deposited at the external face of the plasma membrane, oriented parallel to each other and transverse to the future axis of elongation (the spatial relationship of the cellulose fibrils to the cortical microtubules unfortunately is not known). Treatment of the protoplasts with drugs that prevent microtubule assembly does not inhibit regeneration of the wall; however, in such walls the cellulose microfibrils are oriented at random and the resulting cells are spherical. Conversely, coumarin blocks cellulose synthesis but not the assembly of cortical microtubules; subsequent removal of the coumarin permits deposition of cellulose in the normal parallel arrays (Marchant, 1979). The conclusion from these and many other examples seems plain enough: the cortical microtubules determine the spatial pattern of cellulose fibril deposition and in this manner define the cell's emerging shape.

Some conception of just how this comes about is emerging from studies on the ultrastructural basis of cellulose synthesis (Brown and Wilson, 1977; Giddings et al., 1980). In several unicellular algae cellulose occurs in the wall in the form of domains, each of which consists of a band of perfectly parallel cellulose fibrils (Figure 14.7a). The plasma membrane beneath each cellulose band is rich in intramembrane particles whose arrangement is highly ordered: particles are grouped into rosettes and the latter form regular arrays whose width corresponds to that of the cellulose band (Figure 14.7b). Giddings et al. (1980) suggest that each rosette represents a single complex of cellulose synthase, which spins out a single cellulose fibril (Figure 14.7c and d); spontaneous lateral aggregation of the parallel fibrils produces the cellulose domain. The arrays of

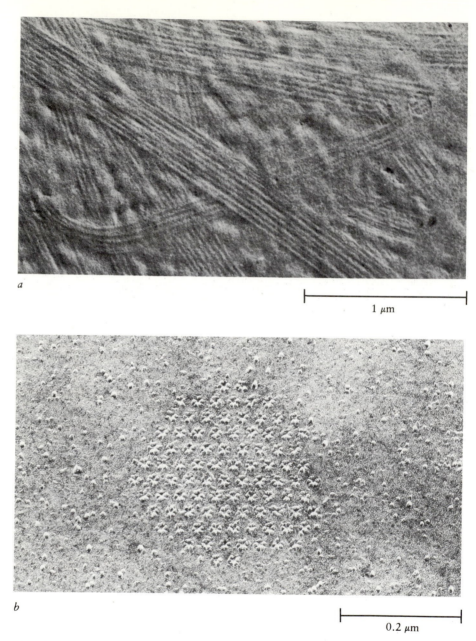

a

1 μm

b

0.2 μm

FIGURE 14.7 Cellulose deposition in the secondary wall of the green alga *Micrasterias denticulata*. (*a*) Bands of cellulose fibrils, each 5 nm wide. (*b*) An array of rosettes embedded in the plasma membrane. Each rosette corresponds to a single complex of cellulose synthase. (*c, d*, opposite page) Model of cellulose deposition. A row of rosettes forms a set of cellulose fibrils, which aggregate laterally to form the larger fibrils of the secondary wall. (From Giddings et al., 1980; reproduced from *Journal of Cell Biology* with permission of Rockefeller University. Electron micrograph courtesy of T. H. Giddings.)

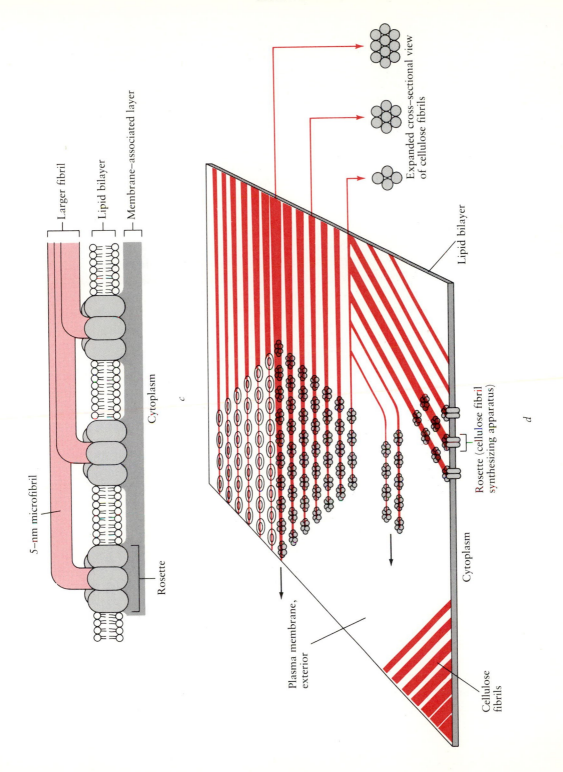

Expanded cross-sectional view of cellulose fibrils

Larger fibril

Lipid bilayer

Membrane–associated layer

5–nm microfibril

Rosette

Cytoplasm

c

Lipid bilayer

Rosette (cellulose fibril synthesizing apparatus)

Cytoplasm

Plasma membrane, exterior

Cellulose fibrils

d

cellulose synthase are apparently produced in the Golgi apparatus by mechanisms quite unknown, transported to the plasma membrane in special vesicles, and inserted at sites determined by the location of cortical microtubules. The cellulose synthase arrays are thought to be mobile along tracks defined by the cortical microtubule net.

Cytoskeletal Fibers Cooperate to Determine Shape.   For another perspective let us glance briefly at animal cells, which again display a strong correlation between cellular shape and the arrangement of cytoplasmic microtubules. Since there is no rigid wall, cell shapes tend to be transitory. Chilling and drugs that inhibit microtubule assembly usually cause the cytoplasmic microtubule network to vanish and the cell to round up. It follows that the microtubules are not secondarily correlated with cell shape but help to maintain that shape, and further, that cytoplasmic microtubules are subject to continuous turnover and constitute a dynamic rather than a fixed scaffold. (It should be mentioned in passing that microtubular structures vary greatly in their stability. Ciliary axonemes, for instance, do not undergo turnover and are quite untouched by cold or colchicine.) Observations with a diversity of animal cells indicate, however, that cell shape is not specifically linked to microtubules but rather depends on the cytoskeleton in its entirety: actin cables are just as vital to biological architecture as are microtubules and in some cells bear the brunt of the burden. A case in point is the role of actin filaments in the morphology and movements of amoebas or fibroblasts (Chapter 12). Quiescent fibroblasts assume a flattened form, elongated in the direction of movement. Thick bundles of actin filaments, aptly called stress fibers, span the cell and obviously serve a structural role. When the cells enter mitosis (or become highly motile), the stress fibers vanish; the cells round up, ceasing to adhere to the substratum; and actin is rearranged into meshworks of short filaments beneath the plasma membrane, including the contractile ring (Figure 14.8). With the completion of mitosis the cells settle down and flatten, and the scaffold of actin bundles is regenerated by a remarkable architectural transformation that includes an intermediate stage resembling a geodome (Figure 12.11).

Form in animal cells is not a matter of *either* microtubules *or* microfilaments but arises by their collaboration, aided by other fibrous structures such as the intermediate filaments and possibly the cytoplasmic matrix. The interactive aspect of shape determination is neatly illustrated by a recent study with nerve cells in culture. Freshly explanted cells send forth axonal processes (neurites) that are modeled on a core of cytoplasmic microtubules. Exposure to cold or to inhibitors of microtubule assembly causes the neurites to retract, and if the microtubules are permitted to reassemble, the axon is reconstituted, essentially in its original form and topography. Significantly, neurite retraction is not a passive structural collapse but an active process that involves both actin and ATP. When both cytochalasin and antitubulin drugs are present, axons retain their form despite the absence of microtubules (Solomon, 1981; Solomon and Magendantz, 1981). Just how the cytoskeleton determines the shape of these and other animal

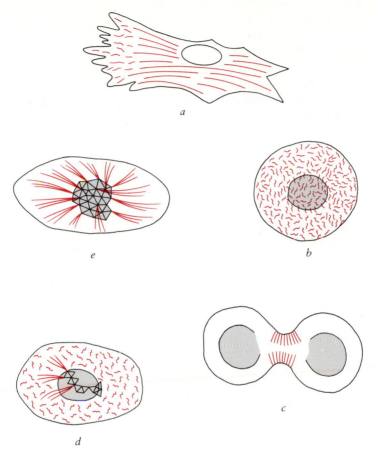

FIGURE 14.8   Shape transformation and the reorganization of actin filaments in
fibroblasts at various stages of their life cycle. (*a*) Bundles of actin filaments (stress fibers)
span a cell that is quiescent, flattened, and attached to the substratum; in the
lamellopodium, actin is found in the form of short filaments. (*b*) A rounded, mobile cell
detached from the substratum, with actin arranged predominantly as a diffuse gel of short
filaments. (*c*) A dividing cell, emphasizing the transient contractile ring. (*d*) and (*e*) The
geodome, an intermediate stage in the conversion of a rounded mobile cell into a quiescent
flattened one. Actin filament bundles have begun to assemble onto the vertices of the
geodome. (After Lazarides and Revel, 1979, and other sources.)

cells is not clear, but the observations suggest that the form of wall-less cells
reflects at least to some degree the balance of forces exerted by contractile
elements, somewhat as muscle tone contributes to holding up our own bodies.

For a last illustration of the close relationship between cytoskeletal form and
function and the overall shape of the eukaryotic cell we turn to the fungi.
Brewer's yeast (*Saccharomyces cerevisae*) multiplies by budding, a process that
superficially resembles cell division in streptococci but is structurally quite

different. A key event is the formation of the primary septum that partitions the growing bud from the mother cell: the cell wall of yeast is made of interlaced glucan and mannan fibrils, but the primary septum alone is constructed of chitin. The division cycle (Figure 14.9) begins with a local bulging of the cell surface, followed by the deposition of a ring of chitin in the neck of the incipient bud. The dimensions of this chitin annulus are quite precisely defined, and it forms at a particular stage of the cell cycle within a discrete period of about 10 minutes. Subsequent wall growth becomes localized beyond the waist formed by the chitin ring, making the cell balloon into the characteristic dumbbell shape. In the meantime the nucleus, centrally located to begin with, has begun to elongate and migrate toward the bud; eventually the nucleus positions itself within the isthmus defined by the chitin ring and then divides. This is followed by closure of the plasma membrane and completion of the primary septum with chitin. Subsequent events include the deposition of the secondary septum, made of glucans and mannans, by both mother and daughter cells, and ultimately their physical separation.

There is no reason to believe that the chemical or physical characteristics of the wall polysaccharides account for the yeast's shape and its manner of division. Nor are the biochemical reactions unique to yeast: filamentous fungi, such as *Neurospora crassa*, also manufacture chitin, but they confine its deposition to the

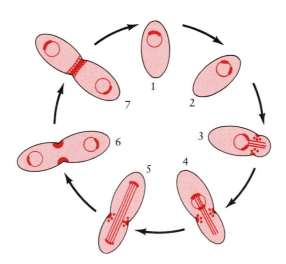

FIGURE 14.9   Landmarks in the division cycle of the yeast *Saccharomyces cereviseae*. As a virgin cell prepares to divide (1), the spindle pole body duplicates (2) and the bud begins to take shape. The bud is set off by a chitin ring (3); microtubules project from the nucleus into the bud. As the bud enlarges (4), the nucleus positions itself across the isthmus that separates the mother cell from the bud. The spindle elongates and the chromosomes are distributed between the two cells (5, 6). Cytokinesis concludes with the completion of the secondary wall by both mother and daughter cells (7). (After a sketch by E. Cabib, with permission.)

hyphal apex. In yeast, as in other walled cells, turgor pressure supplies the driving force for surface expansion, and the shape of the growing zone owes something to surface tension. But the general morphology of the budding yeast is almost certainly engendered by the process that lays down the chitin annulus at a particular stage of the division cycle. Chitin synthase in yeast is embedded in the plasma membrane as an inactive zymogen that is quite uniformly distributed over the cell's surface. Local chitin synthesis is initiated by an activator of the synthase, probably a proteolytic enzyme, which the cell manages to direct to the site of bud emergence packaged in membrane vesicles (Cabib et al., 1982). Localized activation is critical: there is a temperature-sensitive mutant that, at the restrictive temperature, deposits chitin all over its surface and consequently does not divide properly (Sloat et al., 1981). These vesicles are presumably members of that heterogeneous population that carries proteins, lipids, and cell wall precursors from the Golgi region to the growing point (Chapter 12). Cytoplasmic transport in yeast is no better understood than it is in other cells, but its direction and terminus clearly depend on the spatial arrangement of both microtubules and microfilaments (Byers and Goetsch, 1975; Kilmartin and Adams, 1984; Adams and Pringle, 1984). Once again, all we know suggests that the gross form of budding yeast reflects the way their cytoskeleton is organized.

Organizing Centers. So far, so good, but what organizes the cytoskeleton? Despite feverish activity in this field there are no clear answers. Let the biogenesis of the microtubule lattice illustrate the general principle that the cytoskeleton arises by multiple individual steps that are hierarchically ordered and vectorially deployed (Figure 14.10). The process is illustrated by animal cells in tissue culture, whose lattice of cytoplasmic microtubules can be displayed in its entirety by indirect immunofluorescence. When the cells are chilled, the microtubules disappear, apparently by disassembly into monomers. On warming the lattice regrows, recovering its former extent and general form within an hour or so. Both processes are polarized. Disassembly proceeds from the periphery inward to the nucleus; microtubules reconstitute by growing outward from a region near the nuclear membrane, called the centrosome, which contains a pair of centrioles (Tucker, 1979; Weber and Osborn, 1979). One end of each cytoplasmic microtubule is apparently inserted into the centrosome, and growth proceeds by the addition of tubulin monomers to the distal end.

The centrioles from which the lattice grows exemplify a universal class of structures called microtubule-organizing centers, or MTOC. These clearly play a key role in the orderly assembly of all microtubular ensembles but their modus operandi is far from clear. Experiments with cell-free extracts demonstrate that MTOC nucleate the assembly of microtubules and thus greatly enhance its rate. But they also appear to determine the spatial orientation of microtubules that sprout in their vicinity (Figure 14.10) and, according to some authors, even their length. Centrioles are themselves barrel-shaped arrays of microtubules, but the centriolar tubules are not continuous with their cytosolic offspring and there is no obvious mechanism to ensure spatial coherence. Equally mysterious is the

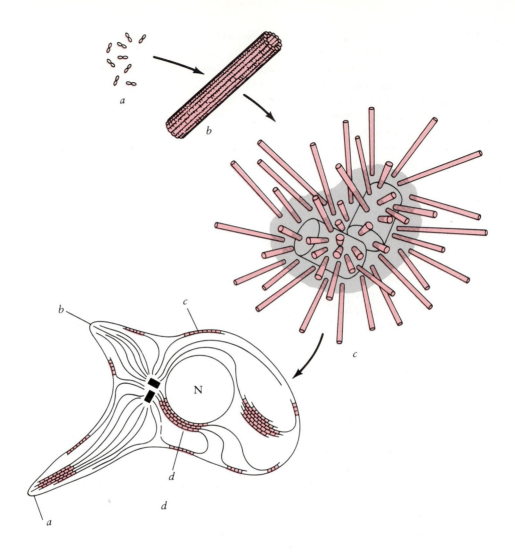

FIGURE 14.10   The hierarchy of microtubule organization in animal cells. (*a*) Tubulin monomers. (*b*) Standard microtubule. (*c*) Microtubules radiating from the environs of a pair of centrioles. (*d*) Possible organization of cytoplasmic microtubules. This cell has a single pair of centrioles (blocked in black) at the centrosome, associated with the nucleus (N). *a*, Microtubules stabilized by cross-links at the tip of a growing cell extension; *b*, conditions at the tip of this pseudopod favor retraction and microtubule disassembly; *c*, tubules run close beneath a specialized region of the surface and help anchor membrane components there; *d*, tubules that followed curved paths around the nucleus become linked by bridges to each other and to components of the nuclear envelope. (From Tucker, 1979, with permission of Academic Press.)

genesis of centrioles. These bodies usually occur in pairs, positioned at right angles to each other. They can arise de novo but generally assemble adjacent to a preexisting centriole and at right angles to it. Centrioles (and, by extrapolation, other kinds of MTOC as well) thus appear to be endowed with the capacity to supply spatial information to neighboring elements, by mechanisms that are still quite obscure (Tucker, 1979; Raff, 1979; Brinkley et al., 1981; McIntosh, 1983).

The significance of microtubule-organizing centers in ordering the construction of the microtubule lattice is particularly well illustrated by the extension of neurites by cultured neuroblastoma cells. As mentioned above, the neurite, or axonal process, is a slender tube packed with microtubules, which together with neurofilaments are thought to determine the form of that process. When a freshly explanted neuroblastoma cell sends forth neurites, several remarkable things happen. First, the cytoplasmic MTOC (which are sometimes multiple in these cells) aggregate into one or two foci, from which the neurite microtubules will sprout. Further, once the neurites have extended, they define a shape that remains a permanent feature of the cell: should the cell be mistreated and induced to retract its neurites, they will be reconstituted on recovery in very nearly the same spatial position that they occupied originally. By the same token, should the original cell divide, the neurites of the two daughter cells emerge as mirror images of each other (Solomon, 1981). Such observations can be rationalized by assuming that MTOC define the spatial organization of microtubules with respect to some fixed plane and that the position and orientation of MTOC are conserved, constituting a kind of spatial memory that endures for several generations.

It is time to call a halt to the blizzard of particulars and consider how far we have come toward understanding the genesis of eukaryotic cell shape. Since every cell and organism is to a considerable degree unique, only the most general conclusions can apply broadly across the biological spectrum. It seems to be established that the shaping of eukaryotic cells is fundamentally bound up with the organization of their cytoplasm, as expressed in its cytoskeletal architecture. A cell that is growing or developing must necessarily model itself upon itself; therefore as the cytoskeleton is bent so will the cell grow. But this relationship is a very elastic one: it is the cytoskeleton as a whole, not a specialized class of morphogenetic macromolecules, that underlies the determination of shape. The cytoskeleton serves to position molecules and subcellular structures, both in the interior and at the periphery, and thereby orchestrates the medley of dynamic processes that collectively support growth. However, just as there is no single process that confers shape on all cells, we cannot assign a single morphogenetic function to the cytoskeleton. The arrangements seem almost infinitely adaptable, compatible with an endless variety of forms, and surely cannot be directly specified by the primary structure of the molecules that make up the cytoskeleton. Finally, the generation of each particular biological form is a singular and exceedingly intricate puzzle, whose solution consists of specifying which molecules are localized when, where, and how.

Localizing Morphogenetic Work

The preceding sections argued that the form of a cell is not determined directly by its macromolecular constituents, as the shape of a virus is, but reflects the manner in which the work of assembly is organized in space. This formulation skirts the thicket of causality: Does the cytoskeleton organize the cell, or the cell its cytoskeleton? The answer, surely, is both. Cells are at all times spatially organized to a high degree. Growth and development involve the progressive and coordinated transformation of one cytoskeletal framework into another with the aid of signals that impinge on the existing structure and encourage extension here, a change in direction there, or even total overhaul, as happens during the division of animal cells. Neither the constituents of the cytoskeleton nor the signals that regulate its activities and construction vary greatly in kind from one organism to another. Instead I believe that the diversity of forms chiefly reflects the manner in which regulatory signals are patterned in time and space.

There seems to be no special class of signals concerned with morphogenesis, but a growing body of observations implicates protein kinases, cAMP, the local proton concentration, and particularly Ca^{2+} ions in the regulation of cytoskeleton assembly and function. The microtubule lattice of animal cells, for instance, tends to disassemble when the concentration of free Ca^{2+} ions rises to the micromolar level. Calcium lability is mediated by calmodulin, which is built into the cytoskeleton as one of the microtubule-associated proteins, and is counteracted by certain other MAPs (Schliwa et al., 1981; Kakiuchi and Sobue, 1983). That these effects may be highly localized was demonstrated recently by Keith et al. (1983), who injected calcium-saturated calmodulin into fibroblasts and observed disruption of both microtubules and stress fibers in the immediate vicinity of the injection site. The importance of Ca^{2+} ions in local contractile activities and in the assembly of actin microfilaments was pointed out in several previous chapters, but one should add here that diverse observations implicate the local pH as an alternative or synarchic signal (Regula et al., 1981; Vacquier, 1981; Begg et al., 1982). Several cytoskeletal proteins are subject to phosphorylation by kinases that may or may not require cAMP; the altered shape of avian cells infected by Rous sarcoma virus may result from the phosphorylation of vinculin (Chapter 13). One can well imagine either endogenous signals or environmental ones (hormones, chemotropic agents) being conveyed in the cytoskeleton via a local phosphorylation event or by local changes in the cytoplasmic pH or the calcium concentration.

Biochemical dissection reveals the kinds of processes that underlie the assembly and activities of cytoskeletal elements but sheds only a fitful light on the genesis of actual forms. The higher levels of order that spell the difference between one form and another are lost when one grinds a cell into a homogenate and can only be approached by methods that respect the integrated workings of the intact organism. One of the most fruitful avenues open to us is the study of localized ion currents, for these often monitor a set of morphogenetic signals without perturbing the course of events.

The germinating zygotes of the marine brown algae *Fucus* and *Pelvetia* were introduced in Chapter 13. It will be recalled that the fertilized egg is spherical and apolar but develops a permanent axis of polarity within about 12 hours. The morphological expression of this polarity is a protuberance that will eventually develop into the rhizoid cell (Figure 13.13), but the axis has actually been fixed an hour or two before local outgrowth begins. Axis fixation is unaffected by inhibitors of protein synthesis or of microtubule assembly but is inhibited by cytochalasins B and D, suggesting that it involves microfilament assembly. The physical basis of the polar axis is uncertain, but an early cytological manifestation is the establishment of a directional transport pathway that carries a special set of vesicles from the Golgi apparatus to the site of rhizoid outgrowth. These vesicles contain fucoidin, a sulfated polysaccharide rich in fucose, which is incorporated into the wall of the developing rhizoid cell and contributes to its adhesive properties. Cytochalasin-treated embryos fail to develop this pathway (Quatrano, 1978; Brawley and Quatrano, 1979).

The generation of polarized form and functions is interwoven with the development of the transcellular ion current (Figure 13.14). As was outlined in Chapter 13, the fertilized egg develops a polarized pattern of ion flow that precedes the fixation of the axis and predicts its polarity. Part of the steady transcellular current is carried by a circulation of Ca^{2+} ions: Ca^{2+} ions are expelled from the major part of the egg's surface and flow into the presumptive rhizoid. Since outgrowth tends to occur in a region of maximum Ca^{2+} influx, it is probable that Ca^{2+} influx helps to localize the axis of polarity and all subsequent development. How does the zygote generate this polarized calcium current, and how may the current in turn amplify the polarity of the zygote? In an illuminating study, Brawley and Robinson (1985) combined measurements of current flow with cytochemical techniques that display actin filaments and discovered that cytochalasin prevents the establishment of both the characteristic ion current and a cortical net of actin filaments. There appears to be an intimate reciprocal relationship between calcium channels that convey spatial information and the cortical actin filaments, which may serve to fix the position of these channels and to orient subsequent work functions. From this and earlier research Brawley and Robinson deduce a very tentative sequence of events for the polarization of *Fucus* eggs, which is summarized here with slight modifications (Figure 14.11).

Immediately after fertilization, membrane patches bearing ion channels are randomly distributed over the egg's surface. In response to an external gradient, such as unilateral light, an asymmetric distribution of ion channels is established; calcium channels aggregate in the shaded half and current begins to flow into the presumptive rhizoid pole. This is axis formation; its mechanism is by no means clear but seems to involve the redistribution of the ion channels in concert with the development of a cortical mesh of actin filaments. Ca^{2+} ions are an important component of the inward current, though they carry but a small fraction of the total charge flux. Calcium ions, perhaps with the aid of calmodulin, stimulate contractile activities and/or polarized assembly of an actin

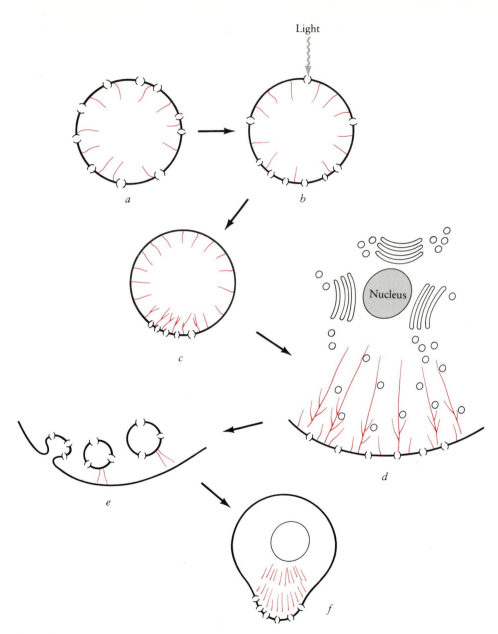

FIGURE 14.11   Polarization of the *Fucus* zygote, a hypothesis. (*a*) In the fertilized egg, Ca^{2+} channels are uniformly distributed. (*b*) After illumination, membrane patches bearing Ca^{2+} channels migrate into the shaded half by a cytochalasin-sensitive process. (*c*) Ca^{2+} channels have clustered at the presumptive rhizoid pole and initiate the growth of a network of actin filaments. (*d*) Golgi vesicles bearing among other things Ca^{2+} channels are guided to the presumptive rhizoid pole, where they fuse with the plasma membrane (*e*). (*f*) Well-developed actin cables extending to the nucleus; rhizoid outgrowth has begun. (Based on an unpublished sketch by Susan Brawley.)

network at the site where outgrowth will occur. Exocytosis of Golgi vesicles carrying precursors for wall and membrane and synthesis occurs at the site of Ca^{2+} influx, permitting that part of the surface to expand in response to turgor pressure. These Golgi vesicles may also bring additional ion channels, whose incorporation amplifies the polarizing effects of Ca^{2+} influx and reinforces the vectorial assembly of microfilaments and the oriented transport of vesicles. The microfilament mesh progressively extends into the cytoplasm until it impinges on the nucleus. Substances of unknown nature, presumably including specific gene products, are deposited in such a manner as to render the polar axis permanent; the cytoplasmic regions corresponding to the future rhizoid and thallus cells have become significantly different. Cytokinesis places identical genomes into this differentiated cytoplasm, in which the two nuclei will express somewhat different sets of genes. The end result is the production of a differentiated, polar embryo containing two distinct cells, one of which is the precursor of the thallus while the other gives rise to the rhizoid.

It is unlikely that the detailed course of events in *Fucus*, whatever it proves to be, is duplicated in any other organism. But the interplay of localized ionic signals from the plasma membrane with architectural elements of the cytoskeleton must be a widespread feature of morphogenesis. During the elongation of pollen tubes, for example, calcium ions are thought to flow into the tip; this may regulate both the exocytosis of apical vesicles and the growth or activities of microfilaments in that region (Jaffe et al., 1975; Picton and Steer, 1982, 1983). More surprisingly, the preceding outline runs broadly parallel to the hypothesis put forward by Bergmann et al. (1983) in an attempt to understand an unrelated morphogenetic sequence, the changing shapes of fibroblasts on the move. Briefly, when a space is opened in a confluent monolayer of fibroblasts, the cells begin to migrate into the wound. Before they do so, they reposition the Golgi apparatus and the microtubule-organizing center forward of the nucleus in the direction of migration. This presumably represents the response to a signal of unknown nature that perturbs the cytoskeleton in a directional manner and leads to its reorientation. Whether ion fluxes and microfilaments play a role in this is not known. In any event, microtubules grow out from the MTOC, guiding Golgi vesicles to the cell's new leading edge, where fresh membrane is inserted and allows the surface to expand. This must be coordinated with the extrusion of cytoplasm into the bulge, perhaps again in response to an ion flux. It would be illuminating to monitor the motile behavior of fibroblasts with a vibrating probe, but their small size precludes such experiments for the present.

The preceding account was intentionally phrased to emphasize the known role of Ca^{2+} ions, and perhaps of protons, as chemical signals that may confer polarity on chemical processes associated with the cytoskeleton. But we do not know for certain how ion currents impress their own polarity on morphogenetic events, and while chemical signaling appears to be the most plausible mechanism, it need not be the only one. For over a decade, Jaffe has emphasized that the separation of calcium pumps from calcium leaks will impose an electric field across the cytoplasm, estimated at about 1 mV across a *Fucus* egg (Jaffe et al.,

1974). This field in itself may provide a direction for morphogenesis by exerting force on charged particles within its range: polarization could in principle be effected by the electrophoresis of cytoplasmic vesicles or by the redistribution of mobile proteins embedded in the plasma membrane. Support for this view comes, among other things, from the well-known fact that applied electric fields sometimes impose their own polarity on developing cells and organisms (Lund, 1947; Jaffee and Nuccitelli, 1977; Brower and Giddings, 1980; Hinkle et al., 1981). It is now established that electric fields can redistribute mobile membrane proteins and perhaps even cytoskeletal elements (Poo, 1981; Luther et al., 1983). Applied fields may mimic or override the actions of endogenous ones; however, they may also promote ion fluxes or protein redistribution in an entirely unphysiological manner, so that the meaning of the biological effects of applied fields is open to question. Nevertheless, there are two instances in which the electric fields generated by endogenous ion currents appear to be implicated in a morphogenetic process. One is the developing oocyte of the moth *Cecropia*. There is a difference in electric potential of about 10 mV across the cytoplasmic bridge that links the oocyte proper to the surrounding nurse cells, and macro-molecules appear to migrate through this bridge by electrophoresis (Woodruff and Telfer, 1980). The other instance is the galvanotaxis of certain embryonic cells and fibroblasts (Stump and Robinson, 1983; Erikson and Nuccitelli, 1984). It is not clear how these cells sense electric fields as small as 10 mV/mm, only 0.2 mV across the cell's width, but we do know that fields of this magnitude are generated by developing embryos and by wounded tissue, so that galvanotaxis of animal cells seems likely to be involved in both embryogenesis and wound healing. As more information accumulates, we may find that living organisms have utilized both chemical signals and electric fields to impose spatial order by means of a localized ion flux.

A final caveat is in order here. Morphogenesis is often associated with transcellular electric currents, in part because ion fluxes are a widespread class of biological signals. But there is no reason to expect an obligatory connection between morphogenesis and bioelectric fields or transcellular ion currents. Morphogenesis requires spatially localized signals to localize work functions, but these signals need not always take the form of a localized ion flux.

Inheritance of Spatial Order

For the past 40 years our understanding of the way living things organize themselves and transmit order from one generation to the next has been formulated almost exclusively in terms of genetic information. We know that this information is encoded in the linear sequence of purine and pyrimidine bases of DNA, transcribed into another linear sequence of nucleotides in RNA, and ultimately translated into the sequence of amino acids in proteins. Its expression is subject to various controls, likewise built into the genome, so that the genetic program embodies not only the kind of information available but also helps to determine the timing and extent of its retrieval. This linear and unidirectional conception of information flow proved so immensely productive and the analogy with everyday actions such as writing so compelling as to generate the prevailing impression that genetic information is all there is to biological order. At least among biologists with a molecular bent, a majority would probably still subscribe to the dictum enunciated 20 years ago by Joshua Lederberg (1966): "The point of faith is this: make the polypeptide sequences at the right time and in the right amounts, and the organization will take care of itself." Is this faith compatible with present knowledge?

Let us define more precisely what is at issue with respect to the relationship between a cell's genome and its form. The question is not whether cell shape is inherited; of course it is. Nor is there any doubt that the transcription and translation of genetic information, at the proper time and to the proper extent, play a crucial role in growth, development, and differentiation. Cellular constituents (simple ions aside) are either themselves nucleic acids or the products of enzymes whose structures and catalytic activities are determined by nucleic acids. In this sense, all cellular constituents and functions can be said to be genetically determined. But this nevertheless leaves open the questions of whether instructions for the growth and organization of a cell are explicitly embodied in its DNA; and whether the linear flow of information from DNA to RNA to protein provides a sufficient framework for inquiring what actually happens when a cell grows and divides. I believe that the answer to the latter questions is no. As we ascend the great chain of being from viruses through bacteria to protists and higher organisms, there is a progressive loosening of the connections between genes and morphology. The structural context in which genetic information is expressed looms progressively larger, to the point that morphology cannot be understood as the direct expression of genetic instructions but must be seen as reflecting the workings of the whole cell or organism. The object of the present section is to justify this opinion, to document it with a few examples, and to explore how it bears on the ways in which living things convert energy into organization. It must be added that the "structuralist" conception of living organisms, to borrow a term from Webster and Goodwin (1981), is today a minority view, but one with a long tradition. It stems from the work of Hans Driesch and D'Arcy Thompson at the turn of the century and includes such illustrious contemporaries as C. H. Waddington and the late Tracy Sonneborn, whose research provided the first insights into the molecular basis of nongenomic heredity.

If it is true that the form of an organism is not explicitly spelled out by its genetic constitution but arises implicitly (or epigenetically) in the context of its growth and development, then it becomes easier to understand why the genetic approach to cellular morphogenesis has proven so disappointing. The obvious procedure, the isolation of morphologically aberrant mutants, is not likely to reveal specific morphogenetic macromolecules because by and large these do not exist as such. An instructive example is provided by a series of *Neurospora crassa* mutants that exhibit a grossly abnormal pattern of branching and are deficient in the synthesis of cyclic AMP; in several of these, exogenous cAMP restores the normal morphology (Scott, 1976; Rosenberg and Pall, 1979). The observations certainly suggest that cAMP is required for the normal branching of hyphae, but it would be an error to infer that the level of this nucleotide controls the location or frequency of branch points. Cyclic AMP is a regulator of many cellular functions, and a deficiency is likely to impinge on cytoskeletal structure, intracellular transport, vesicle fusion, or even ion currents. The mutation is thus best interpreted as one that perturbs the context within which many other genes find expression, and it does not carry us very far forward. Virus assembly aside (King, 1980), I know of no case where a gene product can be assigned a specifically shape-determining function. Examples no doubt exist, but they appear to be uncommon. A more productive approach has been to introduce lesions into the cell's structural framework, leaving the genome intact, and to inquire whether the resultant morphological alterations are inherited, for such

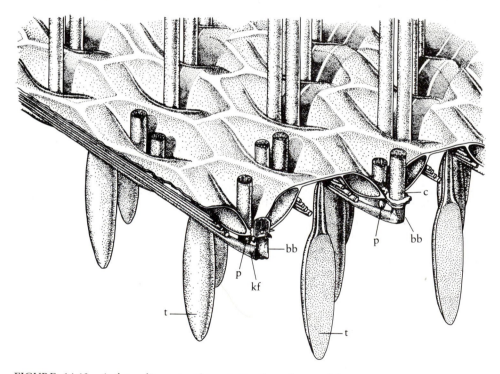

FIGURE 14.12   A three-dimensional reconstruction of part of the cortex of *Paramecium*. The anterior end of the animal is to the upper left, and the kineties run diagonally in that direction; some ciliary units have pairs of cilia. Appendages marked are the cilia (c), basal bodies (bb), kinetodesmal fibers (kf), parasomal sac (p), and trichocysts (t); microtubular bands have been omitted. (After Jurand and Selman, 1969, with permission of Macmillan Publishing, London.)

lesions identify circumstances in which the structural context is crucial to the correct expression of the genetic information. Not many organisms lend themselves to the requisite microsurgical procedures or to the unambiguous identification of their consequences. Most of the information at hand comes from work with ciliated protozoa but is pertinent to cell biology in general. Authoritative reviews of the literature have been published by Sonneborn (1970, 1977), Beisson (1977), Nanney (1977, 1980), de Terra (1978), Aufderheide et al. (1980), and Frankel (1984); let a handful of examples suffice here.

Structural Inheritance.   The surface of *Paramecium* is studded with thousands of cilia, whose synchronized beating underlies the animal's swimming and behavior (Chapter 11). The cilia are placed not at random but in longitudinal rows called kineties. Each cilium springs from a kinetosome, or basal body, inserted into the organism's cortex; this is the focus of an organized territory about 1 μm in diameter, which contains in addition to the kinetosome and cilium a set of accessory structures: the rootlet (or kinetodesmal fiber), a pair of microtubule ribbons, and also the parasomal sac. Within each row, trichocysts alternate with the cilia (Figure 14.12). All these structures are precisely oriented with respect to one another, and successive ciliary units have identical polarities with respect to the body axes of the animal (Figure 14.13). One would expect the maintenance and precise reproduction of this complex unit to require both the production of molecular building blocks and the proper conditions for their insertion, and gene mutations that affect either of these have been obtained. More instructive for present purposes are the remarkable experiments of Beisson and Sonneborn (1965), who succeeded in surgically inverting the polarity of all the kinetosomes along a segment of one or more kineties. This was done, as Figure 14.13 suggests, by grafting onto one animal a small piece of cortex taken from a sister cell of the same genotype, inserting it into the host cell with its orientation reversed. The initial graft contains only a few kinetosomes but in the course of a few divisions gives rise to complete kineties of reversed polarity. Reversed kineties are inherited, as is the twisty mode of swimming that they bestow; both have been maintained without reversion for as long as 800 generations.

Sonneborn (1970) subsequently subjected the animals to extensive backcrosses designed to ensure that normal animals and those bearing reversed kineties were identical with respect to both genotype and any possible informational macromolecules in the cytoplasm. It is hard to evade the conclusion that inheritance of the reversed pattern is independent of changes in either genetic information or gene expression but reflects the continuity of structural order within the cortex from one generation to the next. Evidently cells with identical genes and even with identical molecules can exhibit quite different hereditary patterns. Sonneborn coined the term "cytotaxis" to describe the "ordering and arranging of new cell structure under the influence of preexisting cell structure." This particular example, however, is more aptly described (Beisson, 1977) as structural inheritance.

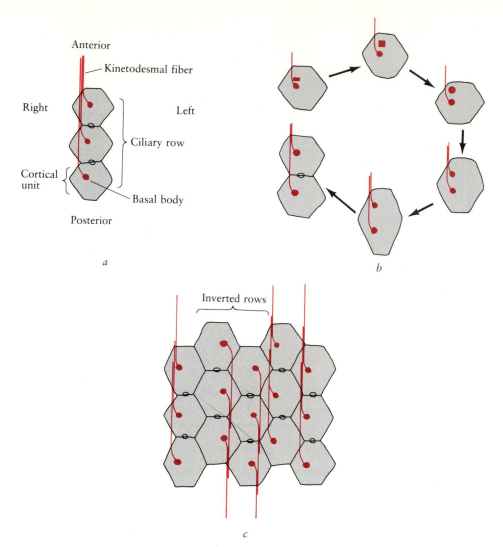

FIGURE 14.13 Normal and inverted ciliary rows. (*a*) Schematic representation of simplified ciliary units in a row. Note the asymmetry of each unit and the polarity of the resultant ciliary row. (*b*) Replication of a ciliary unit. A new basal body (rectangular element) appears within a preexisting unit in a precise spatial relationship to other structures. (*c*) Portion of a ciliate cortex with two inverted ciliary rows. (From Nanney, 1977, with permission of the *Journal of Protozoology.*)

In principle at least we understand quite well how a small graft of reversed cortex gives rise to entire reversed kineties that can persist indefinitely. Cilia, as mentioned above, always sprout from kinetosomes, and the manner of kinetosome duplication holds the key to this instance of structural inheritance. A new kinetosome assembles close to an existing one according to a definite morphogenetic sequence, illustrated in Figure 14.14 for *Tetrahymena* (Allen, 1969). I would emphasize that the old kinetosome does not grow and divide, nor does it replicate itself like a chromosome, but it does unmistakably provide structural guidance for the assembly of the new kinetosome. Assembly takes place at right angles to the old kinetosome and with a definite polarity—anterior in the case of a normal kinetie, posterior in a reversed one. Assembly continues in linear fashion until a whole row of kinetosomes has been produced. In each instance the site where the new kinetosome develops and its polarity are determined only by the microgeography of the cortical unit around the original kinetosome. Like centrioles, which they closely resemble in structure, kinetosomes are microtubular and can be described as microtubule-organizing centers, but just how they do this, in the absence of visible continuity between the old kinetosome and the new, remains a mystery.

Inverted kineties in *Paramecium* are the classic example of structural inheritance but are by no means the only one. Aufderheide et al. (1980) review several others from the protozoan world, including the propagation of inverted ciliary rows in *Tetrahymena,* the positioning of the contractile vacuole with respect to the kineties, the localization of trichocysts and mitochondria, and aspects of the formation of the oral apparatus. The unifying thread is that in every case the site of assembly and development of a new cellular structure is determined by interaction of the molecules entering the new unit with the existing structural framework. The dependence is not absolute: kinetosomes, for instance, can arise de novo. Nevertheless, we appear to be dealing here with a principle of broad significance to growth and development. For structural inheritance is not confined to protozoa. I shall pass over the enormous literature that documents cortical patterning in animal eggs and embryos, but would again draw attention to the discovery that animal cells in culture give rise to daughter cells whose shapes (Solomon, 1981) and movements are related to each other as mirror images. Observations of this kind seem to be best understood in terms of constraints on the positioning of new centrioles, which in turn determine the organization of the daughter cell's cytoskeleton.

At the risk of belaboring the obvious, let me restate what appears to be the general principle in terms of the flow of genetic information. The existing spatial order, or microgeography (Sonneborn, 1970), constrains the expression of a function specified by genes and thus adds a measure of determinacy that was not present in the coding sequence. But it is not really a matter of specific supplementary information that might be attributed to nonnucleic-acid inheritance. Rather, structural inheritance reflects the fact that cell constituents can take up only a limited number of stable positions in space, namely those permitted by the preexisting order. It is in this sense, limited but far from trivial, that we can speak of the modulation of the genetic message by its context.

Global Order.   The recognition that except for some cases of virus replication gene products must generally be inserted into an existing structural framework has profound implications for the relationship between genes and form. But one must question whether structural inheritance, of the local variety described above, suffices to account for the generation and transmission of order at the next higher level, that of the whole cell. Here again protozoa are the most fertile field of inquiry, thanks in part to the very complexity of their organization. Not only do protozoa precisely reproduce their elaborate architecture by growth and division, many are also capable of feats of reorganization and regeneration that smack a little of magic.

Oxytricha, a single cell some 100 μm in length, is one of the more sophisticated ciliates. Figure 14.15a shows the ventral surface, with the oral apparatus anterior and bounded by a band of fused cilia called membranelles. In addition it bears several well-defined groups of cirri, each consisting of some 20 cilia blended together like the hairs of an artist's brush, which are the main organs of locomotion. There are two marginal bands of cirri plus eight frontal cirri, four ventral ones, and six transverse. Division of the organism involves duplication of the entire organization in the posterior half according to the sequence indicated in schematic and simplified form in Figure 14.15a. That some kind of large-scale cytotaxis is involved here, despite the long distance, is inferred from the remarkable fact that under certain circumstances there arise stable, self-perpetuating duplications of the entire structure. Occasionally, either by accident or in consequence of experimental manipulation, daughter cells at division fail to separate. They then proceed to fuse back to back, generating a single cell with two ventral surfaces, each complete with oral apparatus and the full panoply of cirri (Figure 14.15b). Once formed, the doublet is stable and can reproduce itself indefinitely. Doublet formation has nothing to do with mutation, as Sonneborn was able to show in *Paramecium,* which forms doublets that are sexually competent. Rather, it is an alternative and stable pattern of global organization that is brought about by cortical reorganization rather than by genetic change.

Can one then attribute the large-scale ordering of the cortex to structural guidance, as was done in the case of the ciliary rows? Probably not. *Oxytricha* is capable of forming resting cysts in which all visible trace of the original ciliature vanishes, to reappear on excystment: when both singlets and doublets were carried through the encystment procedure, singlets reemerged as singlets and doublets as doublets. Within the limits of present-day electron microscopy, one must conclude that inheritance of the global ciliary pattern does not depend on persistence of the existing ciliature. It does, however, require the structural continuity of the cortex, specifically of a particular portion of the cortex that has the capacity to organize the entire surface.

The extent of this "determinative region" and the degree to which a part of

FIGURE 14.14   Diagram of basal body propagation in *Tetrahymena*, seen from above, from the side, and in cross section. Note that the new basal body arises close to the old one and at right angles to it. As the new basal body grows, it tilts and migrates anteriorly. The cilium and ancillary structures form after the new basal body has assumed its final position. (From Allen, 1969; reproduced from *Journal of Cell Biology* with permission of the Rockefeller University.)

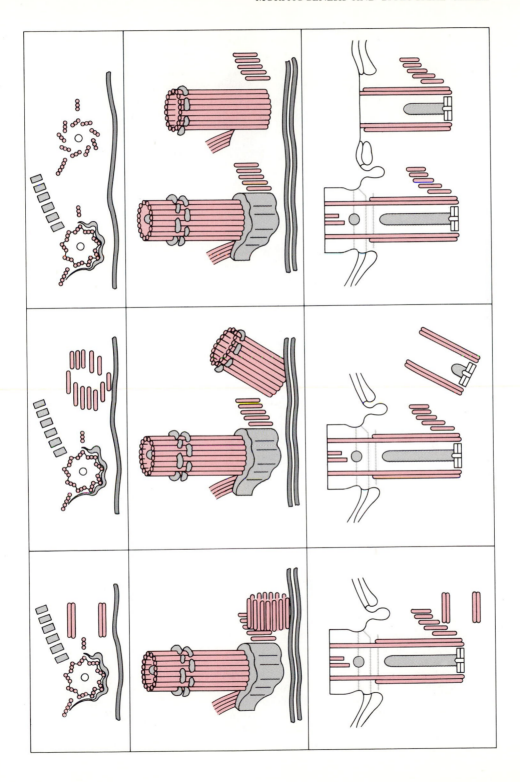

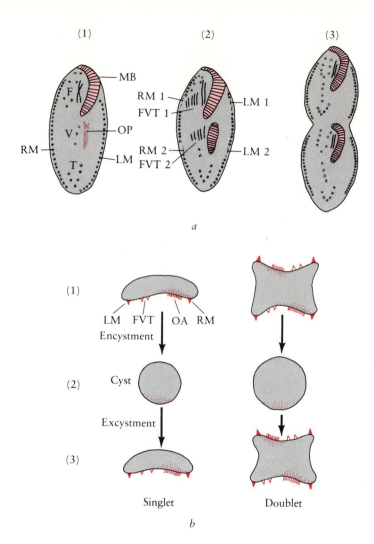

FIGURE 14.15   Transmission of cortical organization in the ciliated protozoan *Oxytricha*. (*a*) Normal development of the ventral ciliature during division. Stage 1, before the initiation of new surface development, shows the position of the oral apparatus with its membranelle band (MB), the oral primordium (OP), which will give rise to the new oral apparatus, and the cirri arranged into the left and right marginal bands (LM, RM) as well as the frontal, ventral, and transverse clusters (F, V, T). In stage 2, just before the onset of division, these structures are being duplicated in the organism's posterior half. The new cirri arise from longitudinal streaks of basal bodies, which have been omitted for clarity. Stage 3, cytokinesis after duplication is complete. (*b*) Schematic and greatly simplified cross sections of singlet (normal) and doublet organisms. The extent of the postulated "determinative region" is roughly indicated by stippling. (After Aufderheide et al., 1980, with permission of the American Society for Microbiology.)

the cell's cortex contains within it the pattern of the whole organism must be inferred primarily from studies on the regeneration of *Oxytricha, Stentor,* and other ciliates following major surgery. There is not space here to recapitulate these astonishing experiments, some a century old; readers who wish to pursue this matter will find an introduction in the reviews by de Terra (1978) and Aufderheide et al. (1980) and in a recent book by Nanney (1980). Suffice it here to say that grossly maimed animals and even fragments containing but one hundredth of the original volume can reorganize themselves to produce small but complete and viable organisms. What is required is only that the fragment include part of the macronucleus and a portion of the cortical surface, the latter corresponding to the determinative region. In *Oxytricha* the determinative region is invisible, but in *Stentor* it is clearly marked by the zone of "stripe contrast," the junction between the broad and the narrow longitudinal stripes that ornament the surface of this organism (Figure 14.16).

What physical principles can account for the inheritance and reconstruction of large-scale patterns? At the least it seems necessary to go beyond local structural guidance to invoke some kind of architectural master plan that governs the spatial organization of the cell as a whole. Joseph Frankel in particular has argued persuasively that the cortex of ciliates has properties traditionally ascribed to the morphogenetic fields of developing embryos (Fran-

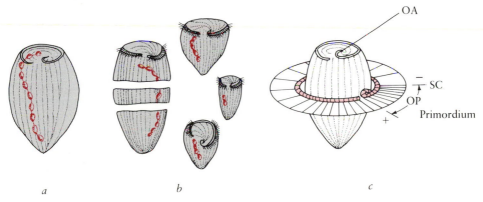

a *b* *c*

FIGURE 14.16   Gradients of positional information in the large ciliate *Stentor*.
(*a*) Schematic drawing of the organism showing the oral apparatus, the elongated nodular macronucleus, and the alternating bands of clear and pigmented stripes. The clear stripes contain rows of pigment granules and are graded in width around the *Stentor*. The widest and narrowest stripes meet at the "locus of stripe contrast." (*b*) Dissection of *Stentor* into parts is followed by regeneration of small but otherwise normal organisms. (*c*) Gradients of positional information. Although superimposed on the form of *Stentor*, this drawing is intended merely to suggest the concept. Two gradients are shown. One is unipolar, emanating from the oral apparatus (OA). Dashed arcs mark lines of constant positional information. The second is circular, emanates from the locus of stripe contrast (SC), and corresponds to the pigment stripes. The position of a new oral primordium (OP) is always on the + side of SC. (After de Terra, 1978, Goodwin, 1980, and Frankel, 1974, with permission.)

kel, 1974, 1982, 1984; Aufderheide et al., 1980; Frankel and Nelsen, 1981). These are (1) a portion of the determinative region can regenerate the entire set of ciliary structures, (2) the number and size of the structures produced are responsive to the size of the whole cell, (3) the potential for development of surface structures is not confined to the place where these normally arise but extends beyond this region, and (4) the developmental potential is itself graded in space. This line of argument suggests that the cortex of a ciliate protozoan constitutes a field of positional information, an idea originally developed by Wolpert (1969) in the context of multicellular embryos and extended here to the level of a single cell. In this view certain regions of the organism, particularly its ends and special zones such as the oral primordium and the locus of stripe contrast, serve as reference points or boundaries. They set up gradients of some condition, possibly but not necessarily a gradient of some diffusible substance, as suggested diagrammatically in Figure 14.16. The "value" of the field at any point on the cell surface, determined by "reading" the various gradients, would specify what may take place at this point—for instance, the positioning of cilia or cirri.

The idea is not without difficulties but seems to be the only one that can, at least in principle, account for that extraordinary capacity of maimed cells and tiny fragments to reconstitute the whole, for a gradient could retain its proportions even on a reduced scale. We know virtually nothing about the physical nature of morphogenetic gradients, nor is it obvious how a single cell might read out and respond to positional information. The extensive mathematical discussion of this subject has been conducted in terms of diffusible morphogens, but the substances themselves remain hypothetical; besides, how could such gradients persist in a streaming cytoplasm? A more appealing alternative is that localized ion pumps and channels, with their dynamic pattern of ionic gradients and electric fields, underlie those global morphogenetic fields that convey positional information (Jaffe, 1979, 1981; Aufderheide et al., 1980). It would be interesting to know whether regions that are determinative in the morphogenetic sense correspond to zones of discontinuity in the pattern of transcellular electric currents—points of current entry, for example.

If there is merit to the proposal that the ciliate cortex constitutes a field of positional information, then we must accommodate additional degrees of freedom between what the genes prescribe and the organisms's form: the value of some gradient, not the shape or specificity of some protein, would determine what goes where. And we may proceed still further, following Goodwin (1980), whose provocative thesis is that the multitude of forms displayed by the various ciliates can all be encompassed by a single mathematical formula, Laplace's general field equation. Each organism represents a particular set of boundary conditions but all conform to the same general rules of order. This is a passage that beckons into uncharted waters, far off the present mainstream of biology. For it implies that we should consider an organism in its totality as an integrated field that exerts over its genome a measure of dominance not dreamt of in molecular philosophy.

Epilogue

Where are we then? We set out at the beginning of this final chapter in the hope that an inquiry into the mechanisms by which cells and simple organisms grow and shape themselves would also illuminate the enigma of biological order. What do we mean when we say that living things convert energy into organization? Or, as Erwin Schrödinger put it 40 years ago, that life feeds on negentropy? Make no mistake about it: cryptic as they are, these pronouncements touch on the essence of life. Living things differ from nonliving ones most pointedly in their capacity to maintain, reproduce, and multiply states of matter characterized by an extreme degree of organization. "Life *is* order, pure and simple"; thus spake Riedl. But an experimental biologist will not be content with such abstractions, however deftly they slip the kernel of truth into the nutshell of epigram: one wants to know just how the trick is done.

In Chapter 1, Harold Morowitz was quoted in explication of Schrödinger to the effect that living things must constantly perform work in order to maintain their highly improbable state; by the same token, work must be done to generate spatial order. What work? That depends on the level of biological organization to which we direct our attention. At the molecular level we recognize the asymmetric and often unique shapes of nucleic acids and proteins as the foundation of all spatial order. The replication, transcription, and translation of genetic information is therefore the most fundamental morphogenetic mechanism, and it suffices to specify the shapes of macromolecules, of simple viruses, and of some organelles that arise by the coming together of prefabricated units. Cells, whether prokaryotic or eukaryotic, represent a higher level of order, for their forms embody, over and above the molecular patterns, traces of dynamic processes. Cellular morphology is the end result of diverse yet integrated work functions: not only the biosynthesis of macromolecules but also their translocation to specific destinations, membrane flow, contractile forces, electrical activities, and others besides. What lends recognizable and reproducible form to the product of all this busyness is that, except for the simpler biosynthetic reactions, energy-linked functions are bound up with the structural matrix of the cell. Work is at all times performed within a framework that already possesses spatial order and transmits this order to the product of its activities. Cell envelope and cytoskeleton serve as a scaffold that positions molecules, guides translocation, and confers direction on growth as a whole. Even that extraordinary *Fucus* egg, imposing polarity on itself where there was none before, draws on an external gradient to instruct its preexistent organized cytoplasm. There is therefore no identifiable mechanism for converting energy into spatial organization, in the same sense that we can speak of the mechanism of photosynthesis or of protein synthesis. All we can find is a variety of patterned signals—chemical as well as electrical, from the genome and from the external environment—that impinge on the spatially ordered procedures of cell physiology. By augmenting one process here, inhibiting another there, conferring a new direction on a third, each cell models itself on itself, and its form is the product of some particular combination of forces and signals.

I assuredly do not mean to imply that it is bootless to study morphogenetic mechanisms. We learn a lot of biology thereby, especially how one organized structure reproduces itself or transforms itself into another organized structure. But we will not learn how spatial order arises unless and until we penetrate that ultimate mystery, the origin of life. For the present, the organization of every cell and all its works stands as a fundamental property of life, one of the givens of biology. All this was already recognized by Schrödinger in his own elliptic fashion: the basic characteristic of life, he said, is the production of order from order, and the only source of biological order is biological order.

Every organism transmits its particular forms, actual and potential, to its offspring. The genome is the sole repository of hereditary information and must ultimately determine form, subject only to limited modulation by the environment. But the inquiry into just how the genome does this leads through another set of Chinese boxes, to show the innermost one empty. There seem to be no "morphogenes," genes uniquely concerned with the specification of shape. Rather, if I may once more paraphrase Gunther Stent, the discovery of how the genes determine form has already been made: genes direct the synthesis of macromolecules, proteins in particular. Gene products come into a preexisting organized matrix consisting of previous gene products, and their functional expression is channeled by the places into which they come and by the signals they receive. Form is not explicitly spelled out in any message but is implicit in its combination with a particular structural context. At the end of the day, only cells make cells.

Struggling with this chapter brought to mind one of the tales recounted by the late Alan Watts in his book *The Way of Zen*. A novice monk comes to the Master and asks, "What is the Buddha?" The Master replies, "Go eat your gruel." "I don't understand," says the perplexed student. "If you don't understand, when you have eaten wash your bowl."

References

ADAMS, A. E. M., and PRINGLE, J. R. (1984). Relationship of actin and tubulin distribution to bud growth in wild-type and morphogenetic-mutant *Saccharyomyces cerevisiae*. *Journal of Cell Biology* 98:934–945.

ALBERTS, B., BRAY, D., LEWIS, J., RAFF, M., ROBERTS, K., and WATSON, J. D. (1983). Op. cit., Chapter 11.

ALLEN, R. D. (1969). The morphogenesis of basal bodies and accessory structures of the cortex of the ciliated protozoan *Tetrahymena pyriformis*. *Journal of Cell Biology* 40:716–733.

AUFDERHEIDE, K. J., FRANKEL, J., and WILLIAMS, N. E. (1980). Formation and positioning of surface-related structures in protozoa. *Microbiological Reviews* **44**:252–302.

BEGG, D. A., REBHUN, L. I., and HYATT, H. (1982). Structural organization of actin in the sea urchin egg cortex: Microvillar elongation in the absence of actin filament bundle formation. *Journal of Cell Biology* **93**:24–32.

BEISSON, J. (1977). Non-nucleic acid inheritance and epigenetic phenomena, in *Cell Biology*, vol. 1, L. Goldstein and D. Prescott, Eds. Academic, New York, pp. 375–421.

BEISSON, J., and SONNENBORN, T. M. (1965). Cytoplasmic inheritance of the organization of the cell cortex in *Paramecium aurelia*. *Proceedings of the National Academy of Sciences USA* **53**:275–282.

BERGMANN, J. E., KUPFER, A., and SINGER, S. J. (1983). Membrane insertion at the leading edge of motile fibroblasts. *Proceedings of the National Academy of Sciences USA* **80**:1367–1371.

BRAWLEY, S. H., and QUATRANO, R. (1979). Sulfation of fucoidin in *Fucus* embryos. IV: Autoradiographic investigation of fucoidin sulfation and secretion during differentiation and the effect of cytochalasin treatment. *Developmental Biology* **73**:193–205.

BRAWLEY, S. H., and ROBINSON, K. R. (1985). Cytochalasin treatment disrupts the endogenous currents associated with cell polarization in fucoid zygotes. *Journal of Cell Biology*, **100**:1173–1184.

BRINKLEY, B. R., COX, S. M., PEPPER, A., WIBLE, L., BRENNER, S. L., and PARDUE, R. L. (1981). Op. cit., Chapter 12.

BROWER, D. L., and GIDDINGS, T. H. (1980). The effects of applied electric fields on *Micrasterias*. II: The distribution of cytoplasmic and plasma membrane components. *Journal of Cell Science* **42**:279–290.

BROWN, R. M., and WILSON, J. H. M. (1977). Golgi apparatus and plasma membrane involvement in secretion and cell surface deposition with special emphasis on cellulose biogenesis, in *International Cell Biology, 1976–1977,* R. B. Brinkley and K. R. Porter, Eds. Rockefeller University Press, New York, pp. 267–283.

BUTLER, P. J. G., and KLUG, A. (1978). The assembly of a virus. *Scientific American* **239**(5):52–59.

BYERS, B., and GOETSCH, L. (1975). Behavior of spindles and spindle plaques in the cell cycle and conjugation of *Saccharomyces cerevisiae*. *Journal of Bacteriology* **124**:511–523.

CABIB, E., ROBERTS, R., and BOWERS, B. (1982). Synthesis of the yeast cell wall and its regulation. *Annual Review of Biochemistry* **51**:763–793.

DANEO-MOORE, L., and SHOCKMAN, G. D. (1977). The bacterial cell surface in growth and division, in *The Synthesis, Assembly and Turnover of Cell Surface Components*, G. Poste and G. L. Nicolson, Eds. Elsevier–North Holland, Amsterdam.

DEROSIER, D. J., TILNEY, L. G., and EGELMAN, E. (1980). Actin in the inner ear: The remarkable structure of the stereocilium. *Nature* **287**:291–296.

DE TERRA, N. (1978). Some regulatory interactions between cell structures at the supramolecular level. *Biological Reviews of the Cambridge Philosophical Society* **53**:427–463.

DUSTIN, P. (1984). Op. cit., Chapter 12.

ERIKSON, C. A., and NUCCITELLI, R. (1984). Embryonic fibroblast motility and orientation can be influenced by physiological electric fields. *Journal of Cell Biology* **98**:296–307.

FRANKEL, J. (1974). Positional information in unicellular organisms. *Journal of Theoretical Biology* **47**:439–481.

FRANKEL, J. (1982). Global patterning in single cells. *Journal of Theoretical Biology* **99**:119–134.

FRANKEL, J. (1984). Pattern formation in ciliated protozoa, in *Pattern Formation, a Primer in Developmental Biology,* G. M. Malacinsky, Ed. Macmillan, New York, pp. 163–196.

FRANKEL, J., and NELSEN, E. M. (1981). Discontinuities and overlaps in patterning within single cells. *Philosophical Transactions of the Royal Society of London* **B295**:525–531.

GIDDINGS, T. H., JR., BROWER, D. L., and STAEHELIN, A. L. (1980). Visualization of particle complexes in the plasma membrane of *Micrasterias denticulata* associated with the formation of cellulose fibrils in primary and secondary walls. *Journal of Cell Biology* **84**:327–339.

GOODWIN, B. C. (1980). Pattern formation and its regeneration in the protozoa. *Symposia of the Society for General Microbiology* **30**:377–404.

GUNNING, B. S., and HARDHAM, A. R. (1982). Op. cit., Chapter 12.

HENNING, U. (1975). Determination of cell shape in bacteria. *Annual Review of Microbiology* **29**:45–60.

HEPLER, P. K., and PALEVITZ, B. A. (1974). Op. cit., Chapter 12.

HIGGINS, M. L., POOLEY, H. M., and SHOCKMAN, G. D. (1971). Reinitiation of cell wall growth after threonine starvation of *Streptococcus faecalis. Journal of Bacteriology* **105**:1175–1183.

HINKLE, L., MCCAIG, C. D., and ROBINSON, K. R. (1981). The direction of growth of differentiating neurones and myoblasts from frog embryos in an applied electric field. *Journal of Physiology* **314**:121–135.

JAFFE, L. F. (1979). Op. cit., Chapter 13.

JAFFE, L. F. (1981). Op. cit., Chapter 13.

JAFFE, L. F., and NUCCITELLI, R. (1977). Op. cit., Chapter 13.

JAFFE, L. F., ROBINSON, K. R., and NUCCITELLI, R. (1974). Op. cit., Chapter 13.

JAFFE, L. A., WEISENSEEL, M. H., and JAFFE, L. F. (1975). Calcium accumulations within the tips of growing pollen tubes. *Journal of Cell Biology* **67**:488–492.

JURAND, A., and SELMAN, G. G. (1969). *The Anatomy of Paramecium aurelia.* Macmillan, New York.

KAKIUCHI, S., and SOBUE, K. (1983). Op. cit., Chapter 13.

KEITH, C., DIPAOLA, M., MAXFIELD, F. R., and SHELANSKI, M. L. (1983). Microinjection of Ca^{2+}-calmodulin causes a localized deploymerization of microtubules. *Journal of Cell Biology* **97**:1918–1924.

KILMARTIN, J. V., and ADAMS, A. E. M. (1984). Structural rearrangements of tubulin and actin during the cell cycle of the yeast *Saccharomyces. Journal of Cell Biology* **98**:922–933.

KING, J. (1980). Regulation of structural protein interactions as revealed in phage morphogenesis, in *Biological Regulation and Development,* vol. 2, R. F. Goldberger, Ed. Plenum, New York, pp. 101–132.

KOCH, A. L. (1983). The surface stress theory of microbial morphogenesis. *Advances in Microbial Physiology* **24**:301–366.

KOCH, A. L., HIGGINS, M. L., and DOYLE, R. J (1981). Surface tension-like forces determine bacterial shapes: *Streptococcus faecium*. *Journal of General Microbiology* **123**:151–161.

KOCH, A. L., HIGGINS, M. J., and DOYLE, R. J. (1982). The role of surface stress in the morphology of microbes. *Journal of General Microbiology* **128**:927–945.

LAZARIDES, E., and REVEL, J. P. (1979). Op. cit., Chapter 12.

LEDERBERG, J. (1966). Remarks. *Current Topics in Developmental Biology* **1**:ix–xiii.

LUND, E. J. (1947). Op. cit., Chapter 13.

LUTHER, P. W., PENG, H. B., and LIN, J. J-C. (1983). Changes in cell shape and actin distribution induced by constant electric fields. *Nature* **303**:61–64.

MARCHANT, H. J. (1979). Microtubules, cell wall deposition and the determination of plant cell shape. *Nature* **278**:167–168.

MARCHANT, H. J., and HINES, E. R. (1978). The role of microtubules and cell-wall deposition in elongation of regenerating protoplasts of *Mougeotia*. *Planta* **146**:41–48.

MCINTOSH, J. R. (1983). The centrosome as an organizer of the cytoskeleton, in *Spatial Organization of Eukaryotic Cells*, J. R. McIntosh, Ed. *Modern Cell Biology* **2**:115–142; Alan R. Liss, New York.

NANNEY, D. L. (1977). Molecules and morphologies: The perpetuation of pattern in ciliated protozoa. *Journal of Protozoology* **24**:27–35.

NANNEY, D. L. (1980). *Experimental Ciliatology: An Introduction to Genetic and Developmental Analysis in Ciliates*. Wiley, New York.

PICTON, J. M., and STEER, M. W. (1982). A model for the mechanism of tip extension in pollen tubes. *Journal of Theoretical Biology* **98**:15–20.

PICTON, J. M., and STEER, M. W. (1983). Evidence for the role of Ca^{2+} ions in tip extension in pollen tubes. *Protoplasma* **115**:11–17.

POO, M. M. (1981). In-situ electrophoresis of membrane components. *Annual Review of Biophysics and Bioengineering* **10**:245–276.

PORTER, K. R. (1966). Op. cit., Chapter 12.

QUATRANO, R. (1978). Op. cit., Chapter 13.

RAFF, E. C. (1979). The control of microtubule assembly *in vivo*. *International Review of Cytology* **59**:1–96.

REGULA, C. S., PFEIFFER, J. R., and BERLIN, R. D. (1981). Microtubule assembly and disassembly at alkaline pH. *Journal of Cell Biology* **89**:45–53.

RIEDL, R. (1978). Op. cit., Chapter 1.

ROSENBERG, G., and PALL, M. (1979). Properties of two cyclic-nucleotide deficient mutants of *Neurospora crassa*. *Journal of Bacteriology* **137**:1140–1144.

SCHLIWA, M., EUTENEUER, U., BULINSKI, J. C., and IZANT, J. G. (1981). Calcium lability of cytoplasmic microtubules and its modulation by microtubule-associated proteins. *Proceedings of the National Academy of Sciences USA* **78**:1037–1041.

SCOTT, W. A. (1976). Biochemical genetics of morphogenesis in *Neurospora*. *Annual Review of Microbiology* **30**:85–104.

SHOCKMAN, G. D., DANEO-MOORE, L., and HIGGINS, M. L. (1974). Problems of cell wall and membrane growth, enlargement and division. *Annals of the New York Academy of Science* **235**:161–196.

SLOAT, B. F., ADAMS, A., and PRINGLE, J. R. (1981). Roles of the CDC24 gene product in cellular morphogenesis during the *Saccharomyces cerevisae* cell cycle. *Journal of Cell Biology* 89:395–405.

SOLOMON, F. (1981). Specification of cell morphology by endogenous determinants. *Journal of Cell Biology* 90:547–553.

SOLOMON, F., and MAGENDANTZ, M. (1981). Cytochalasin separates the disassembly of microtubules from loss of asymmetric morphology. *Journal of Cell Biology* 89:157–161.

SONNEBORN, T. M. (1970). Gene action in development. *Proceedings of the Royal Society of London* B176:347–366.

SONNEBORN, T. M. (1977). Ciliate morphogenesis and its bearing on general cellular morphogenesis. *Cell Surface Reviews* 4:327–355.

STENT, G. S. (1978). Op. cit., Chapter 13.

STEPHENS, R. E., and EDDS, K. T. (1976). Op. cit., Chapter 12.

STUMP, R. F., and ROBINSON, K. R. (1983). *Xenopus* neural crest cell migration in an applied electrical field. *Journal of Cell Biology* 97:1226–1233.

TANFORD, C. (1978). The hydrophobic effect. *Science* 200:1012–1018.

THOMPSON, D. W. (1917). Op. cit., Chapter 13.

TIPPER, D. J., and WRIGHT, A. (1979). The structure and biosynthesis of bacterial cell walls, in *The Bacteria*, vol. 7, J. R. Sokatch and L. N. Ornston, Eds. Academic, New York, pp. 291–426.

TUCKER, J. B. (1979). Spatial organization of microtubules, in *Microtubules*, K. Roberts and J. S. Hyams, Eds. Academic Press, London, pp. 347–357.

VACQUIER, V. D. (1981). Op. cit., Chapter 13.

WEBER, K., and OSBORN, M. (1979). Intracellular display of microtubular structure revealed by indirect immunofluorescence microscopy, in *Microtubules*, K. Roberts and J. S. Hyams, Eds. Academic, New York, pp. 279–313.

WEBSTER, G., and GOODWIN, B. (1981). History and structure in biology. *Perspectives in Biology and Medicine* 25:26–39.

WEISS, P. A. (1967). One plus one does not equal two, in *The Neurosciences, A Study Program*, G. C. Quarton, T. Melnechuk, and F. O. Schmitt, Eds. Rockefeller University Press, New York, pp. 801–821.

WOLPERT, L. (1969). Positional information and the spatial pattern of cellular differentiation. *Journal of Theoretical Biology* 25:1–47.

WOOD, W. B. (1979). Bacteriophage T4 assembly and the morphogenesis of subcellular structure. *Harvey Lectures* 73:203–223.

WOODRUFF, R. I., and TELFER, W. H. (1980). Electrophoresis of proteins in intercellular bridges. *Nature* 286:84–86.

Index